Bacterial Outer Membranes

BACTERIAL OUTER MEMBRANES

Biogenesis and Functions

MASAYORI INOUYE
State University of New York, Stony Brook

A WILEY-INTERSCIENCE PUBLICATION

JOHN WILEY & SONS, New York • Chichester • Brisbane • Toronto

Library of Congress Cataloging in Publication Data

Main entry under title:

Bacterial outer membranes.

"A Wiley-Interscience publication."
Includes index.
1. Bacterial cell walls. 2. Gram negative bacteria. I. Inouye, Masayori.

QR77.3.B3 589.9'08'75 79-13999
ISBN 0-471-04676-0

Printed in the United States of America

10 9 8 7 6 5 4 3 2 1

Contributors

Mark Achtman
Max-Planck Institut für molekulare Genetik, Ihnestrape 63-73, 1000 Berlin 33 (Dahlem), West Germany

Philip J. Bassford, Jr.
Department of Microbiology and Molecular Genetics, Harvard Medical School, Boston, Massachusetts

Manfred E. Bayer
The Institute for Cancer Research, Fox-Chase Cancer Center, 7701 Durholme Ave., Philadelphia, Pennsylvania

Jonathan K. Beckwith
Department of Microbiology and Molecular Genetics, Harvard Medical School, Boston, Massachusetts

Thomas M. Buchanan
Department of Medicine and Pathology, The University of Washington, Seattle, Washington 98195, and Immunology Research Laboratory, U.S. Public Health Service Hospital, 1131 14th Avenue South, Seattle, Washington

John E. Cronan, Jr.
Department of Microbiology, University of Illinois, Urbana, Illinois

Simon Halegoua
Neurobiology Department, The Salk Institute, San Diego, California

Gerald L. Hazelbauer
Membrane Group, Wallenberg Laboratory, University of Uppsala, Uppsala, Sweden

Masayori Inouye
Department of Biochemistry, State University of New York at Stony Brook, Stony Brook, New York

Jordan Konisky
Department of Microbiology, University of Illinois, Urbana, Illinois

Paul A. Manning
Max-Planck Institut für molekulare Genetik, Ihnestrape 63-73, 1000 Berlin 33 (Dahlem), West Germany

David Mirelman
Department of Biophysics, The Weizmann Institute of Science, Rehovoth, Israel

Hiroshi Nikaido
Department of Microbiology and Immunology, University of California, Berkeley, California

Misao Ohki
Biology Division, National Cancer Center Institute, Chuo-ku, Tokyo, Japan

Mary J. Osborn
Department of Microbiology, University of Connecticut, Health Center, Farmington, Connecticut

William A. Pearce
Department of Pathobiology, The University of Washington, Seattle, Washington, and Immunology Research Laboratory, U.S. Public Health Service Hospital, Seattle, Washington

Peter Reeves
Department of Microbiology and Immunology, The University of Adelaide, Adelaide, South Australia

Thomas J. Silhavy
Department of Microbiology and Molecular Genetics, Harvard Medical School, Boston, Massachusetts

Preface

In the last few years research interest in the outer membrane, especially the outer membrane proteins of gram-negative bacteria, has become extremely active. One can easily see this trend in a rapid survey of publications in the *Journal of Bacteriology*: During the years 1971 and 1972, only one paper concerning outer membrane proteins was published, and in the next three years four to five papers appeared per year. However, in 1976 and 1977, the number of publications increased fourfold, and in 1978 more than 20 papers were published.

One of the reasons for this outburst of activity in outer membrane research is that the outer membrane contains several major proteins, which can be easily purified and characterized. Studies on the structure, function, and biogenesis of these proteins have broad implications in many different areas of membrane biochemistry and molecular biology. Furthermore, investigations into the nature of the lipopolysaccharide-covered outer surface of gram-negative bacteria are also important for an understanding of the interaction between bacteria and their environment, such as host animal tissues.

In this book, efforts have been made to illustrate the dynamic aspects of the outer membrane, how its individual constituents are synthesized and assembled in the outer membrane and how they function. It is our good fortune that those scientists in the front line of outer membrane research have contributed chapters to this book. It is my hope that this book will be useful, not only for those who are engaged in outer membrane research, but also for those who have a general interest in membrane biogenesis and function.

Finally, I would like to express my gratitude to Dr. Hiroshi Nikaido for his valuable advice in editing this book and to Dr. Stanley F. Kudzin of State University of New York, New Paltz for his help in the production of this book.

MASAYORI INOUYE

Stony Brook, New York
July 1979

Contents

Chapter 1 What is the Outer Membrane?

MASAYORI INOUYE

Department of Biochemistry, State University of New York at Stony Brook

1 INTRODUCTION

At the end of the last century, Christian Gram, a Danish histologist, developed a method of staining bacteria in tissues that is now one of the most valuable and most widely used procedures for bacterial staining (see review by Bartholemew and Miltwer, 1952). Bacteria that are stained with the I_2–crystal violet complex by this method are classified into two groups; gram positive and gram negative. The affinity of crystal violet for bacteria in this staining procedure appears to be associated with the nature of the bacterial envelope, since gram-negative bacteria are known to be more resistant that gram-positive bacteria to the actions of certain dyes, chemicals, and antibiotics.

In this chapter, I describe the structure and functions of the envelope of gram-negative bacteria and discuss its unique features.

2 STRUCTURE OF THE ENVELOPE OF GRAM-NEGATIVE BACTERIA

As shown in Figure 1, *Escherichia coli*, a gram-negative bacterium, is surrounded by five electron dense tracks in contrast to the typical double-

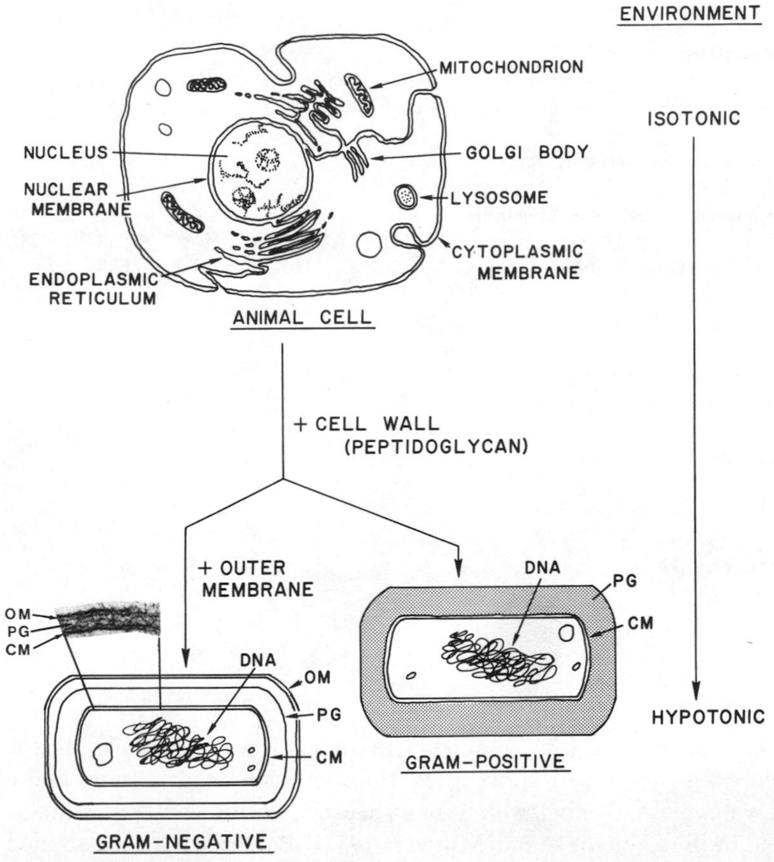

Figure 1 Schematic illustration of the membrane structures of eukaryotic and prokaryotic cells. CM, cytoplasmic membrane; CW, cell wall; OM, outer membrane; PG, peptidoglycan. The ultrastructure of the envelope of the gram-negative cell, in which there are five election dense layers, is shown in the insert.

tracked layer of a unit membrane (Murray et al., 1965; dePetris, 1976). Based on the fact that the central track disappeared after lysozyme treatment of cells, it was concluded that the cell wall or the peptidoglycan layer exists between the two double-track layers, which represent two distinct membrane systems. Both systems consist of a typical unit membrane having a thickness of about 75 Å (Figure 1) (see also review by Glauert and Thornley, 1969). These membranes are called the outer membrane and the inner, or cytoplasmic, membrane. It should be noted that adhesion sites between the outer membrane and the cytoplasmic membrane have been found and their important roles in the assembly of outer membrane components have been demonstrated (see Chapter 6 by Bayer).

On the other hand, gram-positive bacteria such as *Bacillus subtilis* are surrounded by a cytoplasmic membrane and a thick electron dense layer representing the cell wall or the peptidoglycan (about 200 Å), which is exterior to the cytoplasmic membrane (Figure 1).

3 FUNCTIONS OF THE OUTER MEMBRANE

Why is there such a drastic difference in the envelope structure between gram-negative and gram-positive bacteria? Why do gram-negative bacteria have a thin peptidoglycan layer and an outer membrane instead of a thick peptidoglycan layer as in gram-positive bacteria? What are the functions of the outer membrane? To answer these questions let us consider the structure of the envelope of gram-negative bacteria in relation to the membrane systems of animal cells. Also, let us discuss the significance of the cell envelope in regard to the natural habitats.

As shown in Figure 1, animal cells require only the cytoplasmic membrane (plasma membrane) because they live in osmotically isotonic environments. On the other hand, bacteria usually live in hypotonic environments that would cause the lysis of animal cells. To prevent cells from lysing in the hypotonic environment, the bacteria cytoplasmic membranes are surrounded by the cell wall, a rigid network composed of the peptidoglycan. The peptidoglycan is highly cross-linked and the cells are probably surrounded by a single supermacromolecule comprised of peptidoglycan. As can be seen in Figure 1, gram-positive bacteria have very thick multilayered cell walls, whereas gram-negative bacteria have a single layer of peptidoglycan (see Chapter 5 by Mirelman). In both cases, the peptidoglycan layers can be removed by lysozyme, resulting in the formation of spherical cells in an isotonic medium (these spherical cells are called protoplasts and spheroplasts for gram-positive and gram-negative bacteria, respectively). They will lyse, as do animal cells, if they are placed in a hypotonic medium.

The envelope structure of gram-negative bacteria is more differentiated than that of gram-positive bacteria, having an extra membrane system outside the peptidoglycan layer (Figure 1). A similar differentiation of membrane systems can also be seen in animal cells that contain the lysosomal membrane, mitochondrial membrane, endoplasmic reticulum, and nuclear membrane in addition to the cytoplasmic membrane as shown in Figure 1. The functions of the outer membrane are, in part, very similar to those of the eukaryotic lysosomal membrane. The lysosomes are organelles surrounded by a single membrane and are sometimes called "suicide bags" because they contain many hydrolytic enzymes, such as phosphatases, glycosidases, nucleases, proteases, and lipases, are able to hydrolyze all classes of essential components released inside the cells. These enzymes are used to digest that foreign material taken in by the cell, such as bacteria. Bacteria also need these kinds of hydrolytic enzymes for their survival because they have to digest macromolecules or low molecular weight organic compounds that are utilized as nutrients. However, because bacteria do not contain lysosomes, these enzymes must be maintained separately from other components within the cell by other means in order to prevent self-digestion. Gram-positive bacteria simply excrete these enzymes outside the cell, whereas gram-negative bacteria elaborately house these enzymes in the space between the outer and the cytoplasmic membranes. This space is called the periplasmic space and appears to play vital roles in cell growth. The space may include as much as 20–40% of the total cell mass (Stock et al., 1977; see also Chapter 11 by Nikaido) and contains other important proteins, such as specific amino acid and sugar transport binding proteins. These proteins are involved in transport of specific amino acids and sugars through the cytoplasmic membrane. Thus the periplasmic space would appear to be more sophisticated, at least on a functional basis, than the eukaryotic lysosome. As is seen above, one of the functions of the outer membrane is to confine the periplasmic enzymes and proteins to the periplasmic space. At the same time, the outer membrane provides specific and nonspecific channels for those nutrients and ions required for growth (see Chapter 10 by Konisky and Chapter 11 by Nikaido). It should be noted that these nutrients are transported through the outer membrane channels by passive diffusion. After they are passively transported into the periplasmic space they are then actively incorporated into the cytoplasm across the cytoplasmic membrane. The active transport systems for these nutrients are exclusively located in the cytoplasmic membrane. Nonspecific diffusion pores for small hydrophilic molecules have been well characterized and have been shown to consist of a class of major outer membrane proteins (matrix proteins or porins) (see Chapter 11 by Nikaido).

On the other hand, the outer membrane also serves as a selective barrier to the cell exterior. As is described earlier, gram-negative bacteria are more resistant than gram-positive bacteria to the actions of certain dyes, chemicals, enzymes, and antibiotics. This is because gram-negative bacteria have an outer membrane that prevents toxic compounds from entering the cells. It is important, especially for those gram-negative bacteria that comprise the normal intestinal flora of animals (*Enterobacteriaceae*, enteric bacteria), that the outer membrane protects the cytoplasmic membrane from direct exposure to bile salts that would lyse the cells. The most important component in the outer membrane in this regard is the lipopolysaccharide that exists exclusively in the outer membrane (see Chapter 2 by Osborn). Its long polysaccharide chains, extending toward the outside of the cell, not only prevent certain compounds (mostly hydrophobic; see Chapter 11 by Nikaido) from penetrating into the cell, but also play an important role by interacting with natural environments such as animal tissues. Furthermore, lipopolysaccharide is known as endotoxin, the major toxin of pathogenic enteric bacteria.

It should be pointed out that the outer membrane contains many specific receptors for phages and colicins; however, their primary roles have been revealed to be specific receptors and channels for nutrients required for growth (see Chapter 10 by Konisky). Moreover, the outer membrane plays important roles in cell–cell interaction during conjugation (see Chapter 12 by Manning and Achtman) as well as indirect roles in chemotaxis (see Chapter 13 by Hazelbauer).

4 COMPONENTS OF THE OUTER MEMBRANE

Methods to separate the outer membrane and the cytoplasmic membrane have been developed (Miura and Mizushima, 1969; Osborn et al., 1972). In these methods the outer membrane is separated from the cytoplasmic membrane because of its greater density due to the presence of lipopolysaccharide. Other purification methods employ detergents that solubilize the cytoplasmic membrane but not the outer membrane (see review by DiRienzo et al., 1978).

The outer membrane thus isolated is composed of protein, phospholipid, and lipopolysaccharide. In comparison to the cytoplasmic membrane, the outer membrane contains a small variety of proteins in rather large quantities. Therefore, it is easy to purify and characterize these major proteins. The content of phospholipid in the outer membrane is much less than that in the cytoplasmic membrane, and the outer membrane appears to be enriched in phosphatidylethanolamine in comparison with the cytoplasmic membrane (see Chapter 3 by Cronan). In the wild-type *Salmonella typhymurium* the phospholipid content was found not to be large enough to cover even one side of the lipid bilayer (Smit et al., 1975; Kamio and Nikaido, 1976; see Chapter 3 by Cronan and Chapter 11 by Nikaido). On the other hand, the outer membrane contains a specific component, lipopolysaccharide (see the structure in Figure 1 of Chapter 1 by Ocborn and Figure 2 of Chapter 11 by Nikaido). In the case of *S. typhimurium* it has been calculated that there are approximately 2.5×10^6 molecules of lipopolysaccharide per cell (Smit et al., 1975). These molecules are localized exclusively in the outer leaflet of the outer membrane and occupy about 45% of the surface of the outer membrane.

It should be pointed out that the composition of the outer membrane changes rather drastically as a result of mutation. In "deep-rough" mutants of the lipopolysaccharide, the phospholipid content increases significantly, whereas the protein content decreases drastically without changing the number of lipopolysaccharide molecules per cell (Smit et al., 1975).

Besides those components described above, the peptidoglycan is closely associated with the outer membrane (see Chapter 5 by Mirelman). Some outer membrane proteins are shown to have specific interactions with the peptidoglycan (see Chapter 4 by Halegoua and Inouye), and a part of the lipoprotein, one of the major outer membrane proteins, is actually covalently linked to the peptidoglycan, as is described in the next section.

5 OUTER MEMBRANE PROTEINS

Recently, the outer membrane proteins were extensively reviewed by DiRienzo et al. (1978). Therefore, I present here only a brief description of these proteins to help the reader better understand the other chapters of this book.

5.1 Major Proteins

The definition of major proteins is rather arbitrary. A minor protein may become a major protein when its production is fully induced. However, the outer membrane of wild-type *E. coli* K-12 contains at least three classes of major proteins; matrix proteins, ompA protein (tolG protein), and lipoprotein. Although at the present time a universal nomenclature for the outer membrane proteins has not been agreed upon, the designations matrix protein, ompA protein, and lipoprotein are used throughout this book (the reader is referred to DiRienzo et al. (1978) for a cross-reference of other nomenclatures).

5.1.1 Matrix Proteins—Porins Matrix proteins are characterized by their tight, but noncovalent, association with the peptidoglycan. One of these proteins was first isolated by an elegant method described by Rosenbusch (1974). He purified, to homogeneity, matrix protein Ia from *E. coli* B. Its molecular weight was calculated as 36,500 and it consisted of a single polypeptide of 336 amino acid residues. On the other hand, *E. coli* K-12 contains matrix protein Ib in addition to matrix protein Ia (see review by DiRienzo et al. (1978); see also Chapter 8 by Reeves). These two proteins appear to be coded for by independent genes, and their relative amounts vary greatly with growth conditions.

A striking features of matrix proteins is their extremely high content of β-structure in contrast to many other "intrinsic" membrane proteins, which show high contents of α-helix (Rosenbusch, 1974; Nakamura and Mizushima, 1976). Electron micrographs of the negative-stained matrix protein–peptidoglycan complex show that the matrix protein molecules are arranged as a hexagonal lattice layer with a 7.7 nm repeat (Rosenbusch, 1974; Steven et al., 1977; see also Chapter 11 by Nikaido). There are about 1.5×10^5 molecules of matrix proteins per cell and the hexagonal lattice structure covers 60% or more of the outer surface of the peptidoglycan (Steven et al., 1977).

Nakae showed that the incorporation of the matrix protein into artificial lipopolysaccharide–phospholipid vesicles greatly enhances their permeability to sucrose (Nakae, 1976a, 1976b; see Chapter 11 by Nikaido). Those vesicles that were reconstituted with the matrix protein showed almost the same molecular sieving properties as the intact outer membrane, which excludes oligo- and polysaccharides with molecular weights higher than 900 (Nakae and Nikaido, 1975). These results indicate that the function of the matrix proteins is to form passive diffusion pores and because of this property they are also called "porin" (Nakae, 1976b).

Another interesting aspect of the matrix proteins is the regulation of their production. Gene expression of both the *ompF* gene (for matrix protein Ia) and the *ompC* gene (for matrix protein Ib) is controlled by another independent gene called *ompB*. Furthermore, in mutants that lack both matrix proteins Ia and Ib, new proteins having properties similar to those of the matrix proteins are produced (see Chapter 8 by Reeves).

5.1.2 ompA Protein (tolG Protein) This protein also exists in large quantities in the outer membrane and is known to show an anomalous mobility on SDS gels (see review by DiRienzo et al., 1978). The molecular weight is approximately 30,000 and the protein has a high β-structure content (Nakamura and Mizushima, 1976). One important function of the ompA protein is its requirement in F pilus-mediated conjugation (see Chapter 12 by Manning and Achtman). It has been suggested that ompA protein also forms pores (Manning et al., 1977); however, this conclusion may have to be reconsidered (see Chapter 11 by Nikaido).

5.1.3 Lipoprotein The lipoprotein of the outer membrane is one of the most thoroughly investigated membrane proteins. This protein is the most abundant protein in the cell in terms of number of molecules and has many unique features. Recently, an extensive review on the lipoprotein was published (Inouye, 1979).

In 1969, Braun and Rehn first reported the existence of a lipoprotein, covalently linked to the peptidoglycan, with a molecular weight of about 7000. The complete chemical structure of this protein has been determined by Braun and his co-workers as shown in Figure 2 (Braun and Bosch, 1972; Hantke and Braun, 1973). The lipoprotein consists of 58 amino acid residues and lacks histidine, tryptophan, glycine, proline, and phenylalanine. It is linked by the ϵ-amino group of its *C*-terminal lysine to the carboxyl group of every tenth to twelfth *meso*-diaminopimelic acid residue of the peptidoglycan. The *N*-terminal portion of the lipoprotein consists of glycerylcysteine [*S*-(propane-2′,3′diol)-3-thioaminopropionic acid] to which two fatty acids are attached by two ester linkages and one fatty acid is attached by an amide linkage. The fatty acids bound as esters are similar to the fatty acids found in the phospholipids, while the amide-linked fatty acids consist of 65% palmitate, the rest being mainly monounsaturated fatty acids.

Inouye and his co-workers (Inouye et al., 1972; Hirashima et al., 1973) independently found that the lipoprotein also exists in the *E. coli* membrane without covalent bonds to the peptidoglycan (i.e., free instead of bound). The free, as well as the bound, form of the lipoprotein exclusively exists in the outer membrane. There are about 2.4×10^5 molecules of the bound form per cell, and about twice as much of the free form, that is, about 4.8×10^5 molecules/cell. The total free and bound lipoprotein molecules, 7.2×10^5, makes this lipoprotein the most abundant protein, numerically, in the cell (see review by Inouye, 1979).

Thc frcc form lipoprotcin was extensively purified and paracrystallized (Inouye et al., 1976). The purified protein has a very high α-helical content and, from the known sequence of the lipoprotein, a three-dimensional molecular model for the assembly of the lipoprotein have been proposed (see reviews by DiRienzo et al., 1978; Inouye, 1979). Although the exact

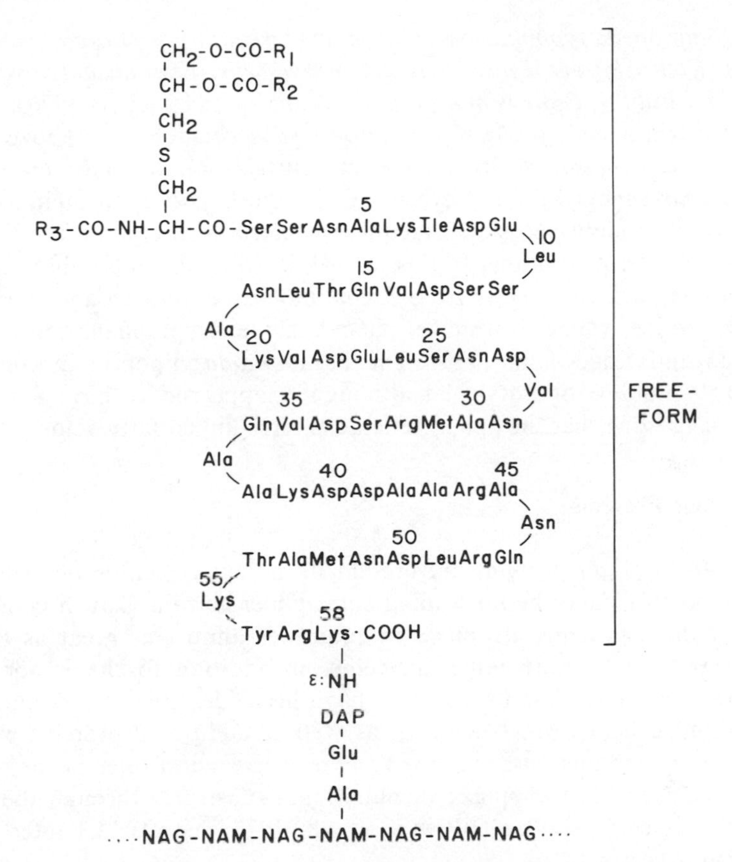

Figure 2 The complete chemical structure of the bound form of the lipoprotein (Braun and Bosch, 1972; Hantke and Braun, 1973). DAP, diaminopimilic acid; NAM, *N*-acetylmuramic acid; NAG, *N*-acetylglucosamine. R_1, R_2, and R_3 represent hydrocarbon chains of fatty acids.

function of the lipoprotein in the outer membrane is still obscure, analysis of lipoprotein mutants has revealed that the lipoprotein at least plays an important role in maintaining the integrity of the outer membrane structure (see reviews by DiRienzo et al., 1978; Inouye, 1979). The mechanism of biosynthesis and assembly of the lipoprotein has been extensively investigated and is discussed in Chapter 4.

An intriguing aspect of lipoprotein research is the examination of the existence of the lipoprotein in various gram-negative bacteria. When the homology of the lipoprotein gene between *E. coli* and other gram-negative bacteria was examined with ^{32}P-labeled purified lipoprotein mRNA from *E. coli* it was found that there are three different groups classified according to the degree of homology with the *E. coli* lipoprotein: (*E. coli*, *Shigella dysen-*

teriae, *Salmonella typhimurium*, *Citrobacter freundii*) > (*Enterobacter aerogenes*, *Klebsiella aerogenes*, *Serratia marcescens*, *Erwinia amvlovora*) > (*Proteus mirabilis*, *Proteus morganii*) (Nakamura and Inouye, 1979). DNAs from bacteria outside the family *Enterobacteriaceae* described above (*Pseudomonas aeruginosa*, *Acinitobacter* sp. H01-N, *Caulobacter crescentus*, *Myxococcus xanthus*) did not hybridize with the *E. coli* lipoprotein mRNA. However, it is highly possible that these bacteria also contain the lipoprotein in the outer membrane. In this regard, it should be noted that the lipoprotein was purified recently from *Pseudomonas aeruginosa* and was found to lack proline, valine, isoleucine, phenylalanine, tryptophan, and cysteine (Mizuno and Kageyama, 1978). The *Pseudomonas* lipoprotein contained only 0.89 mole % of fatty acid although it appeared to have a glycerol group, suggesting that the lipoprotein lacks ester-linked fatty acids.

5.2 Minor Proteins

About 10–20 minor proteins are present in the outer membrane. The term "minor protein" may be misleading since under certain growth conditions some of these proteins are made in quantities almost as great as that of "major protein." Many minor proteins, in addition to the major outer membrane proteins described above, have been identified as receptors for phages and colicins. Most of them, as well as additional proteins with no known receptor functions, are now known to have vital roles in the growth of the cell, such as the uptake of nutritional substrates through the outer membrane (see review by DiRienzo et al., 1978; see also Chapter 10 by Konisky).

6 CONCLUSION

In this chapter, I have attempted to described the structure and functions of the outer membrane as a unique membrane system. One may use the outer membrane as a model system for the study of the biosynthesis, and assembly of the outer membrane components may be directly applicable to eukaryotic membrane systems. Studies on the secretory mechanism of outer membrane proteins across the cytoplasmic membrane may provide valuable information as to how hormones and proteins are secreted across the eukaryotic cytoplasmic membrane (see Chapter 4 by Halegoua and Inouye, and Chapter 7 by Silhavy et al.).

The outer membrane also provides unique and excellent systems for the

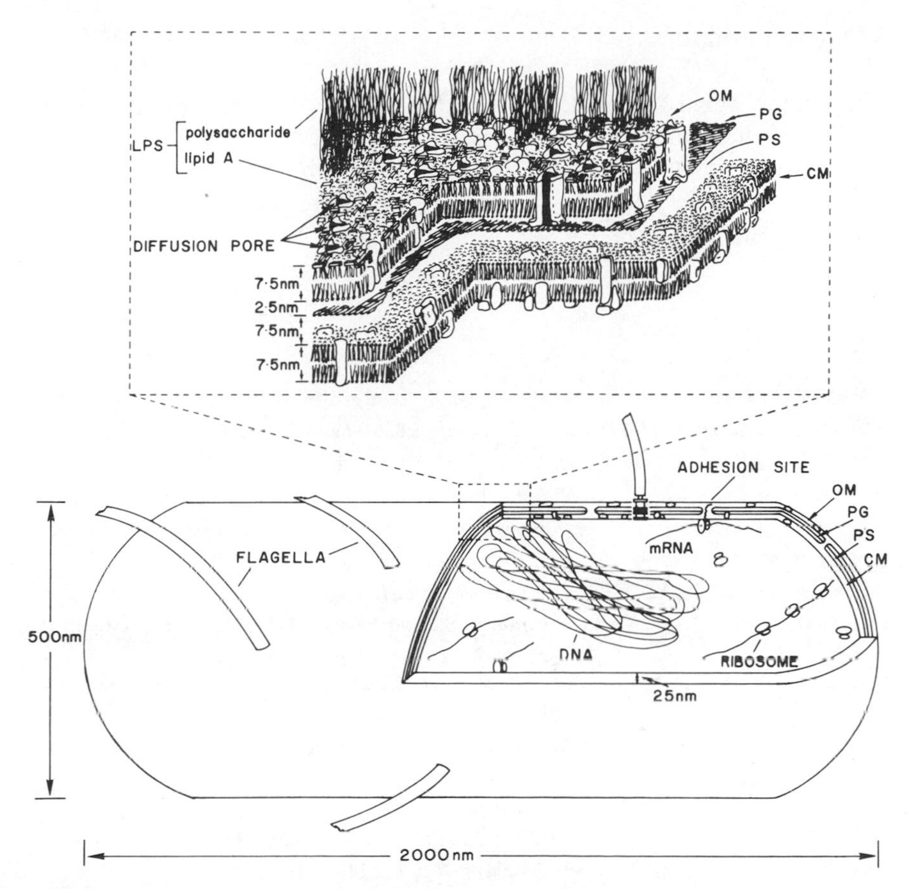

Figure 3 Molecular architecture of the *E. coli* envelope. OM, outer membrane; PG, peptidoglycan; PS, perisplasmic space; CM, cytoplasmic membrane. The structure of the basal end of the flagellum is from Chapter 13 by Hazelbauer.

investigation of many other problems associated with membranes, such as regulation of the biogenesis of membranes (see Chapter 8 by Reeves; Chapter 9 by Ohki), the genetics of membrane proteins (see Chapter 8 by Reeves), membrane receptors and diffusion pores (see Chapter 10 by Konisky and Chapter 11 by Nikaido), cell–cell interactions (see Chapter 12 by Manning and Achtman), and the interaction between pathogenic bacteria and animal tissues (see Chapter 14 by Buchanan and Pearce).

On the basis of the current knowledge described in this chapter, a schematic representation of the possible molecular architecture of the *E. coli* envelope in perspective to the whole cell is shown in Figure 3.

ACKNOWLEDGMENTS

The author would like to thank Dr. Joseph M. DiRienzo for the critical reading of the manuscript and for drawing Figures 2 and 4. This work was supported by grants GM 19043 from the U.S. Public Health Service, PCM 76-07320 from the National Science Foundation, and BC-67D from the American Cancer Society.

REFERENCES

Bartholemew, J. W., and Miltwer, T. (1952), *Bacteriol. Rev.*, **16,** 1–29.

Braun, V., and Bosch, V. (1972). *Eur. J. Biochem.*, **28,** 51–69.

Braun, V., and Rehn, K. (1969), *Eur. J. Biochem.*, **10,** 426–438.

dePetris, S. (1976), *J. Ultrastruct. Res.*, **19,** 45–83.

DiRienzo, J. M., Nakamura, K., and Inouye, M. (1978), *Annu. Rev. Biochem.*, **47,** 481–532.

Glauert, A. M., and Thornley, M. J. (1969), *Annu. Rev. Microbiol.*, **23,** 159–198.

Hantke, K., and Braun, V. (1973), *Eur. J. Biochem.*, **34,** 284–296.

Hirashima, A., Wu, H. C., Venkateswaran, P. S., and Inouye, M. (1973), *J. Biol. Chem.*, **284,** 5654–5659.

Inouye, M. (1979), in *Biomembrane* (L. A. Manson, ed.), Plenum Press, New York, in press.

Inouye, M., Shaw, J., and Shen, C. (1972), *J. Biol. Chem.*, **247,** 8154–8159.

Inouye, S., Takeishi, K., Lee, N., DeMartini, M., and Inouye, M. (1976), *J. Bacteriol.*, **127,** 555–563.

Kamio, Y., and Nikaido, H., (1976), *Biochemistry*, **15,** 2561–2570.

Manning, P. A., Pugsley, A. P., and Reeves, P. (1977), *J. Mol. Biol.*, **116,** 285–300.

Miura, T., and Mizushima, S. (1969), *Biochem. Biophys. Acta*, **193,** 268–276.

Mizuno, T., and Kageyama, M. (1978), *J. Biochem.*, **85,** 115–122.

Murray, R. G. E., Steed, P., and Elson, H. E. (1965), *Can. J. Microbiol.*, **11,** 547–560.

Nakae, T. (1976a), *J. Biol. Chem.*, **251,** 2176–2178.

Nakae, T. (1976b), *Biochem. Biophys. Res. Commun.*, **71,** 877–884.

Nakae, T., and Nikaido, H. (1975), *J. Biol. Chem.*, **250,** 7359–7365.

Nakamura, K., and Inouye, M. (1979), *J. Bacteriol.*, **137,** 595–604.

Nakamura, K., and Mizushima, S. (1976), *J. Biochem.*, **80,** 1411–1422.

Osborn, M. J., Gander, J. E., Parisi, E., and Carson, J. (1972), *J. Biol. Chem.*, **247,** 3962–3972.

Rosenbush, J. P. (1974), *J. Biol. Chem.*, **249,** 8019–8029.

Smit, J., Kamio, Y., and Nikaido, H. (1975), *J. Bacteriol.*, **124,** 942–958.

Steven, A. C., Ten Heggeler, B., Müller, R., Kistler, J., and Rosenbush, J. P. (1977), *J. Cell. Biol.*, **72,** 292–301.

Stock, J. B., Rauch, B., and Roseman, S. (1977), *J. Biol. Chem.*, **252,** 7850–7861.

Part 1
BIOGENESIS

Chapter 2 Biosynthesis and Assembly of the Lipopolysaccharide of the Outer Membrane

M. J. OSBORN

Department of Microbiology University of Connecticut Health Center, Farmington

1 INTRODUCTION

This chapter is not intended as a comprehensive review of lipopolysaccharide biosynthesis, but instead centers on aspects of the field that have come into prominence in the past few years and that, in the author's opinion, merit continued attention as significant unsolved problems. The discussion focuses on biosynthesis of lipid A, the mechanism of translocation of lipopolysaccharide, and the assembly of lipopolysaccharide in the outer membrane and is limited to the enteric bacteria, *Salmonella* and *Escherichia coli*. The pathways of biosynthesis of the polysaccharide core and O-antigen chains are relatively clear in principle if not in detail, although many problems related to molecular organization and mechanism of these enzyme systems remain of interest. Biosynthesis of the polysaccharide is not considered here, since the topic has been the subject of many previous reviewers (Nikaido, 1968; Osborn, 1969; Osborn, 1971; Osborn and Rothfield, 1971; Robbins and Wright, 1971; Rothfield and Romeo, 1971; Wright and Kanegasaki, 1971; Lennarz and Scher, 1972; Nikaido, 1973; Rothfield and Hinckley, 1974). For discussions relevant to the present topics, and for other points of view, the reader is referred to reviews by Nikaido (1973), Leive (1974), Costerton et al. (1974), Bayer (1975), and Luderitz et al. (1978), as well as other chapters in this volume.

2 BIOSYNTHESIS OF LIPID A

2.1 Structure of Lipid A

Lipid A is a unique molecule whose basic structure appears to have been highly conserved (Luderitz et al., 1978). Although the molecule has seemed at times demoniacal in its resistance to structural analysis, the structure evolved in recent years for the lipid A of *Salmonella* (Rietschel et al., 1977) and *E. coli* K-12 (Rosner et al., 1979; Rosner and Khorana, 1979) is now reasonably well defined (Figure 1*A*). The backbone is a β-1,6-linked disaccharide of *N*-β-hydroxymyristoyl-D-glucosamine, which is additionally *O*-acylated by β-hydroxymyristic acid (as the 3-myristoyl diester) and saturated fatty acids (predominantly C_{12} and C_{14} with variable amounts of C_{16}). The polysaccharide chain is attached through ketosidic linkage of 3-deoxy-D-mannooctulosonate (ketodeoxyoctonate, KDO) to the 3′-hydroxyl of the nonreducing glucosamine. The structure is further complicated (and confused) by phosphoryl substituents at the reducing terminus and the 4′-position, the nature of which has been clarified only very recently. The presence of a pyrophosphoryl group at the reducing terminus of the disaccharide is now well established on the basis of both chemical methods and ^{31}P NMR studies (Mühlradt et al., 1977; Rosner and Khorana, 1979, Rosner et al., 1979); however, the original hypothesis that this was engaged in pyrophosphoryl cross-links between disaccharide subunits (Lüderitz et al., 1973) is not in accord with the more recent findings. These indicate that pyrophosphorylethanolamine is present at the reducing terminus of the lipid A of *Salmonella*, while the 4′-phosphate is in diester linkage to the 1-position of 4-aminoarabinose (Mühlradt et al., 1977). The ethanolamine and 4-aminoarabinose substituents appear, however, to be variable and substoichiometric in amount and are entirely absent from the lipid A of a deep-rough mutant of *E. coli* K-12 analyzed by Rosner et al. (1979). The function of these modifying residues and the nature of the biosynthetic control of their

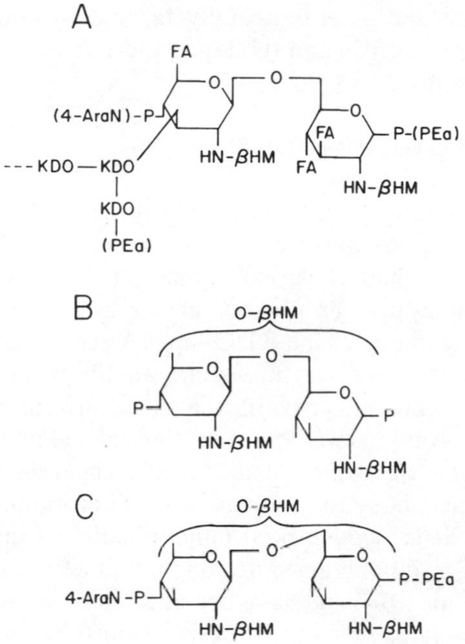

Figure 1 Structure of (*A*) lipid A and the (*B*) acidic and (*C*) neutral forms of lipid A precursor. 4-AraN, 4-aminoarabinose; HM, D-β-hydroxymyristic acid; FA, fatty acid; KDO, 3-deoxy-D-mannooctulosonate (2-keto-3-deoxyoctonate); PEa, phosphorylethanolamine.

addition are still unknown, but it is striking that both contain free amino groups and thus not only reduce the net negative charge, but confer zwitterionic character on the molecule. It is worth noting that the phosphorylethanolamine residue that substitutes to a variable extent the branch KDO of the KDO trisaccharide (Dröge et al., 1970) provides yet another zwitterionic substituent in the polar "head group" region of the lipid.

Early analytical data led to suggestions that monomeric lipopolysaccharide chains were cross-linked by phosphate bridges either in the backbone region of the polysaccharide (Osborn, 1969) or in lipid A (Lüderitz et al., 1973), and molecular weight determinations on presumably disaggregated derivatives (Malchow et al., 1969; Romeo et al., 1970) were consistent with an average trimeric structure. However, recent studies on both *Salmonella* (Mühlradt et al., 1977) and *E. coli* K-12 (Rosner et al., 1979; Rosner and Khorana, 1979) strongly support a non-cross-linked monomeric structure for lipid A, and it seems at present most likely that individual lipopolysaccharide chains are not connected by covalent cross-links. The highly aggregated state of isolated lipopolysaccharide is clearly referable to its amphipathic nature, but is also strongly favored by divalent cations and polyamines that are avidly bound (Galanos and Lüderitz, 1975) and exceedingly difficult to remove.

2.2 Pathway of Biosynthesis of Lipid A

Elucidation of the pathways of biosynthesis of the core and O-antigen polysaccharide chains of lipopolysaccharide was greatly aided by the facile isolation of mutants blocked at various points in the biosynthetic sequence, and the same strategy was an obvious choice in considering experimental approaches to biosynthesis of the KDO–lipid A region. However, the likelihood that this portion of the molecule would prove essential for the structural or functional integrity of the outer membrane suggested that mutants in lipid A synthesis would have to be isolated as conditional lethals, and this supposition was strengthened by the apparent failure of conventional selection techniques to yield nonconditional mutants of this class. Mutants of *Salmonella typhimurium* conditionally defective in synthesis of KDO were indeed readily obtained as conditional lethals (Rick and Osborn, 1972; Lehmann et al. 1977) and have provided considerable insight into late stages in the pathway of lipid A synthesis. Unfortunately, mutants in early steps of lipid A assembly have remained elusive, and this most intriguing segment of the pathway is still entirely unknown.

2.2.1 The Terminal Segment of the Pathway: Incorporation of KDO and Beyond Early studies of Heath et al. (1966) on enzymatic incorporation of KDO into degraded lipid A fractions suggested that addition of KDO might precede transfer of *O*-fatty acyl residues to the glucosamine disaccharide unit and that mutants in KDO synthesis would be, in effect, true lipid A mutants. Biosynthesis of KDO involves three sequential reactions catalyzed respectively by D-ribulose-5-P isomerase (Lim and Cohen, 1966), KDO-8-P synthetase, and KDO-8-P phosphatase (Levin and Racker, 1959; Ghalambor et al., 1966):

$$\text{D-ribulose-5-P} \rightleftarrows \text{D-arabinose-5-P} \quad (1)$$

$$\text{D-arabinose-5-P} + \text{P-enolpyruvate} \rightarrow \text{KDO-8-P} + P_i \quad (2)$$

$$\text{KDO-8-P} \rightarrow \text{KDO} + P_i \quad (3)$$

Free KDO is then converted to the nucleotide sugar CMP-KDO (Ghalambor and Heath, 1966), which serves as a donor of KDO residues in lipopolysaccharide synthesis. The first mutant to be obtained in KDO synthesis was isolated as an auxotroph for D-arabinose-5-P, the obligatory precursor of KDO (Rick and Osborn, 1972). The procedure was based on the observation that exogenous D-arabinose-5-P is an excellent substrate (although not an inducer) for the hexosephosphate transport system (Eidels et al., 1974) and is incorporated efficiently and relatively specifically into KDO residues the lipopolysaccharide. The mutant, *S. typhimurium* PRX2, was dependent on exogenous D-arabinose-5-P for lipopolysaccharide synthesis and growth at 37°C and was found to have an altered KDO-8-P synthetase (reaction 2) with an elevated apparent K_m for D-arabinose-5-P. Interestingly, the K_m defect in the altered enzyme and expression of the mutant phenotype are markedly temperature dependent (Rick and Osborn, 1977). At low growth temperatures (29°C and below), the K_m is increased only moderately (approximately sixfold higher than wild type) and growth and lipopolysaccharide synthesis are essentially independent of exogenous D-arabinose-5-P. At 37°C, the K_m is elevated 25 to 30-fold and the mutant requires exogenous D-arabinose-5-P for maintenance of an effective internal concentration of substrate. At 42°C, the defective enzyme is inactive, and this temperature is nonpermissive for lipopolysaccharide synthesis and growth even in the presence of D-arabinose-5-P. The conditional D-arabinose-5-P auxotrophy of the original mutant was particularly advantageous in establishing the KDO-negative phenotype; however, ts mutants in KDO-8-P synthetase are also readily obtained by simple selection for isolates temperature sensitive in growth and lipopolysaccharide synthesis (Lehman et al., 1977). The KDO-8-synthetase mutation has been mapped at approximately 58 min on the *Salmonella* chromosome (Rick and Osborn, 1977).

Transfer of KDO-8-P synthetase mutants to nonpermissive conditions results in prompt cessation of lipopolysaccharide synthesis and appearance of a new product, identified as an incomplete KDO-deficient precursor of lipid A (Rick et al., 1977; Lehmann, 1977; Lehmann and Rupprecht, 1977). The incomplete product (Figures 1*B* and 1*C*) has the *N*-β-hydroxymyristoylglucosamine disaccharide unit characteristic of lipid A, as well as the O-linked β-hydroxymyristate and the two phosphate groups of the complete molecule, but it lacks both KDO and the saturated C_{12}, C_{14}, and C_{16} *O*-fatty acyl chains of lipid A. The phosphorylethanolamine and 4-aminoarabinose residues found in lipid A may be either absent or present to variable extents depending on growth conditions and strain. At 42°C, PRX2 and its derivatives (Rick et al., 1977) and the independently isolated mutant *ts*1 (Lehmann, 1977) produce the so-called acidic form of the precursor, which lacks these substituents and contains only monoester phosphate. At lower growth temperatures, however, *ts*1 was found to produce a new, "neutral" form of the precursor in which the phosphate residues were in pyrophosphate linkage to phosphorylethanolamine and diester linkage to 4-aminoarabinose, respectively (Figure 1*C*) (Lehmann and Rupprecht, 1977). In other mutants, substitution by phosphorylethanolamine and/or 4-aminoarabinose was observed even at 42°C (Lehmann et al., 1977). The temperature sensitivity of incorporation of these polar substituents appears to be unrelated to the ts defects in KDO synthesis and presumably reflects a normal modification of lipid A structure. The phenomenon is reminiscent of medium-dependent shifts in the relative amounts of the major outer membrane proteins b and c (Ia and Ib) observed in *E. coli* by van Alphen and Lugtenberg (1977) and suggests operation of a general mechanism for subtle modulation of the composition and functional properties of the outer membrane in response to growth conditions.

The structures of the KDO-deficient lipid A precursors suggest a pathway for the final stages of lipid A synthesis as outlined in Figure 2. That both the acidic and neutral forms of the precursor are intermediates in the pathway and not side products of abortive synthesis has been established by pulse-chase experiments that show that pulse labeled precursor is efficiently

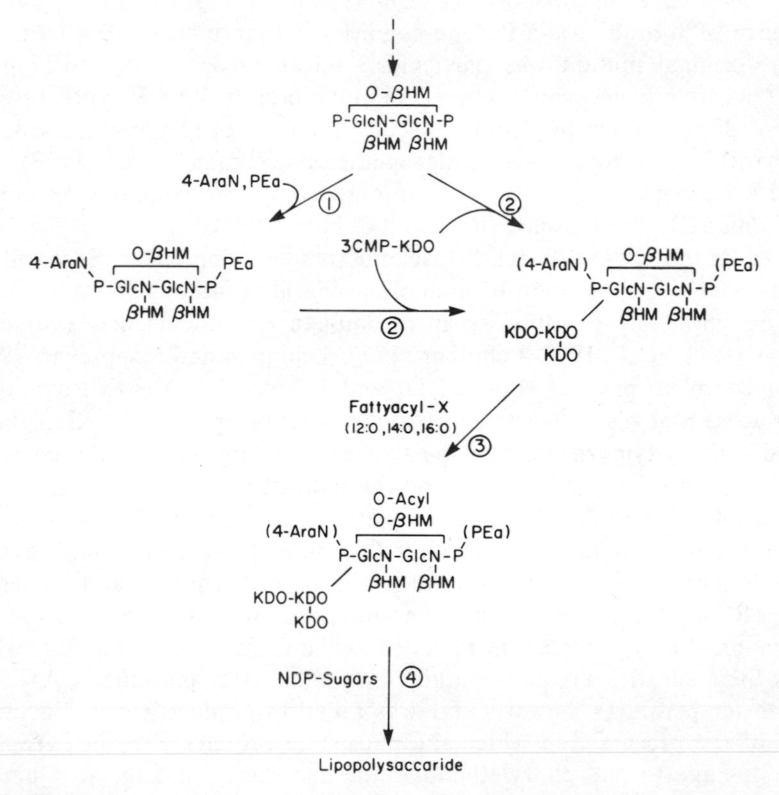

Figure 2 Postulated pathway of lipopolysaccharide synthesis from lipid A precursor. Abbreviations are as in Figure 1.

chased into lipopolysaccharide *in vivo* following shift from nonpermissive to permissive conditions (Lehmann et al., 1978; Osborn et al., 1979). It is clear that addition of KDO (step 2) obligatorily precedes incorporation of ester-linked 12:O, 14:O, and 16:O fatty acids (step 3), and the results of Lehmann and co-workers (Lehmann and Rupprecht, 1977; Lehmann, 1977; Lehmann et al., 1978) strongly suggest that the acidic precursor is the direct acceptor of the optional phosphorylethanolamine and 4-aminoarabinose residues (step 1). The pathways of biosynthesis and mechanism of incorporation of these substituents are not known. Labeling studies suggest that the ethanolamine residues of lipolysaccharide are derived from serine, and this is consistent with observations that ethanolamine-deficient lipopolysaccharide is formed during serine deprivation in auxotophs of *S. typhimurium* (M. J. Osborn, unpublished observations). The fact that lipopolysaccharides of R_e mutants blocked in incorporation of the first heptose residue of the polysaccharide contain the full complement of *O*-acyl chains suggests that the final acylation reactions are completed prior to extension of the polysaccharide chain (step 4). Curiously, however, this is not obligatory (see below).

Evidence that the KDO- and *O*-acyl-deficient precursor is the direct acceptor of KDO residues has been obtained both *in vitro* and *in vivo*. Enzymatic transfer of KDO from CMP-KDO to alkali- and acid-degraded fractions of lipid A was demonstrated in *E. coli* as early as 1966 by Heath et al. (1966), but further characterization of the transferase systems and its relation to the biosynthetic pathway awaited isolation of a chemically and physiologically defined acceptor. Transfer of KDO residues to the acidic form of lipid A precursor is catalyzed by a membrane-bound enzyme system from *S. typhimurium* (Munson et al., 1978) according to the reaction:

$$2\text{CMP-KDO} + [\text{precursor}] \xrightarrow{Mg^{2+}} \text{KDO-KDO-}[\text{precursor}] + 2\text{CMP}$$

In contrast to the glucosyl and galactosyl transferases involved in synthesis of the polysaccharide core, the CMP-KDO:lipid A KDO transferase system is very firmly membrane bound and is solubilized only by nonionic detergents at alkaline pH. In addition, the detergent-soluble KDO transferase system lacks the requirement for added phospholipid characteristic of the soluble core glycosyltransferases (Rothfield and Romeo, 1971), although dispersion of the acceptor in detergent was required for maximal acceptor activity. Purified lipid A precursor was the preferred acceptor of KDO residues, but significant activity was also observed with lipid A prepared by mild acid hydrolysis of wild-type *S. typhimurium* lipopolysaccharide. Intact lipopolysaccharides were inactive. Acceptor activity of the neutral form of lipid A precursor was not tested with the isolated KDO transferase system, but the studies of Lehmann and coworkers (Lehmann et al., 1977; Lehmann et al., 1978) indicate that the neutral precursor is a functional acceptor of KDO *in vivo*.

The observed stoichiometry of the KDO transferase reaction presents some puzzling features. Only one reaction product could be detected with either membrane-bound or solubilized enzyme, and this product contained 2 moles of KDO per mole of lipid A precursor rather than the expected 3 moles (Munson et al., 1978). Although it seems most likely that separate transferases would be required for addition of the first and second KDO residues, no evidence for separation of activities was obtained during a hundred-fold purification of the solubilized enzyme fraction, and efforts to obtain the presumed intermediate product by manipulation of reaction conditions were also unsuccessful. While the possibility that the two KDO residues are added as a disaccharide unit rather than sequentially has not been excluded, this seems unlikely. No requirement for additional cofactors, such as undecaprenyl-P, was detected during partial purification of the solubilized transferase activity, and no evidence could be obtained for formation of lipid-linked or H_2O-soluble intermediates. Unfortunately, the enzyme systems has thus far resisted the further fractionation necessary to resolve the nature of the presumed intermediate steps in the observed overall reaction.

The reasons for the apparent failure of the KDO transferase system to form the complete branched KDO trisaccharide structure *in vitro* also remain unclear. Structural analysis of the enzymatic product yielded somewhat ambiguous results but suggested that the KDO-KDO linkage in the disaccharide corresponded to the branch KDO2-4KDO unit of the normal KDO trisaccharide, rather than the main chain KDO2-7(8)KDO unit to which the polysaccharide is attached (Prehm et al., 1975). While it is possible that incorporation of the third KDO occurs only after completion of the lipid A structure, *in vivo* evidence suggests that neither the 4-amino-arabinose and phosphorylethanolamine residues of lipid A (Lehmann et al., 1978; Rosner et al., 1979) nor the saturated fatty acyl groups (Walenga and Osborn, 1979b) are essential for addition of the core polysaccharide. An alternative possibility is suggested by a deep-rough mutant of *E. coli* (described by Prehm et al., 1975) that makes only the 2,4-linked KDO disaccharide. This mutant also lacks the phosphorylethanolamine residue that normally substitutes the branch KDO, and it was postulated that this substituent may be important for formation of the complete KDO trisaccharide structure.

Formation of a derivative of the lipid A precursor containing two KDO residues has also been established *in vivo* as an intermediate step in the conversion of the precursor to lipopolysaccharide (Walenga and Osborn, 1979a). Transient accumulation of a new KDO-containing product in the inner membrane was observed following shift of pulse-labeled cultures to permissive temperature, and kinetic studies strongly support identification of this material as an intermediate in the synthesis of lipopolysaccharide from lipid A precursor. The intermediate formed *in vivo* was indistinguishable from the enzymatically synthesized product in chromatographic properties and composition. Unfortunately, the amounts of material isolated were too small for detailed structural analysis, and the nature of the KDO-KDO linkage in the *in vivo* intermediate is not yet known.

The nature of the acyl transferase reactions responsible for addition of the ester-linked saturated fatty acids to lipid A remains entirely unknown. Preliminary attempts in the author's laboratory to demonstrate *in vitro* transfer of these fatty acids from acylCoA's to the enzymatic product of the KDO transferase system have been unsuccessful. Whether this reflects a faulty assumption about the nature of the presumed acyl acceptor or an incorrect choice of acyl donor remains to be determined. AcylACP is perhaps a more probable physiological donor, and while the CoA derivatives substitute effectively for acylACPs in phospholipid synthesis (van den Bosch and Vagelos, 1970), the donor specificity for acylation of lipid A may be more stringent. To establish the nature of the immediate acceptor of *O*-acyl residues more definitively, efforts were made to trap this intermediate specifically *in vivo* (Walenga and Osborn, 1979b). For this purpose, cultures containing preformed radioactive lipid A precursor were shifted to conditions permissive for KDO synthesis in the presence of cerulenin. This compound is a potent inhibitor of fatty acid synthesis (Omura, 1976) and abolished incorporation of acetate into phospholipid and lipopolysaccharide under the conditions employed. It was expected that cerulenin would interrupt conversion of preformed lipid A precursor to lipopolysaccharide at the *O*-acylation stage and thereby result in stable accumulation of the preceding intermediate in the pathway. This result was not obtained. Rather, the lipid A precursor (as well as the KDO-containing intermediate) was efficiently converted to an underacylated species of lipopolysaccharide that had the polysaccharide chain of the normal polymer, but the acyl chain composition of the lipid A precursor. The finding that addition of saturated *O*-acyl residues to lipid A is not essential for polysaccharide chain elongation was unexpected, but the phenomenon is not restricted to the KDO mutant or the rather special experimental conditions of temperature shift down. Formation of very similar underacylated lipopolysaccharides has been observed in cerulenin-treated cultures of *Proteus mirabilis* (Rottem et al., 1978) and in a mutant of *E. coli* temperature sensitive in total fatty acid synthesis (Walenga and Osborn, 1979b). Synthesis of *O*-acyl-deficient lipopolysaccharides thus appears to be a general phenomenon under conditions of limited endogenous fatty acid synthesis, in which synthesis of β-hydroxymyristate is preferentially conserved (Silbert et al., 1973). It is of interest that, although phospholipid synthesis can be rescued in general fatty acid mutants of *E. coli* by supplementation with a mixture of saturated and unsaturated fatty acids (Silbert et al., 1973), the acyl chain composition of the lipopolysaccharide is not restored to normal by exogenous saturated acids (Walenga and Osborn, 1979b). The result again suggests that fatty acylCoA's are unable to function as donors for acylation of lipid A.

It seems likely that the underacylated lipopolysaccharides are formed as products of abnormal synthesis resulting from sloppy acceptor specificity of the backbone and core glycosyltransferases, rather than as normal intermediates in the pathway. Direct experimental test of the ability of the underacylated species to accept additional acyl groups is difficult because of its rapid and irreversible translocation to the outer membrane (Walenga and Osborn, 1979b). However, the KDO–lipid A glycolipids of heptose-less (R_e) mutants contain a fully normal complement of saturated fatty acids (Rietschel et al., 1972; Rooney and Goldfine, 1972), consistent with the hypothesis that acylation of lipid A is completed prior to transfer of the first heptose residue of the polysaccharide backbone. While the physiological significance of the underacylated lipopolysaccharides is questionable, their apparently facile synthesis may well explain the failure to isolate mutants specifically blocked in the final *O*-acyl transfer reactions by the usual selection and screening techniques. Such mutants would be of interest for biosynthetic studies, but also in analysis of the role of the lipid A structure in the assembly and functional organization of the outer membrane.

2.2.2 Early Steps in the Pathway: Assembly of Lipid A Precursor The early stages of lipid A synthesis remain an enigma. Are the *N*-β-hydroxymyristoyl residues introduced before or after synthesis of the glucosamine disaccharide? Is the precursor of the glucosamine residues UDP-*N*-acetylglucosamine, UDP-glucosamine, or UDP-*N*-β-hydroxymyristoylglucosamine? Does undecaprenyl-PO_4 or an analogous membrane-bound carrier coenzyme participate in assembly of the disaccharide-1-PO_4 unit? Mutants specifically blocked in early stages of lipid A synthesis would be invaluable in answering these questions, but such mutants have not yet been defined. In theory, the general method of selection employed for isolation of temperature-sensitive KDO mutants, based on conditional phage sensitivities and temperature-sensitive growth (Lehmann, et al., 1977), should also yield early mutants blocked prior to KDO incorporation. However, the assumption of a conditional lethal phenotype may not be valid for all classes of lipid A mutants. The causal relationship between expression of the KDO-negative phenotype and growth stasis is well established (Rick and Osborn,

1977; Lehmann et al., 1977), but the mechanism of the pleiotropic inhibition of growth in these mutants is not well understood. The inhibition of growth appears to be correlated with accumulation of the lipid A precursor in the inner membrane (Osborn et al., 1979), and it is possible that stasis results from some specific toxic effect of the precursor molecule on essential inner membrane functions rather than a defect in functional assembly of the outer membrane.

A further and fundamental question that remains to be answered concerns the synthesis and utilization of D-β-hydroxymyristate. The assumption that lipid A-specific β-hydroxy fatty acids are derived from the normal pathway of fatty acid synthesis is confirmed by analysis of general fatty acid mutants of *E. coli* (Silbert et al., 1973) and *S. typhimurium* (P. McVerry and M. J. Osborn unpublished results), and the D-configuration found in lipid A is the same as that of the 3-hydroxy fatty acylACP intermediates of fatty acid synthesis. However, it has not been established whether β-hydroxymyristate destined for lipid A arises directly from the major pathway of acyl chain elongation or is formed by a product-specific branch pathway. The latter possibility is particularly attractive since indirect evidence suggests that synthesis of fatty acids for lipopolysaccharide and phospholipid may not be totally coregulated. The rate of synthesis of phospholipid-specific fatty acids is immediately and profoundly inhibited during glycerol starvation of a glycerol auxotroph of *E. coli* (Nunn et al., 1977), whereas synthesis of lipopolysaccharide continues at relatively normal rates for approximately 0.5 of a generation time in comparable auxotrophs of *S. typhimurium* (M. J. Osborn and N. S. Rasmussen, unpublished results). The observation that β-hydroxymyristate synthesis is preferentially conserved under partially nonpermissive conditions in general fatty acid mutants of *E. coli* (Silbert et al., 1973) also suggests a mechanism for independent regulation of the end products of fatty acid synthesis. An alternative possible mechanism of β-hydroxymyristate formation—secondary hydroxylation of 14:O either before or after transfer to lipid A—cannot be excluded at present, but is more difficult to reconcile with potential control mechanisms. The high degree of specificity in the fatty acid compositions of lipopolysaccharide and phospholipid raises additional questions about the nature and specificity of the respective acyl donors. Although exogenous fatty acids are readily incorporated into phospholipids, no detectable incorporation of exogenous D-β-hydroxy[^{14}C]myristate into lipopolysaccharide could be observed in β-oxidation-negative mutants (*fadE*) of *E. coli* (M. J. Osborn and K. A. Phan, unpublished results); the result suggests that the acylCoA cannot act as donor in lipid A synthesis. Incorporation of the acyl chain from synthetic D-β-hydroxymyristoylACP into phosphatidylethanolamine has been demonstrated in isolated cell envelope from *E. coli* 0111 (Taylor and Heath, 1969), yet the β-hydroxy fatty acid is effectively excluded from phospholipids *in vivo*. Conversely, long chain and unsaturated fatty acids are excluded from lipopolysaccharide. A possibility that merits consideration is that the compositional specificity is not solely conferred by acyl chain specificity of the relevant acyltransferases, but arises from substrate segregation at the level of the acyl donor. Perhaps the immediate donor of β-hydroxymyristoyl and other acyl chains to lipid A is neither ACP nor CoA, but an as yet unidentified coenzyme specific to the pathway.

3 ASSEMBLY OF LIPOPOLYSACCHARIDE IN THE OUTER MEMBRANE

The outer membrane is a classic example of a type of membrane biogenesis in which all major constituents are synthesized elsewhere in the cell and assembly of the membrane requires translocation of each component from its site of synthesis to sites of insertion and integration into the final membrane structure. Assembly of the outer membrane poses additional special problems, in that the membrane is outside the cell and is presumably restricted in its communication with other cell compartments. The inner (cytoplasmic) membrane is well established as the site of lipopolysaccharide synthesis (Osborn, Gander and Parisi, 1972), and current models for translocation of lipopolysaccharide (Mühlradt et al., 1973; Bayer, 1975), postulate transfer to the outer membrane at limited regions of contact between the two membranes, the zones of adhesion initially described by Bayer (1968). The hypothesis that zones of adhesion serve as specialized sites of intermembrane translocation and outer membrane assembly is an attractive one, and has been invoked to account for translocation of outer membrane phospholipids (Jones and Osborn, 1977) and proteins (Silhavy, Bassford and Beckwith, 1979) as well as lipopolysaccharide. At the present time, however it is possible to describe lipopolysaccharide translocation only at a phenomenological level; the physical and chemical events involved in the process at the molecular level are still poorly defined. Uncertainties include: the nature of the actual translocation events, i.e., intra-, inter- and transmembrane movements of the lipopolysaccharide molecule; whether these events require facilitation and the mechanism by which the process is rendered irreversible; and most importantly, the molecular architecture and functional organization of zones of adhesion.

3.1 Translocational Movement of Lipopolysaccharide

There is agreement that lipopolysaccharide is inserted into the outer membrane at a limited number of randomly distributed sites, although estimates of the number of functional sites vary from 20 to 50 (Bayer, 1975; Kulpa and Leive, 1976) to 250 (Mühlradt et al., 1973). However, it is not known whether there is a corresponding segregation of the sites of lipopolysaccharide synthesis in the inner membrane. Sucrose density gradient fractionation of inner membrane has not revealed significant separation of pulse-labeled lipopolysaccharide or the O-antigen enzyme complex from components of the respiratory chain (Osborn et al., 1972; and unpublished observations), but the crudity of the technique precludes conclusions about structural differentiation of the membrane. Studies on rates of translocation of lipopolysaccharide show a significant time lag (1–2 min at 25°C) between the appearance of pulse-labeled lipopolysaccharide in the inner membrane and its transfer to outer membrane, (M. J. Osborn unpublished observations), but whether this relates to transit time from site of synthesis to site of translocation is not clear. The question of structural coupling between synthesis and translocation of lipopolysaccharide and the extent to which newly synthesized molecules must undergo lateral diffusion in the inner membrane to reach sites of translocation therefore remains open.

Topological considerations suggest that functional integration of lipopolysaccharide into the outer membrane may require transmembrane translocation (flip-flop), as well as transfer between membranes. The distribution of lipopolysaccharide in the outer membrane is asymmetric and is restricted to the outer face of the bilayer (Mühlradt and Golecki, 1975). In addition, it is likely that synthesis of lipopolysaccharide takes place, at least in part, at the inner face of the inner membrane. Although the sidedness of the biosynthetic enzyme of the inner membrane has not been determined, access of the glycosyltransferases to their nucleotide sugar substrates implies exposure at the cytoplasmic interface. Thus integration of lipopolysaccharide into the outer membranes would appear to require transposition across two bilayers. A number of formal mechanisms for effecting such transposition can be envisioned, corresponding to alternative notions about the molecular architecture of zones of adhesion. Four such schemes, all entirely fanciful, are indicated in Figure 3. If the adherent sites are considered simply as points of close contact between the inner and outer membrane (Figure 3*a*), two successive transmembrane flip-flop events would be necessary in addition to intermembrane transfer between the inner and outer membrane. Alternatively, fusion of the two bilayers (Figure 3*b*) would reduce the number of transpositions to a single transbilayer flip-flop, and establishment of direct continuity between the inner and outer membranes (Figure 3*c*) would require only lateral diffusion of lipopolysaccharide within a single leaflet. Another, and radically different, mechanism (Figure 3*d*) postulates that translocation of lipopolysaccharide occurs by a kind of budding or primitive exocytosis of inner membrane and that sites of lipopolysaccharide trans-

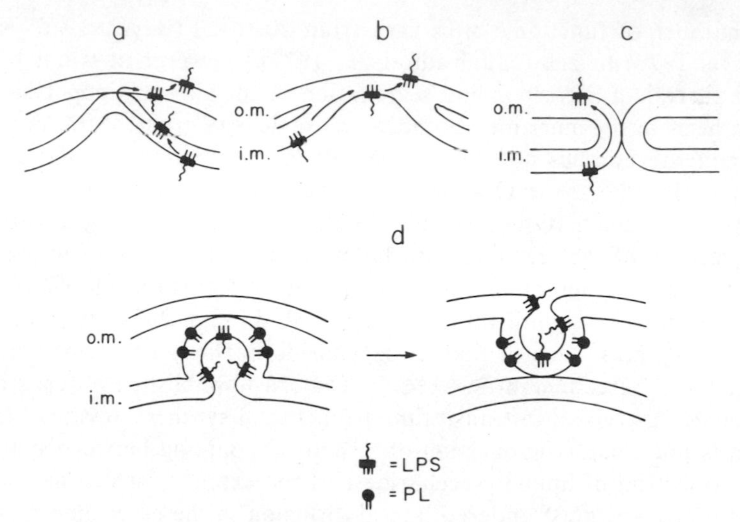

Figure 3 Possible modes of translocation of lipopolysaccharide. LPS, lipopolysaccharide; PL, phospholipid; i.m., inner membrane o.m., outer membrane.

location visualized as zones of adhesion by electron microscopy of plasmolyzed cells represent an intermediate stage of this dynamic process. This mechanism could also account directly and simply for the asymmetric localization of phospholipid in the inner leaflet of the outer membrane postulated by Smit et al. (1975).

Experimental evidence bearing on the occurrence of transmembrane transposition or flip-flop of lipopolysaccharide is fragmentary but provocative. Early studies of Shands (1966) with ferritin-conjugated O-specific antibody showed extensive labeling of the external face of the inner membrane, as well as both faces of the outer membrane in penicillin spheroplasts of *S. typhimurium*. The finding of lipopolysaccharide at the periplasmic face of the inner membrane has apparently not been further investigated, and its relationship to the normal translocation process remains unknown. Using similar ferritin-labeling techniques, Mühlradt and Golecki (1975) found that lipopolysaccharide was restricted to the external face of the outer membrane in "cell wall" (murein–outer membrane) fractions treated with lysozyme at 0°C, but that rapid reorientation to the inner face occurred when such preparations were exposed to higher temperatures. They concluded that maintenance of the asymmetric distribution of lipopolysaccharide in the outer membrane is dependent on the integrity of the underlying murein layer and suggested that discontinuities in murein at zones of adhesion may be important in transmembrane translocation of new lipopolysaccharide to the external face of the outer membrane. The mechanism of the observed reorientation of lipopolysaccharide and the role of murein in stabilizing the asymmetric outer membrane structure are not yet clear. Spontaneous molecular flip-flop of a molecule carrying a bulky polysaccharide chain across an intact membrane bilayer is unexpected, and it has not been established whether other components of the outer membrane also undergo reorientation under these conditions. The phenomena clearly merit further study from the points of view of both assembly and molecular organization of the outer membrane.

3.2 Is Lipopolysaccharide Translocation a Facilitated Process?

The question posed is the following: Is translocation of lipopolysaccharide a spontaneous diffusional process, or does the process require catalysis for thermodynamic or kinetic reasons? Surprisingly, no data are available on energy requirements in lipopolysaccharide translocation. The question is particularly cogent in light of recent findings of Donahue-Rolfe and Schaechter (1978) that the rate of translocation of phosphatidylethanolamine to outer membrane is markedly inhibited by uncouplers and energy poisons in *E. coli*. Whether the effect signifies an energy requirement for phospholipid translocation per se or for maintenance of a required structural organization is not yet clear. A requirement for facilitation, although not necessarily for expenditure of energy, might be expected if translocation of lipopolysaccharide (or phospholipid) involves transmembrane flip-flop or intermembrane transfer between bilayers (see Figure 3*a*). In addition, the possibility must be considered that translocation of lipopolysaccharide is "pulled" by covalent or noncovalent modification at the molecule within the outer membrane. In the author's opinion, covalent modification of lipopolysaccharide is unlikely. Covalent cross-linking to outer membrane protein(s) may occur (Wu and Heath, 1973), but is not obligatory (Garten et al., 1975; Leive, 1974; P. McVerry and M. J. Osborn, unpublished results), and may arise under certain conditions from a relatively nonspecific secondary interaction of lipopolysaccharide with lysine aldehyde residues of outer membrane proteins (Diedrich and Schnaitman, 1978). On the other hand, it is now clear that lipopolysaccharide interacts noncovalently with major outer membrane proteins and that these interactions are extensive and of very high affinity (Datta et al., 1977; van Alphen et al., 1977; Schindler and Rosenbusch, 1978; Mutoh et al., 1978). It is very likely that posttranslocational lipopolysaccharide–protein interactions in the outer membrane account for the functional irreversibility of lipopolysaccharide translocation (Osborn et al., 1972; Mühlradt et al., 1973). It is less likely that complexation with outer membrane proteins is required for translocation per se, since synthesis and integration of lipopolysaccharide into outer membrane continues for several hours in the absence of protein synthesis (Knox et al., 1967; Rothfield and Pearlman-Kothencz, 1969) and the rate of translocation is unaffected under these conditions (M. J. Osborn, unpublished observations).

The existence of proteins that catalyze or facilitate translocation of lipopolysaccharide has not been demonstrated. However, studies on the relation of lipopolysaccharide structure to translocation have raised the possibility that specific recognition of lipopolysaccharide may indeed occur at some point in the overall process. The suggestion is based on the finding that translocation of the incomplete lipid A precursor formed by KDO-8-P synthetase mutants is grossly defective (Osborn et al., 1979). A large fraction (30–50%) of the incomplete lipid is recovered in the inner membrane, and this is accounted for by a five-fold reduction in the rate of translocation compared to that of the lipopolysaccharide of the same strain. However, integration of the incomplete molecule into the outer membrane appears to be stable and functionally irreversible, like that of the parental lipopolysaccharide. As is indicated above (Section 2.2), previously accumulated lipid A precursor is rapidly converted to lipopolysaccharide following shift of mutant cultures to permissive conditions. However, this conversion is restricted to precursor present in the inner membrane at the time of shift (Osborn et al., 1979); material previously translocated to the outer membrane is stable and remains inaccessible to the biosynthetic enzyme system. Similarly, experiments in which exogenous lipid A precursor was introduced from phospholipid vesicles into the outer membrane of intact wild-type cells showed no detectable transfer to inner membrane or conversion to lipopolysaccharide under conditions in which vesicle-derived phospholipids were rapidly translocated back to the inner membrane (Jones and Osborn, 1977). Although it is difficult to distinguish unequivocally by *in vivo* pulse-chase techniques between defective translocation and defective integration into outer membrane, the results suggest that translocation itself is affected.

The nature of the structural features required for effective translocation of lipopolysaccharide is not altogether clear. The polysaccharide chain does not appear to be of major importance, since mutant lipopolysaccharides of R_a, R_c, and R_e chemotypes are translocated at rates similar to that of the wild type in *S. typhimurium* (M. J. Osborn and J. E. Gander, unpublished observations). More surprising, perhaps, is the finding that complete acylation of lipid A is not necessary: the underacylated lipopolysaccharide formed from lipid A precursor in the presence of cerulenin (see above, Section 2.2) is translocated at a normal rate (Walenga and Osborn, 1979b). These data suggest that the KDO trisaccharide region may be important as a structural determinant of translocation, but direct evidence on this point is not yet available. It is of possible interest in this connection that the cluster of KDO carboxyl groups appears to provide a binding site of rather high affinity for divalent cations (Schindler and Osborn, 1979). Studies of cation binding by equilibrium dialysis and fluorescence titration of dansyl-lipopolysaccharide have shown the presence of two classes of binding sites for Ca^{2+} and Mg^{2+} with widely differing K_D's. The high affinity site (K_D = 6–15 M) was observed in all lipopolysaccharides (S, R_c, R_e) but was absent in lipid A and the KDO-deficient lipid A precursor. Low affinity binding of these cations ($K_D \geq 1$ mM) was found in all preparations and was attributed to phosphate and pyrophosphate residues. The role of divalent cations in maintenance of the structural organization of lipopolysaccharide in the outer membrane is well known (Leive, 1974); whether interaction of cations with KDO residues plays an additional role in translocation remains to be determined.

3.3 Isolation and Characterization of Zones of Adhesion

The detailed fine structure of zones of adhesion has not yet been resolved by electron microscopy, and their isolation as a biochemical entity has thus far met with limited success. Potential markers to distinguish zones of adhesion from bulk envelope include enrichment for newly synthesized lipopolysaccharide (Mühlradt et al., 1973; Mühlradt et al., 1974; Bayer, 1974) and sites of phage attachment and injection (Bayer and Starkey, 1972; Bayer, 1975). Kulpa and Leive (1976) effected a partial separation of fragments of outer membrane containing newly inserted lipopolysaccharide by sucrose density gradient fractionation of membranes from *E. coli* J5 (*galE*) pulse labeled with galactose. The existence of compositional differences between these "insertion points" and the bulk membrane has not yet been established but is of obvious interest. The use of phage attachment as a probe for zones of adhesion was initially explored by Tomita et al. (1976), who carried out sucrose gradient centrifugation of membranes obtained from ϵ^{15}-infected *Salmonella anatum*. Under conditions of productive infection, the phage protein was specifically recovered in association with a membrane fraction sedimenting at a density intermediate between inner and outer membrane, and it was suggested that this fraction might represent phage irreversibly attached to zones of adhesion. However, in similar experiments with phage P22 and *S. typhimurium* (Crowlesmith et al., 1978), it was found that the so-called ϕ band was not at equilibrium in sucrose gradient centrifugation and primarily consisted of empty phage beads. No specific association with membranous material (other than bulk outer membrane) was detected.

Physical isolation of zones of adhesion is necessarily predicted on the assumption that these structures are sufficiently stable to survive conditions of cell rupture and comminution of membrane fragments and are sufficiently different in composition from the bulk membranes to permit their separation. Whether the existing methodologies are inadequate or whether the assumption is incorrect is not evident to the author. The question of the molecular architecture and functional organization of zones of adhesion is central to the problem of outer membrane assembly, and new approaches, genetic and ultrastructural as well as biochemical, are clearly needed to answer this question.

REFERENCES

Bayer, M. E. (1968), *J. Gen. Microbiol.*, **53,** 395–404.

Bayer, M. E. (1974), *Am. N.Y. Acad. Sci.*, **235,** 6–28.

Bayer, M. E. (1975), *Membrane Biogenesis* (A. Tzagaloff, ed.), Plenum, New York, pp. 393–427.

Bayer, M. E., and Starkey, T. (1972), *Virology*, **49,** 236–256.

Costeron, J. W., Ingram, J. M., and Cheng, K.-J. (1974), *Bacteriol. Rev.*, **38,** 87–110.

Crowlesmith, I., Schindler, M., and Osborn, M. J. (1978), *J. Bacteriol.*, **135,** 259–269.

Datta, D. B., Ander, B., and Henning, U. (1977), *J. Bacteriol.*, **131,** 821–829.

Diedrich, D. L., and Schnaitman, C. A. (1978), *Proc. Natl. Acad. Sci. U.S.*, **75,** 3708–3712.

Donahue-Rolfe, A., and Schaechter, M. E. (1978), Abstracts, 78th Annual Meeting of the American Society of Microbiology, 154.

Dröge, W., Lehmann, V., Lüderitz, O., and Westphal, O. (1970), *Eur. J. Biochem.*, **14,** 175–184.

Eidels, L., Rick, P. D., Stimler, N. P., and Osborn, M. J. (1974), *J. Bacteriol.*, **119,** 138–143.

Galanos, C., and Lüderitz, O. (1975), *Eur. J. Biochem.*, **54,** 603–610.

Galanos, C., Lüderitz, O., Rietschel, E. T., and Westphal, O. (1977), in *Biochemistry of Lipids II*, Vol. 14 (T. W. Goodwin ed.), University Park Press, Baltimore, pp. 239–335.

Garten, W., Hindennach, I., and Henning, U. (1975), *Eur. J. Biochem.*, **59,** 215–221.

Ghalambor, M. A., and Heath, E. C. (1966), *J. Biol. Chem.*, **241,** 3222–3227.

Ghalambor, M. A., Levine, E. M., and Heath, E. C. (1966), *J. Biol. Chem.*, **241,** 3207–3215.

Heath, E. C., Mayer, R. M., Edstrom, R. D., and Beaudreau, C. A. (1966), *Ann. N.Y. Acad. Sci.*, **133,** 315–333.

Jones, N. C., and Osborn, M. J. (1977), *J. Biol. Chem.*, **252,** 7405–7412.

Knox, R. W., Cullen, I., and Work, E. (1967), *Biochem. J.*, **103,** 192–201.

Kulpa, L. F., and Leive, L. (1976), *J. Bacteriol.*, **126,** 467–477.

Lehmann, V. (1977), *Eur. J. Biochem.*, **75,** 257–266.

Lehmann, V., and Rupprecht, R. (1977), *Eur. J. Biochem.*, **81,** 443–452.

Lehmann, V., Rupprecht, E., and Osborn, M. J. (1977), *Eur. J. Biochem.*, **76,** 41–49.

Lehmann, V., Redmond, J., Egan, A., and Minner, I. (1978), *Eur. J. Biochem.*, **86,** 487–496.

Leive, L. (1974), *Ann. N.Y. Acad. Sci.*, **235,** 109–129.

Lennarz, W. J., and Scher, M. G. (1972), *Biochim. Biophys. Acta*, **265,** 417–441.

Levin, D. H., and Racker, E. (1959), *J. Biol. Chem.*, **234,** 2532–2539.

Lim, R., and Cohen, S. S. (1966), *J. Biol. Chem.*, **241,** 4304–4315.

Luderitz, O., Galanos, C., Lehmann, V., Nurminen, M., Rietschel, E. T., Rosenfelder, G., Simon, M., and Westphal, O. (1973), *J. Infect. Dis.*, **128,** 517–529.

Luderitz, O., Galanos, C., Lehmann, V., Mayer, H., Rietschel, E. T. and Weckesser, J. (1978), *Naturwissenschaften*, **65,** 578–585.

Malchow, D., Luderitz, O., Kickhofen, B., and Westphal, O. (1969), *Eur. J. Biochem.*, **7,** 239–246.

Mühlradt, P. F., and Golecki, J. R. (1975), *Eur. J. Biochem.*, **51,** 343–352.

Mühlradt, P. F., Menzel, J., Golecki, J. R., and Speth, V. (1973), *Eur. J. Biochem.*, **35,** 471–481.

Mühlradt, P. F., Menzel, J., Golecki, J. R., and Speth, V. (1974), *Eur. J. Biochem.*, **43,** 533–539.

Mühlradt, P., Wray, V., and Lehmann, V. (1977), *Eur. J. Biochem.*, **81,** 193–203.

Munson, R. S., Jr., Rasmussen, N. S., and Osborn, M. J. (1978), *J. Biol. Chem.*, **253,** 1503–1511.

Mutoh, N., Furakawa, H., and Mizushima, S. (1978), *J. Bacteriol.*, **136,** 693–699.

Nikaido, H. (1968), *Adv. Enzymol.*, **31,** 77–124.

Nikaido, H. (1973), in *Bacterial Membranes and Walls* (L. Leive, ed.), Dekker, New York, pp. 131–208.

Nunn, W. D., Kelly, D. L., and Stumfall, M. Y. (1977), *J. Bacteriol.*, **132,** 526–531.

Omura, S. (1976), *Bacteriol. Rev.*, **40,** 681–697.

Osborn, M. J. (1969), *Annu. Rev. Biochem.*, **38,** 501–538.

Osborn, M. J. (1971), in *Structure and Function of Biological Membranes* (L. I. Rothfield, ed.), Academic Press, New York, pp. 343–400.

Osborn, M. J., and Rothfield, L. I. (1971), in *Microbial Toxins*, Vol. IV (G. Weinbaum, S. Kadis, and S. I. Ajl, eds.), Academic Press, New York, pp. 331–350.

Osborn, M. J., Gander, J. E., and Parisi, E. (1972), *J. Biol. Chem.*, **247,** 3973–3986.

Osborn, M. J., Rasmussen, N. S., and Rick, P. D. (1979), sumitted to *J. Biol. Chem.*

Prehm, P. K. Stirm, S., Jann, B., and Jann, K. (1975), *Eur. J. Biochem.*, **56,** 41–55.

Rick, P. D., and Osborn, M. J. (1972), *Proc. Natl. Acad. Sci. U.S.*, **69,** 3756–3760.

Rick, P. D., and Osborn, M. J. (1977), *J. Biol. Chem.*, **252,** 4895–4903.

Rick, P. D., Fung, L. W.-M., Ho, C., and Osborn, M. J. (1977), *J. Biol. Chem.*, **252,** 4904–4912.

Rietschel, E. T., Gottert, H., Luderitz, O., and Westphal, O. (1972), *Eur. J. Biochem.* **28,** 166–173.

Rietschel, E. T., Hase, S., King, M. T., Redmond, J., and Lehmann, V. (1977). *Microbiology—1977*. (D. Schlessinger, ed.), American Society of Microbiology, Washington D.C., pp. 262–268.

Robbins, P. W., and Wright, A. (1971), in *Microbial Toxins*, Vol. IV (G. Weinbaum, S. Kadis, and S. J. Ajl, eds.), Academic Press, New York, pp. 351–368.

Romeo, D., Girard, A., and Rothfield, L. (1970), *J. Mol. Biol.*, **53,** 475–490.

Rooney, S. A., and Goldfine, H. (1972), *J. Bacteriol.*, **111,** 531–541.

Rosner, M. R., and Khorana, H. G. (1979), *J. Biol. Chem.*, in press.

Rosner, M. R., Tang, J., Barzilay, I., and Khorana, H. G. (1979), *J. Biol. Chem.*, in press.

Rothfield, L. I., and Hinckley, A. H. III (1974), in *Comparative Biochemistry and Physiology of Transport* (L. Bolis, K. Bloch, S. E. Luria, and F. Lynen, eds.), North Holland, Amsterdam, pp. 102–112.

Rothfield, L., and Pearlman-Kothencz, M. (1969), *J. Mol. Biol.*, **44,** 477–492.

Rothfield, L., and Romeo, D. (1971), *Bacteriol. Rev.*, **35,** 14–38.

Rottem, S., Markowitz, O., and Razin, S. (1978), *Eur. J. Biochem.*, **85,** 451–456.

Schindler, H., and Rosenbusch, J. P. (1978), *Proc. Natl. Acad. Sci. U.S.*, **75,** 3751–3755.

Schindler, M., and Osborn, M. J. (1979), submitted to *Biochemistry.*

Shands, J. W. (1966), *Ann. N.Y. Acad. Sci.*, **133,** 292–298.

Silbert, D. F., Ulbright, T. M., and Honegger, J. (1973), *Biochemistry*, **12,** 164–171.

Silhavy, T. J., Bassford, P. J., Jr., and Beckwith, J. R. (1979), this volume, Chapter 7.

Smit, J., Kamio, Y., and Nikaido, H. (1975), *J. Bacteriol.*, **124,** 942–958.

Taylor, S. S., and Heath, E. C. (1969), *J. Biol. Chem.* **244,** 6605–6616.

Tomita, T., and Kaanegasaki, S. (1976), *Biochem. Biophys. Res. Commun.*, **73,** 807–813.

Van Alphen, W., and Lugtenberg, B. (1977), *J. Bacteriol.*, **131,** 623–630.

Van Alphen, L., Havekes, L., and Lugtenberg, B. (1977), *FEBS Lett.*, **75,** 285–290.

Van den Bosch, H., and Vagelos, P. R. (1970), *Biochim. Biophys. Acta*, **218,** 233–248.

Walenga, R. W., and Osborn, M. J. (1979a), submitted to *J. Biol. Chem.*

Walenga, R. W., and Osborn, M. J. (1979b),

Wright, A., and Kanegasaki, S. (1971), *Physiol. Rev.*, **51,** 748–784.

Wu, M. C., and Heath, E. C. (1973), *Proc. Natl. Acad. Sci. U.S.*, **70,** 2572–2576.

Chapter 3 Phosphoolipid Synthesis and Assembly

JOHN E. CRONAN, JR.

Department of Microbiology, University of Illinois, Urbana

1 INTRODUCTION

The biosynthesis of the membrane lipids of enterobacteria is quite well understood. The lipids consist of phospholipids, lipid A, and the lipoprotein.

The phospholipids, all of which are membrane bound, are the major (>90%) fatty acid-containing structure in the enterobacteria. This chapter concentrates on the biogenesis of the phospholipid component of the membranes and is concerned with the topology of synthesis and movement of phospholipids during membrane biogenesis. I largely bypass the bulk of the data on the enzymology and regulation of fatty acid and phospholipid synthesis because these subjects were recently reviewed by the present author (Cronan and Vagelos, 1972; Cronan and Gelmann, 1975; Cronan, 1978), by C. H. Raetz (1978), and by others (for references see Cronan, 1978). The status of the work on lipid A and the lipoprotein is reviewed from different perspectives elsewhere in this volume.

The phospholipids of *Escherichia coli* are phosphatidylethanolamine (PE), phosphatidylglycerol (PG), and cardiolipin (CL), all of which have similar fatty acid compositions consisting principally of palmitic, myristic, palmitoleic and *cis*-vaccenic acids. PE is the major phospholipid (70–80% of the lipid phosphorus). The relative amounts of PG and CL depend on the growth phase of the cells, as do the relative amounts of the palmitoleic and *cis*-vaccenic acyl groups and their cyclopropane derivatives. The phospholipids of *Salmonella tymphimurium* are identical to those of *E. coli* (Ames, 1968). The phospholipids of the other enterobacteria and of other gram-negative organisms are, in general, quite similar to those of *E. coli* (Goldfine, 1972). The structures of the major fatty acids and phospholipids of *E. coli* are given in Figures 1 and 2. The phospholipid biosynthetic

Structural type	*General structure*	*Systematic names*	*Trivial names*
Saturated	$CH_3-(CH_2)_x-COOH$	Dodecanoic acid ($x = 10$)	Lauric acid
		Tetradecanoic acid ($x = 12$)	Myristic acid
		Hexadecanoic acid ($x = 14$)	Palmitic acid
		Octadecanoic acid ($x = 16$)	Stearic acid
Unsaturated	$CH_3-(CH_2)_5\overset{H}{C}=\overset{H}{C}-(CH_2)_x-COOH$	*cis*-9-Hexadecenoic acid ($x = 7$)	Palmitoleic acid
		cis-11-Octadecenoic acid ($x = 9$)	*cis*-Vaccenic acid
Cyclopropane	$CH_3-(CH_2)_5-\underset{H}{C}\overset{CH_2}{\frown}\underset{H}{C}-(CH_2)_x-COOH$	*cis*-9, 10-Methylene hexadecanoic acid ($x = 7$)	none
		cis-11, 12-Methylene octadecanoic acid ($x = 9$)	Lactobacillic acid
Hydroxy	$CH_3-(CH_2)_x-CHOH-CH_2-COOH$	D(−)-3-Hydroxytetradecanoic acid ($x = 10$)	β-Hydroxymyristic acid

Figure 1 Structures and nomenclature of the fatty acids of *E. coli*. All of the above fatty acids are components of the phospholipids of *E. coli*, except β-hydroxymyristic acid is only found in the lipid A component of LPS. For detailed information see Cronan and Vagelos (1972).

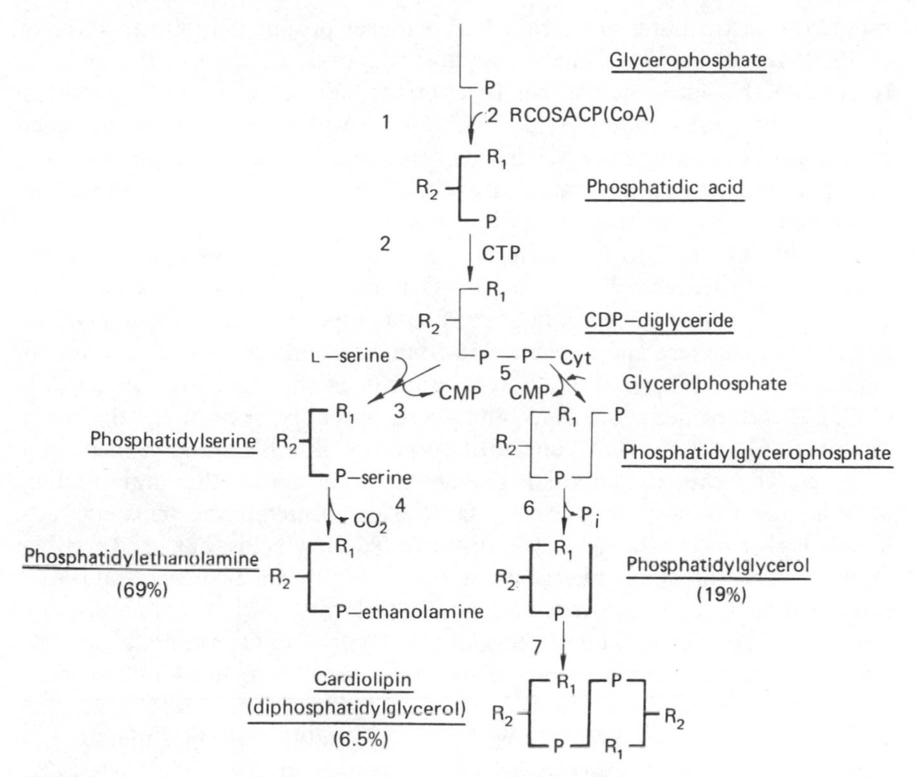

Figure 2 Structures and biosynthesis of the phospholipids of *E. coli*. R_1 and R_2 denote the fatty acyl moieties of the phospholipids, R_1 generally being palmitic acid and R_2 being an unsaturated or cyclopropane fatty acid. The vertical line denotes the glycerol moiety. The values in parentheses denote the percentage of lipid phosphorus in that lipid in exponentially growing cells. The biosynthesis is catalyzed by the following enzymes: (*1*) *sn*-glycerol 3-phosphate and monoacylglycerol 3-phosphate acyltransferases; (*2*) CTP-diglyceride synthetase; (*3*) phosphatidylserine synthetase; (*4*) phosphatidylserine decarboxylase; (*5*) phosphatidylglycerolphosphate synthetase; (*6*) phosphatidylglycerol phosphate phosphatase; (*7*) cardiolipin synthetase (catalyzes the conversion of two phosphatidylglycerol molecules to cardiolipin plus glycerol). The major phospholipids and the important intermediates are given in boldface. All the intermediates have been demonstrated *in vivo* and together comprise $<1\%$ of the total lipid phosphorus. For detailed information consult Cronan and Vagelos (1972) and Raetz (1978).

pathway is also given in Figure 2. The reader is referred to the reviews cited above for more detailed information on the lipid structures and pathways.

2 LIPID COMPOSITION OF INNER AND OUTER MEMBRANES

The lipid compositions of the two membranes were uncertain for several years since the first two papers to report the lipid composition of inner and outer membranes presented conflicting data. Osborn and coworkers (1972) reported that the outer membrane had a higher proportion (80 versus 60%) of PE than did the inner membrane and was thus deficient in PG and CL relative to the inner membrane. In contrast, White et al. (1972) reported that the outer membrane was deficient in PE when compared to the inner membrane. The latter workers also reported that the outer membrane phospholipids had a higher content of saturated fatty acids (45% versus 35%) than was found in the inner membrane phospholipids.

The bulk of the data in the literature since those first two papers (Joseleau-Petit and Kepes, 1975; Jones and Osborn, 1977b; Goodell et al., 1974, R. M. Bell, personal communication), and especially the recent detailed paper by Lugtenberg and Peters (1976), indicate that the outer membrane is enriched in PE. The phospholipid compositions reported by White et al. (1972) therefore seem spurious and were probably skewed by the large amounts of lyso PE and of "unidentified phospholipids" found in their lipid analyses. This explanation seems reasonable because the other main finding of White and co-workers (1972), that the outer membrane phospholipids have a higher ratio of saturated to unsaturated fatty acids, has been verified in wild-type cells (Lugtenberg and Peters, 1976) and also in unsaturated fatty acid auxotrophs supplemented with unsaturated fatty acids not usually found in enterobacteria (Lugtenberg and Peters, 1976; Overath et al., 1975). Some of the increased proportion of unsaturated fatty acid in the inner membrane is probably due to its greater content of PG enriched in *cis*-vaccenic acid (Kito et al., 1975). Most fatty acid analyses of inner membranes show an increased proportion of *cis*-vaccenic acid (White et al., 1972; Koplow and Goldfine, 1974; Lugtenberg and Peters, 1976). However, Lugtenberg and Peters (1976) have reported that the inner membrane PE has a somewhat higher proportion of unsaturate than does the PE of the outer membrane.

All the above data on the phospholipid composition of inner and outer membranes were obtained by phospholipid analysis of the isolated membranes. However, the preparative procedures usually involve sonication or shearing in a French pressure cell. Such treatments can cause the intermembrane transfer of phospholipids (Tsukagoski and Fox, 1971; Devor et al., 1976). Indeed, under some conditions, cocentrifugation of two membrane populations can result in intermembrane transfer of phospholipids (Devor *et al.*, 1976). Therefore, it, seems quite possible that the phospholipid compositions of the inner and outer membranes may differ by a greater degree than is usually observed. Thus the more extreme values in the literature may more accurately reflect the *in situ* compositions. The most extreme values are those of Lugtenberg and Peters (1976) and of Bell (R. M. Bell, personal communication), who have obtained outer membrane preparations from *E. coli* K-12 in which 90–95% of the phospholipid was PE.

If a major portion of the outer membrane is PE, it is possible that the phospholipid of this membrane may not be packed into the standard lamellar arrangement. This possibility should be considered since pure PE readily forms an H_{II} phase, an arrangement in which the PE molecules form a hydrophobic sheet containing cylindrical holes (Luzzati, 1968), although PE can also be packed into a lamellar structure (Stollery and Vial, 1977). The holes in the H_{II} phase consist of water and the polar head groups of the phospholipids, and such holes could be of obvious physiological relevance. It would be of interest therefore to examine the phospholipids composition of outer membranes prepared by methods that prevent lipid exchange (Tsukagoski and Fox, 1971). The amount of H_{II} phase in these membranes might be assayed by ^{31}P NMR (Cullis and de Kruijff, 1978).

Another possible approach to the determination of the relative PE contents of inner and outer membranes is by means of *in situ* labeling of phospholipids using reagents that penetrate the outer but not the inner membrane. Marinetti and Love (1977) have presented such labeling data using two amino group reagents. However, these experiments are extremely unsatisfactory. The specificity of the probes was only assumed and the cells examined had previously been frozen.

3 PHYSICAL PROPERTIES OF OUTER MEMBRANE PHOSPHOLIPIDS

3.1 Phase Transitions

The lipid phase properties of the phospholipids of the outer membrane are very similar to those of the inner membrane. Most of the work on order–disorder phase transitions in *E. coli* have used either purified inner membranes or membrane vesicles prepared by the method of Kaback (1971) (which contain a definite although variable amount of outer membrane). The agreement within the field is quite good and does not depend on the type of preparation used (Cronan and Gelmann, 1975), which strongly suggests that the shapes and maxima of the phase transitions of inner and outer membranes are quite similar. The only direct study of the transitions of separated inner and outer membranes (Overath et al., 1975) supports this conclusion. These workers found very similar transitions for both membranes, the only difference being that a smaller portion of the outer membrane phospholipids participates in the transition. This last point is discussed below (see Section 3.4).

3.2 Lateral Diffusion

The movement of lipids in the plane of the membrane is quite rapid. Using a spin-labeled fatty acid, Sackman et al. (1973) found a coefficient of lateral diffusion of 3.2×10^{-8} cm^2/sec in *E. coli* "Kaback vesicles" (Kaback, 1971) above the phase transition. Based on this value an average lipid molecule moves about 3.5 μm in 1 sec. This is a value similar to that observed in bilayers of pure synthetic lipids (Lee, 1975).

In apparent conflict, Dupont et al. (1972) using an X-ray technique capable of kinetic resolution, reported that the time required for the lipid molecules of Kaback vesicles to complete the disorder–order transition was on the order of 10 min. Based on these data the authors suggested that the diffusion of lipids in the *E. coli* membrane is considerably slower (ca. ten- to hundred-fold) than the diffusion rate seen in pure lipid bilayers. However, more recent data (Caron et al., 1974; Mateu et al., 1978) suggest that a major portion of the time needed for completion of the disorder–order transition is consumed by nucleation rather than by diffusion and thus the results of Dupont et al. (1972) probably do not conflict with the spin-label results. In this regard it should also be noted that on theoretical grounds (Keith et al., 1973), the line-width method of evaluation of spin-labeling results can be expected to give an overestimate of the difussion rate. In view of these considerations a diffusion constant of about 10^{-8} cm^2/sec seems a reasonable estimate for the lipids of the *E. coli* cell envelope. This estimate is similar to that found in a variety of cells using a variety of methods (for review see Lee, 1975) and indicates that a given lipid molecule could move from one end of a bacterium to the other in less than a second. The fluorine NMR data on *E. coli* vesicles containing ^{19}F-labeled phospholipids (Gent et al., 1978; Gent and Ho, 1978) also seem consistent with a diffusion rate of about 10^{-8} cm/sec^2.

Recently Nikaido et al. (1977) reported spin-labeled experiments quite similar to those of Sackman et al. (1973) and from which a diffusion constant of 2.5×10^{-8} cm/sec^2 can be calculated. These workers find identical diffusion rates for the probes in isolated inner and outer membranes.

In conclusion, the diffusion rates of the phospholipids of both the inner and outer membranes are very rapid and the rates are typical of the fluidity observed in biological and artifical membranes.

3.3 Is the Outer Membrane Phospholipid in a Bilayer?

As is discussed above, the outer membrane of wild-type *S. tymphimurium* and *E. coli* seems enriched in PE when compared to inner membranes. The location of the PE in the outer membrane has been the subject of considerable work. It has been reported that the PE molecules of the outer membrane are formed into a lipid monolayer (Kamio and Nikaido, 1976), the polar groups of the PE molecules facing towards the inner membrane. The external leaflet of the outer membrane would then lack lipid and would be composed of only LPS and protein. The other possibility to be considered is that the phospholipids of the outer membranes are formed into a conventional lipid bilayer, although other organizations (see above) also seem possible. Unfortunately, the various physical techniques used to examine the phospholipid in the outer membrane are unable to resolve this question. None of the techniques can discriminate between a lipid bilayer and a lipid monolayer (Lee, 1975). Several groups have therefore approached this problem by chemical labeling with various impermeant amino group reagents.

Kamio and Nikaido (1976) reported that *S. tymphimurium* wild-type cells and R_c (galactose-less) strains were resistant both to phospholipase C and to labeling of PE by CNBr-activated dextran. In contrast, strains lacking the portion of the LPS internal to the galactose residues had an increased (ca. two-fold) content of outer membrane phospholipid that was sensitive to phospholipase C and to activated dextran labeling. Smit et al. (1975) had calculated that the wild-type outer membrane contains only half the phospholipid needed to form a continuous bilayer around the cell. This calculation coupled with the phospholipase and activated dextran labeling data led Nikaido and co-workers (Smit et al., 1975; Kamio and Nikaido, 1976) to posit that the wild-type outer membrane contains not a lipid bilayer, but a single monolayer on the interior face of the membrane. The mutants lacking a major portion of the LPS were then postulated to contain a lipid bilayer as a result of the substitution of phospholipid for LPS. The presence of this postulated bilayer would thus confer sensitivity to phospholipase C digestion and activated dextran labeling (Kamio and Nikaido, 1976).

Van Alphen et al. (1977) have reported analogous experiments using various *E. coli* mutants and phospholipases A_2 and C. Although their results are quite similar to those of Kamio and Nikaido (1976), these authors have interpreted their results as a demonstration of the shielding of the phospholipids by other membrane components rather than in terms of lipid location or structure. They have shown that mutants lacking various outer membrane proteins have increased phospholipid contents similar to those seen in deep-rough LPS mutants. However, one of these mutants is insensitive to phospholipase despite an increased lipid content. Phospholipase sensitivity can be conferred on another strain by EDTA treatment (LPS is not released). Neither of these results is consistent with the Nikaido hypothesis (see above) and thus van Alphen et al. (1977) favor a shielding hypothesis. In this regard Schindler and Teuber (1978) have shown that a *S. tymphimurium galE* strain cannot be labeled (in either protein or lipid) by a dansyl chloride impermeant reagent. However, addition of Tris-Cl, EDTA, divalent ions, and various cationic antibiotics allowed appreciable labeling of phospholipids and outer membrane proteins. These authors attribute the resistance of the native membrane towards dansylation to ionic interaction in the membrane.

The freeze-fracture technique suggests that the outer membrane does not have a typical lipid bilayer structure (Bayer et al., 1975; Smit et al., 1975; Verkleij et al., 1977; Schweizer et al., 1976). In wild-type strains, the frequency and extent of fracture within the outer membrane is quite low compared to the behavior of the inner membrane (Bayer et al., 1975; Smit et al., 1975). However, in mutant outer membranes that are enriched in phospholipid, cleavage often occurs within the outer membrane (Bayer et al., 1975; Smit et al., 1975). The morphology of the cleavage face of the outer membranes appears similar in wild-type and in phospholipid-enriched mutant strains. A large variety of bacterial strains have been examined and thus the observations seem quite firm.

To this reviewer, it seems that all the above data can be accommodated by a model in which the phospholipid in the wild-type outer membrane is in a lipid bilayer, but the lipid bilayer occurs only in small patches. In phospholipid-enriched mutants, the increased phospholipid content (and the general concomitant decrease in other components) leads to larger patches of bilayer. The proposal of small patches of phospholipid bilayer in the wild-type membrane could explain the freeze fracture results since the target size for cleavage would be smaller in the outer membrane than in the inner membrane. Also, the phospholipids by virtue of their patchiness would generally closely neighbor other membrane components that could shield the lipids from external probes. This model is also consistent with the results of Jones and Osborn (1977a, b), who have incorporated exogenous lipids into intact cells. This incorporation has the properties expected of a direct fusion of two phospholipid bilayers (Pagano and Weinstein, 1978). Jones and Osborn (1977a) found that a defective LPS in the recipient cell was required for fusion. However, only a twofold difference was found between a heptose-less strain (whose LPS contains only ketodeoxyoctonate and lipid A) and a strain missing only the galactose residues. The heptose-less strain has an increased outer membrane phospholipid content and often is freeze-fractured through the outer membrane (Smit et al., 1975), whereas the galactoseless mutant is wild type in both phospholipid content (Jones and Osborn, 1977a) and its freeze fracture behavior (Smit et al., 1975). In regards to fusion with lipid vesicles, the outer membrane of the galactoseless strain therefore behaves as though it has a lipid bilayer with half the surface area of the heptose-less strain rather than a monolayer on the inner face of the outer membrane. The fusion of synthetic phospholipid vesicles with a lipid monolayer would not be expected to occur (Pagano and Weinstein, 1978).

In principle, a direct test of the bilayer versus monolayer question in the outer membrane should be possible by experiments analogous to those done by Engleman (1971) on *Acholeoplasma laidlawaii*. Using X-ray diffraction (see Levine and Wilkens, 1971; Wilkens et al., 1971), Engleman (1971) was able to demonstrate that the thickness of the bilayer increased with increasing chain length of the phospholipid fatty acids (chain length was controlled by exogenous supplementation). The thickness increases in proportion to the average length of the hydrocarbon chains as expected for a bilayer rather than a monolayer structure (Engleman, 1971; Wilkens et al., 1971). A similar experiment could be done by growing fatty acid auxotrophs of *E. coli* (Cronan, 1978) on media supplemented with fatty acid mixtures of differing average chain length.

3.4 Boundary Lipid

Overath et al. (1975) reported that substantially less than half of the outer membrane phospholipid takes part in the order–disorder transition. The values for the inner membrane are about twice as high. Overath and coworkers (1975) suggested that a major portion of the lipids in the outer membrane is unable to undergo the lipid–lipid interactions characteristic of the order–disorder transition because of interaction with protein. A similar interpretation has been put forth for the lipids of Kaback vesicles by Träuble and Overath (1973).

Data similar to those of Overath and co-workers (1975) have been reported in a variety of model systems (see Papahadjopoulos et al., 1975) and have led to the concept of boundary lipid. Boundary lipid is defined as lipid bound to protein that, as a result of this binding, has properties different from those of the bulk lipid phase. The current notion of boundary lipid, although consistent with X-ray diffraction and colorimetric data, is based primarily on ESR spin label and ^{2}H NMR data.

Jost and co-workers (Jost et al., 1973, 1977) have found that a spin labeled phospholipid is immobilized by cytochrome oxidase and that the immobilized lipid exchanges slowly on the ESR time scale with the bulk membrane phospholipid. However, using ^{2}H NMR, Oldfield et al. (1978) have found no evidence for immobilized lipid and have reported that the presence of cytochrome oxidase (or of a number of other proteins) introduces more disorder into the phospholipid structure.

The resolution of the apparent conflict between the ESR and ^{2}H NMR data seems to lie in the differing time scales of the two techniques. ESR spin-labeled experiments have a time scale of about 10^8/sec, whereas the ^{2}H NMR time scale is about 3 orders of magnitude slower. Molecular motions that are fast on the ^{2}H NMR time scale (10^5–10^7/sec) would be very slow on an ESR time scale. In fact, the ESR technique would classify the molecules undergoing rapid (ca. 10^6/sec) movement as immobile.

In regard to the outer membranes, other workers (Rottem and Leive, 1977; Cheng et al., 1974) have interpreted the results of Overath and coworkers (1975) as indicating that the phospholipids of the outer membrane are less fluid than the inner membrane phospholipids. However, the current ideas of boundary lipid suggest the opposite conclusion. The result of Overath et al. (1975) can be explained by an increased outer membrane lipid fluidity due to the high protein content. The fluidizing effects of the protein would prevent a sizable fraction of the lipid from becoming ordered at low temperature and thus less of the phospholipid would undergo the order–disorder transition, exactly the result observed by Overath et al. (1975).

The finding that much of the outer membrane phospholipid does not participate in the lipid phase transition is therefore consistent with the current notions concerning boundary lipid. However, as noted by Overath et al. (1975), the presence of lipopolysaccharide could also alter the properties of the outer membrane phospholipids. Two recent spin-label studies have examined the relationship between the motion of a spin label (presumably located in the lipid phase) and LPS content of the outer membrane. Both Nikaido et al. (1977) and Rottem and Leive (1977) report that the motion of 5-doxylstearate in outer membranes with a galactose-deficient LPS is similar to that seen in inner membranes. The two reports differ, however, on the behavior of 12-doxylstearate. Rottem and Leive (1977) found differing behavior for the probe in inner and outer membranes whereas Nikaido et al. (1977) found identical spectra. Upon addition of galactose residues to the LPS *in vivo*, Rottem and Leive (1977) found a restriction of spin-label motion, whereas using a model system, Nikaido et al. (1977) found no effect of LPS on spin-label motion. A resolution of this difference may lie in the fact that Nikaido et al. (1977) used an LPS that lacked galactose residues.

4 SPACIAL ORGANIZATION OF MEMBRANE LIPID SYNTHESIS

4.1 Location of the Phospholipid Biosynthetic Enzymes

The enzymes of fatty acid synthesis are cytosolic enzymes, whereas the enzymes of phospholipid synthesis are (with one exception) integral proteins of the inner membrane. It seems reasonable that the fatty acid synthetic enzymes might be loosely associated with the membrane, but the only evidence favoring this notion is the autoradiographic evidence of van den Bosch et al. (1970), who showed that the acyl carrier protein (ACP) component of pantothenate-deprived cells is localized at the outer periphery of the cytoplasm.

Bell et al. (1971) and White et al. (1971) showed that the phospholipid biosynthetic enzymes of *E. coli* are inner membrane enzymes. These enzymes, PS decarboxylase, PGP synthetase, PGP phosphatase, cardiolipin synthetase, CDP-diglyceride synthetase, and the *sn*-glycerol-3-P and monoacylglycerol 3-P acyltransferases (Figure 2) were definitely shown to be inner membrane-bound enzymes. The trace of activity found in outer membranes could be attributed to contamination of the outer with the inner membrane. Similar results have been reported for diglyceride kinase (Schneider and Kennedy, 1976) CDP-diglyceride hydrolyase (Raetz et al., 1976) and monoacylphosphatidylglycerol synthetase (Nishijima et al., 1978).

All these proteins are integral membrane proteins; they can only be solubilized by detergent treatment. Another lipid biosynthetic enzyme, cyclopropane fatty acid synthase, is also bound to the inner membrane, but since the enzyme can be solubilized by high salt, it seems to be an extrinsic protein (Taylor and Cronan, 1979).

The only enzyme involved in phospholipid synthesis that is not an integral inner membrane protein is phosphatidylserine (PS) synthetase. A large fraction of this enzyme is found bound to ribosomes in cell extracts. This surprising finding, first reported by Raetz and Kennedy (1972), was challenged by Machtiger and Fox (1973). However, the total activity of phosphatidylserine synthetase observed by these latter authors was only about 10% of that observed by others and thus it seems probable that Machtiger and Fox (1973) lost or inactivated the ribosome-bound activity. Indeed, the recovery of ribosomes reported by these workers was extremely low. Ishinaga and Kito (1974) reported that about 80% of the phosphatidylserine synthetase was ribosome bound; most of the remainder was soluble, although a small amount was membrane bound. The soluble enzyme was readily bound to ribosomes, membranes, or phospholipids. The genetic evidence argues strongly that *E. coli* contains only a single phosphatidylserine synthetase (for reviews see Raetz, 1978; Cronan, 1978). The most straightforward intracellular location for this enzyme, considering its properties and its role in phospholipid synthesis, is loosely associated with the inner membrane of *E. coli*. The soluble and ribosomal activities would then be artifacts due to cell disruption. In this regard, it should be noted that the association of phosphatidylserine synthetase with ribosomes is not stochiometric (Raetz et al., 1977). Although no lipid biosynthetic enzymes are found in the outer membrane, several lipid catabolic activities are specific components of the outer membrane. Scandella and Kornberg (1971) reported the purification of a phospholipase A from the outer membrane, and Albright et al. (1973)

reported phospholipase A1, A2, and 1-acyl lysophospholipase activities to be located in the outer membrane. Bell et al. (1971) reported a "phospholipase plus lysophospholipase" activity in the outer membrane. Recent work by Nishjima et al. (1977) indicates that all these activities plus 2-acyl lysophospholipase activity reside in the same outer membrane protein. This enzyme, which is (and should properly be named as) a phospholipase B, can hydrolyze both fatty acids from a given diacyl phospholipid molecule. An additional 2-acyl lysophospholipase has been localized to the inner membrane (Albright et al., 1973; Nishijima et al., 1978). Another lipid catabolic activity, a phosphatase acting on phosphatidic acid and lysophosphatidic acid (van den Bosch and Vagelos, 1970) has also been reported to be localized in the outer membrane of *E. coli* (Bell et al., 1971).

The only exception to the specific localization of membrane-bound lipid metabolic enzymes to either the inner or the outer membrane is an activity that acylates 1-acyl lysophosphatidylethanolamine to the diacyl lipid. This activity has been reported to be equally distributed between the inner and outer membranes (Vos et al., 1978). Machtiger and Fox (1973) reported that PS decarboxylase was distributed between the inner and outer membranes. However, three other laboratories (Bell et al., 1971; White et al., 1971; Schneider and Kennedy, 1976) have found that virtually all the PS decarboxylase is located in the inner membrane. The specific activities found by Machtiger and Fox (1973) are very low and suggest that PS decarboxylase was inactivated during purification of the inner membrane fraction.

4.2 Topology of Membrane Lipid Synthesis

The location of the phospholipid biosynthetic enzymes in the inner membrane and presumably in the inner face of the inner membrane raises the question of the mechanism whereby newly synthesized phospholipid molecules are transferred to the outer face of the inner membrane and to the outer membrane. Unfortunately, the presence of two membranes makes answering the question for gram-negative organisms technically difficult. For this reason Rothman and Kennedy (1977a, b) have studied the gram-positive organism *Bacillus megaterium*. These workers (Rothman and Kennedy, 1977a) have reported that the distribution of PE is asymmetric in *B. megaterium*. In a convincing series of experiments using membrane impermeant labeling reagents, these workers showed that about one-third of the phosphatidylethanolamine can be rapidly labeled by two different reagents specific for amine groups (Rothman and Kennedy, 1977a). The reagents were shown not to enter the cell under these conditions. When conditions that allowed slow entry of a labeling reagent were used, one-third of the cellular PE reacted quickly with the reagent, whereas the remaining PE reacted much more slowly. In inverted membrane vesicles, two-thirds of the PE was found to be readily labeled. Since bacilli have only a single cellular membrane, these workers interpret their results as showing that the outer leaflet of the lipid bilayer, in contact with the external milieu of this bacterium, has half as much PE as the inner leaflet. Assuming that both leaflets have equal amounts of lipid, it follows that PG, the other major lipid of this organism, must be largely in the outer leaflet of the bilayer. These results indicate that bacteria, as well as eucaryotic cells (for review see Rothman and Lenard, 1977), have membranes with an asymmetric lipid distribution.

Rothman and Kennedy (1977b) then coupled these labeling techniques with "pulse-chase" incorporation experiments using phospholipid precursors to approach the topology of PE biosynthesis. These workers found that newly synthesized phosphatidylethanolamine molecules are resistant to modification by impermeant reagents. Upon a chase with the nonradioactive phospholipid precursor, the radioactive PE becomes accessible to the reagents that label the external surface of the cells. Rothman and Kennedy (1977b) interpret these data as showing that the newly synthesized PE is first located in the inner face of the plasma membrane lipid bilayer and is later rotated ("flip-flopped") through the lipid bilayer to become part of the lipid leaflet at the external face of the membrane. This interpretation is certainly reasonable and consistent with the data. However, it should be pointed out that these workers have not actually localized the new lipid in the inner face of the membrane. It is only known that the lipid is not accessible to the labeling reagent.

An added complexity in these experiments is the recent very novel finding that PG seems to be a precursor of PE in bacilli. Two groups (Lombardi et al., 1978; Langley and Kennedy, 1978) have found that upon addition of radioactive lipid precursors (palmitic acid or $^{32}P_i$), PG is labeled immediately, whereas PE is labeled only after a lag of 15 min. These results indicate that phosphatidic acid units can be transferred from the PG pool into the PE pool. The precursor to "outside" PE in the experiment of Rothman and Kennedy (1977b) could as well be PG (thought to be mostly external) rather than "internal" PE. Another complication is the possibility of conversion of PG to PE during exposure to the impermeant labeling reagents. Clearly much additional work is needed to see if a straightforward interpretation of the results of Rothman and Kennedy (1977b) is justified. In this context it should be noted that *B. amyloliquifaciens* is reported to have 90% of its PE in the outer leaflet (Paton et al., 1978), a result in startling contrast to the 33% observed by Rothman and Kennedy (1977a) in *B. megaterium*.

Several laboratories (Lin et al., 1971; Tsukagoski et al., 1971; Green and

Schaechter, 1972) have tested if phospholipids are synthesized at discrete sites on the cell envelope. In general, these studies have asked if phospholipids are preferentially synthesized at either the polar ends or the lateral sides of the cell envelope, but no site of preferential synthesis has been found. However, the time scale of such experiments is very much (> hundred-fold) greater than the time required for the lateral diffusion of a lipid molecule from one end of an *E. coli* cell to another (see above) and therefore, the experiments cannot discriminate between a disperse mode of synthesis and a few discrete sites of synthesis followed by rapid lateral diffusion.

However, Green and Schaechter (1972) have reported that the phospholipids of *E. coli* segregate among daughter cells as though the phospholipids are present in units of about 10^{13} molecules rather than as single molecules. These data do not conflict with the data indicating a rapid lateral diffusion of lipids in *E. coli*, since the size of the subunits (Green and Schaechter estimate 250 units/cell) is sufficiently large so that the random diffusion of a given lipid molecule would not be seriously constrained by the subunit boundaries. Another possibility is lateral diffusion of the subunits.

Goodell et al. (1974) have examined the phospholipid compositions of inner and outer membranes of minicells produced from the poles of various mutant *E. coli* strains. These workers found that both the inner and outer membranes of minicells have about one-fifth less PE than the membranes of wild-type or of the multinucleate filaments that result from minicell production.

4.3 Movement of Phospholipids Between Inner and Outer Membranes

4.3.1 Vesicle Fusion Experiments Jones and Osborn (1977a, b) reported that incubation of intact cells of *S. typhimurium* with small phospholipid bilayer vesicles resulted in a significant transfer of intact vesicular phospholipid to the cells. The transfer required a cell with an altered lipopolysaccharide (see above), Ca^{2+} or spermidine, and was temperature dependent. The transfer was shown to be unidirectional and all the vesicle components including lipids not normally found in enterobacteria (phosphatidylcholine and cholesteryl oleate) were transferred to the cell in a ratio similar to that of the donor vesicles. The vesicle lipid was found in both the inner and outer membranes and resulted in no apparent physiological damage to the cell. Vesicle donated phosphatidylserine (PS) was decarboxylated after transfer to cells, and both the PS and its decarboxylation product, PE, were found in the inner and the outer membranes.

In a related study, McIntyre and Bell (1978) showed that *E. coli* will bind lysophosphatidic acid and convert a portion of this precursor to cellular phospholipids. Although this process was without apparent physiological effect, a marked accumulation of intracellular multilamellar (liposome-like) structures were observed in cells exposed to the lipid. Baldassare et al. (1977) extended the work of George-Nascimento et al. (1974) and showed that incubation of isolated inner membranes deficient in a given enzyme activity (because of lipids of unnatural composition) with phospholipid vesicles of normal composition allowed restoration of enzymatic activity.

The most straightforward interpretation of these data is that phospholipids fuse with the other membrane and by fusion enter the process whereby phospholipids are exchanged between the inner and outer membranes. The pitfalls of these experiments are obvious but are difficult to conquer. The possibility of a passive binding of the phospholipids to the exterior of the cell followed by randomization of the lipid to the inner and outer membrane during membrane isolation is very difficult to rule out. A second possibility is that the vesicle lipids enter the cell but do not mix with *de novo* synthesized lipids. The intracellular lamellae of McIntyre and Bell (1978) could represent such a situation. Subsequent apparent fractionation of these lipids with the inner and outer membranes could then be an artifact of isolation. By far the strongest evidence that the vesicle lipids mix freely with the lipids of the inner and outer membranes are the findings that the PS and lysophosphatidic acid are metabolized by membrane lipid biosynthetic enzymes (Jones and Osborn, 1977a; McIntyre and Bell, 1978). Since these enzymes are believed to residue solely on the inner membrane (see above), it follows that the vesicular lipids have been transported to the inner membrane. The finding that PE formed by decarboxylation of PS is found in the outer membrane is therefore interpreted as movement to the inner membrane, followed by decarboxylation and movement to the outer membrane. This interpretation is, strictly speaking, only as strong as the data localizing the phosphatidylserine decarboxylase and 1-acyl lysophosphatidic acid acyltransferase to the inner membrane (see above). In both cases about 90–95% of the enzyme activity is found associated with the inner membrane (Bell et al., 1971; White et al., 1971; Schneider and Kennedy, 1976). The remaining 5–10% of the activity found with the outer membrane can be explained by contamination with inner membrane. However, it should be noted that the amount of enzyme activity found in the outer membrane is sufficient to account for the vesicle lipid conversion that was observed and thus further tests are needed. Some possible additional tests are incubation of intact cells with vesicles of various phospholipids to see if these lipids could (*a*) restore function to inactive enzymes (Baldassare et al., 1977) (to be observed *in vivo*) and (*b*) change the lipid phase transition in intact cells. Other possible experiments would be to check if vesicle-donated PG can be converted to CL, to the *sn*-glycerol-1-phosphate moiety of the membrane-derived oligosaccharide (MDO), or to the diacyl glycerol moiety of the lipoprotein, or if the PG unsaturated moieties could be converted to their cyclopropane derivatives. Since several lipid containing bacteriophages incorporate existing membrane lipids into their capsids (see below), it would also be interesting to see if these phages incorporate vesicle-donated phospholipids into the phage capsids.

4.3.2 Other Approaches Osborn et al. (1974) have reported pulse-chase experiments that are consistent with PE being made in the inner membrane and being subsequently translocated to the outer membrane.

Another line of evidence stems from cyclopropane fatty acid synthesis. As enterobacteria progress from exponential to stationary phase, most of the phospholipid unsaturated acyl moieties are converted to their cyclopropane derivatives. For instance, *E. coli K-12* shows virtually complete ($> 95\%$) conversion of palmitoleyl residues to *cis*-9,10-methylenehexadecanoate residues in a few hours. Both *in vitro* (Thomas and Law, 1966; Taylor and Cronan, 1979) and *in vivo* (Cronan, 1968; Cronan et al., 1974; Cronan et al., 1979) experiments indicate that conversion to the cyclopropane derivative occurs by transfer of the methyl group of *S*-adenosyl methionine to an intact phospholipid molecule. Genetic data (Taylor and Cronan, 1976) indicate that this conversion is catalyzed by a single enzyme. This enzyme appears to be loosely bound to the inner membrane; no detectable activity is found in the outer membrane or the periplasm (Taylor and Cronan, 1979). Since this inner membrane enzyme can modify most of the phospholipids of the inner and outer membranes, this is strong, albeit circumstantial, evidence that phospholipids are translocated between the inner and outer membranes.

4.3.3 Are Enzymes Required for Translocation? The most straightforward interpretation of the existing data is that the outer membrane phospholipids arise from and continually mix with the inner membrane phospholipids. The differing lipid composition of the two membranes could be due to a selectivity dictated by interaction of the lipids with other membrane components, such as protein and LPS. It seems probable that the sites of phospholipid translocation are the zones of adhesion between the inner and outer membranes (Bayer, 1974) at which LPS is translocated (Mühlradt et al., 1973; Leive, 1977), but what is the mechanism of translocation?

Translocation would seem to require at least two different processes: (*a*) lateral diffusion of phospholipids and (*b*) transbilayer movement (flip-flop) of phospholipid molecules from the inner to the outer leaflets of membrane bilayers. As is discussed above, the lateral diffusion of phospholipid molecules is very rapid in both inner and outer membranes. The rate of diffusion is much more rapid than that needed for membrane lipid translocation and does not require enzymatic catalysis (see above).

In model membranes composed of homogeneous lipids transbilayer movement (flip-flop) of phospholipids is very slow ($t_{1/2}$ = >10 days) (for reviews see Rothman and Lenard, 1977; Bergelson and Barsukov, 1977). In the process of membrane biogenesis, it seems probable that flip-flop would need to be much more rapid than that observed in model systems (Rothman and Lenard, 1977). If the experiments of Rothman and Kennedy (1977a, b) discussed above do give an accurate value for flip-flop in *B. megaterium*, then the rate of flip-flop in the organism is at least 10^4-fold faster than that observed in model systems and thus *a priori* it would seem that flip-flop is a catalyzed reaction *in vivo*.

However, a series of recent papers have pointed out various factors that could increase the rate of flip-flop *in vivo* without resort to enzymatic catalysis or to loss of the barrier properties of the membrane. In model systems the rate of flip-flop can be increased significantly by (*a*) temperatures within the phase transition (de Kruijff and Van Zoelen, 1978), (*b*) asymmetry of the bilayers with respect to head group (de Kruijff and Baken, 1978) or fatty acid composition (de Kruijff and Wirtz, 1977), and (*c*) addition of a transmembrane protein to the vesicles (Van Zoelen et al., 1978a; de Kruijff et al., 1978). This last effect is one of the largest and was seen when the rate of flip-flop of lysophosphatidylcholine (Van Zoelen et al., 1978a) or of phosphatidylcholine (de Kruijff et al., 1978) was examined in phosphatidylcholine vesicles that contained about 5 molecules of the transmembrane sialoprotein, glycophorin, per vesicle. In the presence of glycophorin, the rate of flip-flop was increased about a hundredfold. Since glycophorin is thought to have only a structural role in the erthryocyte membrane, the effect has been attributed to the presence of a transmembrane protein in the lipid bilayer (de Kruijff et al., 1978). It has been proposed that by altering the structure of the lipid bilayer (see Section 3.4), the protein will promote phospholipid flip-flop. However, it should be noted that the glycophorin used contained several molecules of phospholipid per molecule of protein (Armitage et al., 1977; Van Zoelen et al., 1978b). Since this phospholipid is associated with the intramembrane portion of the protein (Armitage et al., 1977), an effect of the glycophorin-bound phospholipid rather than of the protein per se cannot be ruled out at the present time. However, it appears that a combination of the influences of membrane proteins, phospholipid heterogeneity, and fluidity above the phase transition could result in a rate of uncatalyzed flip-flop sufficient for membrane biogenesis.

Soluble proteins capable of the exchange of phospholipid molecules between synthetic and biological bilayer membranes have been known for several years in animal tissues (for recent reviews see Zilversmit and Hughes, 1976; Wirtz, 1974). A similar protein, perhaps localized at the adhesion zones, could facilitate phospholipid exchange between the inner and outer membranes. However, no such protein had been demonstrated in bacterial systems until the recent report of Cohen et al. (1979). These authors report that the photosynthetic bacterium *Rhodopseudomonas sphaeroides* contains a soluble protein fraction that catalyzes transfer of a variety of phospholipids between synthetic phospholipid vesicles and the intracytoplasmic membrane vesicles (a vesicle involved in photochemical reactions) of the organism. The transfer activity is also found in heterotrophically grown cells and thus does not appear to be solely involved in biogenesis of the photosynthetic apparatus. The assay developed by Cohen et al. (1979) should allow the assay of other bacteria for phospholipid transfer proteins.

5 PHOSPHOLIPIDS AND THE INSERTION AND FUNCTION OF MEMBRANE PROTEINS

The literature on this subject is very extensive and most has been reviewed previously (Cronan and Gelmann, 1975; Cronan, 1978). The emphasis in this section is on the effects of phospholipid synthesis and structure on outer membrane protein insertion. However, the lack of a well-understood functional role for most outer membrane proteins necessitates some discussion of inner membrane and periplasmic proteins.

5.1 Requirement for Phospholipid Synthesis?

Phospholipid synthesis in *E. coli* and *S. tymphimurium* can be inhibited over tenfold by glycerol starvation of a glycerol auxotroph (for review see Cronan, 1978). Two other methods are addition of the antibiotic cerulenin and use of mutant strains defective in early steps of the fatty acid biosynthetic pathway (Cronan and Gelmann, 1975; Cronan, 1978). The glycerol auxotrophs are preferred because of the possible side effects of a block in fatty acid synthesis on the synthesis of the lipid A moiety of LPS (for review see Cronan and Gelmann, 1975).

McIntyre and co-workers (McIntyre and Bell, 1975; McIntyre et al., 1977) reported that when phospholipid synthesis was halted in a glycerol (*pls*B) auxotroph of *E. coli*, both the inner and outer membranes of the cells became enriched with protein. The protein to phospholipid ratio of both membranes increased approximately 60%. This work demonstrated that the outer membrane is not normally saturated with protein and that the synthesis of membrane phospholipid is not required for the synthesis and insertion of bulk outer membrane protein. The protein-enriched membranes are not lethal to the cell and the enrichment is quickly normalized upon the restoration of phospholipid synthesis (McIntyre et al., 1977). Labeling experiments by Hsu and Fox (1970) and Lin and Wu (1976) and a number of experiments with inner membrane proteins (for review see Cronan and Gelmann, 1975; Cronan, 1978) also indicate that bulk protein insertion does not require concomitant phospholipid synthesis. However, specific outer membrane and periplasmic proteins have only been examined quite recently.

Lin and Wu (1976) reported experiments in which a glycerol auxotroph of *S. typhimurium* was deprived of glycerol and the rate of synthesis of envelope protein and of the lipoprotein were measured. An appreciable rate of synthesis for these protein fractions remained even several hours after the cessation of phospholipid synthesis (no data for earlier times of glycerol deprivation were reported). However, only a small fraction of the lipoprotein from the glycerol-deprived cultures contained a glycerol moiety. It therefore seems that the synthesis of the apoprotein protein of the lipoprotein proceeds independently of the attachment of the diglyceride moiety.

Randall (1975) has recently reported that insertion of the lamB protein (which functions in maltose transport and is the receptor for phage lambda) into the outer membrane proceeds normally during cerelenin inhibition (>90%) of phospholipid synthesis. In a similar study Beacham et al. (1976) examined the induction and subsequent secretion of alkaline phosphatase during the absence of phospholipid synthesis. These workers avoided the complication of derepression by phosphate limitation by commencing alkaline phosphatase formation using a combination of a temperature-sensitive suppressor and a nonsense mutation in the alkaline phosphatase structural gene. Under these conditions alkaline phosphatase activity could be formed normally in the absence (<10% of normal) of phospholipid synthesis.

Webster and co-workers (Chamberlain and Webster, 1976; Cashman and Webster, 1977) have studied the interactions of phospholipid synthesis and the association of the major coat protein of a filamentous bacteriophage with the inner membrane. An increase in the CL content of the inner membrane coincides with membrane insertion of the coat protein (Chamberlain and Webster, 1976). However, Pluschke et al. (1978), using an *E. coli* mutant defective in CL synthesis, showed that increased CL content is not a necessity for insertion of the coat protein.

Cashman and Webster (1977) examined the effects of an inhibition of phospholipid synthesis on the insertion of the major coat protein into the inner membrane of the host cell. Unexpectedly these workers found that the rate of coat protein synthesis was inhibited under these conditions. Since the major coat protein is made with a signal sequence that is subsequently clipped by a membrane protease (Chang et al., 1978), it seems possible that phospholipid synthesis is somehow required for the protease action. However, the lamB protein of *E. coli* is also the protease cleavage product of a larger precursor (Randall et al., 1978), and this protein seems to be inserted normally into the outer membrane in the absence of phospholipid synthesis (Randall, 1975).

5.2 Influence of Lipid Phase Transition

A copious literature on the insertion into and function of the lactose transport protein in the inner membrane has been reviewed previously (Cronan and Gelmann, 1975; Cronan, 1978). Briefly, if suitable precautions against membrane damage are taken, the insertion of the lactose transport protein is not dependent on the order–disorder phase properties of the lipid bilayer. The phase properties do, however, have an effect on the rate versus temperature profile of the lactose transport system. These results (see Cronan, 1978) are currently interpreted as denoting the partition of the transport protein between lipid phases. Although the physical evidence for discrete phases under the condition of these experiments is not strong (Linden et al., 1977), the distinctions between phage separations and phase transitions overlap sufficiently to make the argument semantic (Cronan, 1978).

Recently, Shechter and co-workers (Therisod et al., 1977; Letellier et al., 1977) have contested the lipid phase separation interpertation of the function of the lactose transport system. These workers report that the number of lactose carriers changes during phase separation (transition) and thus the assumptions of the partition hypothesis are untenable. However, the data of Shechter and co-workers (Therisod et al., 1977; Letellier et al., 1977) are based on a measurement of the amount of functional lactose transport protein by the fluorescence of dansyl galactoside upon energization of membrane vesicles. Recently, the basic parameters of the dansyl galactoside technique as developed by Kaback and coworkers (for review see Schuldiner and Kaback, 1977) were attacked by Overath, Simoni, and co-workers (Overath et al., 1979), who reported that contrary to the original reports, dansyl galactosides are transported by *E. coli* and the fluorescence increase observed upon energization is an experiment artifact due to nonspecific binding of the galactoside to phospholipid. Therefore, until the validity of the dansyl galactoside technique is clear, the partition hypothesis remains a reasonable working model.

In regard to outer membrane proteins, Ito and co-workers (Ito et al., 1977) reported that the rate of incorporation of a radioactive amino acid into the protein of the inner membrane proceeds more rapidly than incorporation into the outer membrane proteins. Similar results had been observed by Lee and Inouye (1974). More strikingly Ito et al. (1977), reported that the rate of appearance of label in the outer (and to some extent the inner) membrane was very slow when the membranes were incubated at a temperature below the lipid phase transition. These experiments used an elaidate grown unsaturated fatty acid auxotroph. However, the kinetics of the incorporation into outer membranes were also slow in the elaidate grown cells even at a temperature above the phase transition. The latter result raises the possibility of the presence of elaidate and/or cellular damage. Another complication is that the inner and outer membrane preparations of the elaidate grown cells appear to be mutually contaminated. The contamination is probably greater than it appears, since the autoradiographic procedure used accentuates major protein bands (Laskey and Mills, 1975). Indeed, because of extensive cell lysis, Overath et al. (1975) were unable to prepare pure outer membranes from an elaidate grown *E. coli* strain under conditions similar to those of Ito et al. (1977). For these reasons the data of Ito and co-workers must be considered preliminary in nature. Cells supplemented with a variety of fatty acids should be examined. A set of physiological controls such as those used by Kimura and Izui (1976) (see below) would also aid in interpretation.

Kimura and Izui (1976) have shown that the induction of alkaline phosphatase is much slower at 25 or 30°C in an unsaturated fatty acid auxotroph grown on elaidate than that of the same strain grown on oleate. Induction is similar in the elaidate and oleate grown cells at higher temperatures. These authors showed that the cells were viable and that the elaidate grown cells derepressed alkaline phosphatase normally upon shift from low to high temperature. These experiments therefore seem physiologically relevant. However, the interpretation of the experiment is difficult since the mechanism of alkaline phosphatase derepression is not well understood. Several gene products are involved, some of which may also be involved in phosphate transport (Willsky et al., 1973). For example, Gallant and Stapleton (1963) have found that derepression of alkaline phosphate commences before any effect on the intracellular phosphate pool can be detected. Another problem is that Kimura and Izui (1976) assayed only enzyme activity rather than alkaline phosphate protein. A precursor of the alkaline phosphatase monomer is the primary transcription product of the *phoA* gene (H. Inouye et al., 1977; H. Inouye and Beckwith, 1977). This precursor must be proteolytically cleaved by an outer membrane enzyme (H. Inouye and Beckwith, 1977) before the monomers can be dimerized into the active enzyme. Since this cleavage is probably the rate-limiting step in the formation of active enzyme (Torriani, 1968), many more experiments are needed before a molecular interpretation of the results of Kimura and Izui (1976) can be made. A repeat of these experiments using an immunological assay for *phoA* gene product might clarify the molecular level at which elaidate supplementation and temperature act.

6 PHOSPHOLIPIDS AS PRECURSORS TO MEMBRANE MACROMOLECULES

Membrane phospholipids recently have been found to be metabolic precursors to two water soluble compounds, the membrane derived oligosaccharides (MDO) of the periplasmic space and the lipoprotein of the outer membrane.

6.1 Membrane-Derived Oligosaccharides

Kennedy and co-workers (van Golde et al., 1973; Kennedy et al., 1976) have isolated a novel class of oligosaccharide from *E. coli* containing components that appear to be derived from the membrane phospholipids. The compounds contain glucose as the sole sugar and consist of at least three species of similar molecular weight (4000–5000). The glucose residues on one of the components are linked to succinic acid by an O-ester linkage and to glycerol-P and phosphoethanolamine by phosphodiester links to C-6 of the glucose residue (Kennedy et al., 1976). The other fraction that has been characterized lacks the succinate and ethanolamine residues (Kennedy et al., 1976).

The glycerol phosphate moiety of the oligosaccharide is *sn*-glycerol 1-phosphate, and thus has the same configuration as the polar group of phosphatidylglycerol (Kennedy et al., 1976). This finding, coupled with the results of "chase" experiments where the label lost from glycerol labeled lipids could be recovered in the oligosaccharides, led Kennedy and co-workers (van Golde et al., 1973; Kennedy et al., 1976) to propose that phosphatidylglycerol and/or cardiolipin was the source of the oligosaccharide glycerol phosphate residues. Strong support for this pathway has recently been obtained by Schulman and Kennedy (1977), who find that mutants of *E. coli* containing only traces of glucose (as a result of a defect in phosphoglucose isomerase) do not contain the membrane-derived oligosaccharides. In such mutants, although growth is normal, the turnover of both phosphatidylglycerol and cardiolipin is very slow. However, upon addition of glucose to the medium, a dramatic stimulation of turnover of the two lipids is seen. The phosphoethanolamine residue is also thought to be derived from the membrane lipids; however, this has not been demonstrated. It would be interesting to see if the mutants lacking the oligosaccharide have phosphoethanolamine residues in their LPS molecules. The oligosaccharides are reported to be localized in the periplasmic space (Schulman and Kennedy, 1977).

6.2 Lipoprotein

Shulman and Kennedy (1977) have also investigated the relationship of the membrane-derived oligosaccharides to the synthesis of the diglyceride moiety of the lipoprotein. Shulman and Kennedy (1977) showed that this diglyceride moiety is derived from a lipid of long half-life. The membrane-derived oligosaccharides do not seem required for transfer of the diglyceride moiety to the lipoprotein. Based on these data, either cardiolipin or phosphatidylglycerol could be the source of the diglyceride moiety of the lipoprotein.

The studies of Schulman and Kennedy (1977) have been confirmed by Chattopadhkyay and Wu (1977), who have also shown that the cardiolipin-deficient strain of Pluschke et al. (1978) forms the diglyceride moiety of the lipoprotein normally. Another important finding is that the diglyceride is linked to the protein by a thioether link to C-1 of the *sn*-glycerol moiety not to C-3. This finding indicates that precursor of the diglyceride is not the diglyceride moiety of a phospholipid molecule (linkage to C-3 would be expected). The only likely sources of *sn*-1-glycerol-phosphate for transfer to the protein are the nonacylated glycerol of PG and the monoacylglycerol moiety of 3′-monoacylphosphatidylglycerol. This latter lipid is a minor (~1%) component of the lipids of *E. coli* (Cho et al., 1977) and of *S. tymphimirium* (Olsen and Ballou, 1971) and is formed by transacylation from other phospholipid molecules such as PG and PE (Cho et al., 1977). A recent paper (Nishijima et al., 1978) reports that this reaction proceeds in extracts by transfer from 2-acyl lysophospholipids formed by the phospholipase B activity of the outer membrane. However, the authors have not yet shown that the phospholipase B-deficient strains (*pldA*) lack monoacylphosphatidylglycerol. The monoacyl group of the monoacylphosphatidylglycerol formed *in vitro* has a very high content of unsaturated fatty acids (Nishijima et al., 1978), whereas the *in vivo* product is very deficient in unsaturated fatty acids (Olsen and Ballou, 1971). This difference indicates that a transacylation from 2-acyllysophospholipids is probably not the mechanism whereby monoacyl PG is formed *in vivo*.

A mechanism involving transfer of the monoacylglycerol moiety of monoacyl PG is attractive because it would rationalize the presence of this lipid. However, PG seems more likely to be the precursor of the lipoprotein glycerol moiety since a lipoprotein mutant lacking both ester-linked fatty acids has been described (Rotering and Braun, 1977). The mutant retains the amide-linked fatty acid (Rotering and Braun, 1977; S. Inouye et al., 1977).

The origin of the amide-linked fatty acid moiety of the lipoprotein is unknown. It seems reasonable that the amide-linked fatty acids of the lipoprotein and of lipid A are incorporated by similar donors. The donors could arise as intermediates of the fatty acid biosynthetic pathway and hence could be thioesters of acyl carrier protein. The development of procedures to enzymatically synthesize and purify acyl-acyl carrier protein substrates should permit a test of this hypothesis (Spencer et al., 1978; Rock and Garwin, 1979; Rock and Cronan, 1979).

The transfer of the polar portion of PG or monoacyl PG to protein to form the lipoprotein would result in the formation of phosphatidic acid or (depending on the reaction) perhaps 1,2-diglyceride. Likewise, the transfer of the glycerol-1-phosphate residue of PG to MDO would result in 1,2-diglyceride. It seems probable that the 1,2-diglyceride would then be recycled into phospholipid synthesis as phosphatic acid by diglyceride kinase, an inner membrane enzyme. Mutants defective in diglyceride kinase have recently been isolated (Raetz and Newman, 1978) and should allow a test of this hypothesis. The outer membrane enzyme, phosphatidic acid phosphatase (van den Bosch and Vagelos, 1970), might also play a role in this salvage pathway.

The functions of the membrane-derived oligosaccharides and the lipoprotein are unclear, but the synthesis of these molecules together with that of cardiolipin accounts for the turnover of PG observed during the growth of *E. coli*. It would seem that the oligosaccharides and the lipoprotein may act to preserve the integrity of the periplasmic space and in this regard, it is interesting that the rate of PG turnover is increased in cells adapted to growth on media of low osmolarity (Munro and Bell, 1973a, b). Growth of the diglyceride kinase mutant is also inhibited by media of low osmolarity (Raetz and Newman, 1978).

7 PHOSPHOLIPID-CONTAINING BACTERIOPHAGES

Bacteriophages containing phospholipid as a capsid component have been known for a number of years (for review see Franklin, 1974, 1977); however, only recently have a number of R factor-dependent lipid-containing phages been described (Stanisich, 1974; Bradley and Rutherford, 1975; Olsen et al., 1974; Wong and Bryan, 1978). Prior to the isolation of these new phages, the known lipid-containing phages had ill-characterized bacterial hosts (usually pseudomonads) (Franklin, 1974). However, the R factor-dependent phages are able to grow on most bacteria that can harbor an appropriate R plasmid. These hosts include *E. coli K-12* and *S. tymphimurium* LT2. The ability to grow these bacteriophages on well-defined and genetically manipulatable hosts will greatly increase the utility of these phages in membrane research.

Two general types of lipid-containing phages are known. The first type to be extensively studied is typified by phage PM2 (Franklin, 1974, 1977). This phage contains DNA in an icosahedron of about 600 Å in diameter in which the lipid is an internal component of the capsid. The second phage type has an RNA genome and the lipid is exterior to the phage capsid (Franklin, 1977). This type of phage, of which $\phi 6$ is the best studied, greatly resembles enveloped animal viruses in morphology. Unfortunately, only phages of the PM2 type have been found among the R factor-dependent lipid-containing phages. These plasmid-dependent phages are called PR3, PR4, PR5, and PRD1. Phages PR3, PR4, and PR5 (Bradley and Rutherford, 1975; Wong and Bryan, 1978) appear related, whereas phage PRD1 has a much larger genome and a lower relative lipid content (Olsen et al., 1974).

Phages PR3, PR4, and PR5 are morphologically similar to PM2 except that the PR phages probably have a short tail (Bradley and Rutherford, 1975; Bradley and Cohen, 1977; Wong and Bryan, 1978; Muller and Cronan, 1979). However, phage PR4 has a linear double stranded DNA of 9.6×10^6 daltons (Muller and Cronan, 1980), whereas phage PM2 has a smaller (6.2×10^6 daltons) super-coiled circular genome (Franklin, 1974). Phage PR5, which is serologically related to phages PR3 and PR4, has been reported to have a linear double stranded DNA of about 8×10^6 daltons (Wong and Bryan, 1978). It seems likely that the genomes of PR4 and PR5 are of identical size and the lower value of Wong and Bryan (1978) can be attributed to the lack of an internal standard in their electron micrographs. The phospholipid content of these phages is about 10–15% of the particle weight.

The lipids of phage PR4 are all phospholipids and are those normally found in large quantities in the host (Muller and Cronan, 1980). The phospholipid composition is about 55% PE, 36% PG, and 6% CL. The increased PG content over that in the host is characteristic of phospholipid-containing phages (Franklin, 1977). It should be noted that the lipid composition differs significantly from those reported by Sands (1976) and by Sands and Cadden (1975), who reported the presence of large amounts of PS and other unidentified lipids in phage PR4. It seems clear that the PS reported was actually lyso PE formed by the rather harsh methods used by these workers for extraction of the phage lipids. In this regard, phage PR5 was reported to have a phospholipid composition very similar to that reported for phage PR4 by Muller and Cronan (1980). The fatty acid composition of the phospholipids of phage PR4 are indistinguishable from those of the host phospholipids (Muller and Cronan, 1980).

The origin of the phospholipids of the lipid-containing bacteriophages appears to be the cell envelope of the host cell. Pulse-chase experiments with lipid precursors indicate that both the phospholipids present at the time of infection and the phospholipid molecules made after infection appear in phage particles (Snipes et al., 1974; Tsukagoski and Franklin, 1974; Diedrich and Cota-Robles, 1976; Sands, 1976; Muller and Cronan, 1980). Although early results (Snipes et al., 1974; Tsukagoshi and Franklin, 1974) indicated that phage PM2 altered the host phospholipid metabolism, more recent results with phages PM2 (Diedrich and Cota-Robles, 1976) and PR4 (Sands, 1976; Muller and Cronan, 1980) indicate that infection has no effect on host phospholipid synthesis until the onset of lysis.

The enrichment of PG in the lipid-containing phages therefore cannot be explained by an increased rate of PG synthesis during phage infection. The increased content of PG in the phage must then arise from a selective incorporation of PG over PE during phage morphogenesis. The most straightforward selective mechanism could be a preferential interaction of PG with the phage capsid proteins. Indeed, Schafer and Franklin (1975) isolated a basic capsid protein from phage PM2 that binds PG by electrostatic interaction. A second possibility is that the lipid-containing phages obtain their phospholipid from a portion of the cell membrane that is enriched in PG, such as the poles of the cell (Goodell et al., 1974).

The requirements of phage PR4 for specific lipids for maturation and infection can be tested by the use of various mutants of *E. coli* blocked in fatty acid and phospholipid synthesis (see Cronan, 1978 for review). Experiments along these lines have been reported by Sands and Auperin (1977). However, most of the data presented in that paper were expressed in only relative terms and the absolute burst sizes (when presented) are quite low compared to those usually obtained (Stanisich, 1974; Muller and Cronan, 1980). Also, no analytical data on the lipid composition of either the host or the phage were presented. The data of Sands and Auperin, 1977) therefore do not allow interpretation.

In conclusion, because of their ability to grow on well-characterized hosts and their small genome size, the lipid-containing bacteriophages should be a useful system in the study of protein–lipid interaction and of membrane biogenesis.

ACKNOWLEDGMENTS

The preparation of this manuscript and our unpublished results cited herein were supported by grants AI10186 and AI15650 from the National Institute of Allergy and Infectious Diseases. I thank Mrs. M. Gillett for her assistance with the manuscript and the members of my laboratory for their advice.

REFERENCES

Albright, F. R., White, D. A., and Lennarz, W. J. (1973), *J. Biol. Chem.*, **248,** 3968–3977.

Ames, G. F. (1968), *J. Bacteriol.*, **95,** 833–843.

Armitage, I. M., Shapiro, D. L., Furthmayer, H., and Marchesi, V. T. (1977), *Biochemistry*, **16,** 1317–1320.

Baldassare, J. J., Brenckle, G. M., Hoffman, M., and Silbert, D. F. (1977). *J. Biol. Chem.*, **252,** 8797–8803.

Bayer, M. E. (1974), *Ann. N.Y. Acad. Sci.*, **235,** 6–28.

Bayer, M. E., Koplow, J., and Goldfine, H. (1975), *Proc. Natl. Acad. Sci. U.S.*, **72,** 5145–5149.

Beacham, I. R., Taylor, N. S., and Youell, M. (1976), *J. Bacteriol.*, **128,** 522–527.

Bell, R. M., Mavis, R. D., Osborn, M. J., and Vagelos, P. R. (1971), *Biochim. Biophys. Acta*, **249,** 628–635.

Bergelson, L. D., and Barsukov, L. I. (1977). *Science*, **197,** 224–230.

Bradley, D. E., and Cohen, D. R. (1977). *J. Gen. Microbiol.*, **98,** 619–623.

Bradley, D. E., and Rutherford, E. L. (1975), *Can. J. Microbiol.*, **21,** 152–163.

Caron, F., Mateu, L., Rigsig, P., and Azerad, R. (1974), *J. Mol. Biol.*, **85,** 279–300.

Cashman, J. S., and Webster, R. E. (1977), *J. Bacteriol.* **129,** 1245–1249.

Chamberlain, B. K., and Webster, R. E. (1976), *J. Biol. Chem.*, **251,** 7739–7745.

Chang, C. N., Blobell, G., and Model, P. (1978), *Proc. Natl. Acad. Sci. U.S.*, **75,** 361–365.

Chattopadhkyay, P. K., and Wu, H. C. (1977), *Proc. Natl. Acad. Sci. U.S.*, **74,** 5318–5322.

Cheng, S., Thomas, J. K., and Kulpa, C. F. (1974), *Biochemistry*, **13,** 1135–1139.

Cho, K. S., Hong, S. D., Cho, J. M., Chang, C. S., and Lee, S. L. (1977), *Biochim. Biophys. Acta*, **486,** 47–54.

Cohen, L. K., Lueking, D. R., and Kaplan, S. (1979), *J. Biol. Chem.*, **254,** 721–728.

Cronan, J. E., Jr. (1968), *J. Bacteriol.*, **95,** 2054–2060.

Cronan, J. E., Jr. (1978), *Annu. Rev. Biochem.*, **47,** 163–189.

Cronan, J. E., Jr., and Gelmann, E. P. (1975), *Bacteriol. Rev.*, **39,** 232–256.

Cronan, J. E., Jr., and Vagelos, P. R. (1972), *Biochim. Biophys. Acta*, **265,** 25–65.

Cronan, J. E., Jr., Nunn, W. D., and Batchelor, J. G. (1974), *Biochim. Biophys. Acta*, **348,** 63–75.

Cronan, J. E., Jr., Reed, R., Taylor, F. R., and Jackson, M. B. (1979), *J. Bacteriol.*, **138,** 118–121.

Cullis, P. R., and de Kruijff, B. (1978), *Biochim. Biophys. Acta*, **507,** 207–218.

de Kruijff, B., and Baken, P. (1978), *Biochim. Biophys. Acta*, **507,** 38–47.

de Kruijff, B., and van Zoelen, E. J. J. (1978), *Biochim. Biophys. Acta*, **511,** 105–115.

de Kruijff, B., and Wirtz, K. W. A. (1977), *Biochim. Biophys. Acta*, **468,** 318–326.

de Kruijff, B., van Zoelen, E. J. J., and van Deenan, L. L. M. (1978), *Biochim. Biophys. Acta*, **509,** 537–542.

Devor, K. A., Teather, R. M., Brenner, M., Schwarz, H., Wirz, H., and Overath, P. (1976), *Eur. J. Biochem.*, **63,** 459–467.

Diedrich, D. C., and Cota-Robles, E. M. (1976), *J. Virol.*, **19,** 446–456.

Dupont, Y., Gabriel, A., Chabre, M., Gulik-Krzywicki, T., and Schecter, E. (1972), *Nature (Lond.)*, **238,** 331–333.

Engleman, D. M. (1971), *J. Mol. Biol.*, **58,** 153–165.

Franklin, R. M. (1974), *Curr. Top. Microbiol. Immunol.*, **6,** 107–159.

Franklin, R. M. (1977), in *Synthesis, Assembly, and Turnover of Cell Surface Components* (G. Poste and G. L. Nicolson, eds.), Elsevier, Amsterdam.

Gallant, J., and Stapleton, R. (1963), *Proc. Natl. Acad. Sci. U.S.*, **50,** 348–352.

Gent, M. P. N., and Ho, C. (1978), *Biochemistry*, **17,** 3023–3036.

Gent, M. P. N., Cottam, P. C., and Ho, C. (1978), *Proc. Natl. Acad. Sci. U.S.*, **75,** 630–634.

George-Nascimento, C., Zehner, Z. E., and Wakil, S. J. (1974), *J. Supramol. Struct.*, **2,** 646–669.

Goldfine, H. (1972), *Adv. Microbiol. Physiol.*, **8,** 1–57.

Goodell, E. W., Schwarz, U., and Teather, R. M. (1974), *Eur. J. Biochem.*, **47,** 567–572.

Green, E. W., and Schaechter, M. (1972), *Proc. Natl. Acad. Sci. U.S.*, **69,** 2312–2316.

Hsu, C. C., and Fox, C. F. (1970), *J. Bacteriol.*, **103,** 410–416.

Inouye, H., and Beckwith, J. (1977), *Proc. Natl. Acad. Sci. U.S.*, **74,** 1440–1444.

Inouye, H., Pratt, C., Beckwith, J., and Torriani, A. (1977), *J. Mol. Biol.*, **110,** 75–87.

Inouye, S., Lee, N., Inouye, M., Wu, H. C., Suzuki, H., Nishimura, Y., Iketoni, H., and Hirota, Y. (1977), *J. Bacteriol.*, **132,** 308–313.

Ishinaga, M., and Kito, M. (1974), *Eur. J. Biochem.*, **42,** 483–487.

Ito, K., Sato, T., and Yura, T. (1977), *Cell*, **11,** 551–559.

Jones, N. C., and Osborn, M. J. (1977a), *J. Biol. Chem.*, **252,** 7398–7404.

Jones, N. C., and Osborn, M. J. (1977b), *J. Biol. Chem.*, **252,** 7405–7412.

Joseleau-Petit, D., and Kepes, A. (1975), *Biochim. Biophys. Acta*, **406,** 36–49.

Jost, P. C., Griffith, O. H., Capaldi, R. A., and Vanderkooi, G. (1973), *Proc. Natl. Acad. Sci. U.S.*, **70,** 480–485.

Jost, P. C., Nakakaoukaren, K. K., and Griffith, O. H. (1977), *Biochemistry*, **16,** 3110–3114.

Kaback, H. B. (1971), *Methods Enzymol.*, **22,** 99–127.

Kamio, Y., and Nikaido, H. (1976), *Biochemistry*, **15,** 2561–2570.

Keith, A. D., Sharroff, M., and Cohn, G. (1973), *Biochim. Biophys. Acta*, **300,** 379–419.

Kimura, K., and Izui, K. (1976), *Biochem. Biophys. Res. Commun.*, **70,** 900–906.

Kito, M., Ishinaga, M., Nishihara, M., Kata, M., Sawada, S., and Hata, T. (1975), *Eur. J. Biochem.*, **54,** 55–63.

Kennedy, E. P., Rumley, M. K., Schulman, H., and van Golde, L. M. G. (1976), *J. Biol. Chem.*, **251,** 4208–4213.

Koplow, J., and Goldfine, H. (1974), *J. Bacteriol.*, **117,** 527–543.

Langley, K. E., and Kennedy, E. P. (1978), *Fed. Proc.*, **37,** 1691 (Abstr. 2323).

Laskey, R. A., and Mills, A. D. (1975), *Eur. J. Biochem.*, **56,** 335–341.

Lee, A. G. (1975), *Prog. Biophys. Mol. Biol.*, **29,** 3–56.

Lee, N., and Inouye, M. (1974), *FEBS Lett.*, **39,** 167–170.

Leive, L. (1977), *Proc. Natl. Acad. Sci. U.S.*, **74,** 5065–5060.

Letellier, L., Weil, R., and Shechter, E. (1977), *Biochemistry*, **16,** 3777–3780.

Levine, Y. K., and Wilkens, M. H. F. (1971), *Nature, New Biol.*, **230,** 69–72.

Lin, E. C. C., Hirota, Y., and Jacob, F. (1971), *J. Bacteriol.*, **108,** 375–385.

Lin, J. J.-C., and Wu, H. C. P. (1976), *J. Bacteriol.*, **125,** 892–904.

Linden, C. D., Blasie, J. K., and Fox, C. F. (1977), *Biochemistry*, **16,** 1621–1625.

Lombardi, F. J., Chen, S. L., and Fulco, A. J. (1978), *Fed. Proc.*, **37,** 1494 (Abstr. 1236).

Lugtenberg, E. J. J., and Peters, R. (1976), *Biochem. Biophys. Acta*, **441,** 38–47.

Luzzati, V. (1968), in *Biological Membranes* (D. Chapman, ed.), Academic Press, London and New York, pp. 71–123.

Machtiger, N. A., and Fox, C. F. (1973), *J. Supramolecular Struct.*, **1,** 545–564.

Marinetti, G. V., and Love, R. (1977), *Chem. Phys. Lipid*, **18,** 170–180.

McIntyre, T. M., and Bell, R. M. (1975), *J. Biol. Chem.*, **250,** 9053–9059.

Mateu, L., Caron, F., Luzzati, V., and Villecocq, A. (1978), *Biochim. Biophys. Acta*, **508,** 109–121.

McIntyre, T. M., and Bell, R. M. (1978), *J. Bacteriol.*, **135,** 215–226.

McIntyre, T. M., Chamberlain, B. K., Webster, R. E., and Bell, R. M. (1977), *J. Biol. Chem.*, **252,** 4487–4493.

Mühlradt, P. F., Mengel, J., Golecki, J. R., and Speth, V. (1973), *Eur. J. Biochem.*, **35,** 471–481.

Muller, E., and Cronan, J. E., Jr. (1980), *Virology*, (submitted).

Munro, G. F., and Bell, C. A. (1973a), *J. Bacteriol.*, **116,** 257–262.

Munro, G. F., and Bell, C. A. (1973b), *J. Bacteriol.*, **116,** 1479–1481.

Nikaido, H. Takeuchi, Y., Ohnishi, S.-I., and Nakae, T. (1977), *Biochem. Biophys. Acta*, **465,** 152–164.

Nishijima, M., Nakaido, S., Tamori, Y., and Nojima, S., (1977), *Eur. J. Biochem.*, **73,** 115–124.

Nishijima, M., Sa-eki, T., Tamori, Y., Doi, O., and Nojima, S. (1978), *Biochem. Biophys. Acta*, **528,** 107–118.

Oldfield, E., Gilmore, H., Glaser, M., Gutowsky, H. S., Hshung, J. C., Kang, S. Y., King, T. E., Meadows, M., and Rice, D. (1978), *Proc. Natl. Acad. Sci. U.S.*, **75,** 4657–4660.

Olsen, R. H., Siak, J. S., and Gray, R. H. (1974), *J. Virol.*, **14,** 689–699.

Olsen, R. W., and Ballou, C. E. (1971), *J. Biol. Chem.*, **246,** 3305–3313.

Osborn, M. J., Gander, J. E. Parisi, E., and Carson, J. (1972), *J. Biol. Chem.*, **247,** 3962–3972.

Osborn, M. J., Rick, P. D., Lehmann, V., Rupprecht, E., and Singh, M. (1974), *Ann. N.Y. Acad. Sci.*, **235,** 52–65.

Overath, P., Brenner, M., Gulik-Krzywicki, T., Shechter, E., and Letellier, L. (1975), *Biochim. Biophys. Acta*, **389,** 385–369.

Overath, P., Teather, R. M., Simoni, R. D., Gichele, G., and Wilhelm, V. (1979), *Biochemistry*, **18,** 1–11.

Pagano, R. E., and Wernstein, S. N. (1978), *Annu. Rev. Biophys. Bioeng.* **7,** 435–468.

Papahadjopoulos, D., Moscorello, M., Eylar, F. H., and Isac, T. (1975), *Biochim. Biophys. Acta*, **401,** 317–335.

Paton, J. C., May, B. C., and Elliott, W. H. (1978), *J. Bacteriol.*, **135,** 393–401.

Pluschke, G., Hirota, Y., and Overath, P. (1978), *J. Biol. Chem.*, **253,** 5048–5055.

Raetz, C. H. (1978), *Microbiol. Rev.*, **42,** 614–659.

Raetz, C. R. H., and Kennedy, E. P. (1972), *J. Biol. Chem.*, **247,** 2008–2014.

Raetz, C. R. H., and Newman, R. F. (1978), *J. Biol. Chem.*, **253,** 3882–3887.

Raetz, C. R. H., Dowhan, W., and Kennedy, E. P. (1976), *J. Bacteriol.* **125,** 855–863.

Raetz, C. R. H., Larson, T. J., and Rowhon, W. (1977), *Proc. Natl. Acad. Sci. U.S.*, **74,** 1412–1416.

Randall, L. L. (1975), *J. Bacteriol.*, **122,** 347–351.

Randall, L. L., Hardy, S. J. S., and Josefsson, L.-G. (1978), *Proc. Natl. Acad. Sci., U.S.* **75,** 1209–1212.

Rock, C. O., and Cronan, J. E., Jr. (1979), *J. Biol. Chem.*, **254,** in press.

Rock, C. O., and Garwin, J. L. (1979), *J. Biol. Chem.*, **254,** in press.

Rotering, H., and Braun, V. (1977), *FEBS Lett.*, **83,** 41–44.

Rothman, J. E., and Kennedy, E. P. (1977a), *J. Mol. Biol.*, **110,** 630–648.

Rothman, J. E., and Kennedy, E. P. (1977b), *Proc. Natl. Acad. Sci. U.S.*, **74,** 1821–1825.

Rothman, J. E., and Lenard, J. (1977), *Science*, **195,** 743–753.

Rottem, S., and Leive, L. (1977), *J. Biol. Chem.*, **252,** 2077–2081.

Sackmann, E., Träuble, H., Galla, H.-J., and Overath, P. (1973), *Biochemistry*, **12,** 5360–5369.

Sands, J. A. (1976), *J. Virol.*, **19,** 296–301.

Sands, J. A., and Auperin, D. (1977), *J. Virol.*, **22,** 315–320.

Sands, J. A., and Cadden, S. P. (1975), *FEBS Lett.*, **58,** 43–46.

Scandella, C. J., and Kornberg, A. (1971), *Biochemistry*, **10,** 4447–4456.

Schäfer, R., and Franklin, R. M. (1975), *J. Mol. Biol.*, **97,** 21–34.

Schindler, P. R. G., and Teuber, M. (1978), *J. Bacteriol.*, **135,** 198–206.

Schneider, E. G., and Kennedy, E. P. (1976), *Biochim. Biophys. Acta*, **441,** 201–212.

Schuldiner, S., and Kaback, H. R. (1977), *Biochim. Biophys. Acta*, **472,** 399–478.

Schulman, H., and Kennedy, E. P. (1977), *J. Biol. Chem.*, **252,** 4250–4255.

Schweizer, M., Schwarz, H., Sonntag, I., and Henning, V. (1976), *Biochim. Biophys. Acta*, **448,** 474–491.

Smit, J., Kamio, Y., and Nikaido, H. (1975), *J. Bacteriol.*, **124,** 942–958.

Snipes, W., Douthwright, J., Sands, J., and Keith, A. (1974), *Biochim. Biophys. Acta*, **363,** 340–350.

Spencer, A. K., Greenspan, A. D., and Cronan, J. E., Jr. (1978), *J. Biol. Chem.*, **253,** 5922–5926.

Stanisich, V. A. (1974), *J. Gen. Microbiol.*, **84,** 332–343.

Stollery, J. G., and Vail, W. J. (1977), *Biochim. Biophys. Acta*, **471,** 372–390.

Taylor, F. R., and Cronan, J. E., Jr. (1976), *J. Bacteriol.*, **125,** 518–525.

Taylor, F. R., and Cronan, J. E., Jr. (1979), *Biochemistry*, **18,** in press.

Theresod, H., Letellier, L., Wert, R., and Shechter, E. (1977), *Biochemistry*, **16,** 3772–3776.

Thomas, J., and Law, J. H. (1966), *J. Biol. Chem.*, **241,** 5013–5018.

Torriani, A. (1968), *J. Bacteriol.*, **96,** 1200–1207.

Träuble, H., and Overath, P. (1973), *Biochim. Biophys. Acta*, **307,** 491–512.

Tsukagoski, N., and Fox, C. F. (1971), *Biochemistry*, **17,** 3309–3313.

Tsukagoski, N., and Franklin, R. M. (1974), *Virology*, **59,** 408–417.

Tsukagoski, N., Fielding, P., and Fox, C. F. (1971), *Biochim. Biophys. Res. Commun.*, **44,** 497–502.

Van Alphen, W., Lugtenberg, B., van Boxtel, R., and Rerhoef, K. (1977), *Biochim. Biophys. Acta*, **466,** 257–268.

Van den Bosch, H., and Vagelos, P. R. (1970), *Biochim. Biophys. Acta*, **218,** 233–248.

Van den Bosch, H., Williamson, J. R., and Vagelos, P. R. (1970), *Nature*, **228,** 338–342.

Van Golde, L. M. G., Schulman, H., and Kennedy, E. P. (1973), *Proc. Natl. Acad. Sci. U.S.*, **70,** 1368–1372.

Van Zoelen, E. J. J., de Kruijff, B., and van Deenan, L. L. M. (1978a), *Biochim. Biophys. Acta*, **508,** 97–108.

Van Zoelen, E. J. J., Verkleij, A. J., Zwall, R. F. A., and van Deenan, L. L. M. (1978b). *Eur. J. Biochem.*, **86,** 539–546.

Verkleij, A., van Alphen, L., Bijvelt, J., and Lugtenberg, B. (1977), *Biochim. Biophys. Acta*, **466,** 269–282.

Vos, M. M., Op den Kamp, J. A. F., Beckerdite-Quagliata, S., and Elsbach, P. (1978), *Biochim. Biophys. Acta*, **508,** 165–173.

White, D. A., Albright, F. R., Lennerz, W. J., and Schnaitman, C. A. (1971), *Biochim. Biophys. Acta*, **249,** 636–643.

White, D. A., Lennarz, W. J., and Schnaitman, C. A. (1972), *J. Bacteriol.*, **109,** 686–690.

Wilkens, M. H. F., Blaurock, A. E., and Engleman, D. M. (1971), *Nature, New Biol.*, **230,** 72–76.

Willsky, G. R., Bennet, R. L., and Malomey, M. H. (1973), *J. Bacteriol.*, **113,** 529–539.

Wirtz, K. W. A. (1974), *Biochim. Biophys. Acta*, **344,** 95–117.

Wong, F. H., and Bryan, L. E. (1978), *Can. J. Microbiol.*, **24,** 875–882.

Zilversmit, D. B., and Hughes, M. E. (1976), *Methods Membrane Biol.*, **7,** 211–259.

Chapter 4 Biosynthesis and Assembly of the Outer Membrane Proteins

SIMON HALEGOUA* AND MASAYORI INOUYE

Department of Biochemistry, State University of New York at Stony Brook

* Present address: Neurobiology Department, The Salk Institute, San Diego, California

1 INTRODUCTION

The modes of biosynthesis and assembly of the outer membrane proteins in *Escherichia coli* offers very challenging and perhaps some of the most perplexing questions in protein biosynthesis and membrane biogenesis. In considering the biosynthesis and assembly of the outer membrane proteins one is faced with two major problems. As is shown in Chapter 1, the *E. coli* envelope consists of two independent membranes, the cytoplasmic membrane and the outer membrane, and a space between the two membranes, the periplasmic space. Although the periplasmic space is considerably large, no ribosomes exist within this enclosed area. Therefore, the outer membrane proteins must be synthesized on polysomes from the cytoplasm. Hence the first problem facing proteins destined for the outer membrane is the traversal of the cytoplasmic membrane barrier. This problem is not specific to the outer membrane proteins but is shared by proteins of the periplasmic space, as well as cytoplasmic membrane proteins, which are at least in part exposed to the external side of that membrane. In fact, the question of protein translocation across a membrane is a major basic question in molecular biology, since at least all secretory proteins in eukaryotic systems face this same problem. Upon translocation of the outer membrane protein across the cytoplasmic membrane, the outer membrane protein is faced with its second major problem, specific insertion and assembly into the outer membrane structure. This again poses a very general question in membrane molecular biology: How is membrane localization of a protein determined?

As complex and formidable as it may appear, the biosynthesis and assembly of the outer membrane proteins occur quickly and efficiently, so that the synthesis and assembly are complete even after very short pulse-label times (see Section 3.1). As is seen later, the synthesis and assembly of the outer membrane proteins show interesting distinct differences from the synthesis of cytoplasmic proteins. These differences include the existence of precursors for the outer membrane proteins containing amino-terminal peptide extensions with unique features, synthesis on membrane-bound polysomes, and a close relationship to and dependence on the properties of the membrane, such as the fluid state of the membrane lipids. Although many intriguing questions remain to be answered and many details need to be worked out, it is shown in this chapter that a basic framework of the requirements for the synthesis, translocation, and assembly of the outer membrane protein, as well as some factors that regulate these processes, have been established over the last decade.

Among the outer membrane proteins, the biosynthesis and assembly of the lipoprotein have been most extensively investigated and recently reviewed (Inouye, 1975; Inouye, 1979). Furthermore, the reader is referred to reviews for general properties and functions of the outer membrane proteins (DiRienzo et al., 1978) and for more general aspects of secretion and membrane localization of proteins in *E. coli* (Inouye and Halegoua, 1979).

2 UNIQUE FEATURES

2.1 Differential Inhibition by Antibiotics

In a pioneering study to explore the characteristics of membrane and cytoplasmic protein synthesis *in vivo*, the effects of a number of antibiotic protein synthesis inhibitors were examined (Hirashima et al., 1973). The effects of five ribosome-directed antibiotics (kasugamycin, tetracycline, chloramphenicol, sparsomycin, and puromycin) on the synthesis of cytoplasmic and individual envelope proteins were examined. In general, when the antibiotic was added to a growing culture labeled by a radioactive amino acid, the biosynthesis of envelope proteins was found to be strikingly more resistant to kasugamycin and puromycin than the biosynthesis of cytoplasmic proteins. Conversely, envelope protein synthesis was more sensitive to tetracycline and sparsomycin.

When the labeled envelope proteins were further examined by SDS–polyacrylamide gel electrophoresis, differential sensitivites to the antibiotics were also found among individual envelope proteins. Synthesis of ompA protein was more resistant to kasugamycin, sparsomycin, and chloramphenicol than the other envelope proteins, while lipoprotein was more sensitive to chloramphenicol. In the case of puromycin, a much more severe differential effect was observed. Although the synthesis of envelope proteins was in general more resistant than that of cytoplasmic proteins, lipoprotein synthesis was strikingly more resistant to puromycin than the synthesis of the other envelope proteins. Even at 600 μg/ml puromycin, lipoprotein was synthesized at 60% its normal rate, while all other envelope protein synthesis was almost completely inhibited. This puromycin-resistant biosynthesis was also shown under conditions of exclusive lipoprotein synthesis. Since lipoprotein lacks five amino acids, including histidine, during histidine starvation of cells, lipoprotein was produced at 90% of its normal rate after 1 hr

(Hirashima and Inouye, 1973). Under this condition the lipoprotein synthesis was again found to be resistant to puromycin. Furthermore, lipoprotein produced in the presence of puromycin contained the same amount of fatty acids as lipoprotein produced under normal conditions.

The studies described above indicate that the mode of synthesis of envelope proteins differs from that of cytoplasmic proteins. Determination of the reasons for these differential inhibitions by antibiotics will provide insights into the specific mode(s) of biosynthesis and assembly of the envelope proteins. It was suggested by the authors that the differential effects may be due to compartmentalization of polysomes for specific proteins or specific differences in the biosynthetic machinery for the different proteins. The reasons for puromycin resistance of lipoprotein and other envelope protein synthesis are discussed in a later section.

In addition to the five ribosome-directed antibiotics tested, the effects of rifampicin, an inhibitor of RNA synthesis, on the synthesis of envelope and cytoplasmic proteins *in vivo* was examined. It was found that the envelope protein synthesis was more resistant to rifampicin than that of cytoplasmic proteins. Since rifampicin is known to block initiation of mRNA synthesis, the stabilities of mRNAs can be estimated by measuring the capacity for protein synthesis after the addition of rifampicin to the culture (Hirashima et al., 1973). It was found that the average half-lives of the mRNAs for cytoplasmic and membrane proteins were 2.1 and 5.5 min, respectively. This indicates that mRNAs for the membrane proteins are, on the average, about 2.5 times more stable than those for the cytoplasmic proteins. The half-lives for individual envelope proteins were calculated to be 3.2, 4.0 and 11.5 for matrix protein, ompA protein (tolG protein), and lipoprotein, respectively. Hence, aside from the unusual puromycin resistance of the lipoprotein biosynthesis, the mRNA coding for lipoprotein is extraordinarily stable; mRNA stability appears to be characteristic for outer membrane proteins since no differences in stability were observed between cytoplasmic membrane protein mRNA and those of cytoplasmic proteins (Lee and Inouye, 1974). Aspects of stable mRNA for outer membrane proteins are discussed in a later section.

2.2 Membrane-Bound Polysomes

2.2.1 ***Existence*** The idea that proteins that are secreted across a membrane are made on membrane-bound ribosomes was first demonstrated in eukaryotic systems. Palade and co-workers found that there is a correlation between the amount of ribosomes bound to the endoplasmic reticulum and the secretion of proteins (see review, Palade, 1975). It was later found that puromycin released incomplete chains to the interior of microsomes, suggesting that the secreted protein passes through the membrane while the peptide chain is elongated (Redman and Sabatini, 1966). From these and other results it was proposed that membrane-bound ribosomes in animal cells are responsible for protein secretion across the membrane (Blobel and Sabatini, 1971; Blobel and Dobberstein, 1975).

By analogy, secreted proteins in prokaryotic cells might also be synthesized on membrane-bound polysomes. However, the prokaryotic membrane-bound polysomes could not be easily detected by electron microscopy, and methods for the isolation of a functional class of ribosomes bound to membrane were quite cumbersome. Cancedda and Schlessinger (1974) first separated polysomes of *E. coli* into two classes, membrane-bound and free polysomes. They demonstrated that alkaline phosphatase, a periplasmic protein, was preferentially synthesized (70–90%) on the membrane-bound polysomes. The existence of membrane-bound polysomes was more recently confirmed and extended with the use of new and better techniques for their isolation. After sonic disruption of the cells, the membrane-bound polysomes isolated on a sucrose density gradient were found to synthesize two classes of proteins, one that remained membrane bound and another that became soluble upon release from the ribosomes (Randall and Hardy, 1977). The former class was identified immunologically as outer membrane proteins and the latter class was identified as periplasmic proteins. Synthesis of the cytoplasmic protein EFTu could not be detected by the membrane-bound polysomes. The synthesis of alkaline phosphatase on membrane-bound polysomes also was confirmed recently using somewhat different procedures for isolation of membrane-bound and free polysomes (Smith et al., 1977; W. P. Smith, P. C. Tai and B. D. Davis, personal communication; Varienne et al., 1978).

2.2.2 Role for Translocation The existence of membrane-bound polysomes would suggest that the nascent peptide is translocated across the membrane during the protein synthesis. If this is the case, the nascent peptide should at least in part be exposed to the outside of the cell and would thereby be accessible to chemical modification by an external agent. This possibility was tested (Smith et al., 1977) using the surface labeling agent [^{35}S]acetylmethionyl methylphosphate sulfone, which labels amino groups. This agent was reacted with spheroplasts (only the outer membrane is permeabilized to the agent) in which protein synthesis was ongoing. When the membrane-bound polysomes were isolated from these cells, the radioactivity was recovered in peptidyl-tRNA. A portion of the labeled product after *in vitro* chain completion was identified as alkaline phosphatase. Hence this result has provided the first evidence that proteins are secreted across the cytoplasmic membrane as growing chains. The nature of membrane binding of the polysomes is discussed in Section 3.2.

2.2.3 ***Puromycin-Resistant Biosynthesis*** As is discussed earlier, the biosynthesis of *E. coli* membrane proteins show differential sensitivities to antibiotics compared to cytoplasmic proteins. The most striking example of this are the differential effects of puromycin on the synthesis of these proteins. Membrane protein synthesis was generally more puromycin resistant than the synthesis of cytoplasmic proteins. In particular, lipoprotein was specifically produced in the presence of puromycin, which inhibited all other protein synthesis. As is discussed later, the mRNA for the lipoprotein is abundant in the cell, and the size of the mRNA is relatively smaller than that of other cellular mRNAs. However, these facts are probably not the reason for the puromycin resistance of the lipoprotein biosynthesis, since (*a*) the lipoprotein synthesis is as sensitive to other ribosome-directed antibiotics such as tetracycline, kasugamycin, and sparsomycin as the biosynthesis of many other membrane proteins, and (*b*) the lipoprotein synthesis is more sensitive to chloramphenicol than the synthesis of other membrane proteins. As was discussed previously (Hirashima et al., 1973), the puromycin-resistant biosynthesis could be due to compartmentation of specific polysomes or differentiation of components required for protein synthesis. Based on these considerations, lipoprotein synthesis was examined in three different systems in which the permeability barrier to puromycin was progressively disrupted by EDTA treatment, toluene treatment, and in an *in vitro* system using isolated polysomes (Halegoua et al., 1976). Puromycin sensitivity of overall protein synthesis was increased by about tenfold in each permeabilized system: 50% inhibitions were obtained at 330, 35, 2.7, and 0.22 μg/ml puromycin for intact cells, EDTA-treated cells, toluene-treated cells, and the polysome system, respectively. In each of the above systems, lipoprotein synthesis remained puromycin resistant relative to all other protein synthesis, indicating that puromycin resistance was an intrinsic property of the lipoprotein biosynthetic machinery. However, when the ribosomal proteins of polysomes engaged in the synthesis of outer membrane proteins (rifampicin-resistant polysomes) were compared to those engaged in overall protein synthesis, no quantitative differences could be detected (Randall and Hardy, 1975). Later, Randall and Hardy (1977) found that protein synthesis directed by membrane-bound polysomes that were shown to synthesize both periplasmic and outer membrane proteins was much more resistant to puromycin than free polysomes synthesizing predominantly cytoplasmic proteins.

Based on the above data, it is quite possible that puromycin resistance may be a function of membrane binding of the polysomes, and it may therefore be a useful probe for determination of the extent of membrane binding of a particular polysome or class of polysomes. It is interesting to point out that the biosynthesis of membrane proteins in rabbit reticulocytes was also found to be more resistant to puromycin than the synthesis of the cytoplasmic protein globin (Koch et al., 1975).

3 TRANSLOCATION ACROSS THE CYTOPLASMIC MEMBRANE

3.1 Precursors of Secreted Proteins

3.1.1 Accumulation by Membrane Perturbants Since outer membrane protein biosynthesis and assembly must undergo a multistage process consisting at least of (*a*) synthesis, (*b*) translocation across the cytoplasmic membrane, and (*c*) assembly into the outer membrane, it was thought that examination of precursors or intermediate forms of these proteins would help to elucidate the mechanisms involved in these processes. However, efforts to find these intermediates by short pulse-chase experiments have failed (Lee and Inouye, 1974; Leij, et al., 1978), suggesting that processing and assembly of precursor molecules, if any, occur very rapidly. Therefore, another approach has been taken to examine the intermediate steps in biosynthesis and assembly, namely, making use of perturbants of the envelope structure, which allows accumulation of intermediates by inhibition at different stages of biosynthesis and assembly. The first perturbant used was toluene treatment of cells, which is known to result in breakdown of the cellular permeability barrier with partial dissolution of the cytoplasmic membrane. Since the toluene-treated cells are permeable to molecules such as nucleoside triphosphates they have provided an important and easily obtainable system, partially *in vitro* to study DNA synthesis (Moses and Richardson, 1970), RNA synthesis (Peterson et al., 1971), and peptidoglycan synthesis (Schrader and Fan, 1974). Conditions were established in which cells may be treated with toluene and retain the ability for protein synthesis (Halegoua et al., 1976a; Halegoua et al., 1976b). This system was found to be entirely dependent on the addition of ATP and to exclusively synthesize membrane proteins. As can be seen in Figure 1, the SDS–gel electrophoresis pattern of the membrane proteins produced by toluene-treated cells differs substantially from the *in vivo* pattern. When the membrane proteins of toluene-treated cells were treated with anti-lipoprotein antiserum, two distinct peaks were observed in the immunoprecipitate, one

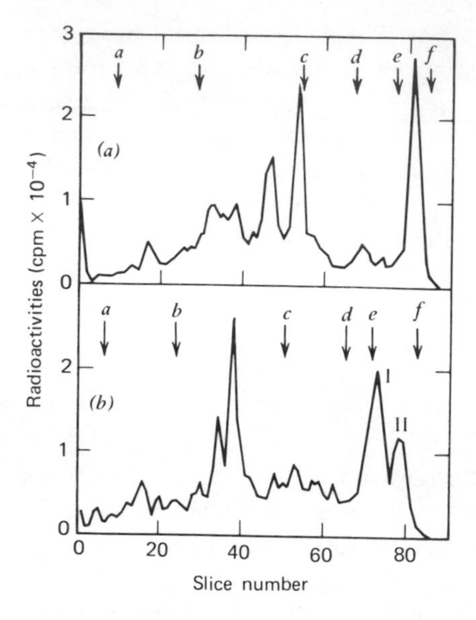

Figure 1 Sodium dodecyl sulfate–gel electrophoresis of membrane proteins from intact and toluene-treated cells. Proteins were labeled with [^{3}H]arginine and the membrane fractions were prepared and analyzed by SDS–gel electrophoresis. (*A*) Membrane proteins of intact cells; (*B*) membrane proteins of toluene-treated cells. Peaks labeled *I* and *II* indicate the new form of lipoprotein and lipoprotein, respectively. Arrows with letters indicate internal molecular weight standards: (*a*) dimer, (*b*) monomer of DANS-bovine serum albumin, (*c*) dimer, (*d*) monomer of DANS-egg-white lysozyme, (*e*) cytochrome c, (*f*) DANS-insulin. Reprinted by permission of the American Society of Biological Chemists, Inc. From Halegoua et al. (1977).

(peak II in Figure 1) comigrated with the *in vivo* lipoprotein and the other (peak I) appeared to correspond to a new form of lipoprotein with an apparent molecular weight of about 15,000 (Halegoua et al., 1976b; Halegoua et al., 1977). This new form of lipoprotein shared common biosynthetic characteristics as lipoprotein, such as relative puromycin and rifampicin resistance and chloramphenicol sensitivity. Peptide mapping of the new protein revealed that it had the same carboxyl-terminal structure as lipoprotein (Halegoua et al., 1977). However, the amino-terminal end was found to be methionine (Halegoua et al., 1977) instead of the glycerylcysteine found in the case of lipoprotein. Thus, based on the assumption that the new protein contained an amino-terminal peptide extension on the lipoprotein sequence, its amino acid composition was determined by double-labeling experiments (Halegoua et al., 1977). It was found that the peptide extension contained at least 18–19 extra amino acid residues enriched in hydrophobic residues (63% in contrast to 42% in the case of lipoprotein). It should also be noted that it contained three glycine residues, which lipoprotein lacks. The new form of lipoprotein, designated prolipoprotein, was believed to play a major role in the biosynthesis and assembly of lipoprotein. Thus toluene treatment appeared to block processing of the prolipoprotein to lipoprotein. As is discussed in the next section, the prolipoprotein accumulated in toluene-treated cells was identical to the *in vitro* translation product directed by the purified lipoprotein mRNA.

Milder conditions for treatment of cells with toluene, that is, treatment for 1.5 min instead of 10 min, enhanced the incorporation of [^{35}S]methionine into new membrane proteins of higher molecular weights (Sekizawa et al., 1977). The production of two new membrane proteins was clearly evident, and based on the results obtained by immunoprecipitation, SDS–polyacrylamide gel electrophoresis, and autoradiography it was shown that the two new proteins were putative precursors of matrix and ompA proteins (promatrix and pro-ompA proteins), with molecular weights about 2000 higher than those of the two major outer membrane proteins. Since all three major outer membrane proteins may be formed from precursors that contain about 20 amino acid residues, a common mechanism for their synthesis seems to be involved. The amino terminus of both promatrix protein and prolipoprotein was methionine (Halegoua et al., 1977; Sekizawa et al., 1977). The distribution of leucine residues at the amino terminal region of promatrix protein differs from that of the matrix protein, which suggests that the peptide extension is located at the amino-terminal end as in prolipoprotein.

The use of phenethyl alcohol (PEA) instead of toluene as a membrane perturbant in order to accumulate precursors of secreted proteins proved to be a valuable tool in studying outer membrane protein synthesis and assembly (Halegoua and Inouye, 1979). Although similar in structure to toluene, PEA is water soluble up to 2%; thus the extent of PEA administration to the cells was quite easily controlled. In addition, unlike toluene, the effects of PEA were reversible and less damaging in terms of macromolecular synthesis and membrane structure. At 0.3% PEA, among the major outer membrane proteins (matrix protein, ompA protein, and lipoprotein), synthesis of matrix protein was specifically lost with the concomitant accumulation of promatrix protein (Figure 2*A*). Under this condition, unlike the case with toluene, the accumulation of promatrix protein was reversible. Upon removal of PEA and regrowth of the cells in culture, promatrix protein was processed and the resulting matrix protein was normally assembled into the outer membrane. This first demonstrated a direct precursor–product relationship between the proprotein and processed outer membrane protein. The promatrix protein specifically accumulated under this condition was also subcellularly localized, as is discussed in a later section.

When higher concentrations of PEA were administered to the cells, differential effects on the synthesis and assembly of the major outer membrane proteins were observed (Halegoua and Inouye, 1979). From 0.5 to 1.0% PEA, both matrix protein and ompA protein were missing from the SDS–gel pattern of the membrane fraction and instead both promatrix and pro-ompA protein were produced (Figures 2*B*, 2*C*, and 2*D*). At these con-

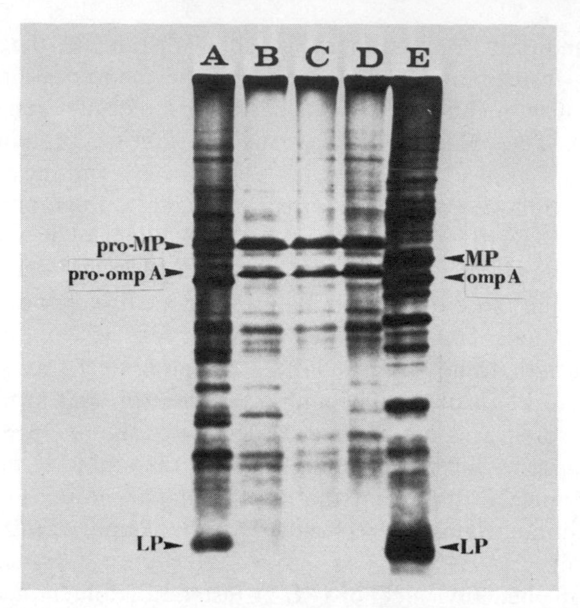

Figure 2 Differential effects of PEA on the synthesis and assembly of outer membrane proteins. Protein synthesis was carried out in PEA-permeabilized cells or in culture without PEA with [^{35}S]methionine. SDS–polyacrylamide gel electrophoresis of isolated envelope fractions was also carried out. (*A*) 0.3% PEA, (*B*) 0.5% PEA, (*C*) 0.8% PEA, (*D*) 1.0% PEA, (*E*) *in vivo*. Arrows designate positions of identified outer membrane proteins; MP, matrix protein; ompA, ompA protein; LP, lipoprotein. Reprinted by permission from Halegoua and Inouye (1979).

centrations of PEA processing of lipoprotein was not inhibited. Therefore, PEA could be used to differentially inhibit processing and assembly of each major outer membrane protein by manipulating the PEA concentration. Other differential effects of PEA on the assembly of the outer membrane proteins are discussed in following sections.

3.1.2 **In vitro *Synthesis*** A final goal in the study of envelope protein biosynthesis and assembly is the establishment of cellfree systems for these processes. The first major step toward this goal has been the purification of the lipoprotein mRNA from the cell. Since *E. coli* mRNA is generally extremely unstable, usually it is not an easy task to purify a specific mRNA from cells. However, purification of the lipoprotein mRNA offers distinct advantages over the other cellular mRNAs as follows: (*a*) as is discussed earlier, the lipoprotein mRNA is extremely stable with a half-life of 11.5 min, (*b*) the mRNA is assumed to be very abundant in the cell, since lipoprotein is the most abundant protein based on molecular numbers (Inouye, 1975; Inouye, 1979), and (*c*) the mRNA is assumed to be smaller than most other cellular mRNAs since lipoprotein has a very small molecular weight (7200). Based on these criteria, the biologically active lipoprotein mRNA was isolated (Hirashima et al., 1974). The *in vitro* translation product was cross-reactive with anti-lipoprotein serum, and peptide mapping of the product indicated that it had the identical carboxyl-terminal structure to that of lipoprotein (Hirashima et al., 1974).

As is discussed above, prolipoprotein, a putative precursor of lipoprotein was produced in toluene-treated cells. If prolipoprotein is indeed a precursor of lipoprotein, it would be expected to be seen as the primary translation product of the lipoprotein mRNA. When the immunoprecipitated prolipoprotein from toluene-treated cells was mixed with the cellfree product directed by the purified lipoprotein mRNA, the two products were found to comigrate on SDS–gel electrophoresis (Inouye et al., 1977), suggesting that the two proteins may be the same.

The amino acid sequence of the peptide extension at the amino-terminal end of prolipoprotein was believed to give essential clues to solving the mechanism of translocation and assembly of lipoprotein. The amino acid sequence of this region was obtained using cell-free translation of the purified lipoprotein mRNA. To determine the length of the peptide extension, the cell-free product directed by the purified mRNA was labeled with [^{3}H]leucine and was subjected to 46 consecutive Edman degradations in a sequenator that automatically removes amino acid residues one at a time from the amino-terminal end of the protein. Figure 3 shows the release of radioactivity at each cycle of the Edman degradation. As can be seen from Figure 3, radioactivity peaks appeared at cycles 6, 8, 13, 17, 18, 30, 37, and 44. As expected from sequenator analysis, the recovery of radioactivity decreases with increasing cycles, and there is always some trailing of

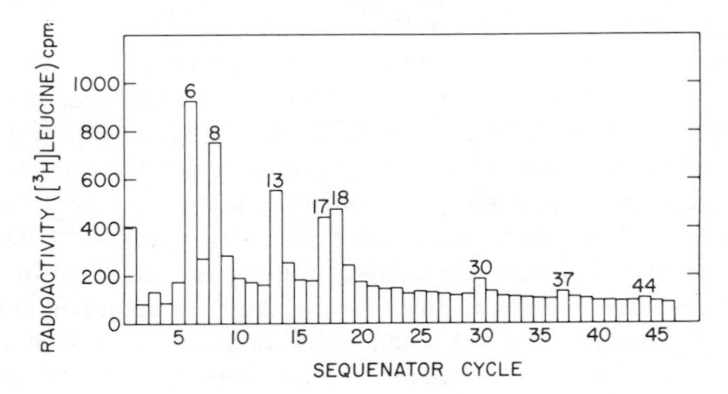

Figure 3 Forty-six consecutive Edman degradations of the cell-free product labeled with [^{3}H]leucine (Inouye et al., 1977).

radioactivity from the previous cycle because of incomplete degradation at the previous cycle. Therefore, the radioactivity at cycle 18 is slightly higher than that at cycle 17 because of the trailing of the radioactivity from cycle 17.

Lipoprotein has leucine residues at the tenth, seventeenth, twenty-fourth, and forty-ninth positions from its amino terminus (Braun and Bosch, 1972). Thus one notices that the distribution of leucine residues in the cell-free product at positions 30, 37, and 44 coincide very well with that of positions 10, 17, and 24 of lipoprotein if prolipoprotein is assumed to have 20 extra amino acid residues at its amino-terminal end. It is also important to point out that there are no leucine residues between positions 21 and 29 in Figure 3. The results in Figure 3 also indicate that there are five leucine residues in the extended region at positions 6, 8, 13, 17, and 18. This is also in very good agreement with the data obtained from toluene-treated cells, in which 4–5 leucine residues were predicted to exist in the extended region of prolipoprotein (Halegoua et al., 1977).

If the extended region has 20 amino acid residues, one can predict the appearance of amino acids after cycle 21 of the Edman degradation, because the amino acid sequence of lipoprotein is known. For example, serine residues are known to be at positions 2, 3, 11, 12, 25, and 33 of lipoprotein (Braun and Bosch, 1972). Thus one can predict that serine residues should appear at cycles 22, 23, 31, 32, 45, and 53 of the Edman degradation of the cell-free product. Figure 4*A* shows the Edman degradation of the cell-free product labeled with [^{3}H]serine up to cycle 34. The radioactivities clearly appear at cycles 22, 23, 31 and 32 as predicted. Figure 4*A* also shows that there is one serine residue at position 15 of the extended region. Figures 4*B*, 4*C*, and 4*D* provide further confirmation of the length of the extended region. Alanine residues appear at cycle 25 of the Edman degradation as predicted as well as at positions 3, 10, and 19 (Figure 4*B*). Lysine residues are detected at position 26 as predicted, as well as at positions 2 and 5 of the extended region (Figure 4*C*). Furthermore, isoleucine residues appeared at cycle 27 of the Edman degradation as predicted, as well as at position 12 of the extended region of the cell-free product (Figure 4*D*).

To determine the complete amino acid sequence of the extended region, Edman degradation was carried out with the cell free product labeled with the following amino acids, which were predicted to exist in the peptide extension of prolipoprotein produced in toluene-treated cells: methionine, glycine, threonine, and valine. In this way, the final complete amino acid sequence of prolipoprotein was obtained as shown in Table 1. It should be pointed out that the first methionine residue seems to be formulated in the cell-free product, since a hot TCA treatment of the product improves the recovery of the radioactivity in the Edman degradation. This indicates that

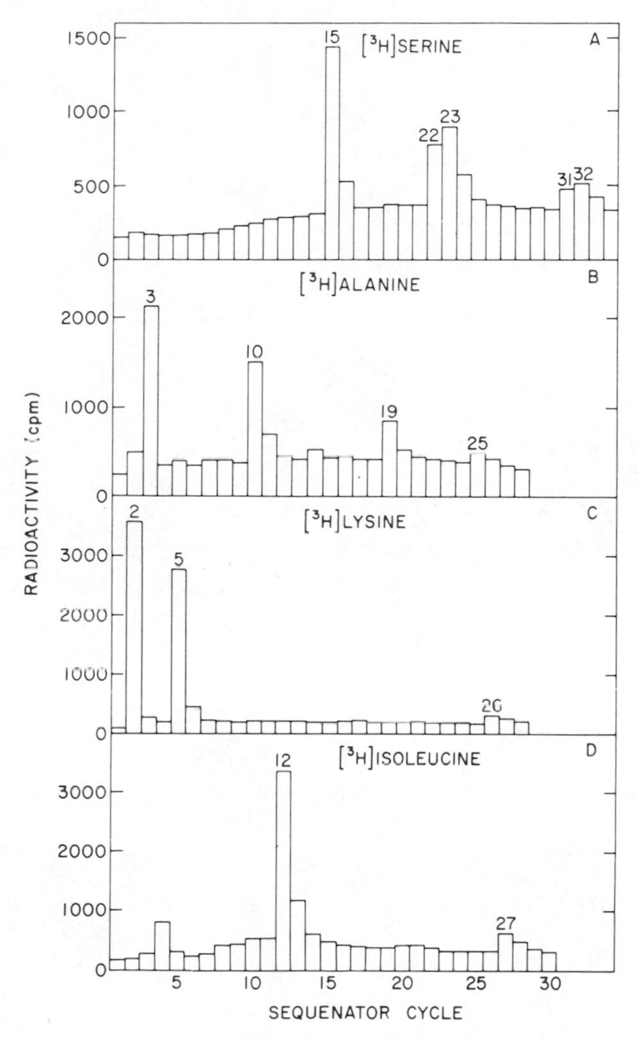

Figure 4 Sequential Edman degradation of the cell-free products labeled with [^{3}H]serine, [^{3}H]alanine, [^{3}H]lysine, and [^{3}H]isoleucine (Inouye et al., 1977).

prolipoprotein shown in Table 1 is the primary product in the cell-free system.

By use of the membrane-bound polysome system of Randall and Hardy (1977), a higher molecular weight form of the λ receptor (*lamB* gene product) was found to be made along with λ receptor by chain completion *in vitro*. The new form had molecular weight about 2000 higher and was presumed to be the precursor form of the λ receptor, similar to those of

Table 1 Amino Acid Sequences of the Peptide Extension of Precursor Proteins of *E. coli*

	23	22	21	20	19	18	17	16	15	14	13	12	11	10	9	8	7	6	5	4	3	2	1	
Prolipoprotein				Met	Lys	Ala	Thr	Lys	Leu	Val	Leu	Gly	Ala	Val	Ile	Leu	Gly	Ser	Thr	Leu	Leu	Ala	Gly	Cys[a]
Pro-β-lactamase:	Met	Ser	Ile	Gln	His	Phe	Arg	Val	Ala	Leu	Ile	Pro	Phe	Phe	Ala	Ala	Phe	Cys	Leu	Pro	Val	Phe	Ala	His[b]
Pro-fd-major coat protein (gene 8):	Met	Lys	Lys	Ser	Leu	Val	Leu	Lys	Ala	Ser	Val	Ala	Val	Ala	Thr	Leu	Val	Pro	Met	Leu	Ser	Phe	Ala	Ala[c]
Pro-fd-minor coat protein (gene 3):						Met	Lys	Lys	Leu	Leu	Phe	Ala	Ile	Pro	Leu	Val	Val	Pro	Phe	Tyr	Ser	His	Ser	Ala[d]

[a] Inouye et al., 1977.
[b] Sutcliffe, 1978.
[c] Sugimoto et al., 1977.
[d] Schaller et al., 1978.

other outer membrane proteins described above. In a similar manner, putative precursors of two periplasmic proteins, maltose binding protein and arabinose binding protein, were identified (Randall et al., 1978). A higher molecular weight precursor of alkaline phosphatase, a periplasmic enzyme, was first synthesized and processed using an *in vitro* coupled transcription–translation system directed by the DNA from a transducing phage containing the gene for alkaline phosphatase (*phoA*) (Inouye and Beckwith, 1977; see Chapter 7). Hence the existence of protein precursors appears to be a general phenomenon for secreted proteins in *E. coli* (see review, Inouye and Halegoua, 1979).

3.1.3 Accumulation by Nonfluid Membrane When an unsaturated fatty acid auxotroph of *E. coli* is grown on elaidate, a shift in growth temperature to 20°C results in a nonfluid state of the membrane lipids. Under such a nonfluid condition of membranes the biosynthesis and assembly of outer proteins were found to be greatly affected (DiRienzo and Inouye, 1979). For instance, pro-ompA protein is accumulated that can be chased into the mature ompA protein in the outer membrane upon shifting the temperature back from 20 to 30 or 37°C. Effects of lipid fluidity on the assembly of outer membrane proteins are discussed later.

3.2 Role of Precursor Peptide Extension

3.2.1 Structure As is shown in the previous section, prolipoprotein provided the first amino acid sequence of a peptide extension on an envelope protein precursor. From the amino acid sequence of the prolipoprotein peptide extension (Table 1), Inouye et al. (1977) pointed out several unique features of the sequence that were believed to be important in the functioning of the peptide extension in translocation of the protein across the cytoplasmic membrane. These are features (*a*) the extended region is basic and positively charged at neutral pH because it contains two lysine but no acidic amino acid residues; (*b*) the peptide extension contains three glycine residues, an amino acid that is not present in the lipoprotein sequence; (*c*) the extended peptide is much more hydrophobic than lipoprotein, since 60% of the amino acid residues in this region are hydrophobic in contrast to 38% in lipoprotein; (*d*) the distribution of the hydrophobic amino acids in the peptide extension is completely different from their periodical distribution in lipoprotein.

More recently, amino acid sequences of peptide extensions on other envelope protein purecursors have been determined. The amino acid sequence of the β-lactamase precursor (pro-β-lactamase) was determined from the nucleotide sequence of the plasmid gene coding for the protein

(Table 1) (Sutcliffe, 1978). Although β-lactamase is a periplasmic enzyme, the sequence of the pro-β-lactamase peptide extension of 23 amino acid residues is remarkably similar to that of prolipoprotein. As in the case with β-lactamase, two cytoplasmic membrane proteins were deduced to be made from precursors containing amino-terminal peptide extensions, based on nucleotide sequences. In one case the sequence of the peptide extension of 23 amino acid residues of bacteriophage fd major coat protein precursor was deduced from the nucleotide sequence of the mRNA for the protein (Sugimoto et al., 1977) (Table 1). From the DNA sequence of gene 3 of bacteriophage fd, the peptide extension of 18 amino acids of the minor coat protein precursor was deduced (Schaller, et al., 1978) (Table 1). Again, the amino acid sequences of these peptide extensions bear an uncanny resemblance to each other and to that of prolipoprotein and pro-β-lactamase.

Although obtained from proteins with very different locations in the envelope, the parity among the sequences of the peptide extensions shown in Table 1 indicate that the common features must reflect a common function of the peptide extensions. With this in mind, the common features among the peptide extensions of envelope protein precursors were derived from their sequences shown in Table 1 by examination of certain classes of amino acids along the sequence of the peptide extension (Inouye and Halegoua, 1979). The distribution of three classes of amino acids, basic amino acids, hydrophobic amino acids, and proline + glycine residues, along the peptide extension are plotted in Figure 5*A*. In the figure, the average content of the particular class of amino acid at position N is calculated as the average content of the amino acids at positions $N - 1$, N, and $N + 1$ to avoid the fluctuations seen in the contents calculated for a single position. From the analysis demonstrated in Figure 5 and the sequences given in Table 1, the peptide extension can be divided into three main sections containing the following general features. It should be noted that in both Table 1 and Figure 5 the amino acid residues in the peptide extensions are numbered from the processing or cleavage site residue of the peptide extension (numbered 1) to the amino terminus.

Rule 1 The amino-terminal section (section *I*) has a basic character. (*a*) The amino-terminal section has a basic nature as a result of a contingency of at least two basic, but no acidic, amino acid residues. (*b*) The basic section *I* is located about 55 Å from the cleavage site. The first basic residue (lysine or arginine) appears at the sixteenth or seventeenth position from the cleavage site without fail.

Rule 2 There exists a section (section *II*) 50 Å in length of high hydrophobicity between the basic section *I* and the cleavage site. As clearly seen in

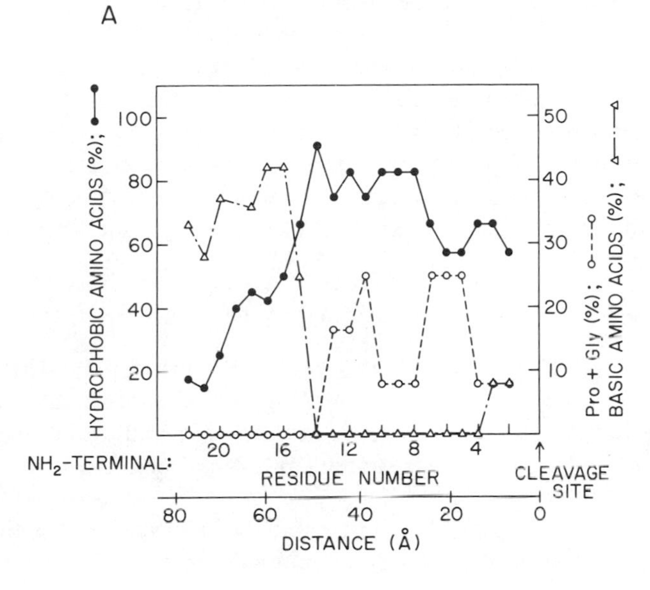

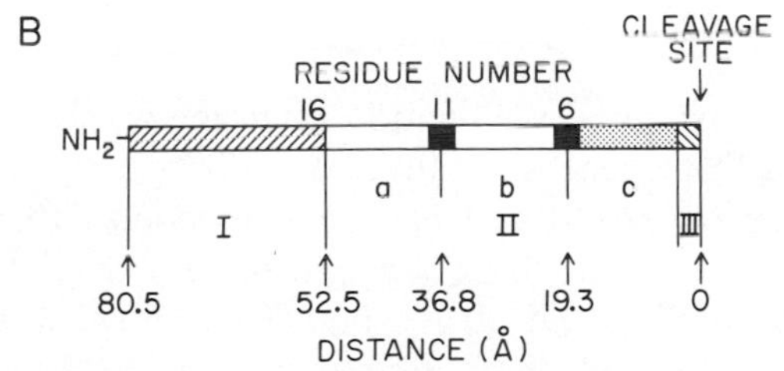

Figure 5 Amino acid distribution in the peptide extensions of precursor proteins in *E. coli* (from Inouye and Halegoua, 1979). (*A*) Using amino acid sequences given in Table 1, the content (*C*) of amino acid residues of a particular class at each position (*N*) is calculated as follows: $C = 100X$ (total number of amino acid residues of a particular class at positions $N - 1$, N, and $N + 1$)/(total number of amino acid residues at positions $N - 1$, N, and $N + 1$). *C* is then plotted against residue number from the cleavage point or the distance (Å) from the cleavage point. (●) Contents of hydrophobic amino acid residues (Ala, Cys, Val, Met, Ile, Leu, Tyr, Phe, and Trp except for Met at the amino-terminal ends); (Δ) contents of basic amino acid residues (Lys, His, and Arg); (O) contents of Pro and Gly. (*B*) The general structure of the peptide extension deduced from *A*. (*I*) basic region; (*II*) hydrophobic region, and (*III*) recognition site for cleavage of the peptide extension. Section *II* is divided into three subsections, *a*, *b*, and *c*, by pro or gly at positions 6 and 11. Distances are calculated using 3.5 Å for the distance between two adjacent amino acid residues.

Figure 5*A*, the content of hydrophobic amino acids in the peptide extension sharply increases immediately after the basic section *I*. Hydrophobicity is particularly high between residue positions 6 and 15.

Rule 3 Within the hydrophobic section *II*, there appears either a proline or glycine residue at least 15–20 Å from the cleavage site. It is evident from Figure 5*A* that there are two peaks of proline + glycine residues along the peptide extension. One peak occurs between positions 11 and 13 (based on the average) and the other occurs between positions 5 and 7. All the sequences shown in Table 1 contain a proline or glycine residue between positions 4 and 7 (Gly^{7} for 1, Pro^{4} for 2, Pro^{6} for 3, and Pro^{6} for 4) and three out of those four sequences contain a proline or glycine residue between positions 10 and 12 (Gly^{12} for 1, Pro^{12} for 2 and Pro^{10} for 4). This class of residue divides the hydrophobic section *II* into three subsections, *a*, *b*, and *c* (see Figure 5). As seen in Figure 5, sections *IIa* and *IIb* are extremely hydrophobic, while section *IIc* is somewhat less hydrophobic.

Rule 4 The amino acid residue existing at the cleavage site (section *III*) is one whose side chain at the α-carbon contains none or only one carbon (i.e., Gly, Ala, Ser, and Cys).

Based on the above rules, the peptide extension is schematically drawn in Figure 5, in which the general features described above are demonstrated. The peptide extension consists of (*a*) a basic section *I* that is 52.5 Å away from the cleavage site and whose length can vary from 10 to 25 Å, (*b*) a hydrophoblic section *II* about 50 Å long, that is divided into three subsections, *a*, *b*, and *c* by proline or glycine residues, and (*c*) an amino acid residue with a short side chain at the cleavage site (section *III*).

As is mentioned in the preceding section, secretory proteins in eukaryotic systems are also made from precursors containing amino-terminal peptide extensions similar in length to the ones from *E. coli*. It is therefore very interesting to note that the rules derived from the *E. coli* precursor peptide extensions were found to be applicable to some peptide extensions of eukaryotic secretory protein precursors (see review by Inouye and Halegoua, 1979). This suggests that there may be a common mechanism of secretion of proteins across the membrane for both prokaryotic and eukaryotic systems.

3.2.2 Function: Models for Translocation What is the function(s) of the peptide extension on precursors of envelope proteins? The general features of the precursor peptide extension are derived above from proteins having completely different locations in the envelope, the cytoplasmic membrane, the periplasmic space, and the outer membrane. However, as is mentioned above, the remarkable similarities of their precursor peptide extensions would indicate a common function for the peptide extensions. That common function would undoubtedly be in the translocation of the protein in whole (in the case of some outer membrane proteins and periplasmic proteins) or in part (in the case of cytoplasmic membrane proteins) across the cytoplasmic membrane. It is indeed astounding that the peptide extensions on *E. coli* envelope protein precursors bear a profound similarity to those on secretory protein precursors in eukaryotic cells. These latter extensions must also be translocated across a membrane, thus demonstrating a process that is highly conserved through evolution.

Based on the general features for the peptide extension outlined above, a possible role for the peptide extension in the translocation process has been outlined (Inouye et al., 1977; DiRienzo et al., 1978; Inouye and Halegoua, 1979) as follows. As the precursor protein in synthesized on the polysome, the positively charged section *I* allows the initial attachment of the peptide extension and consequently the polysome to the negatively charged inner surface of the cytoplasmic membrane through ionic interactions (see Figure 6). The *E. coli* membrane surface is negatively charged at neutral pH as a

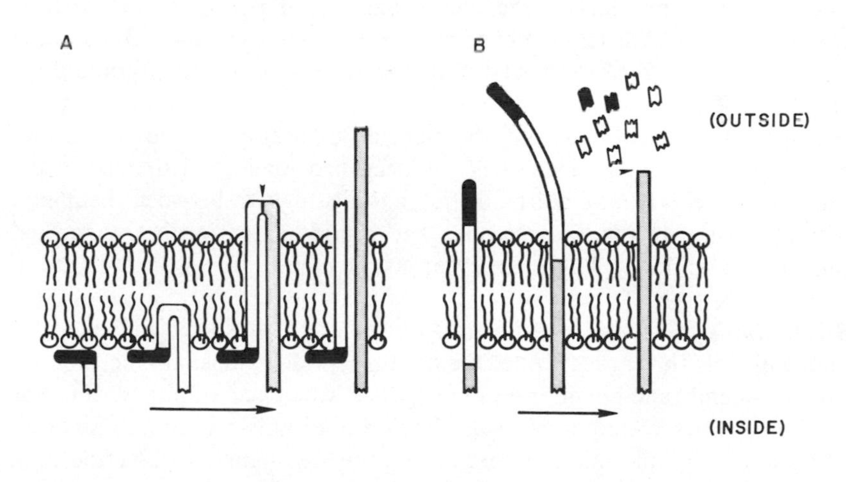

Figure 6 Loop model and linear model for the translocation of secretory proteins across the membrane (Inouye and Halegoua, 1979). (*A*) Loop model and (*B*) linear model. Solid portions represent the basic regions of the first several amino acid residues of the peptide extension (section *I* in Figure 5). The following blank portions represent the hydrophobic regions of the peptide extension (Section *II*) and the cleavage site (section *III*) followed by shaded portions of mature proteins. Small arrows show the cleavage sites of the precursor molecules.

result of the presence of phosphatidyl glycerol. The next hydrophobic section, section *II*, is then progressively inserted into the cytoplasmic membrane by hydrophobic interaction with the lipid bilayer (Figure 6). As the peptide further elongates, the loop formed by section *II* is further extended into the lipid bilayer. The processing or cleavage site of the peptide extension (section *III*) is eventually exposed to the outside surface of the cytoplasmic membrane (see Figure 6). The length of sections *II* and *III*, about 53 Å, is long enough to extend across the lipid bilayer to expose section *III* to the outside surface of the membrane where processing at the cleavage site may take place. The role of the proline or glycine residues in section *II* may be important in bending the peptide at the position to form a loop. This proposed mechanism of protein translocation across a membrane is called the loop model (Inouye et al., 1979; Inouye and Halegoua, 1979).

An alternative mechanism is the linear model (signal hypothesis) proposed by Blobel and Dobberstein (1975) for the transfer of eukaryotic secretory proteins across membranes. In this model it is proposed that the hydrophobic nature of the peptide extension of secretory protein precursors allows for initial interaction of the polysome with the membrane followed by interaction of the peptide extension with a specific receptor protein in the membrane. The result is activation of the receptor protein to mediate aggregation of proteins at the membrane-bound polysome site to form a transmembrane hydrophilic channel. The peptide extension and the succeeding secretory protein are linearly translocated across the membrane through channel during synthesis of the latter. This model does not take into account the specific characteristics of the peptide extensions as described above. However, no direct evidence has yet been provided to distinguish between the two models. It may not be difficult to distinguish between them experimentally, since in the linear model, the peptide extension is transported to the outside, while it is left in the cytoplasmic membrane in the loop model.

3.2.3 Basis for Membrane-Bound Polysomes It has been indicated in the above models that at least one role of the peptide extension is in the formation of membrane-bound polysomes. The existence of membrane-bound polysomes for secreted proteins has been well established, and these proteins traverse the cytoplasmic membrane as growing chains (see Section 2.2). A major question to be answered experimentally is, What is the basis for membrane binding of the polysomes? Evidence has been presented recently that suggests there is no specific attachment of the membrane-bound ribosome to the membrane other than by way of the nascent peptide chain (Smith et al., 1978). When membrane-bound polysomes were isolated and treated with puromycin, it was observed that 75% of the ribosomes were released from the membrane by the puromycin treatment. Since puromycin releases the ribosomes from the nascent chain by formation of peptidyl-puromycin, it was concluded that the released ribosomes had no specific membrane attachment of their own but were attached to the membrane by way of the nascent peptide chain. Presumably, the peptide extension would play an intimate role in the membrane binding. The mRNA could also be eliminated as a mediator of membrane binding since no ribosome release factor was present in the *in vitro* polysome system, and therefore the ribosomes released by puromycin remained in complex with the mRNA. However, it is interesting that in is study, 25% of the ribosomes in the membrane-bound polysome preparation was not released by puromycin. Among this 25% it was shown that 10–15% contained peptide that did not react with puromycin (even at 500 μg/ml). Since, as is discussed in Section 2, secreted proteins in *E. coli* were shown to be more resistant to puromycin than to cytoplasmic proteins, it is possible that the 25% of the polysomes that remained membrane bound after puromycin treatment were engaged in the production of a class of secreted proteins such as lipoprotein (see Section 2). Thus it cannot be ruled out that these polysomes have an, as yet, undetermined functional attachment of ribosome to membrane. At any rate, it would be of interest to determine which secreted proteins are produced in puromycin-sensitive membrane-bound polysomes and which are produced in puromycin-resistant membrane-bound polysomes.

Another question that has thus far eluded explanation concerns the nature of the binding of the polysome to the membrane. As is discussed above, at least for most membrane-bound polysomes, the nascent peptide may be the sole mode of attachment of the polysome to the membrane, and the peptide extension of precursor proteins is generally believed to be the mediator of such binding. However, the precise mode of interaction of the peptide extension with the membrane is as yet undetermined. In the previous section, two models for the polysome attachment to the membrane are described. In the loop model, the peptide extension itself has characteristics, as demonstrated in Figure 6, such that it can interact directly with the membrane structure. In the linear model (signal hypothesis), the membrane is proposed to contain a specific receptor protein that is recognized by the precursor peptide extension. If such a receptor protein exists in the cell it should be possible to isolate a mutant with a defect in this protein in which the outer membrane protein(s) could be traverse the cytoplasmic membrane. No such mutant has yet been isolated. In addition, if each polysome of the secreted envelope proteins has its own receptor protein as implied in the signal hypothesis, or if secreted proteins have a common receptor, there should be as many as 10,000 receptor proteins per cell judging from the number of membrane-bound ribosomes per cell (ca. 10,000–40,000, Smith et al., 1977). If this is the case, it should be possible to identify and isolate this protein from the cytoplasmic membrane.

An approach toward elucidating the interaction of the peptide extension with the cytoplasmic membrane would be to isolate mutants in the peptide extension that do not allow binding and hence translocation of the outer membrane protein across the cytoplasmic membrane. One system that is being used for such mutant isolation makes use of gene fusion between the genes for an outer membrane protein (λ receptor) and a cytoplasmic protein (β-galactosidase), described fully in Chapter 6 of this book. Mutants of the resulting hybrid protein that can no longer traverse the cytoplasmic membrane are providing further insight into the interaction of the peptide extension with the membrane.

3.3 Energy for Translocation

The data discussed above lead to yet another question regarding the mechanism of protein translocation across the membrane, that is, What provides the energy for translocation? It has been suggested that the energy for movement of the growing peptide chain through the membrane is derived from protein synthesis. If this were the case, one would expect that the ribosome should be anchored to the membrane apart from the growing chain so that as the peptide was further elongated, it would be "pushed" through the membrane. Such a specific attachment of the ribosome to the membrane occurs in eukaryotic cells since the membrane-bound polysome remains membrane bound even after release of the nascent chain by puromycin, and these ribosomes can then be released by high salt concentrations (Adelman et al., 1973; Rosbash and Penman, 1971). Since, as is discussed above, at least the majority of *E. coli* membrane-bound polysomes do not seem to be held to the membrane by way of ribosome–membrane interaction, sources of energy other than protein synthesis must be considered for the translocation of proteins across the cytoplasmic membrane. One possibility is the stabilization of the protein transfer across the membrane by protein folding of the peptide as it crosses the membrane. Alternatively, the energy may be derived from the membrane by an as yet undetermined process. In this regard, it is interesting to note that in a preliminary experiment outer membrane protein synthesis and assembly in glucose-grown cells was not inhibited by addition of carbomylcyanide-*m*-chlorophenylhydrazone (CCCP) (Halegoua, unpublished data), which is known to eliminate the membrane potential.

3.4 Is Precursor Processing Required For Translocation?

As is seen in the previous sections, the peptide extension on outer membrane protein precursors appears to be required for translocation of the proteins across the cytoplasmic membrane. It is mentioned in Section 1 that the precursor is processed by removal of the peptide extension during or immediately after the synthesis of the protein. A question that must be answered to understand the role of the precursor peptide extension in the sequence of assembly of the outer membrane proteins is whether processing is necessarily coupled to the translocation process of the protein across the cytoplasmic membrane.

One way to answer this question would be to somehow accumulate the precursor protein in the cell and determine its location as interior or exterior to the cytoplasmic membrane. This has been achieved by two independent means. As is discussed earlier, promatrix protein could be specifically accumulated in cells in the presence of 0.3% PEA. Promatrix protein thus accumulated was subcellularly localized by a variety of methods (Halegoua and Inouye, 1979). The results are summarized in Figure 7 together

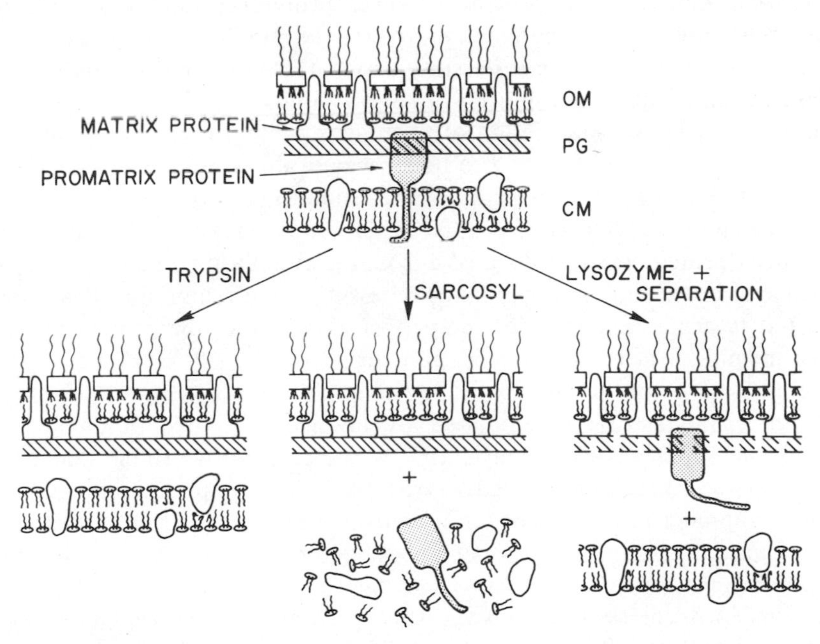

Figure 7 Possible localization in the envelope of promatrix protein produced in the PEA-treated cells (Halegoua and Inouye, 1979). OM, outer membrane, PG, peptidoglycan; CM, cytoplasmic membrane.

with a schematic interpretation of the promatrix protein location in the envelope. Promatrix protein was found totally with the cell membrane fraction. When cells containing promatrix protein were treated with trypsin in the absence of Mg^{2+}, promatrix protein was completely digested, but it was not affected when trypsin was added in the presence of Mg^{2+}. These data demonstrate that the bulk of promatrix protein is located between the cytoplasmic and outer membranes as shown in Figure 7, since trypsin penetrates the outer membrane but not the cytoplasmic membrane when it is added to cells only in the absence of Mg^{2+}. Promatrix protein was also solubilized from the envelope by sodium sarcosinate, indicating it was not properly inserted into the outer membrane, and raised the possibility of a cytoplasmic membrane association. However, when the cells were lysed with EDTA–lysozyme and the cytoplasmic and outer membranes were separated by sucrose density gradient centrifugation, promatrix protein was only found with the outer membrane fraction. These results indicate that promatrix protein has a strong affinity for the outer membrane fraction even if it is associated with the cytoplasmic membrane, and therefore remains with the outer membrane upon cytoplasmic–outer membrane separation. Unlike matrix protein, however, promatrix protein was released from this outer membrane fraction by 0.5 *M* NaCl. Thus, as shown in Figure 7, promatrix protein is possibly associated with the peptidoglycan by ionic interactions since matrix protein binds to the peptidoglycan (Rosenbusch, 1974). Association of promatrix protein with the cytoplasmic membrane is shown in Figure 7 to be by way of its peptide extension, based on the properties of the peptide extension discussed in Section 3.2. However, direct evidence for this is presently lacking. The fact that promatrix protein was translocated across the cytoplasmic membrane in the absence of processing indicates that processing is not required for translocation but rather is necessary for the proper assembly of the protein into the outer membrane after the translocation step.

In addition to the results above, the ompA protein precursor, pro-ompA protein, could also be accumulated by lowering the lipid fluidity of the membrane (DiRienzo and Inouye, 1979). The pro-ompA protein thus accumulated was also found to have traversed the cytoplasmic membrane. Hence, although processing of the precursor may occur during its synthesis under normal conditions, it is not an obligatory coupling to translocation across the membrane.

An independent study to answer this question was carried out using a mutant of prolipoprotein (Lin et al., 1978; Wu et al., 1978). In this mutant prolipoprotein was accumulated as a result of a mutation in the peptide extension in which the glycine residue at position 14 of prolipoprotein (position 7 in Table 1) was changed to aspartic acid. The prolipoprotein accumulated in these cells was localized by lysozyme–EDTA treatment followed by separation of the cytoplasmic and outer membranes by sucrose density gradient centrifugation. It was found that the mutant prolipoprotein was distributed as 60% in the outer membrane fraction, 20% in the cytoplasmic membrane fraction, and 20% in the soluble cell fraction (Lin *et al.*, 1978). This result also indicates that the mutant prolipoprotein was translocated across the cytoplasmic membrane without being processed.

4 MODIFICATION AND ASSEMBLY

4.1 Precursors and Processing

As is discussed above, the peptide extensions of the precursors of outer membrane proteins appear to be essential for translocation across the cytoplasmic membrane. The next question is how the peptide extensions are processed and how the processed proteins assembled in the outer membrane. PEA (phenethyl alcohol) was found to be very useful in the study of the assembly intermediates of outer membrane proteins (Halegoua and Inouye, 1979). At the higher concentrations of PEA, processing of three outer membrane protein precursors was differentially inhibited. Processing of promatrix protein was the most sensitive to PEA followed by pro-ompA protein and lipoprotein, the most resistant. On the other hand, when PEA was added at concentrations lower than the above critical concentrations for each protein, the precursor was properly processed, but the processed protein (ompA protein and lipoprotein) was not found in the membrane fraction. Instead, the processed proteins appeared to be accumulated in the periplasmic space, since they were found in the soluble cell fraction and were releasable from the cells by osmotic shock. It could not, however, be ruled out that the processed proteins were very loosely bound to the outer membrane. The above data suggest that the proper assembly of the processed protein into the outer membrane was inhibited by PEA, resulting in the apparent accumulation of the processed protein in the periplasmic space.

Since the effects of PEA are reversible for accumulation of ompA protein (Halegoua and Inouye, 1979), a pulse-chase experiment could be performed. Thus ompA protein accumulated in the periplasmic space at 0.4% PEA was inserted into the outer membrane upon removal of PEA and regrowth of the cells in culture; ompA protein was found to be properly assembled into the outer membrane because it was not solubilized by sarcosyl treatment of the membrane fraction and was found in the outer membrane fraction upon cytoplasmic–outer membrane separation. These data indicate that the periplasmic ompA protein is an intermediate form of the outer membrane protein that can be accumulated by PEA. Three conclusions may be drawn from these experiments. First, the peptide extension on the precursor form of the outer membrane protein is not essential for the assembly of the protein into the outer membrane. Second, another step after removal of the precursor peptide extension is required for proper assembly of the protein into the outer membrane. Last, ongoing synthesis is not necessary for the protein assembly into the outer membrane.

The differences in the degree of sensitivity to PEA of insertion of each outer membrane protein into the outer membrane is the same as the sensitivity to PEA inhibition of the precursor processing for each protein, with matrix protein being the most sensitive, followed by ompA protein, and then by lipoprotein. This result may be explained if the inhibition of processing of the precursor protein and assembly of the processed protein into the outer membrane occur at the same sites for each protein and if the different outer membrane proteins are processed and assembled at different sites (each protein may also have a different processing enzyme). Alternatively, the differential inhibition by PEA may be due to the different structural configurations of each protein and different affinities for other components of the outer membrane (i.e., lipopolysaccharide).

As is mentioned in Section 3, the precursor forms of the proteins have very short lifetimes and the peptide extensions are processed very quickly. Also, the precursor peptide extension appears to be unnecessary for protein insertion into the outer membrane after translocation. Instead, it is possible that removal of the peptide extension may be necessary for completion of the assembly process of the outer membrane protein. In fact, it is suggested in the loop model (Section 3.2) that the peptide extension interacts specifically with the cytoplasmic membrane and that the cleavage of the peptide extensions is required for outer membrane proteins to be released from the cytoplasmic membrane. Therefore, this model predicts that precursor proteins with peptide extensions would be accumulated on the outer surface of the cytoplasmic membrane if the cleavage of the peptide extensions is inhibited. This may be the case for promatrix protein accumulated during PEA treatment (Halegoua and Inouye, 1979; see Figure 7).

It should be noted that in the case of the mutant prolipoprotein (see Section 3.4; Lin et al., 1978) 60% of the mutant prolipoprotein was found bound to the outer membrane, 20% was bound to cytoplasmic membrane, and 20% was in the soluble cell fraction. However, since this prolipoprotein is not bound to the peptidoglycan as is normal lipoprotein, the mutant prolipoprotein bound to the outer membrane is probably not assembled properly into the outer membrane. It is therefore premature to conclude from the data that the processing of the precursor molecules is not required for assembly of the outer membrane protein in the final location.

Identification, localization, and characterization of the enzymes required for processing of the outer membrane protein precursors would provide valuable information and approaches toward understanding the role of processing in the assembly of the outer membrane proteins. However, at present it is not known how many different processing enzymes there are for the outer membrane proteins, whether each protein has its own specific enzyme or, whether groups of proteins share common enzymes. Based on the known amino acid sequences of peptide extensions, it was pointed out in Section 3.2 that even for precursors of different classes of envelope proteins, an amino acid residue with a short side chain is found at the cleavage site of the peptide extension (position 1 residue, Table 1). This is also found with secretory protein precursors in eucaryotic systems (see review, Inouye and Halegoua, 1979). However, the amino acids found on either side of the cleavage site residue of the peptide extension (position 2 residue and amino terminus of processed protein, Table 1), do not show any common feature. The processing enzymes may have similar specificities that may depend on the conformation of the cleavage site region. For instance, the short side chain amino acid residues at the cleavage site of the peptide extension may allow bending at that position (see loop model, Section 3.2), making the cleavage site accessible to the processing enzyme. It is interesting to note that in eukaryotic systems the processing enzyme shows a very wide range of species specificity (see Inouye and Halegoua, 1979), indicating that the precise mode of processing of secretory proteins is highly conserved during evolution.

No processing enzyme has yet been purified and characterized because of the difficulty of purification and of preparation of the proper enzyme substrates. Addition of several types of protease inhibitors to cells [tosyl-lysine chloromethylketone (TLCK), antipain, leupectin, and diisopropylfluorophosphate] did not result in the accumulation of unprocessed precursor molecules of the envelope proteins (Ito, 1977, 1978). However, the appearance of matrix protein in the envelope was severely inhibited by some of the inhibitors and TLCK resulted in the formation of abnormal proteins. The reasons for these results are not presently known. It is also an open question whether the processing enzyme is exo- or endoproteolytic. However, when bacteriophage f1 procoat protein (a cytoplasmic membrane protein) was cleaved by a crude enzyme fraction, a fragment of the peptide extension was isolated containing at least eight amino acids from the amino terminus of the precursor protein (Chang et al., 1978). This indicates endoproteolytic enzyme activity, although a definitive finding awaits processing enzyme purification.

Thus far, only two precursor protein processing enzyme activities have been tentatively localized. In the case of proalkaline phosphatase, the processing enzyme activity was found to be associated with an outer membrane fraction (Inouye and Beckwith, 1977). For the procoat protein of bacteriophage f1, the processing activity was in the cytoplasmic membrane fraction. In all systems, eukaryotic, as well as prokaryotic (one possible exception in the case of M13 procoat protein; G. Mandel and W. Wickner, personal communication), processing of the precursor by membrane vesicles has always been required to occur only during the protein synthesis (Lingappa et al., 1977; Lingappa et al., 1978; Chang et al., 1978). This implies that proper conformation of the precursor peptide extension and cleavage site can occur with the outer membrane fraction, although proper insertion of the precursor *in vivo* occurs in the cytoplasmic membrane. It is also curious that the processing of proalkaline phosphatase in the presence of total envelope was very slow and even after 4 hr at 35°C a substantial amount of unprocessed precursor was left in the reaction mixture. It should also be pointed out that the enzyme activity for processing appears to be latent. Processing of pro-f1 coat protein by cytoplasmic membrane vesicles was greatly stimulated when the membrane was present together with the nonionic detergent Nikkol (Chang et al., 1978).

Peptidases and proteases of *E. coli* have been described (see review, Miller, 1975) and recently two periplasmic enzymes, aminoendopeptidase (Lazdunski et al., 1975; Murgier et al., 1976) and protease I (Pacand et al., 1976) have been reported. Aminoendopeptidase is specific to L-alanine-*p*-nitroamilide and is inhibited by *N*-ethylmaleimide but not by EDTA and diisopropylphosphoryl fluoride. These enzymes may be used for digestion of exogenous nutrients rather than for processing of the envelope protein precursors.

4.2 Determinants of Outer Membrane Localization

4.2.1 Protein Specificity The factors that would determine the localization of the outer membrane proteins should clearly involve the specific properties of the outer membrane proteins and their interactions with envelope components. The interactions that maintain protein assembly in the outer membrane probably play an important role in the localization of the protein to the outer membrane. Particularly, the interaction with the lipopolysaccharides is considered to be most important. If outer membrane proteins have specific affinities for the lipopolysaccharides, which are exclusively located in the outer leaflet of the outer membrane, newly formed outer membrane proteins can be easily self-assembled in the outer membrane. In fact, the self-assembly of the outer membrane protein with the lipopolysaccharides (Nakamura and Mizushima, 1975; Yu and Mizushima, 1977) and the specific interaction of outer membrane proteins with the lipopolysaccharides (Henning et al., 1973; Datta et al., 1977; Schweizer and Henning, 1977; van Alphen et al., 1977; Yu and Mizushima, 1977) have been shown. Recently, Schweizer et al. (1978) showed that the amino-terminal moiety of ompA protein is active as a phage receptor and is protected from protease digestion in the presence of the lipopolysaccharides. Furthermore, they showed that the lipid part of the lipopolysaccharides (lipid A) essentially exhibits the same activity as the complete lipopolysaccharides. These results indicate that the amino-terminal portion of ompA protein has specific affinity to the lipid A part of the lipopolysaccharides. Beside the interaction between the lipopolysaccharides and the outer membrane proteins, specific interactions between matrix protein and the peptidoglycan (Rosenbusch, 1974) and between λ receptor and the peptidoglycan (Endermann et al., 1978) also have been reported. Matrix protein appears to form a highly periodic array that exhibits threefold symmetry (Steven et al., 1977) and also has been shown to interact with lipoprotein in the outer membrane (DeMartini and Inouye, 1978). However, the interaction between matrix protein and lipoprotein may not be essential for the determination of the final location of these proteins in the outer membrane, since mutants that lack only one of these proteins have been isolated (see review by DiRienzo et al., 1978).

An elegant genetic approach has been taken in order to define the outer membrane protein determinants for final localization in the outer membrane (Silhavy et al., 1977; see Chapter 7). In their study gene fusions were achieved between the *lacZ* gene for β-galactosidase, a cytoplasmic protein, and the *lamB* gene for λ receptor, an outer membrane protein. As a result of the gene fusions, hybrid proteins were produced that contained the amino-terminal sequence of the λ receptor protein, the major portion of the carboxyl-terminal sequence of β-galactosidase, and β-galactosidase activity. Details of these gene fusion experiments are described in Chapter 7. When the location of one of these hybrid proteins in the cell was examined, it was found that 40–70% of the β-galactosidase activity was in the crude membrane fraction, and about 50% of this membrane-bound activity was located in the outer membrane fraction after separation of the cytoplasmic and outer membranes.

From this result it is apparent that the amino-terminal segment of the receptor that is present in the above hybrid protein is responsible for the localization of the hybrid protein to the outer membrane. Since the hybrid protein is about 10,000–15,000 daltons larger than β-galactosidase (Silhavy et al., 1977), the hybrid protein must contain at least 90–140 amino acids from the amino-terminal sequence of the λ receptor protein (the λ receptor has about 500 amino acids in total). Therefore, this fragment of the λ receptor protein must contain the properties necessary for directing, and to some extent maintaining, the protein in the outer membrane. For example, specific interactions of this fragment with outer membrane components, such as the lipopolysaccharide, might be needed for outer membrane localization to occur. The genetic approach described above provides an excellent system to further characterize the determinants both of outer membrane protein localization and of protein translocation across the cytoplasmic membrane by providing a system in which mutants in these processes are easily isolated (see Chapter 7).

4.2.2 Bayer's Junctions A major question concerning the assembly of outer membrane proteins is, once the protein is translocated across the cytoplasmic membrane, how is it transferred from the point of translocation on the cytoplasmic membrane to the point of its insertion into the outer membrane. Are the precursors processed, thereby releasing the processed protein to the periplasmic space to find its own way to the proper site on the outer membrane, or is there an additional mechanism for this transfer process? An intriguing envelope structure that may provide an answer, at least in part, to this question is the Bayer junction (see Chapter 6 by Bayer). These junctions are sites of adhesion or apparent fusion zones between the cytoplasmic and outer membranes, which occur at about 200–400 sites/cell, as seen under the electron microscope. The transfer of capsular polysaccharide (Bayer and Thurow, 1977) and lipopolysaccharide (Mühlradt et al., 1973) from the cytoplasmic membrane through the Bayer junctions to the outer membrane has been established. Recently, it was reported that newly synthesized matrix proteins in *Salmonella typhimurium* can be found at the cell surface preferentially at these junction sites by pulse labeling with ferritin-labeled anti-matrix protein serum (Smit and Nikaido, 1978). It is possible that other outer membrane proteins also may be transferred from the cytoplasmic membrane to the outer membrane through the Bayer's junctions.

4.3 Importance of Membrane Lipid Fluidity

An important factor in understanding the assembly processes of outer membrane proteins is the fluid state of the envelope membranes. An approach that has been taken toward a definition of the role of lipid fluidity in these processes is examination of the effects of altering the membrane fluid state on the biosynthesis and assembly of the envelope proteins. With the use of an unsaturated fatty acid auxotroph the fatty acid composition of the membrane lipids can be dictated. By growing the cells on specific unsaturated fatty acids the fluid state of the membrane can be controlled by altering the growth temperature. The inhibition of alkaline phosphatase induction

(Kimura and Izuki, 1976) under conditions of a crystalline membrane state has been shown to be a result of the membrane binding of the synthesized alkaline phosphatase (Pages et al., 1978). However, whether the membrane-bound form was unprocessed precursor was not determined nor was its precise location in the envelope.

Similar experiments were carried out for examination of the effects of a nonfluid membrane state on the synthesis and assembly of cytoplasmic membrane and outer membrane proteins (Ito et al., 1977). It was found that in general the assembly of outer membrane proteins was more drastically adversely affected than that of cytoplasmic membrane proteins. However, when the individual outer membrane proteins were examined more carefully, it was found that there were differential effects of a nonfluid state of the membrane lipids on the synthesis and assembly of the individual outer membrane proteins (DiRienzo and Inouye, 1979). Under these conditions, the assembly of lipoprotein was hardly affected, whereas the biosynthesis and assembly of matrix protein was completely inhibited. On the other hand, the processing of pro-ompA protein was inhibited, resulting in the accumulation of pro-ompA protein in the envelope fraction. Pro-ompA protein thus accumulated was found to be very loosely associated with the envelope fraction so that it could be easily released from the envelope fraction by sonication. However, the accumulated pro-ompA protein was as sensitive to trypsin as mature ompA protein assembled in the outer membrane, suggesting that pro-ompA protein is partially assembled in the outer membrane. This pro-ompA protein could be chased into mature ompA protein in the outer membrane fraction upon shifting back the fluid state of the membrane to normal (DiRienzo and Inouye, 1979). These results indicate that outer membrane proteins are synthesized and assembled by several different mechanisms in which the fluidity of membrane lipids plays an important role. At present it is not known at which step the assembly of matrix protein is inhibited.

It is quite interesting to note that the differential effects on outer membrane protein assembly by PEA as described previously are remarkably similar to the effects of lipid fluidity above, with synthesis and assembly of matrix protein being the most easily affected, followed by ompA protein and then by lipoprotein. In this regard it is interesting that PEA is structurally a type II local anesthetic. Since these compounds are known to increase the fluid state of the membrane, the effects of PEA on fluidity of *E. coli* envelope was examined by using 5-doxylstearate as an ESR probe (Halegoua and Inouye, 1979). It was found that PEA drastically increased the fluid state of the envelope membranes over the range in which it exerted its effects on outer membrane protein assembly. Therefore, it appears that not only low, but also high, fluid states of the membrane cause similar inhibitory effects on the assembly of the outer membrane proteins. This indicates that a proper fluidity of the membrane is required to maintain the machinery for processing of precursor proteins and for transferring the outer membrane proteins from the cytoplasmic membrane to the outer membrane. It is possible that the change in lipid fluidity might cause severe effects on the structure and function of the Bayer's junctions, which then result in accumulation of precursors of the outer membrane proteins in the cytoplasmic membrane, or of the processed proteins in the periplasm. In this case, the differential effects seen on assembly among the outer membrane proteins may then reflect the different dependencies of each outer membrane protein's assembly on the Bayer's junctions.

4.4 Lipoprotein

Of all the outer membrane proteins, the biosynthesis and assembly of the lipoprotein is the most complex and has been the most extensively investigated. As is discussed in Section 3.1, the lipoprotein was the first envelope protein shown to be produced from a precursor, prolipoprotein, and the lipoprotein mRNA is the only biologically active mRNA purified from *E. coli* to date. The complete structure of lipoprotein and prolipoprotein is known, and the partial nucleotide sequence of the lipoprotein mRNA has been determined (see reviews; Inouye, 1975; Braun, 1975; Inouye, 1979).

After translation of the lipoprotein mRNA, a series of posttranslational modification reactions must take place to yield the final lipoprotein product. The posttranslational modification of lipoprotein may conceptually be divided into stages, including five independent enzymatic reactions as demonstrated in Figure 8. The first stage, similar to the other outer membrane proteins is the translocation and processing of the precursor, prolipoprotein, which results in the removal of the peptide extension (see Section 3). In the second stage the newly formed amino-terminal cysteine residue of lipoprotein is extensively modified. At least three independent reactions appear to be involved in modification of the cysteine residue (Figure 8). In one reaction, acylation of the free amino group occurs to yield an amide-linked fatty acid at this position. From the analysis of the lipoprotein structure (Hantke and Braun, 1973), the fatty acid at this position was found to consist of palmitic acid (65%), palmitoleic acid (11%), and *cis*-vaccinic acid (11%). The other modification results in the attachment of a diglyceride group to the SH group of the amino-terminal cysteine residue. The composition of the ester-linked fatty acids is similar to the fatty acid composition of the phospholipids of the same cells, indicating that this diglyceride moiety is derived from a stage in the pathway of phospholipid metabolism (Hantke and Braun, 1973). From a study using a glycerol-requiring *E. coli*

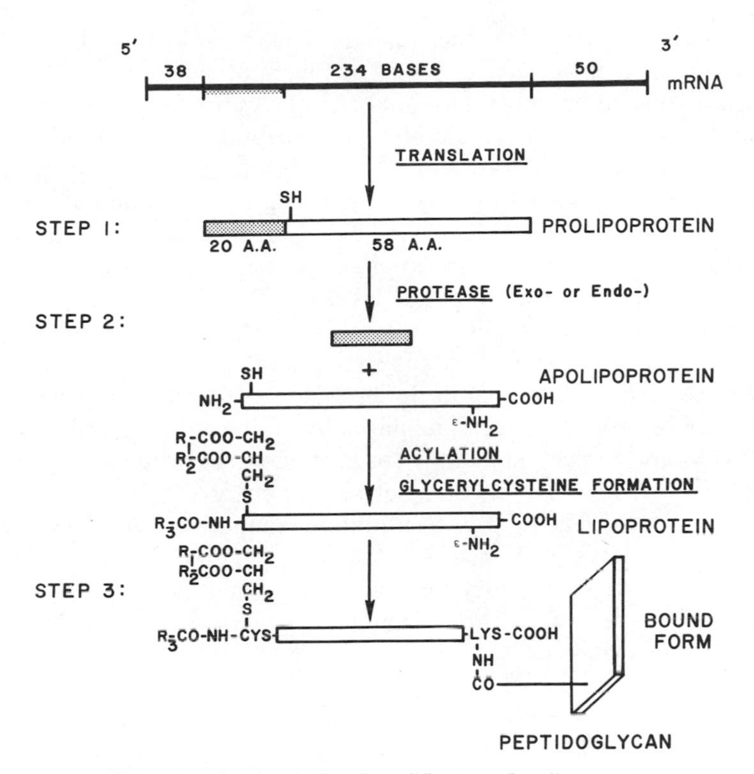

Figure 8 Posttranslational modification of prolipoprotein.

strain it was suggested (Lin and Wu, 1976) that synthesis of the protein moiety of lipoprotein can occur independently of the diglyceride modification. Schulman and Kennedy (1977) have reported that the diglyceride of lipoprotein contains glycerol derived from a metabolic pool with a relatively long half-life, thus ruling out CDP-diglyceride as the diglyceride donor originally predicted (Hantke and Braun, 1973). Using a mutant that lacked phosphoglucose isomerase, glucose-6-phosphate dehydrogenase, and glycerophosphate dehydrogenase, thus controlling the synthesis of both phosphatidylglycerol and cardiolypin, these authors suggested that the lipoprotein's diglyceride moiety may be derived from cardiolipin (Schulman and Kennedy, 1977). However, it has been shown more recently that phosphatidylglycerol is the most probable donor of the lipoprotein's glycerol moiety (Chattopadhyay and Wu, 1977). In cerulenin-treated cells, in which fatty acid synthesis is inhibited, nonacylated [2-^{3}H]glycerol could be exclusively incorporated into phosphatidylglycerol and subsequently into lipoprotein. In addition, it was found that the carbon 1 rather than the carbon 3 of *sn*-glycerol is involved in the thioether linkage (Chattopadhyay and Wu, 1977). This glycerol is then acylated as shown in Figure 8.

The last stage of lipoprotein assembly involves the covalent attachment of one-third of the lipoprotein molecules to the peptidoglycan. In this reaction the ϵ-amino group of the carboxyl-terminal lysine residue on lipoprotein is bound in peptide linkage to the carboxyl group of the diaminopimelic acid in the peptidoglycan. That this bound form of lipoprotein is derived from the free form was shown by a pulse-chase experiment (Inouye et al., 1972) in which the free form was first pulse labeled with [^{14}C]arginine. Upon chasing with nonradioactive arginine, it was found that one-third of the pulse-labeled free form molecules were converted to the bound form. This conversion of free form to bound form lipoprotein was found to proceed even in the absence of protein synthesis inhibited by amino acid starvation or addition of chloramphenicol, and in the absence of energy production (Inouye et al., 1972). Since this reaction does not require energy, it was suggested that the reaction proceeds by transpeptidation similar to the formation of peptide cross bridges in the cell wall (Inouye et al., 1972; Inouye, 1975). A new penicillin derivative, bicyclomycin, has been reported to specifically inhibit the biosynthesis of lipoprotein, particularly the formation of the bound form (Tanaka et al., 1976). The mode of inhibition by this drug has yet to be determined.

In view of the extensive posttranslational modification of lipoprotein described above, it is of great interest which stages, if any, are involved in the assembly of lipoprotein into the outer membrane. As is described above, the second stage of lipoprotein modification involves the modification by fatty acids. The fatty acids impart to lipoprotein a very high hydrophobic nature, suggesting a role of this modification in the assembly of lipoprotein into the outer membrane. Clearly, the amide-linked fatty acid can only be attached after processing of the prolipoprotein (removal of the peptide extension). However, it is not known whether the attachment of diglyceride occurs before or after processing. In the case of the prolipoprotein mutant (Wu et al., 1978, see Section 3), the accumulated mutant prolipoprotein lacks the covalent diglyceride. One line of data that suggests lipoprotein fatty acid modification may be essential for the proper lipoprotein assembly in the outer membrane comes from the PEA-permeabilized cell system. As is shown in Section 4, assembly intermediates of ompA protein and lipoprotein may be accumulated by PEA as processed proteins in the periplasmic space. In the case of the periplasmic lipoprotein, it was found that the protein was sensitive to trypsin treatment (Halegoua and Inouye, 1979) contrary to the case of the outer membrane lipoprotein. Since the trypsin resistance of lipoprotein is considered to be due to the presence of the covalently linked fatty acids, the periplasmic lipoprotein must be lacking in one or more of these fatty acids even though the peptide extension on prolipoprotein was removed. It is believed that complete analysis of the amino-terminal structure of the periplasmic lipoprotein will provide insights into the precise requirements of the fatty acid modifications for assembly of lipoprotein.

The last stage of lipoprotein modification, the conversion to bound form, is apparently not essential to the insertion of lipoprotein into the outer membrane, since free form is first inserted into the outer membrane and a part of free form is converted into bound form. The bound form is considered to be important for maintenance of the structural integrity of the outer membrane, since in a mutant that is deficient in bound-form lipoprotein periplasmic enzymes were found to leak out of the cell and the ultrastructure of the envelope of the mutant cell was found to be abnormal (Weigand et al., 1976).

4.5 Other Outer Membrane Proteins

It has been reported that both matrix protein and ompA protein contain sugar. Matrix protein possibly contains 1 mole of glucosamine per mole of protein (Garten and Henning, 1974), whereas ompA protein contains 2 moles of reducing sugar per mole (Chai and Foulds, 1977). Besides post-translational modification by sugar, the existence of α-aminoadipic acid δ-semialdehyde (allysine) in matrix protein has been reported (Diedrich and Schnaitman, 1978). The positions of lysine residues converted to allysine were found to be nonspecific. The function of allysine residues in matrix protein is not understood at present, although it has been suggested that it may be involved in cross-linking of matrix protein to other outer membrane components (Diedrich and Schnaitman, 1978).

5 REGULATION

Regulation of the production of outer membrane proteins entails a potentially intricate design. The outer membrane proteins not only may undergo regulation at the stages of transcription and translation as the cytoplasmic proteins, but may also be regulated at the levels of translocation across the cytoplasmic membrane and insertion into the outer membrane. An understanding of the regulation of these processes would provide valuable insights into the complex issue of membrane biogenesis, including the coordination of production of the different membrane constituents.

At present, very little is known about the regulatory processes of outer membrane proteins. However, a number of phenomena have been described that not only provide some insight, but also offer direction for further studies.

5.1 Protein Composition and Gene Dosage

An intriguing aspect of outer membrane protein composition is the extreme variation in the abundance in the outer membrane among the different proteins. In the case of some outer membrane proteins only a few hundred molecules exist in the outer membrane, while for the major proteins such as matrix protein, ompA protein, and lipoprotein, there exist more than 10^5 molecules of each per cell (see Chapter 1).

There exist 7.5×10^5 lipoprotein molecules per cell, making this the most abundant protein in the cell. The content of lipoprotein can be further increased by histidine starvation (see Section 2; Hirashima and Inouye, 1973) or in a strain merodiploid for the structural gene for lipoprotein. In these cells, the lipoprotein free form was doubled in the outer membrane as compared to the haploid strain (Movva et al., 1978). Interestingly, the amount of bound form lipoprotein was the same in both strains, which suggests that the assembly sites for the bound-form lipoprotein are saturated in a haploid strain in contrast to the presence of excess sites for free-form lipoprotein.

In contrast to the constitutive synthesis and assembly of lipoprotein described above, the syntheses of the other major outer membrane proteins, matrix protein and ompA protein, show different characteristics. The production of matrix protein and ompA protein are somehow coordinated such that when one of the proteins is absent, there is a compensatory production of the other. For example, in mutants that lack matrix protein, the amount of ompA protein is increased to the level of matrix protein plus ompA protein in the wild-type outer membrane (Schnaitman, 1974). Conversely, mutants lacking ompA protein partially compensated for this loss by synthesizing increased amounts of matrix protein (Lugtenberg et al., 1976). On the other hand, in a merodiploid strain for the *ompA* gene, no increase in production of the protein was seen contrary to the case of the haploid strain (Datta et al., 1976). One possible reason for both the compensatory production of matrix protein and ompA protein and the absence of the gene dosage effect of the *ompA* gene may be due to a limited number of sites for the synthesis and assembly of the two proteins. It is possible that the assembly sites in the outer membrane are shared by the two proteins and that the space to accommodate the two proteins in the outer membrane is physically limited. Alternatively, the number of sites for polysome binding to the membrane or for protein insertion into the outer membrane may be limited. In any case, the total amount of matrix protein plus ompA protein may be maintained constant.

Another intriguing aspect is the control of gene expression of matrix proteins (see Chapter 8 by Reeves). It has been recently shown that two forms of matrix proteins are produced from two independent genes, located at 21 min for matrix protein Ia and 48 min for matrix protein Ib on the *E. coli* chromosome map (Ichihara and Mizushima, 1978; T. Sato and T. Yura, personal communicaion). The relative amounts of matrix proteins Ia and Ib in the outer membrane vary greatly depending on the composition of the growth medium (Hasegawa et al., 1976). Furthermore, a mutation at a third locus at 74 min (*ompB*) resulted in a complete elimination of both matrix proteins Ia and Ib (Sarma and Reeves, 1977). Both matrix proteins Ia and Ib can also be replaced in strains lysogen for the lambdoid phage PA2 by a large amount of a new protein coded for by the phage (Pugsley and Schnaitman, 1978a). It is also interesting to note that in mutant strains lacking both matrix proteins Ia and Ib, new proteins with properties similar to those of matrix proteins are produced (Foulds and Chai, 1978; Pugsley and Schnaitman, 1978b). The mechanism of this induction of new proteins is not known at present.

5.2 Cell Cycle-Dependent Synthesis and Assembly

A number of reports have appeared describing the synthesis of envelope proteins only, or mostly, at specific times during the cell cycle. This is a very intriguing aspect of regulation of envelope protein biosynthesis and assembly and is reviewed separately by Ohki (Chapter 9) and Nikaido Chapter (11) in this book.

5.3 Gene and mRNA Structure

Analysis of base sequences of genes and mRNAs for the outer membrane protein offers a promising approach toward understanding the regulatory mechanisms involved in the synthesis and assembly of these proteins. A major step toward these goals has been the purification and characterization of the lipoprotein mRNA. Important characteristics of the lipoprotein mRNA that permit its isolation include its remarkable stability, its abundance, and its small size. The purified mRNA was biologically active and could direct the production of cross-reactive material with antilipoprotein serum not only in an *E. coli* cell-free system (Hirashima et al., 1974), but also in a eukaryotic (wheat germ) cell-free system (Wang et al., 1976). ^{32}P-Labeled lipoprotein mRNA also has been purified and its partial base sequence has been determined as shown in Figure 9A (Takeishi et al., 1976;

A

1
Met - Lys - Ala
5'-end: G-C-U-A-C-A-U-G-G-A-G-A-U-U-A-A-C-U-C-A-A-U-C-U-A-G-A-G-G-G-U-A-U-U-A-A-U-A-A-U-G-A-A-A-G-C-U-
1 10 20 30 40

5 10 15 20
Thr - Lys - Leu - Val - Leu - Gly - Ala - Val - Ile - Leu - Gly - Ser - Thr - Leu - Leu - Ala - Gly -
A-C-U-A-A-A-C-U-G-G-U-A-C-U-G-G-G-C-G-C-G-G-U-A-A-U-C-C-U-G-G-G-U-U-C-U-A-C-U-C-U-G-C-U-G-G-C-A-G-G-U-
50 60 70 80 90

25 30 35
Cys - Ser - Ser - Asn - Ala - Lys - Ile - Asp - Gln - Leu - Ser - Ser - Asp - Val - Gln - Thr - Leu -
U-G-C-U-C-C-A-G-C-A-A-C-G-C-U-A-A-A-A-U-C-G-A-U-C-A-G-C-U-G-U-C-U-U-C-U-G-A-C-G-U-U-C-A-G-A-C-U-C-U-G-
100 100 120 130 140

40 45 50
Asn - Ala - Lys - Val - Asp - Gln - Leu - Ser - Asn - Asp - Val - Asn - Ala - Met - Arg - Ser - Asp -
A-A-C-G-C-U-A-A-A-G-U-U-G-A-C-C-A-G-C-U-G-A-G-C-A-A-C-G-A-C-G-U-G-A-A-C-G-C-A-A-U-G-C-G-U-U-C-C-G-A-C-
150 160 170 180 190 200

55 60 65 70
Val - Gln - Ala - Ala - Lys - Asp - Asp - Ala - Ala - Arg - Ala - Asn - Gln - Arg - Leu - Asp - Asn -
G-U-U-C-A-G-G-C-U-G-C-U-A-A-A-G-A-U-G-A-C-G-C-A-G-C-U-C-G-U-G-C-U-A-A-C-C-A-G-C-G-U-C-U-G-G-A-C-A-A-C-
210 220 230 240 250

75 78
Met - Ala - Thr - Lys - Tyr - Arg - Lys
A-U-G-G-C-U-A-C-U-A-A-A-U-A-C-C-G-C-A-A-G-U-A-A-U-A-G-U-A-C-C-U-G-U-G-A-A-G-U-G-A-A-A-A-A-U-G-G-C-G-C-
260 270 280 290 300

A-C-A-U-U-G-U-G-C-G-C-C-A-U-U-U-U-U-U-U$_{OH}$:3'-end
310 320

B

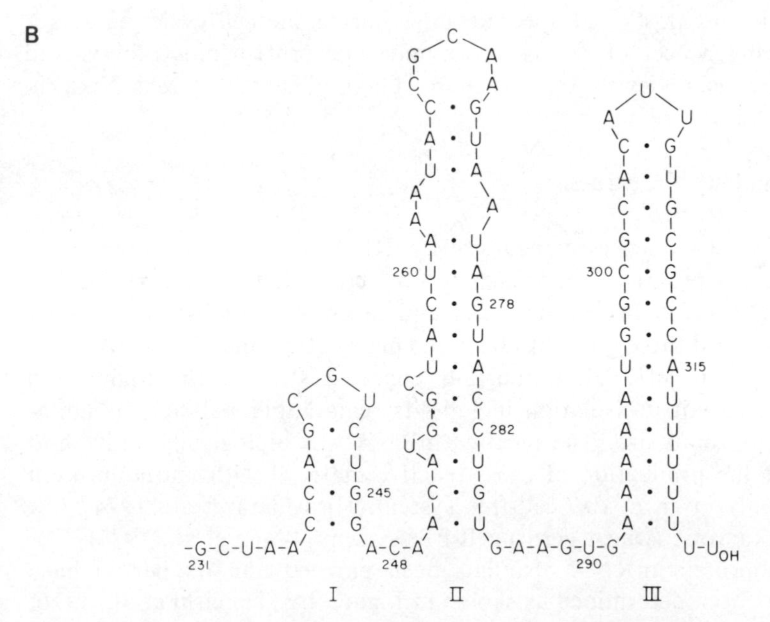

Figure 9 Nucleotide sequence of the lipoprotein mRNA.

Pirtle et al., 1978). The base sequence of the 5′-end of the mRNA confirmed the amino acid sequence of the peptide extension of prolipoprotein.

The lipoprotein mRNA consists of 322 bases. Out of about 88 bases that are untranslated, 38 bases are at the 5′-end before the initiation codon, AUG, and 50 bases are at the 3′-end (see Figure 9A). Remarkably, the sequence of 12, nucleotides, including the initiation codon AUG, GUAUUAAUA*AUG*, was exactly the same as is found at the 80 S ribosome binding site for in brome mosaic virus RNA4, a eukaryotic mRNA (Pirtle et al., 1978). The nucleotide sequence of the untranslated 3′-end contains very stable stem and loop structures as shown in Figure 9B (Pirtle et al., 1979). The stability of the lipoprotein mRNA may be due to these stable secondary structures.

Recently a clone carrying the structural gene for lipoprotein was isolated (K. Nakamura and M. Inouye, manuscript in preparation). There is no doubt that the determination of the base sequence of the lipoprotein gene will provide further important information concerning the synthesis and assembly of lipoprotein and the regulatory mechanisms of its production.

6 CONCLUSION

As is seen in the preceding sections, the processes governing the biosynthesis and assembly of the outer membrane proteins have begun to be dissected. Assembly of outer membrane proteins requires translocation across the cytoplasmic membrane and subsequent transfer and insertion into the outer membrane. The use of membrane perturbants such as toluene or phenethyl alcohol or lowering the fluidity of membrane liquids allowed the uncoupling of these processes, resulting in the accumulation of precursor forms of the outer membrane proteins. The precursors contain amino-terminal peptide extensions about 20 amino acid residues long. The complete amino acid sequence of the peptide extension on the lipoprotein precursor, prolipoprotein, was determined using cell free translation of the purified lipoprotein mRNA and revealed a number of unique features. The existence of precursors was found to be a general feature of the envelope proteins that must traverse the cytoplasmic membrane regardless of their final locations. Thus the outer membrane proteins, periplasmic proteins, and some proteins of the cytoplasmic membrane contain precursors with very similar amino-terminal peptide extensions.

From the amino acid sequences of the peptide extensions of four envelope protein precursors, a set of four common features or rules unique for the peptide extension was deduced (see Section 3.2). Remarkably, the rules are also applicable to the structures of many of the peptide extensions of eukaryotic secretory protein precursors whose sequences have thus far been determined (Inouye and Halegoua, 1979). These four features include, (*a*) a basic amino-terminal section, (*b*) a stretch of hydrophobic amino acids following the basic section, (*c*) a proline or glycine residue within the hydrophobic section, and (*d*) an amino acid with a short side chain at the cleavage site.

The striking homology among the peptide extensions, regardless of the final location of the processed protein product, implies that the peptide extensions are required only for the translocation of the secreted protein across the cytoplasmic membrane and possibly not for the determination of its final location in the envelope. The loop model for translocation of a protein across the membrane was proposed and accommodates all four features of the peptide extension (see Section 3.2). The model includes a detailed account of the peptide extension-mediated, membrane-bound polysome synthesis of the secreted protein. Another important aspect for the translocation of proteins across the cytoplasmic membrane is how ribosomes are bound to the membrane and what energy is needed for the translocation. Smith et al. (1979) have recently reported that ribosomes from *Bacillus subtilis* are attached to membrane solely by way of their secreted chain and that after loss of the extracellular segment of the chain its attachment is maintained at 37°C, as well as at 0°C. Based on these data they suggested that the chain does not passively slip through a membrane but is actively held within a channel. If such a channel is formed by a specific protein(s), its identification and characterization is vital for understanding of the molecular mechanism of the translocation of proteins across the cytoplasmic membrane.

A second major question is how the proteins translocated across the cytoplasmic membrane find their final location in the envelope. The mechanism for the determination of the final location of the secreted proteins is not well understood at present. However, as shown in Figure 10, the final location for the periplasmic proteins and cytoplasmic membrane proteins is possibly determined simply by the nature of the peptide that comes out through the cytoplasmic membrane. If the peptide is hydrophilic as a whole, the final product stays in the periplasmic space, whereas if the hydrophilic amino-terminal region is followed by a very hydrophobic segment, this segment interacts strongly with the lipid bilayer of the cytoplasmic membrane and is forced to stay in the cytoplasmic membrane. If a carboxyl-terminal part of this protein is hydrophilic, this last portion remains in the cytoplasm. Thus this protein becomes a transmembrane protein.

It is most likely that the outer membrane proteins are assembled exclusively in the outer membrane because of their specific affinity to the outer membrane components. As is discussed in Section 4.2, the most probable

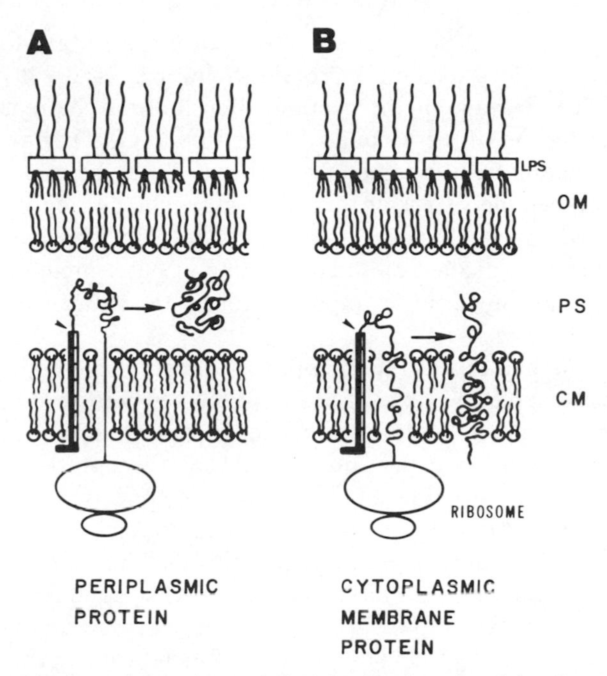

Figure 10 Mechanism of secretion and final localization of periplasmic proteins and cytoplasmic membrane proteins in *E. coli*. (*A*) Periplasmic proteins. When the cleavage site is exposed to the outside surface of the cytoplasmic membrane, the peptide extension is cleaved off and the resultant nascent peptide begins to fold. Because of the hydrophilic property of the proteins, the part that extrudes to the outside of the cytoplasmic membrane remains soluble in the periplasmic space. (*B*) cytoplasmic membrane proteins. After the peptide extension is cleaved off as in the case of the outer membrane proteins and periplasmic proteins, the new amino-terminal part of the protein stays in the periplasmic space, while a suceeding region begins to assemble in the lipid bilayer of the cytoplasmic membrane as the peptide is further synthesized, as a result of the hydrophobic property of this portion of the protein. A part of the carboxyl-terminal portion of the protein as a transmembrane protein is exposed in the cytoplasm. Solid portions and the following thick portions indicate basic sections *I* (see Figure 5) and hydrophobic sections *II* plus section *III*, respectively. Large arrows show the sequence of assembly. LPS, lipopolysaccharides; OM, outer membrane; PS, periplasmic space; and CM, cytoplasmic membrane. For simplicity the peptidoglycan layer between the outer membrane and the cytoplasmic membrane is not illustrated.

candidate for this component, are the lipopolysaccharides, which specifically exist in the outer membrane. The peptidoglycan, the outer membrane proteins, and the complexes formed by these components are also considered to trigger the self-assembly of the outer membrane proteins in the outer membrane. As soon as the amino-terminal section of an outer membrane protein is secreted across the cytoplasmic membrane, it may start to interact with the lipopolysaccharides. The specific interaction between the amino-terminal section of the outer membrane proteins and the lipopolysaccharides has been shown for ompA protein (Schweizer et al., 1978; see also Section 4.2). The cleavage of the peptide extension may be required for a subsequent interaction of the protein with the lipopolysaccharides. Alternatively, the interaction with the lipopolysaccharides may be essential for the cleavage of the peptide extension or there may not be any coordination between the two events. Further investigation of the effects of the lipid fluidity and phenethyl alcohol on the processing of outer membrane protein precursors will possibly provide important clues to distinguish the possibilities described above.

An intriguing hypothesis of the assembly of the outer membrane proteins is that it may be coupled with the insertion of the newly formed lipopolysaccharide molecules into the outer membrane. In this case, outer membrane proteins are transferred into the outer membrane through the Bayer junctions. This view is supported by the recent finding concerning the assembly of matrix proteins (Smit and Nikaido, 1978; see also Section 4.2). However, it is possible that some outer membrane proteins may be assembled without being transferred through the Bayer junctions. One candidate for Bayer junction-independent assembly is lipoprotein, the synthesis and assembly of which were not affected by the nonfluid state of membrane lipids in contrast to other outer membrane proteins (see Section 4.2). In this case, the unique modification at the amino-terminal cysteine residue may be essential for the proper assembly in the outer membrane.

Cell cycle-dependent assembly of some outer membrane proteins (see Chapter 10) is also another very intriguing aspect. There may be a specific structure at the site of septum formation that is essential for transferring these outer membrane proteins into their final location. It should be noted that beside the lipopolysaccharides the interaction of outer membrane proteins with other outer membrane components may also be important for the determination of the final location, although the interaction may not be primarily required for transferring of proteins into the outer membrane.

Figure 11 shows three possible mechanisms by which proteins are assembled in the outer membrane: (*A*) Bayer junction-dependent assembly in which the outer membrane protein assembly is coupled with the assembly of the lipopolysaccharides; (*B*) Bayer junction-indendent assembly (type 1), in which the nascent amino-terminal section interacts with the outer membrane and starts to be assembled in the membrane while the peptide is elongated; and (*C*) Bayer junction-independent assembly (type 2), in which the completion or modification of the peptide is required before assembly. In all cases, the translocation of the proteins is shown to occur by way of the loop model (see Section 3). There also may be another completely different assembly mechanism without a precursor protein, since it has been

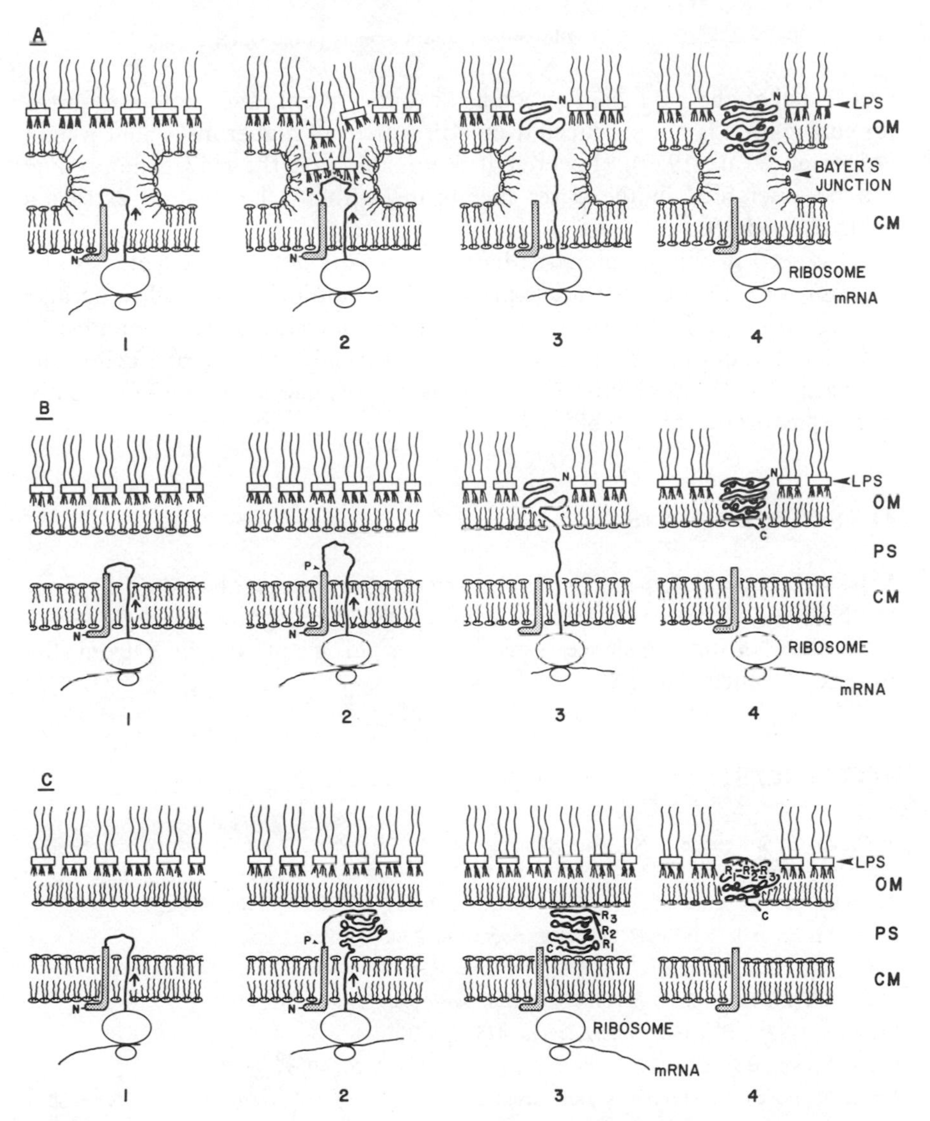

Figure 11 Mechanisms of secretion and final localization of outer membrane proteins in *E. coli*. (*A*) Bayer junction-dependent assembly, (*B*) Bayer junction-independent assembly (type 1), (*C*) Bayer junction-independent assembly (type 2). In case C the protein produced is the lipoprotein and R_1, R_2, and R_3 represent fatty acids linked to the amino-terminal end of the lipoprotein (see Figure 8). Shaded portions designate the peptide extensions of precursor molecules. In all cases proteins are secreted according to the loop model (see Figure 5). Small arrows indicate the direction of the lipopolysaccharide assembly and large arrows indicate the direction of protein secretion. Small arrows with P show the cleavage sites of the peptide extensions. LPS, lipopolysaccharide; OM, outer membrane; PS, periplasmic space; CM, cytoplasmic membrane; N, amino-terminal end; C, carboxyl terminal end. For simplicity the peptidoglycan layer between the outer membrane and the cytoplasmic membrane is not illustrated.

reported that the *traT* gene product, an outer membrane protein required for conjugation is not produced from a precursor of higher molecular weight (Achtman et al., 1977). However, it is possible that the *traT* gene product can be assembled in the outer membrane without the peptide extension being cleaved off.

As is seen in this chapter, the basic framework for the biosynthesis and assembly of the outer membrane proteins has culminated in the last decade. On the basis of this framework it is now possible to investigate a number of more specific questions on these processes. It should not be long before the precise molecular mechanisms of the biosynthesis and assembly of the outer membrane proteins is established.

ACKNOWLEDGMENTS

This work was supported by Public Health Service grant GM19043 from the National Institute of General Medical Sciences, grant PCM 76-07320 from the National Science Foundation, and grant BC-67D from the American Cancer Society.

REFERENCES

Achtman, M., Kennedy, N., and Skurray, R. (1977), *Proc. Natl. Acad. Sci. U.S.*, **74,** 5104–5108.

Adelman, M. R., Sabatini, D. D., and Blobel, G. (1973), *J. Cell Biol.*, **56,** 206–229.

Bayer, M. E., and Thurow, H. (1977), *J. Bacteriol.*, **130,** 911–936.

Blobel, G., and Dobberstein, B. (1975), *J. Cell Biol.*, **67,** 835–851.

Blobel, G., and Sabatini, D. D. (1971), in *Biomembranes*, **2,** 193–195.

Braun, V. (1975), *Biochim. Biophys. Acta*, **415,** 335–377.

Braun, V., and Bosch, V. (1972), *Proc. Natl. Acad. Sci. U.S.*, **69,** 970–974.

Braun, V., Bosch, V., Hantke, K., and Schaller, K. (1974), *Ann. N.Y. Acad. Sci.*, **235,** 66–82.

Chai, T., and Foulds, J. (1977), *Biochim. Biophys. Acta*, **493,** 210–215.

Chang, C. N., Blobel, G., and Model, P. (1978), *Proc. Natl. Acad. Sci. U.S.*, **75,** 361–365.

Chattopadhyay, P. K., and Wu, H. C. (1977), *Proc. Natl. Acad. Sci. U.S.*, **74,** 5318–5322.

Cancedda, R., and Schlessinger, M. J. (1974), *J. Bacteriol.*, **117,** 290–301.

Datta, D. B., Kramer, C., and Henning, J. (1976), *J. Bacteriol.*, **128,** 834–841.

Datta, D. B., Arden, B., and Henning, U. (1977), *J. Bacteriol.*, **131,** 821–829.

DeMartini, M., and Inouye, M. (1978), *J. Bacteriol.*, **133,** 329–335.

DeMartini, M., Inouye, S., and Inouye, M. (1976), *J. Bacteriol.*, **127,** 564–571.

Diedrich, D. L., and Schnaitman, C. A. (1978), *Proc. Natl. Acad. Sci. U.S.*, **75,** 3708–3712.

DiRienzo, J. M., and Inouye, M. (1979), *Cell*, **17,** 155–161.

DiRienzo, J. M., Nakamura, K., and Inouye, M. (1978), *Annu. Rev. Biochem.*, **47,** 481–532.

Endermann, R., Hindenbach, I., and Henning, U. (1978a), *FEBS Lett.*, **88,** 71–74.

Endermann, R., Krämer, C., and Henning, U. (1978b), *FEBS Lett.*, **86,** 21–24.

Foulds, J., and Chai, T. (1978), *J. Bacteriol.*, **133,** 1478–1483.

Garten, W., and Henning, U. (1974), *Eur. J. Biochem.*, **47,** 343–352.

Halegoua, S., and Inouye, M. (1979), *J. Mol. Biol.*, **130,** 39–61.

Halegoua, S., Hirashima, A., and Inouye, M. (1976a), *J. Bacteriol.*, **126,** 183–191.

Halegoua, S., Hirashima, A., Sekizawa, J., and Inouye, M. (1976b), *Eur. J. Biochem.*, **69,** 163–167.

Halegoua, S., Sekizawa, J., and Inouye, M. (1977), *J. Biol. Chem.*, **252,** 2324–2330.

Hantke, K., and Braun, V. (1973), *Eur. J. Biochem.*, **34,** 284–295.

Hasegawa, Y., Yamada, H., and Mizushima, S. (1976), *J. Biochem.*, **80,** 1401–1409.

Henning, U., Höhn, B., and Sonntag, I. (1973), *Eur. J. Biochem.*, **47,** 343–352.

Hirashima, A., and Inouye, M. (1973), *Nature*, **242,** 405–407.

Hirashima, A., and Inouye, M. (1975), *Eur. J. Biochem.*, **60,** 395–398.

Hirashima, A., Childs, G., and Inouye, M. (1973), *J. Mol. Biol.*, **79,** 373–389.

Hirashima, A., Wang, S. S., and Inouye, M. (1974), *Proc. Natl. Acad. Sci. U.S.*, **71,** 4149–4153.

Ichihara, S., and Mizushima, S. (1978), *J. Biochem.*, **83,** 1095–1100.

Inouye, H., and Beckwith, J. (1977), *Proc. Natl. Acad. Sci. U.S.*, **74,** 1440–1444.

Inouye, M. (1974), *Proc. Natl. Acad. Sci. U.S.*, **71,** 2396–2400.

Inouye, M. (1975), in *Membrane Biogenesis* (A. Tzagoloff, ed.), Plenum Press, New York, pp. 351–391.

Inouye, M. (1979), in *Biomembranes* (L.A. Manson, ed.), Plenum Press, New York, in press.

Inouye, M., and Halegoua, S. (1979), *CRC Crit. Rev. Biochem.*, in press.

Inouye, M., Shaw, J., and Shen, C. (1972), *J. Biol. Chem.*, **247,** 8154–8159.

Inouye, S., Wang, S., Sekizawa, J., Halegoua, S., and Inouye, M. (1977), *Proc. Natl. Acad. Sci. U.S.*, **74,** 1004–1008.

Inouye, M., Pirtle, R., Pirtle, I., Sekizawa, J., Nakamura, K., DiRienzo, J., Inouye, S., Wang, S., and Halegoua, S. (1979), *Microbiology*, Microbiology-1979, 34–37.

Ito, K. (1977), *J. Bacteriol.*, **132,** 1021–1023.

Ito, K. (1978), *Biochem. Biophys. Res. Commun.*, **82,** 99–107.

Ito, K., Sato, T., and Yura, T. (1977), *Cell*, **11,** 551–559.

Izuki, K., Matsuhashi, M., and Strominger, J. L. (1966), *Proc. Natl. Acad. Sci. U.S.*, **55,** 656–663.

Kimura, K., and Izuki, K. (1976), *Biochem. Biophys. Res. Commun.*, **70,** 900–906.

Koch, P. A., Gardner, F. H., Gartrell, J. E., Jr. and Carter, J. E., Jr., (1975), *Biochim. Biophys. Acta*, **389,** 117–187.

Lazdunski, C., Busuttil, J., and Lazdunski, A. (1975), *Eur. J. Biochem.*, **60,** 363–369.

Lee, N., and Inouye, M. (1974), *FEBS Lett.*, **39,** 167–170.

Lee, N. Tu, S., and Inouye, M. (1977). *Biochemistry*, **16,** 5026–5030.

Leij, L., Kingma, J., and Witholt, B. (1978), *Biochim. Biophys. Acta*, **512,** 365–376.

Lin, J. J., and Wu, H. C. P. (1976), *J. Bacteriol.*, **125,** 892–904.

Lin, J. J. C., Kanazawa, H., Ozols, J., and Wu, H. C. (1978), *Proc. Natl. Acad. Sci. U.S.*, **75,** 4891–4895.

Lingappa, V. R., Devillers-Thiery, A., and Blobel, G. (1977), *Proc. Natl. Acad. Sci. U.S.*, **74,** 2432–2436.

Lingappa, V. R., Lingappa, J. R., Prasad, R., Ebner, K. E., and Blobel, G. (1978), *Proc. Natl. Acad. Sci. U.S.*, **75,** 2338–2342.

Lugtenberg, B., Peters, R., Bernheimer, H., and Berendsen, W. (1976), *Mol. Gen. Genet.*, **147,** 251.

Miller, C. G. (1975), *Annu. Rev. Microbiol.*, **29,** 485–504.

Moses, R. E., and Richardson, C. C. (1970), *Proc. Natl. Acad. Sci. U.S.*, **67,** 674–681.

Movva, N. R., Katz, E., Asdourian, P. L., Hirota, Y., and Inouye, M. (1978), *J. Bacteriol.*, **133,** 81–84.

Mühlradt, P. F., Menzel, J., Golecki, J. R., and Speth, V. (1973), *Eur. J. Biochem.*, **35,** 471–481.

Murgier, M., Pelissier, C., Lazdunski, A., and Lazdunski, C. (1976), *Eur. J. Biochem.*, **65,** 517–520.

Nakamura, K., and Mizushima, S. (1975), *Biochim. Biophys. Acta*, **413,** 371–393.

Pacand, M., Sibilli, L., and LeBras, G. (1976), *Eur. J. Biochem.*, **69,** 141–151.

Pages, J. M., Piovant, M., Varenne, S., and Lazdunski, C. (1978), *Eur. J. Biochem.*, **86,** 589–602.

Palade, G. E. (1975), *Science*, **189,** 347.

Peterson, R. L., Radcliffe, C. W., and Pace, N. R. (1971), *J. Bacteriol.*, **107,** 585–588.

Pirtle, R. M., Pirtle, I. L., and Inouye, M. (1978), *Proc. Natl. Acad. Sci. U.S.*, **75,** 2190–2194.

Pirtle, R. M., Pirtle, I. L., and Inouye, M. (1979), submitted for publication.

Pugsley, A. P., and Schnaitman, C. (1978a), *J. Bacteriol.*, **133,** 1181–1189.

Pugsley, A. P., and Schnaitman, C. (1978b), *J. Bacteriol.*, **135,** 1118–1129.

Randall, L. L., and Hardy, S. J. S. (1975), *Mol. Gen. Genet.*, **137,** 151–160.

Randall, L. L., and Hardy, S. J. S. (1977), *Eur. J. Biochem.*, **75,** 43–53.

Randall, L. L., Hardy, S. J. S., and Josefsson, L. G. (1978), *Proc. Natl. Acad. Sci. U.S.*, **75,** 1209–1212.

Redman, C. M., and Sabatini, D. D. (1966), *Proc. Natl. Acad. Sci. U.S.*, **56,** 608–615.

Rosbash, M., and Penman, S. (1971), *J. Mol. Biol.*, **59,** 227–241.

Rosenbusch, J. R. (1974), *J. Biol. Chem.*, **249,** 8019–8029.

Sarma, V., and Reeves, P. (1977), *J. Bacteriol.*, **132,** 23–27.

Schaller, H., Beck, E., and Takanami, M. (1978), in *The Single-Stranded DNA Phages*, Cold Spring Harbor Laboratory, New York, pp. 139–163.

Schnaitman, C. A. (1974), *J. Bacteriol.*, **118,** 454–464.

Schrader, W. P., and Fan, D. P. (1974), *J. Biol. Chem.*, **249,** 4815–4818.

Schulman, H., and Kennedy, E. P. (1977), *J. Biol. Chem.*, **252,** 4250–4255.

Schweizer, M., and Henning, U. (1977), *J. Bacteriol.*, **129,** 1651–1657.

Schweizer, M., Hindennach, I., Garten, W., and Henning, U. (1978), *Eur. J. Biochem.*, **82,** 211–217.

Sekizawa, J., Inouye, S., Halegoua, S., and Inouye, M. (1977), *Biochem. Biophys. Res. Commun.*, **77,** 1126–1133.

Silhavy, T. J., CasaDaban, M. J., Shuman, H. A., and Beckwith, J. (1976), *Proc. Natl. Acad. Sci. U.S.*, **73**, 3423–3427.

Silhavy, T. J., Shuman, H. A., Beckwith, J., and Schwartz, M. (1977), *Proc. Natl. Acad. Sci. U.S.*, **74**, 5411–5415.

Smit, J., and Nikaido, H. (1978), *J. Bacteriol.*, **135**, 687–702.

Smith, W. P., Tai, P.-C., and Davis, B. D. (1978), *Proc. Natl. Acad. Sci. U.S.*, **75**, 814–817.

Smith, W. P., Tai, P.-C., Thompson, R. C., and Davis, B. D. (1977), *Proc. Natl. Acad. Sci. U.S.*, **74**, 2830–2834.

Smith, W. P., Tai, P.-C., and Davis, B. (1978), *Proc. Natl. Acad. Sci. U.S.*, **75**, 5922–5925.

Steven, A. C., Ten Heggeler, B., Muller, R., Kistler, J., and Rosenbusch, J. P. (1977), *J. Cell Biol.*, **72**, 292–301.

Sugimoto, K., Sugisaki, H., Okamoto, T., and Takanami, M. (1977), *J. Mol. Biol.*, **110**, 487–507.

Sutcliffe, J. G. (1978), *Proc. Natl. Acad. Sci. U.S.*, **75**, 3737–3741.

Takeishi, K., Yasumura, M., Pirtle, R., and Inouye, M. (1976), *J. Biol. Chem.*, **251**, 6259–6266.

Tanaka, N., Iseki, M., Miyoshi, T., Aoki, H., and Imanaka, H. (1976), *Antibiotics*, **29**, 155–168.

Van Alphen, L., Havekes, L., and Lugtenberg, B. (1977), *FEBS Lett.*, **75**, 285–290.

Varienne, S., Piovant, M., Pages, J. M., and Lazdunski (1978), *Eur. J. Biochem.*, **86**, 603–606.

Wang, S., Marcu, K. B., and Inouye, M. (1976), *Biochem. Biophys. Res. Commun.*, **68**, 1194–1200.

Weigand, R. A., Vinci, K. D., and Rothfield, L. I. (1976), *Proc. Natl. Acad. Sci. U.S.*, **73**, 1882–1886.

Wu, H. C., Hou, C., Lin, J. J. C., and Yem, D. W. (1978), *Proc. Natl. Acad. Sci. U.S.*, **74**, 1388–1392.

Yamada, H., and Mizushima, S. (1977), *J. Biochem.*, **81**, 1889.

Yu, F., and Mizushima, S. (1977), *Biochem. Biophys. Res. Commun.*, **74**, 1394–1402.

Chapter 5 Biosynthesis and Assembly of Cell Wall Peptidoglycan

DAVID MIRELMAN

Department of Biophysics, The Weizmann Institute of Science, Rehovoth, Israel

1 INTRODUCTION

Modern study of the bacterial cell wall peptidoglycan, the main supporting structure of the cell, is approaching the end of its third decade. Tremendous progress has been achieved in this extremely complex field ever since the pioneering work of M. R. J. Salton (1953), who in the early fifties isolated pure bacterial cell walls from mechanically disrupted cells, and of J. T. Park (1952), who isolated and identified the first nucleotide precursor of peptidoglycan from cells grown in the presence of penicillin G.

Peptidoglycan has been investigated from many angles, all of which have begun to become integrated in recent years. While emphasis during the first two decades focused on the chemical structure of peptidoglycan and the diversities among many species, as well as on identifying the key reactions in biosynthesis, recent research has been more concentrated on the elucidation of the molecular mechanism of its assembly and growth. In the last few years a considerably large number of excellent reviews and monographs dealing with many aspects of peptidoglycan structure, biosynthesis, growth, and expansion have appeared (Strominger, 1970; Rogers, 1970; Schleifer and Kandler, 1972; Ghuysen and Shockman, 1973; Fiedler and Glaser, 1973; Glaser, 1973; Ghuysen, 1977a; Daneo-Moore and Shockman, 1977).

Many of these reviews deal with questions related mainly to gram-positive organisms. Consequently, in this chapter I deal primarily with what we know today about the late stages of peptidoglycan biosynthesis and assembly, especially as it applies to gram-negative bacteria. The basic concepts and ideas pertaining to structure and biosynthesis are outlined only briefly and the reader is referred to other monographs for more detailed information on these subjects.

The mode of assembly of macromolecular components such as peptidoglycan into a three-dimensional, supramolecular surface structure is one of the most complex processes to be investigated thus far. Attention is given in this chapter to an analysis of the available evidence on the specific enzymatic processes involved in the mechanism of cell wall peptidoglycan synthesis and assembly during growth of a bacterium.

2 CHEMICAL COMPOSITION AND STRUCTURE OF PEPTIDOGLYCAN

Thanks to the vast work done during the last two decades, we know today a great deal about the overall chemical composition and primary structure of peptidoglycans in a wide variety of organisms. Since peptidoglycans are relatively easy to isolate and purify, numerous comparisons of composition and structure have been made, and peptidoglycans from different species have been classified in a series of chemotypes (Schleifer and Kandler, 1972).

All gram-negative species examined so far, and several gram-positive species, have peptidoglycans of chemotype I (Figures 1 and 2) containing *meso*-diaminopimelic acid (DAP). This structure prevails in both rod and coccal shaped species such as *Escherichia coli* and *Neisseria*, and in a number of gram-positive organisms such as *Bacilli* and *Lactobacillus plantarum*.

In gram-positive organisms, peptidoglycan comprises from 50 to 90% of the mass of the cell envelope and acid or base extraction of the isolated envelopes usually yields an insoluble fraction of peptidoglycan of reasonable purity. In gram-negative organisms, peptidoglycan is one of the minor components of the envelope, yet it is fairly easy to isolate the insoluble peptidoglycan matrix by heating cell envelopes in sodium dodecyl sulfate (SDS) (4%) solutions followed by trypsin digestions of the covalently linked lipoprotein (Boy de la Tour et al., 1965; Mardarowicz, 1966; Braun, 1975).

2.1 The Glycan

The glycan moiety of all peptidoglycans examined consists of linear strands of alternate residues of *N*-acetylglucosamine and *N*-acetylmuramic acid [(2-*N*-acetylamino-2-deoxycarboxyethyl)-2-deoxy-D-glucose] (Figure 1). On the basis of extensive chemical and enzymatic investigations, it is firmly believed that the $\beta(1 \rightarrow 4)$ linkages between the above amino sugars are uniform in all bacteria and resemble those present in the chitin-like structures (Sharon et al., 1966, Chipman et al., 1968).

The glycan chains of most bacteria terminate with free-reducing ends of muramic acid (Figure 3 and see Section 2.3) (Hartmann et al., 1972, Höltje chains of *E. coli*, which terminate in an unusual anhydro $(1 \rightarrow 6)$ structure of muramic acid (Figure 3 and see Section 2.3) (Hartmann et al., 1972, Höltje et al., 1975).

The average glycan chain length in walls appears to be substantially longer than previously thought (Rogers, 1974). By measuring the ratio of total amino sugar alcohol (obtained after reduction of glycan chains with sodium borohydride) to total hexosamine, Ward (1973) estimated the chain length to be about 80 disaccharide units in walls and 140 units in newly synthesized chains of a lytic defective mutant of *Bacillus licheniformis*. Because of the presence of peptidoglycan autolysins, nearly all estimates of either glycan chain length or extent of peptide cross-linking in isolated cell walls must be considered to be minimal values. Thus it seems that long glycan chains are hydrolyzed to shorter lengths when or after they are inserted into the wall. Because of the absence of free reducing glycan ends in the peptido-

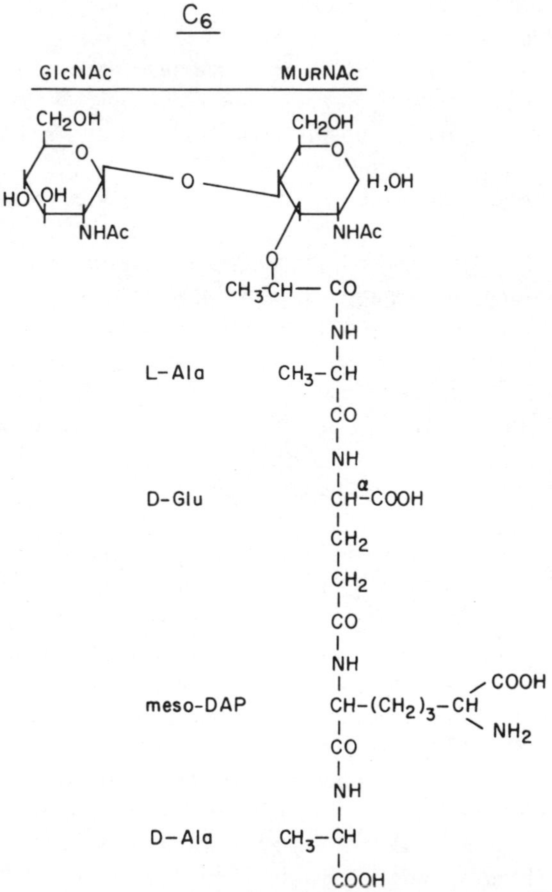

C_6

GlcNAc MurNAc

CH_2OH CH_2OH

HO OH O H,OH

NHAc NHAc

CH_3–CH— CO

NH

L-Ala CH_3–CH

CO

NH

D-Glu CH–COOH (α)

CH_2

CH_2

CO

NH

meso-DAP CH–$(CH_2)_3$–CH(COOH)(NH_2)

CO

NH

D-Ala CH_3–CH

COOH

Figure 1 Chemical structure of the *E. coli* peptidoglycan subunit (monomer) C_6 (Weidel and Pelzer, 1964). GlcNAc = *N*-acetylglucosamine, MurNAc = *N*-acetylmuramic acid. This peptidoglycan fragment is one of the compounds obtained after lysozyme digestion of the *E. coli* cell wall sacculus (see also Figure 2).

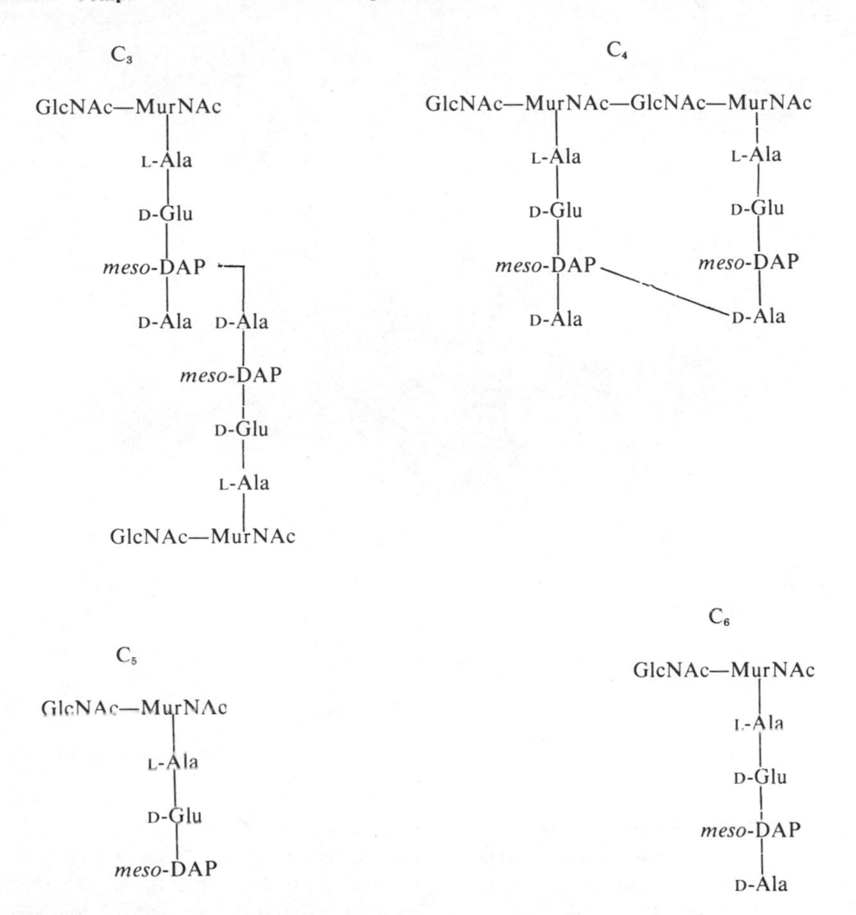

Figure 2 Chemical structures of the four main peptidoglycan fragments obtained upon lysozyme digestion of the cell wall peptidoglycan (Weidel and Pelzer, 1964). C_3 and C_4 are representatives of the cross-linked structures and C_5 and C_6 are representives of the uncross-linked monomers.

glycan of *E. coli*, no estimation of chain lengths could be made by the muramic acid/muramicitol ratio. The average chain lengths were calculated by determining the amount of [^{14}C]galactose that was transferred from UDP[^{14}C]galactose to *N*-acetylglucosamine residues at the nonreducing end of the chains by a galactosyl transferase from human milk (Schindler et al., 1976). The calculated chain lengths (30–60 disaccharide units) were in good agreement with those obtained for gram-positive organisms and indicate that glycans have an average length of 50–100 nm.

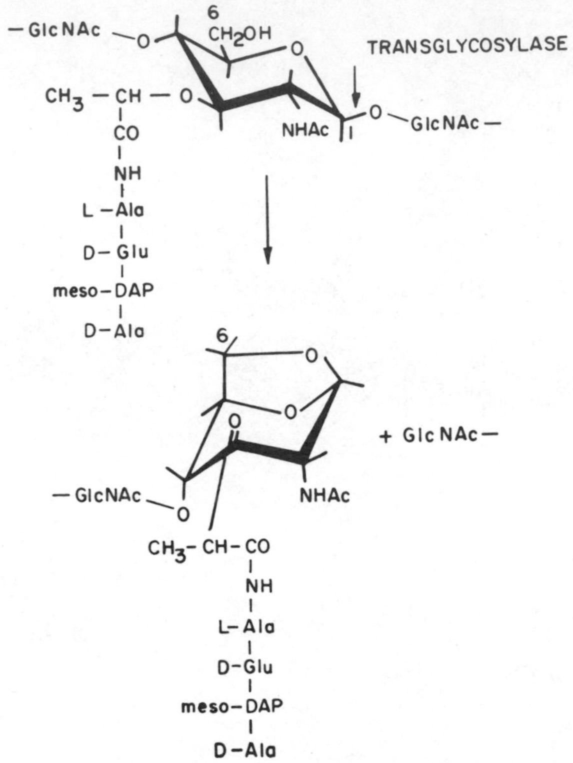

Figure 3 Formation and structure of *E. coli* peptidoglycan fragment X^1. Fragment X which is also obtained, lacks the terminal D-alanine moiety (Hartman et al, 1972 and Höltje et al, 1975). The scheme shows the conversion of the muramic acid moiety from the $^4C_1$ into the $^1C_4$ conformation upon the *E. coli* transglycosylase action. Reprinted with permission of the American Society for Microbiology from Höltje et al, 1975.

2.2 The Peptide

In most peptidoglycans, especially those of gram-positive species, every D-lactyl group of *N*-acetylmuramic acid is substituted by a short tetrapeptide composed of L-alanine-D-isoglutamic acid–*meso*-diaminopimelic acid–D-alanine (Figure 1). In some cases a tripeptide lacking the terminal D-alanine is found (Figure 2). Since the L-configuration of *meso*-DAP is usually involved in the peptide backbone linkage, all tetrapeptides exhibit the unique L-D-L-D alternating sequence.

Peptide units substituting adjacent glycan chains are covalently linked by means of "bridges." On the basis of the composition and location of these interpeptide bridges, peptidoglycans have been classified into four main types. The bridge between two peptide units usually extends between the *C*-terminal D-alanine residue of one tetrapeptide and the ω-amino group of the diamino acid of another (tri- or tetra-) peptide unit. In peptidoglycan of type I, the bridging consists of a direct bond between N of the D-asymmetric center of DAP and the carboxyl of a D-alanine residue (Figure 2). The proportion of peptide units that are crosslinked varies from species to species but is invariably rather low (Schleifer and Kandler, 1972; Ghuysen and Shockman, 1973). The most highly cross-linked peptide is that of *Staphylococcus aureus*, in which more than 90% of the peptide side chains are cross-linked. The peptidoglycans of most gram-negative bacteria contain approximately equimolar amounts of uncross-linked peptide monomer (usually tetrapeptides-C_6 but also tripeptides-C_5) and cross-linked peptide dimers C_3 and C_4 (Figure 2; Weidel and Pelzer, 1964).

The minor differences in composition and chemical structure found in the numerous species do not appear to reflect the variety of shapes or changes in shape of cells in which they are found. Although there is no direct evidence that the chemical structure of peptidoglycan plays a role in determining the shape of the wall, it should be stated here that all analyses of peptidoglycan made so far have mainly considered the major constituents and general structural properties. Minor, but perhaps important, variations in secondary and tertiary structures of peptidoglycan that have not yet been studied in sufficient detail may have considerable influence on the cell shape.

2.3 Enzymes That Degrade Peptidoglycan

The unraveling of the primary structure of the wall peptidoglycan was made possible by the isolation and characterization of a whole series of enzymes selectively active upon different, well-defined linkages within the peptidoglycan (Ghuysen, 1968, Schleifer and Kandler, 1972, Hartmann et al., 1974, Daneo-Moore and Shockman, 1977). Peptidoglycans can be solubilized by enzymes that hydrolyze bonds (1) in the glycan strands (*endo-N*-acetylmuramidases or *endo-N*-acetylglucosaminidases), (2) in the peptide moiety (endopeptidases that hydrolyze peptide bonds of the bridge or at the *C*-terminal ends), or (3) at the junction between the glycan strands and the peptide units (*N*-acetylmuramyl-L-alanine amidase) (Figure 4). A variety of lytic enzymes having specificities for most of the unusual types of peptidoglycan linkages found, have been isolated. All bacteria possess their own peptidoglycan degrading enzymes, autolysins, which are sometimes localized in the cytoplasm, the membrane, or the wall itself. When such enzymes are permitted to act from within, the cells lose their osmotic protection and lyse. The nature and complexity of the autolysin system differs markedly from species to species and this may reflect differences in peptido-

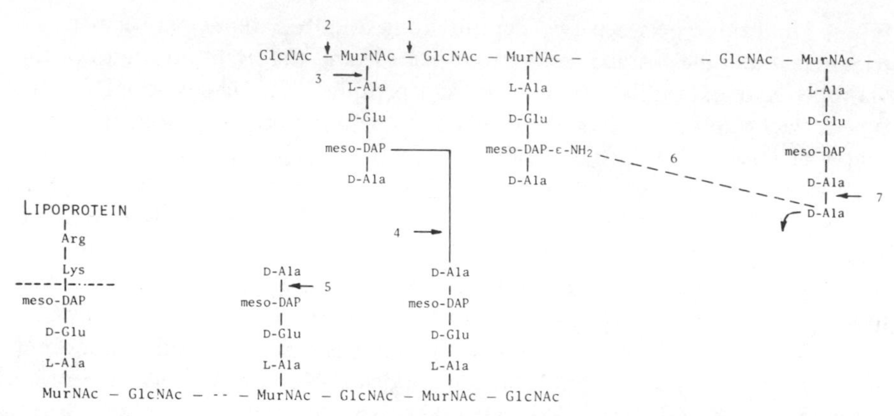

Figure 4 Schematic representation of the sites of enzymatic activity performed by the *E. coli* peptidoglycan metabolizing system: (*1*) *N*-acetylmuramidase and also apparently the transglycosylase, (*2*) *N*-acetylglucosaminidase, (*3*) *N*-acetylmuramyl-L-alanyl amidase, (*4*) DD-endopeptidase, (*5*) LD-carboxypeptidase (or carboxypeptidase II), (*6*) DD-transpeptidase with the concomitant release of the *C*-terminal D-alanine, (*7*) DD-carboxypeptidase. Lipoprotein is bound by an ε-amino lysine group to the α-carboxyl of the L-asymetric center of *meso*-DAP. For further details on the enzymatic activities and locations see Ghuysen, 1978; Ghuysen and Shockman, 1973; Hartmann et al. 1974; Ghuysen, 1977a; and Daneo-Moore and Shockman, 1977.

glycan structure and in the role that these enzymes may play in the regulation of peptidoglycan metabolism during growth and division of the bacteria. These points are also discussed in later sections.

At least six different hydrolases have been found in *E. coli* (Figure 4). They include *N*-acetylglucosaminidase, *N*-acetylmuramidase, muramyl-L-alanylamidase, a cross-bridge splitting endopeptidase, as well as what appear to be several variations of a D-alanine carboxypeptidase I (Tamura et al., 1976a). An interesting and apparently unique autolysin that exhibits an unusual specificity was also discovered in *E. coli* (Hartmann et al., 1972, Höltje et al., 1975; Taylor et al., 1975). The isolated pure enzyme degrades insoluble peptidoglycan into disaccharide-peptide fragments that have an internal anhydro (1→6) structure at the terminal muramic acid residue (Figure 3). This enzyme behaves therefore as a peptidoglycan: peptidoglycan 6-muramyltransferase and may be responsible for the total absence of reducing moieties in intact peptidoglycan strands of *E. coli*. The role of this enzyme in peptidoglycan metabolism is not yet understood. Mutants lacking this activity appear to have normal growth characteristics (U. Schwarz, personal communication).

Solubilization of the netlike structure of the wall peptidoglycan can be brought about by the hydrolysis of any single type of linkage but is achieved only if a sufficient number of bonds are actually hydrolyzed. In view of the large number of autolysins, there is always a suspicion that many of the terminal groups found in all peptidoglycans as they are isolated, are artifacts caused by the sudden, uncontrolled action of these enzymes during isolation. In *E. coli*, where the peptidoglycan matrix can be isolated simply by heating intact log phase bacteria in boiling 4% SDS (Boy de la Tour et al., 1965, Mardarowicz, 1966, Braun, 1975), the autolysins are almost incapable of causing any hydrolysis; yet its peptidoglycan contains numerous terminal groups (Schwarz et al., 1969). It is clear today that most of the terminal groups present in the peptidoglycan reflect significant properties of the peptidoglycan biosynthetic machinery and their careful study may lead to important information on the mechanisms of secondary modifications that occur after assembly of the wall (Daneo-Moore and Shockman, 1977).

2.4 Three-Dimensional Structure

In all the three-dimensional atomic models of peptidoglycan proposed (Kelemen and Rogers, 1971; Oldmixon et al., 1974; Formanek et al., 1974), the glycan chains are assumed to have the chitin-like lattice arrangement. X-Ray diffractions of dried foils of isolated peptidoglycan from both gram-negative *Spirillum serpens* and the gram-positive *L. plantarum*, showed Debye-Scherrer rings with periodicities of about 1.00 and 0.46 nm (Formanek et al., 1974). In addition, the comparison of the infrared spectra of chitin and peptidoglycan showed that the amide absorption bands of both compounds have the same frequencies. This experimental evidence strongly suggests that the structure of the glycan chains in the peptidoglycan is similar to that of chitin (and cellulose). In order to fit the periodic structure of glycan into a lattice structure like that of chitin, the amino sugar residues must form a parallel, twofold screw axis and the peptide chains must be very flat. The unique γ-glutamyl bond of the peptide backbone necessarily directs all the peptide chains to the same side of the glycan stack and imparts on them the hypothetical 2.2_7 helix conformation. In this structure the peptide units become stabilized by forming two hydrogen bonds with the sugar residues and the *N*-acetyl moieties, and these in turn form additional hydrogen bonds to *N*-acetyl residues of adjacent glycan chains (Formanek et al., 1974).

The relative fragility of the *E. coli* peptidoglycan matrix is well demonstrated by the solubilizing effect that 6 *M* guanidine hydrochloride has on it (Leduc and Van Heijenoort, 1975). Hydrogen bonding and perhaps electrostatic interactions must contribute significantly to the maintenance of the special three-dimensional structure of the nondenatured form of this polymer.

A very dense, continuous packing of peptidoglycan in the chitin-like arrangement, as proposed by Formanek et al. (1974), can exist only in certain areas of the wall, since the calculated number of repeating units per cell surface area would mean that only about 30% of the cell surface is covered by peptidoglycan (Braun, 1975).

No information is available on the maximal dimensions that a nonelastic fiber such as peptidoglycan can attain when it is in the fluid or gel-like environment of the intact envelope. The native structure of peptidoglycan in the envelope may be flexible enough and may have a number of degrees of freedom, especially in the zones where the tetrapeptides are not cross-linked without losing the restrictions of its overall shape as suggested by Prević (1970). Furthermore, the peptidoglycan lattice may be regarded as a paracrystalline foil with amorphous regions and crystalline mosaic blocks (Formanek, 1978). Such an extended peptidoglycan network could then perhaps fulfill its structural role and cover most, if not all, of the cell surface area. Changes in growth conditions of *E. coli* cells have been shown to result in different cell sizes and volumes (Rosenberger et al., 1978). These changes seem to have no effect on the amount of peptidoglycan per cell surface area (Zaritsky, et al. 1979).

More work, however, needs to be done before a clear understanding of the three-dimensional structural arrangement of peptidoglycan in the intact gram-negative cell envelope is obtained.

2.5 Location of the Peptidoglycan and Its Interaction With Other Cell Wall Components

Qualitative inspections of electron micrographs such as thin sections or freeze fractures indicate that the cell envelope accounts for a substantial fraction of the cell volume and mass, especially in gram-positive bacteria. In gram-negative bacteria, the cell envelope is composed of a multilayered structure in which a thin peptidoglycan layer, 20–30 Å thick and most likely made up of a single or bimolecular layer, is sandwiched between an outer membrane and the cytoplasmic membrane immediately underlying it (Figure 5). The outer membrane is linked to the peptidoglycan sacculus by way of a covalent attachment between some lipoprotein molecules and the peptidoglycan (Braun, 1975). Approximately every tenth peptide subunit is linked through the α-carboxyl group of the L-asymmetric carbon of diaminopimelic acid to an ϵ-amino group of a *C*-terminal lysine residue of the lipoprotein (Figure 4). Braun (1975) has calculated that there are approximately 2.5×10^5 lipoprotein attachment sites per cell. Two-thirds of the total lipoprotein of the cell envelope exists as free molecules in the outer membrane (Hirashima et al., 1973, Lee and Inouye, 1974). Part of the free lipoprotein becomes covalently attached to peptidoglycan and a dynamic equilibrium is believed to exist between the bound and free forms. The mechanism of attachment of the lipoprotein to the peptidoglycan is also not known. It

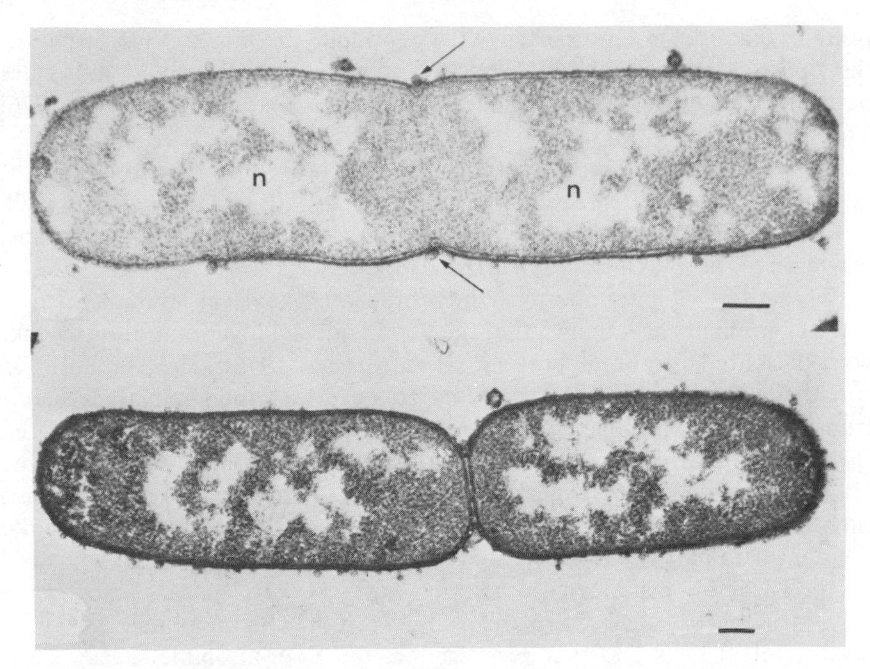

Figure 5 Sections of *E. coli* B/R synchronized by membrane elution at 30 min (upper section) and 45 min (lower section) fixed with acrolein–glutaraldehyde. Note blebs of outer membrane and cell envelope layers (arrows, upper section), as well as the nucleoid (*n*) partitioned to each half of the cell. Lower section shows almost completed septum. Reprinted from Burdett and Murray (1974b) with permission of the American Society for Microbiology.

may proceed by a transpeptidation reaction in which the carboxyl group at the L center of the *meso*-DAP is transferred to the ε-amino group of the *C*-terminal lysine residue of a lipoprotein molecule in the outer membrane. It has been speculated for a number of years that an LD transpeptidase (perhaps the LD-alanine carboxypeptidase II) (Figure 4) catalyzes the attachment with a concomitant release of the D-alanine residue. Unfortunately, no evidence has yet been obtained for such a reaction.

A number of other peptidoglycan-associated outer membrane proteins have been isolated and investigated (Rosenbusch, 1974; Hasegawa et al., 1976; Lugtenberg et al., 1977). These major outer membrane proteins remain attached to the peptidoglycan after treatment with SDS at 60°C but are removed by heating above 60°C. The linkage between the peptidoglycan and proteins is suspected to be of an ionic or hydrophobic nature, although some linkages may be of a Schiff-base type as a result of the presence of allysine (α-amino adipic acid δ-semialdehyde) residues in outer membranes of *E. coli* (Diedrich and Schnaitman, 1938; Mirelman and Siegel, 1979). The proteins that are loosely attached to the peptidoglycan may form or be part of the "pores" or hydrophilic channels through which numerous molecules permeate the outer membrane into the cell (Nakae and Nikaido, 1976; VanAlphen et al., 1978; see also Chapter 11 by H. Nikaido.

The complex arrangement of the gram-negative cell envelope has the advantage of not only providing mechanical strength to the whole cell wall, but also of imparting further protection to the cytoplasmic membrane, which serves as a permeability barrier to the cell.

Separations of outer and inner membranes by physical methods such as sucrose gradients have shown that the peptidoglycan layer migrates together with the outer membrane portion (Osborn et al., 1972). The only location where the peptidoglycan layer seems to be diffuse and indistinguishable from the other envelope components is at the septum region of the cell (Figure 5). During nascent cross-wall formation, Burdett and Murray (1974a, b) and Weigand et al. (1976) observed the presence of a gap separating the two transverse lamellae of peptidoglycan. The outer membrane did not appear to participate in septation until cells began to separate.

Not all of the peptidoglycan of gram-negative cells appears to be in the SDS-insoluble sacculus. Comparisons between the diaminopimelic acid content of intact cells and that recovered in the SDS-insoluble sacculi show considerable differences. In *E. coli* the succulus accounts for 80–90% of the total DAP, whereas in *Pseudomonas aeruginosa* it accounts for only 70–80% (D. Mirelman, unpublished observations). Very little information is available on the structure, location, and function of this SDS-soluble peptidoglycan in the cell envelope (see also Sections 4.3 and 4.4).

3 BIOSYNTHESIS AND ASSEMBLY OF PEPTIDOGLYCAN

Since the shape and rigid integrity of the bacterial cell are maintained largely by the peptidoglycan, detailed understanding of the biosynthesis of peptidoglycan requires knowledge not only of how the component amino sugars and amino acids are joined together, but also of the complex process by which the macromolecule is assembled in an orderly manner external to the cytoplasmic membrane and then modified during growth and cell division.

Present knowledge of the biosynthesis of this highly complex and unusual macromolecule is based mainly on extensive investigations carried out with *Micrococcus luteus* (*lysodeikticus*), *S. aureus*, and *E. coli*. The contributions of the research groups of Park, Strominger, Ghuysen, Perkins, Rogers, and Neuhaus were crucial for the unraveling of the key enzymatic steps leading to the synthesis of peptidoglycan. The findings obtained clearly showed that all known peptidoglycans are synthesized by essentially common mechanisms. The process of biosynthesis is usually divided into three stages (Figure 6).

1 Synthesis of the two uridylic sugar nucleotides, UDP-*N*-acetylglucosamine (UDP-GlcNAc) and UDP-*N*-acetylmuramyl-L-Ala-D-isoGlu-*meso*-DAP-D-Ala-D-Ala (UDP-MurNAc-pentapeptide), that serve as the low molecular weight precursors of peptidoglycan. The synthesis of these precursors is catalyzed by soluble cytoplasmic enzymes (Figure 6, reactions 1–9).
2 Conversion or transfer of the water-soluble nucleotides into lipid-soluble bactoprenyl (or undecaprenyl phosphate) derivatives with sub-

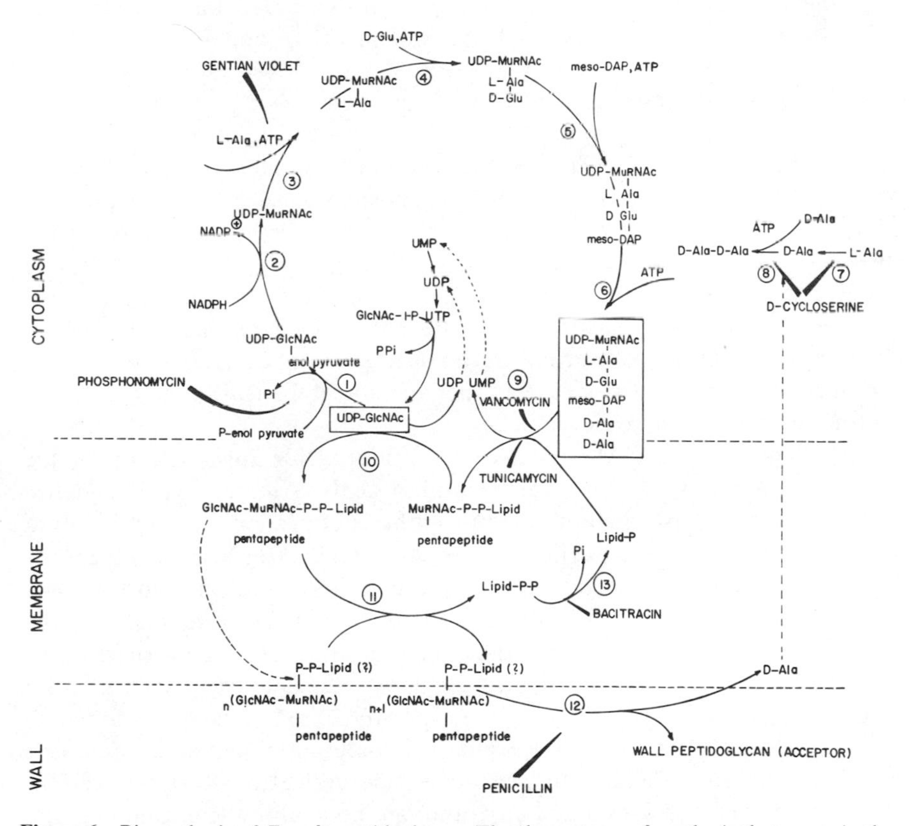

Figure 6 Biosynthesis of *E. coli* peptidoglycan. The three stages of synthesis that occur in the cytoplasm, in the membrane, and outside the membrane are separated by dashed lines. The sites at which various antibiotics have been shown to act are also indicated. For further details on the reactions see text.

sequent regeneration of the uridylic acid carriers. A second step in this stage is the polymerization of the activated derivatives by a transfer reaction to form linear peptidoglycan strands (Figure 6, reactions 10–12).

3 Transfer of polymerized subunits to acceptor sites outside the cytoplasmic membrane outer face and crosslinking of peptidoglycan by a transpeptidation reaction resulting in the insolubilization of the newly synthesized peptidoglycan (see also Figure 7).

In most peptidoglycans other than those of type I of gram-negative organisms, the peptide cross-linking reaction requires the incorporation of additional amino acid residues. The addition of these residues onto the principal pentapeptide takes place either during stage 1 or, more often, during stage 2. (For details on the mechanisms of these reactions in other types of peptidoglycans see Ghuysen and Shockman, 1973, and Ghuysen, 1977a.)

3.1 Stage 1: Synthesis of Precursors—The Nucleotide Cycle

The condensation of *N*-acetylglucosamine α-1-phosphate with UTP catalyzed by a uridyltransferase (or pyrophosphorylase) leads to the formation of UDP-*N*-acetylglucosamine. A specific transferase then catalyzes the transfer of the three-carbon unit, 2-phosphoenolpyruvate, to UDP-GlcNAc to give the 3-pyruvate-enol ether of the latter (Figure 6, reaction 1). This reaction is irreversibly inactivated by phosphonomycin (Kahan et al., 1974). Reduction of the pyruvyl group by an NADPH-linked reductase leads to the formation of the 3-*O*-D-lactyl ether of *N*-acetylglucosamine (*N*-acetylmuramic acid) (Figure 6, reaction 2).

Conversion of UDP-MurNAc to its pentapeptide form occurs by the sequential addition of the requisite amino acids. Amino acyl adenylates, which serve as intermediates in the biosynthesis of proteins, are not involved in the formation of the peptide sequence of UDP-MurNAc-pentapeptide. Synthesis of the pentapeptide is thus quite distinct from protein synthesis and the sequence of amino acids in the peptide is determined solely by the specificities of the synthetic enzymes. The enzymes are highly specific for both the amino acid and the nucleotide substrate. Each step requires ATP and a divalent cation (Mg^{2+} or Mn^{2+}). The *meso*-DAP-adding enzyme of *E. coli* fails to add L-lysine and the purified D-glutamic acid-adding enzyme does not add glutamic acid to any nucleotide derivative other than UDP-MurNAc-L-Ala, nor does it add any other amino acid to the nucleotide.

One exception in the synthesizing sequence is the addition of preformed dipeptide D-alanine-D-alanine to complete the pentapeptide structure. The synthesis of this dipeptide has aroused considerable interest because it is a unique procaryotic pathway for which a number of antagonists have been found. Three enzymes are involved in this synthesis: alanine racemase (Figure 6, reaction 7), D-alanyl-D-alanine synthetase (Figure 6, reaction 8), and the UDP-MurNAc-L-Ala-D-isoGlu-*meso*-DAP, that is, D-Ala-D-Ala ligase or adding enzyme (Figure 6, reaction 6). Details about the enzymatic properties of these enzymes have been reviewed by Neuhaus et al. (1972). In *E. coli* and *Bacillus subtilis*, both *meso*-DAP-containing bacteria, the alanine racemase is inhibited by both D- and L-cycloserine (Johnston et al., 1966; Neuhaus et al., 1972). The synthetase (Figure 6, reaction 8) is inhibited by D-cycloserine, D-Ala-D-Ala, and analogs of D-Ala-D-Ala (Neuhaus et al., 1972). Amino acid exchanges at the amino terminal of this dipeptide decrease its inhibitory activity, but peptides such as D-norvalyl-D-alanine cannot be incorporated by the adding enzyme to the nucleotide tripeptide. Modifications at the *C*-terminal can enhance the effectiveness of peptides as inhibitors of the synthetase (for example, D-alanyl-D-norvaline) but allow the incorporation of these to the tripeptide nucleotide by the ligase (Neuhaus et al., 1972).

3.2 Regulation of Precursor Synthesis

The amount of D-Ala-D-Ala synthesized by the cell seems to be strictly regulated. D-Alanine is a specific inducer of L-alanine dehydrogenase and hence limits the amount of L-alanine that is available to the racemase (Berberich et al., 1968) and in *E. coli* the synthesis of the racemase is also repressed by high concentrations of alanine (Neuhaus et al., 1972).

D-Cycloserine causes considerable accumulation of UDP-*N*-acetylmuramyl-tripeptide in *E. coli*, but penicillin, which inhibits the peptide cross-linking reactions of stage 3 of the biosynthesis, fails to cause the accumulation of UDP-*N*-acetylmuramyl-pentapeptide. These observations suggest that in *E. coli* UDP-*N*-acetylmuramyl-pentapeptide might regulate its own biosynthesis by feedback inhibition (Lugtenberg, 1972; Lugtenberg and Van Dam, 1972a, b; Lugtenberg et al., 1972). This phenomenon is certainly not general. *S. aureus* apparently lacks such a regulation, since in this organism penicillin causes a large accumulation of nucleotides. This accumulation was one of the key initial observations leading to our current knowledge on the mode of action of penicillin (Park, 1952).

Osmotically fragile, temperature-sensitive mutants of *E. coli* impaired at the levels of UDP-*N*-acetylglucosamine-enolpyruvate reductase (*murB*), of L-alanine adding enzyme (*murC*), of *meso*-DAP adding enzyme (*murE*) and of D-Ala-D-Ala adding enzyme (*murF*) were isolated (Matsuzawa et al., 1969; Lugtenberg, 1972; Lugtenberg and Van Dam, 1972a, b; Lugtenberg et al., 1972; Wijsman, 1972; Miyakawa et al., 1972). The *murC*, *murE*, and *murF* genes are located extremely close to each other (at 1–1.5 min of the chromosome map) and might form or be part of an operon. The *murB* gene is located at 77 min. For details on the genetic map location of other enzymes that metabolize peptidoglycan, see Table 1.

3.3 Stage 2: The Bactoprenyl Cycle

The undecaprenyl (lipid) phosphate glycosyl carrier is one of the key compounds in peptidoglycan synthesis as it transports the *N*-acetylmuramyl-pentapeptide and the *N*-acetylglucosamine precursors from the intracellular sites of synthesis through the cytoplasmic membrane to the exocellular sites of polymerization. The undecaprenyl phosphate has been isolated with a very high degree of purity, allowing its structure to be ascertained by mass spectrometry (Wright et al., 1967; Higashi et al., 1967, 1970). In order for it to be functional, the lipid carrier must be phosphorylated. The first reaction of the cycle is the translocation of phospho-*N*-acetylmuramyl-pentapeptide from UDP-MurNAc-pentapeptide to the undecaprenyl phosphate (Figure 6, reaction 9) with the generation of UMP. The translocase has a high specificity for the uracyl moiety and the *C*-terminal D-Ala-D-Ala sequence. Both forward and reverse reactions require Mg^{2+}. This reaction is inhibited by vancomycin (Neuhaus et al., 1972; Perkins and Nieto, 1972).

Tunicamycin, which bears some structural resemblance to the undecaprenyl-phosphate glycosyl carrier (Takatsuki et al., 1977), also prevents the formation of the bactoprenyl disaccharide peptide by blocking the transfer of *N*-acetylglucosamine (Bettinger and Young, 1975). Recent experiments, however, have shown that tunicamycin also inhibits the transfer of phospho-*N*-acetylmuramyl-pentapeptide (Tamura et al., 1976b).

The second step is the transglycosylation of *N*-acetylglucosamine from UDP-GlcNAc to undecaprenyl-P-P-*N*-acetylmuramyl-peptapeptide with the concomitant formation of the activated disaccharide-peptide, a mechanism that ultimately is responsible for the alternating sequence of GlcNAc and MurNAc in the glycan strands of the completed peptidoglycan (Ghuysen and Shockman, 1973). The transfer of disaccharide-pentapeptide units from their undecaprenyl pyrophosphate carrier to an appropriate acceptor occurs at the interface between membrane and the wall and generates bactoprenyl pyrophosphate (Figure 6, reaction 11). This in turn becomes dephosphorylated by a membrane-bound pyrophosphatase that yields P_i and bactoprenyl phosphate, which can begin a new cycle (Figure 6, reaction 13). Bacitracin inhibits the pyrophosphatase by complexing with the substrate (Siewert and Strominger, 1967; Storm, 1974) and this prevents the regeneration of the bactoprenyl phosphate carrier.

Table 1 Properties of Pencillin-Binding Proteins (PBPs) in *E. coli* Membranes

PBP	Mol. Wt.	[^{14}C] Penicillin Binding (%)	Genetic Map Location (min)	β-Lactam Affinity	Morphological Effect	Presumed Main Enzymatic Activity of PBP	Role in Peptidoglycan Expansion
IA	91,000	6	73.5	Cephalosporins			
IB's	91,000	6	3.3	Penicillins Cephalosporins	Lysis	Transpeptidation	Cell elongation
2	66,000	0.7	14.4 and 70	Mecillinam Thienamycin	Spherical cells		Cell shape
3	60,000	1.8	1.8	Cephalexin	Nonseptated filaments	Endopeptidase–DD-Carboxypeptidase	Septum formation
4	49,000	4.0	68	Benzylpenicillin		DD-Carboxypeptidase Ib	
5	42,000	65	13.7	Benzylpenicillin		DD-Carboxypeptidase Ia	
6	40,000	20	13.7	Benzylpenicillin		DD-Carboxypeptidase Ia	

3.4 Stage 3: Insolubilization of Nascent Peptidoglycan

It is crucial to our understanding of bacterial wall assembly to know whether monomeric lipid precursors are incorporated directly into the wall or are first incorporated into oligomeric intermediates, which are then transferred as units to the preformed sacculus. The transfer of disaccharide units from the bactoprenyl-P-P-carrier to a growing peptidoglycan chain can follow two pathways. One is the transglycosylation or polymerization mechanism that enables the elongation of a glycan chain and the other is the transpeptidation reaction that cross-links adjacent peptide side chains.

3.4.1 Glycan Elongation The *in vitro* synthesis of peptidoglycan from the two nucleotide precursors UDP-GlcNAc and UDP-MurNAc-pentapeptide, performed with particulate membrane enzyme fractions of *S. aureus* or *M. luteus*, leads to the incorporation of precursors into an acid-precipitable, chromatographically immobile, linear un-cross-linked peptidoglycan (Anderson et al., 1965, 1966). Ward and Perkins (1973), have shown that chains can grow and elongate up to 140 units long by sequential addition of the new disaccharide-peptide at the reducing terminal (i.e., *N*-acetylmuramic acid) of the growing chain (Figure 6, reaction 11). In this process, the reducing terminal end of the lengthening chain is transferred from its link with the membrane carrier to the nonreducing end (i.e., *N*-acetylglucosamine) of the new disaccharide-peptide unit, which itself is linked somehow to the membrane. The nature of the linkage to the membrane acceptor is not entirely clear. An oligomeric peptidoglycan intermediate isolated from *in vivo* experiments with a double auxotroph of *Bacillus megaterium* has been identified as the [disaccharide-pentapeptide]$_{12}$-pyrophosphoryl-undecaprenol (Cleveland and Gilvarg, 1976). The concept that the initial step after the transfer of the first disaccharide precursor is the emergence of an oligomeric nascent peptidoglycan strand exterior to the plasma membrane is supported also by electron micrographs of isolated protoplasts of *B. megaterium* (Fitz-James, 1974) and of *B. licheniformis* (Elliottet al., 1975a, b).
1975a, b).

When incubated in the presence of β-lactam antibiotics, cells of a number of gram-positive organisms, release un-cross-linked peptidoglycan into the medium (Mirelman et al., 1974b; Keglevic et al., 1974; Tynecka and Ward, 1975). These observations show that transglycosylation is an apparently independent process, and in cases where transpeptidation does not occur, either because of inhibition by β-lactam antibiotics or because of lack of a suitable acceptor, glycan elongation can continue by transglycosylation of disaccharide-peptide units and can result in chains of considerable lengths.

In contrast to gram-positive organisms, particulate membrane preparations of *E. coli* were found to catalyze the synthesis of cross-linked peptidoglycan, which was detected and defined as chromatographically immobile material (Izaki and Strominger, 1968; Izaki et al., 1966, 1968). In their studies with *E. coli* W7, Braun and co-workers (Braun, 1975) observed that the bactoprenyl-P-P-disaccharide peptide unit was the only detectable compound to which ^{3}H-labeled DAP incorporated apart from the peptidoglycan rigid layer; these results suggested that glycan extension in *E. coli* might proceed by direct insertion of newly synthesized disaccharide-peptide monomer units. Recent observations by Van Heijenoort et al. (1978) indicate that a linear un-cross-linked peptidoglycan can be obtained with *E. coli* membrane preparations by using bactoprenyl-P-P-disaccharide-peptide as substrate in the presence of penicillin G and 0.1% sodium deoxycholate. Furthermore, our studies with ether-treated cells of *E. coli* PAT 84 (see Section 3.1) show that an SDS-soluble peptidoglycan that is only partially cross-linked is obtained by heating the trichloroacetic acid (TCA) precipitates of newly synthesized peptidoglycan in 4% SDS (Mirelman et al., 1976). Our interpretation of these results is that TCA precipitates of newly synthesized peptidoglycan consist of both nascent peptidoglycan that has not yet been incorporated into the SDS-insoluble peptidoglycan sacculus (and which in our system accounted for 40–70% of the material) and nascent peptidoglycan that has became covalently linked to the preexisting sacculus (Figure 7). A soluble, chromatographically immobile peptidoglycan that is almost totally un-cross-linked was obtained recently by the present authors also from intact *E. coli* cells that incorporated [^{3}H]DAP (D. Mirelman and R. Bracha, manuscript in preparation). These results confirm other preliminary observations of Braun (1975).

The question of whether the SDS-soluble peptidoglycan in *E. coli* is a precursor or a turnover product of the sacculus has not yet been resolved and it is quite possible that a portion of the peptidoglycan of the envelope may be free and that not all of it is covalently linked to the sacculus matrix, similar to the case for the lipoprotein (Braun, 1975).

A larger amount of the SDS-soluble peptidoglycan was detected in [^{3}H]DAP incorporation experiments carried out with a lipoprotein-lacking mutant of *E. coli* (JE 5510) isolated by Hirota et al. (1977). Gel filtration and paper chromatography of the material revealed that it is a mixture of macromolecular, nondialyzable and chromatographically immobile compounds and of low molecular weight fragments, some of which migrate as oligomers of peptidoglycan and some as the monomers (Figure 1) (D. Mirelman and R. Bracha, manuscript in preparation).

Furthermore, an SDS-soluble peptidoglycan of similar composition was obtained recently in experiments carried out with ether-treated cells of *P. aeruginosa* X-48 (Mirelman and Nuchamowitz 1979a, b). Pulse-chase experi-

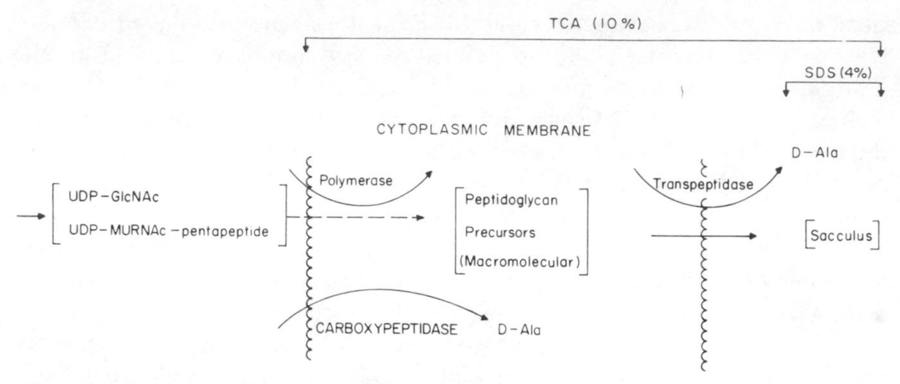

Figure 7 Schematic model of principal reactions that participate in the synthesis and attachment of newly synthesized peptidoglycan strands to preexisting ones in the *E. coli* cell wall. The sodium dodecyl sulfate insoluble material consists of newly synthesized peptidoglycan bound covalently to the preexisting cell wall. The macromolecular peptidoglycan intermediates remain in the cell upon treatment with TCA and are predominantly un-cross-linked in their peptide side chains. Reprinted with permission from D. Mirelman, Y. Yashouv-Gan, and U. Schwarz, *Biochemistry*, **15**, 5045–5053 (1976). Copyright by the American Chemical Society.

ments using unlabeled nucleotide precursors revealed that all the labeled SDS-soluble material could be chased into the SDS-insoluble sacculus within 20 min of incubation, suggesting that, in this organism, it may indeed be a macromolecular intermediate in the biosynthesis of peptidoglycan (Figures 7 and 8).

All these results indicate that the initial steps of peptidoglycan expansion in gram-negative organisms may not be very different from those found for gram-positive ones. In both cases an SDS-soluble, mostly un-cross-linked peptidoglycan strand is the intermediate that becomes subsequently inserted and covalently linked to the preformed, insoluble sacculus. In *E. coli* and *P. aeruginosa* this oligomer is already partially cross-linked at its peptide side chains; upon digestion of this material with lysozyme what appears to be peptidoglycan monomer and dimer fragments are released (Weidel and Pelzer, 1964) (Figures 1 and 2).

3.4.2 The Peptide-Cross-linking System The last and most important reaction in peptidoglycan biosynthesis is the formation of cross-links between the peptide moieties. Based on the fact that the nucleotide precursor of peptidoglycan, UDP-MurNAc-pentapeptide, contains two D-alanine residues at the carboxyl terminus of the peptide whereas thc subunit of peptidoglycan usually contains only a single D-alanine residue (Figures 1 and 6), it was originally postulated by Martin (1964) that the cross-linking reaction might involve transpeptidation between the D-alanyl-D-alanine terminus of one peptide unit and the free amino group of a second peptide with concomitant release of one free D-alanine residue (Figure 4). This hypothesis was particularly attractive in that it provided a mechanism for isoenergetic formation of the cross-linking peptide bond in the wall itself, external to the cytoplasmic membrane and independent of energy donors such as ATP. Studies of Wise and Park (1965), of Tipper and Strominger (1965), and of Strominger et al. (1967) with *S. aureus* provided support not only for the existence of such a transpeptidation reaction, but also for identification of this reaction as the penicillin-sensitive step in cell wall synthesis. Direct demonstration of transpeptidation *in vitro* and its sensitivity

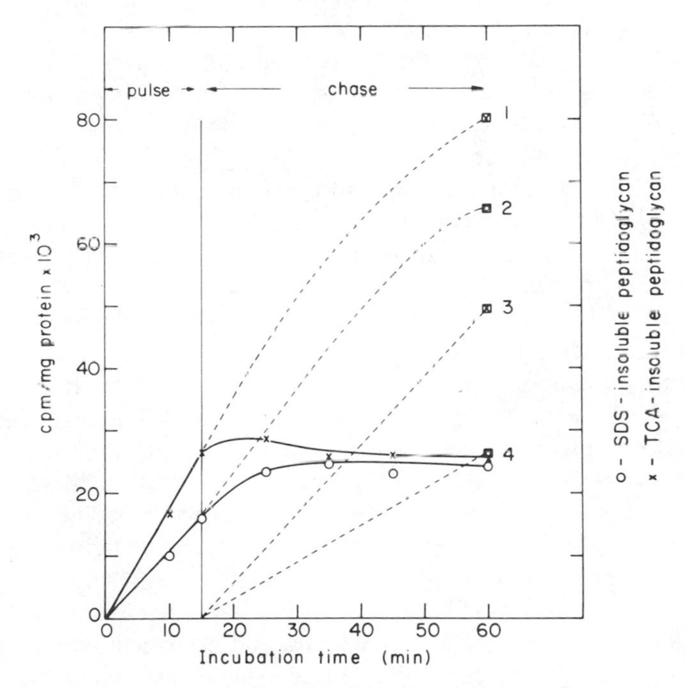

Figure 8 Pulse-chase experiment of peptidoglycan biosynthesis performed with ether-treated *Pseudomonas aeruginosa* X-48 cells. Cells were incubated for 15 min. with UDP[^{14}C]GlcNAc and UDP-MurNAc-pentapeptide as described (Mirelman et al., 1976) and the peptidoglycan synthesized was determined as TCA-insoluble material (×) or SDS-insoluble material (O). After 15 min of incubation, the cells were sedimented, the radioactive nucleotide precursor was removed, an excess of nonlabeled UDP-GlcNAc and UDP-MurNAc-pentapeptide was added, and the incubation was resumed. Aliquots were removed at predetermined times during the incubation and synthesis of peptidoglycan was determined as before. The difference between the SDS- and TCA-insoluble peptidoglycan almost completely disappeared after 20 min of chase incubation. Controls for peptidoglycan synthesis during the chase period (points *1*, *2*, *3*, *4*) indicate that the difference between SDS-and TCA-insoluble material is usually considerably large even after 60 min of incubation.

to penicillin was first achieved with particulate membrane preparations of *E. coli* (Izaki et al., 1966, 1968). Further investigations of the bridge closure reaction soon revealed that it involves a complex enzymatic system and this system varies considerably depending on the bacteria (Ghuysen, 1977a, b, c, 1979).

Fractionation of the cross-linking enzyme complex of *E. coli* revealed that at least three distinct enzymatic activities are involved in the formation of the bridge structure (Strominger, 1970; Ghuysen, 1977b, 1979) (see also Figure 4): (1) the transpeptidase, whose main function is to catalyze the formation of cross-linkages between two peptide units, (2) the DD-carboxypeptidase that hydrolyzes the *C*-terminal D-Ala-D-Ala peptide bond of pentapeptide units, and (3) the endopeptidase, which hydrolyzes the cross-linked dimers, formed by transpeptidation, into monomers. The endopeptidase has been shown to cause hydrolysis of *C*-terminal D-alanine peptide bonds and is therefore assumed to be also a DD-carboxypeptidase.

The complexity of the *E. coli* enzyme system and the difficulties encountered in studying in detail the different enzymes and their kinetic and molecular properties has prompted the search for simpler model systems, where the activities as well as their mode of interaction with β-lactam antibiotics could be studied in detail. Based on this concept, the group led by J. M. Ghuysen in Liège has for the past decade made an outstanding contribution to our understanding of the mechanism of the transpeptidase/DD-carboxypeptidase reaction at the molecular level. Most of their numerous findings have been thoroughly reviewed (Ghuysen 1977a, b, c, 1979) and only a brief synopsis of their work is presented here. Ghuysen and his group found that enzymes secreted during growth of some actinomycetes, in particular *Streptomyces* R61 and *Actinomadura* R39, are able to catalyze *in vitro* both DD-carboxypeptidase and transpeptidase activities. Whenever the enzymes perform transpeptidation, concomitant hydrolysis of the peptide donor also occurs (Figure 9). *C*-Terminal D-Ala is thus released by two pathways, namely straight hydrolysis and transpeptidation, the two processes competing with each other. In aqueous media and in the presence of low concentrations of a peptide acceptor, the enzymes act mostly as DD-carboxypeptidases and their transfer activities are low. However, changing of the environmental conditions by adding solvents of low polarity and higher pH, as well as higher acceptor concentrations, causes a considerable increase in the transpeptidation reaction. In addition to the soluble enzymes, Ghuysen and his group have also isolated several membrane-bound enzymes that act mainly as transpeptidases (Dusart et al., 1975). The kinetics for the membrane-catalyzed transpeptidation between $(acetyl)_2$-L-Lys-D-Ala-D-Ala and Gly-Gly fit the general equation for an enzyme catalyzed bimolecular reaction:

$$E + S \xrightarrow{k} ES \xrightarrow{k_2} E - P \xrightarrow{k_3} E + P.$$

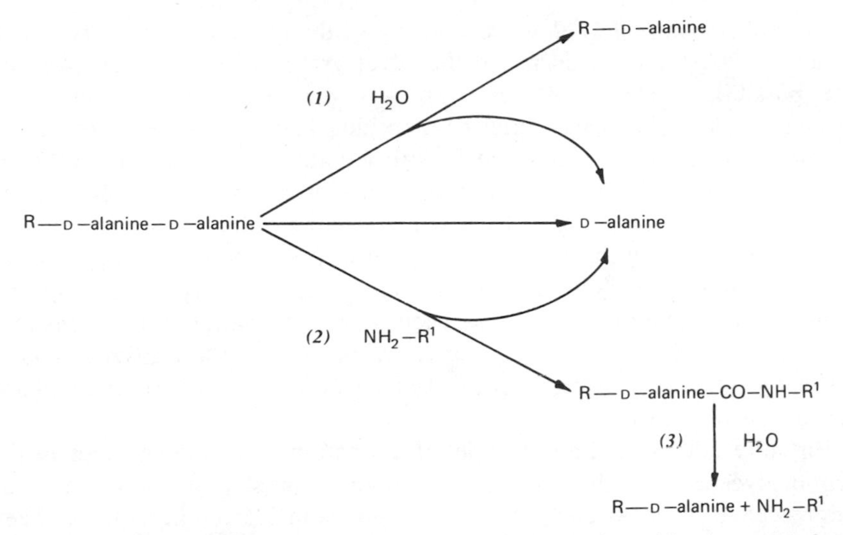

Figure 9 Schematic representation of the enzyme activities performed by the peptidoglycan cross-linking system of *E. coli* as described by Ghuysen (1977a): (*1*) DD-carboxypeptidase activity, (*2*) DD-transpeptidase activity, (*3*) DD-endopeptidase activity. In the natural substrate, R represents *N*-acetylmuramyl-L-Ala-D-Glu-*meso*-DAP, and R′ is the α-carbon linked to the amino group of DAP (D asymetric center) in *N*-acetylmuramyl-L-Ala-D-Glu-*meso*-DAP-D-Ala.

Evidence for acyl–enzyme intermediates of DD-carboxypeptidases from *B. subtilis* and *B. stearothermophilus* have been obtained recently (Nishino et al., 1977; Rasmussen and Strominger, 1978; see also Section 4.1).

There are reports in the literature where enzymes are designated as transpeptidases on the basis of their ability to replace the *C*-terminal D-Ala residue of a peptide donor by glycine or a single D-amino acid residue. Such an exchange does not yet prove the occurrence in the enzyme of an "acceptor site" concerned with peptide dimerization (by transpeptidation) and caution should be taken in designating activities. A number of DD-carboxypeptidases from bacilli (Wickus and Strominger, 1972; Nishino et al., 1977; Chase et al., 1977), *Streptococcus faecalis* (Coyette et al., 1977), *S. aureus* (Kozarich and Strominger, 1978), and *Proteus mirabilis* L-forms (Martin et al., 1976) are unable to catalyze natural model transpeptidation reactions, and their transfer capabilities are limited to the artificial transfer of single amino acid residues.

3.5 Cellular Site of Synthesis and Assembly of Peptidoglycan

In their interesting autoradiographic studies, Ryter et al. (1973) and Schwarz et al. (1975) have shown that pulses for one-eighth of a generation time of very high specific activity ^{3}H-labeled DAP incorporates into the equatorial region of *E. coli* cells. Chasing of the radioactive precursor with unlabeled DAP quickly dispersed the silver grains all over the cell, raising the possibility of the existence of a distribution or randomization mechanism for the newly inserted peptidoglycan. Because of the great experimental difficulties involved, it still cannot be ascertained with complete confidence whether nascent peptidoglycan is first exclusively inserted into the central growth zone of the cell. Since the whole process of cell wall growth requires post-insertion modifications, it is quite likely that peptidoglycan fragments are displaced from their original incorporation site and reinserted into other locations of the cell by a combination of reactions that are catalyzed by specific hydrolytic enzymes such as the transglycosylase, the endopeptidase (Hartmann et al., 1974; Höltje et al., 1975), and perhaps also a transpeptidase.

Virtually nothing is known about the functioning of the enzymes in the peptidoglycan assembly centers. New and promising methods, such as fluorescence spectroscopy and electron spin resonance, have been introduced recently for the study of effects of microenvironment on the synthesis of peptidoglycan (Johnston and Neuhaus, 1975, 1977; Weppner and Neuhaus, 1977, 1978). The quenching efficiencies of a series of nitroxyl stearate derivatives that differ with respect to the depth of the nitroxide within the membrane suggest that the dansyl moiety attached to the ϵ-amino group of the *N*-acetylmuramyl-pentapeptide precursor molecule is close to the surface of the membrane and within 4–6 Å of the lipid matrix.

4 THE EXPANSION AND GROWTH OF THE PEPTIDOGLYCAN

4.1 Role of the Enzymes of the Cross-linkage System

As mentioned above, incubations of particulate membrane preparations of gram-positive organisms with nucleotide precursors led to hardly any transpeptidation. Only the introduction of a novel system consisting of crude cell wall preparations as the enzyme source for peptidoglycan biosynthesis enabled the study of transpeptidation and cross-linkage reactions in gram-positive organisms (Mirelman and Sharon, 1972; Mirelman et al., 1972; Ward and Perkins, 1974a, b,; Hammes and Kandler, 1976). These crude cell wall preparations, which contained strongly adhered membrane fragments, were found not only to catalyze the polymerization of the sugar nucleotide precursors, but also to attach the newly synthesized peptidoglycan to the preexisting one in the cell wall. Such an attachment apparently occurs because the enzymes involved, which are membrane bound, are intimately associated with the growing region of the preexisting peptidoglycan.

Artificial mixtures of cytoplasmic membranes and exogenously added denatured cell walls, that have been prepurified with detergents of any membrane contaminants, do not transfer nascent peptidoglycan to the preformed wall (D. Mirelman, unpublished results).

The crude cell wall systems developed for gram-positive organisms and the ether-treated cells developed later for gram-negative organisms (see below) enabled us for the first time to study the mechanism of wall peptidoglycan expansion under *in vitro* conditions that closely resembled those existing in the intact cell. Studies with *S. aureus* (Mirelman et al., 1972), *M. luteus* (Mirelman et al., 1972, 1974a; Weston et al., 1977), *B. licheniformis* (Ward and Perkins, 1974a, b), *E. coli* (Mirelman et al., 1976, 1977, 1978) and *P. aeruginosa* (Mirelman and Nuchamowitz, 1979a, b) have revealed that low concentrations of penicillin that were similar to the minimal concentrations that inhibit growth of the organism not only inhibited the transpeptidation and bridge closure *in vitro*, but also markedly inhibited the incorporation of newly synthesized peptidoglycan into the preexisting wall. We postulated therefore that the transpeptidation reaction leads to insolubilization of nascent peptidoglycan strands by covalently linking them to preexisting wall peptidoglycan acceptors (Figure 10).

4.2 Direction of Nascent Peptidoglycan Insertion

In *B. licheniformis* the direction of insertion appears to be such that the penultimate *C*-terminal D-alanine of the newly synthesized pentapeptide unit acts as a donor and becomes bound to a free amino group acceptor on the existing peptidoglycan. This was established by using, as a nucleotide precursor, a UDP-MurNAc-pentapeptide in which the free amino group of *meso*-DAP (i.e., the potential acceptor group for transpeptidation) had been blocked by a [^{14}C]acetyl group (Ward and Perkins, 1974a). The [^{14}C]acetylated nascent peptidoglycan was incorporated into the preexisting wall, and labeled peptidoglycan dimer fragments were obtained by lysozyme digestion. This demonstrated that the preexisting peptidoglycan can act as acceptor and the nascent peptidoglycan can act as donor. Quantitation of attachment, however, showed that incorporation of the acetylated nascent peptidoglycan amounted to only about 23% of that of the nonacetylated nascent peptidoglycan. It was suggested that incorporation was reduced because acetylation of the free amino group prevents attachment of additional nascent peptidoglycan strands to acceptors on the newly inserted peptidoglycan.

Transpeptidation in *Gaffkya hommari* has been found to proceed in the opposite direction, that is, the nascent peptidoglycan strands serve as the acceptor and the preexisting peptidoglycan of the wall serves as the donor.

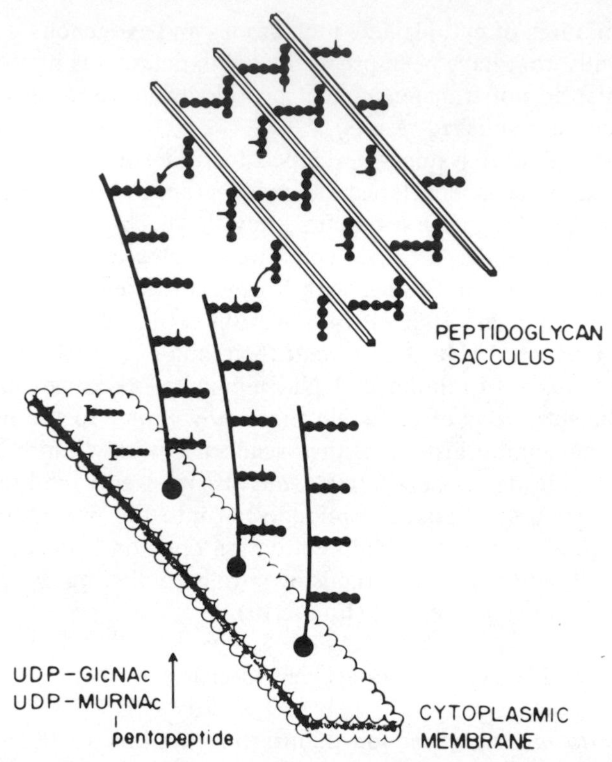

Figure 10 Cross linking of nascent peptidoglycan to the cell wall. UDP-GlcNAc and UDP-MurNAc-pentapeptide are joined to a membrane-bound carrier and are transported to the outer surface of the cytoplasmic membrane where they form linear, mostly non-cross-linked, nascent peptidoglycan strands. Although some cross-linking occurs between the nascent chains, the insolubilization of peptidoglycan is achieved by cross-linking between the peptide side chains of nascent and preexisting glycan chains. In *E. coli*, cross-linking is between the *meso*-DAP of preexisting peptidoglycan and the penultimate D-Ala of the nascent glycan with the release of the terminal D-Ala (see also Figure 4). Those terminal D-Ala residues that are not released during cross-linking are removed at some stage of peptidoglycan synthesis by the enzyme D-alanine carboxypeptidase.

= GlcNAc
|
MurNAc—L—Ala—D—Glu—meso—DAP—D—Ala—D—Ala

UDP-MurNAc-peptides that can function only as acceptors and not as donors were used in the detailed studies of Hammes and Kandler (1976). UDP-MurNAc-L-Ala-D-isoGlu-L-Lys-D-Ala (tetrapeptide) obtained as the product of the action of a DD-carboxypeptidase on the pentapeptide substrate was effectively utilized by membrane preparations of *G. hommari* for the synthesis of soluble, non-cross-linked peptidoglycan (Hammes and Neuhaus, 1974). Crude wall preparations of the same organism (Hammes and Kandler, 1976) incorporated the tetrapeptide substrate into the wall by a transpeptidation reaction in which D-alanyl-D-alanine sequences present in the preexisting cell wall functioned as donors and the ϵ-amino groups of the lysine residues in the newly synthesized peptidoglycan strands functioned as acceptors. Substitution of the ϵ-amino group of lysine with an acetyl group, in either the UDP-MurNAc-pentapeptide or UDP-MurNAc-tetrapeptide substrates, almost completely blocked the incorporation into the preformed cell wall. Based on these and other results the authors proposed that in *G. hommari* a "reverse" mechanism operates and the normal sequence of events is such that a penicillin-sensitive DD-carboxypeptidase first removes the *C*-terminal D-alanine, converting the substrate into a tetrapeptide that is then used as an acceptor for transpeptidation in a penicillin-insensitive reaction (Figure 11). Thus the DD-carboxypeptidase in *G. hommari* (see also paragraph below) appears to be essential in cell wall synthesis in that it controls the ratio between pentapeptide and tetrapeptide units in the newly synthesized peptidoglycan and thus the extent of incorporation and peptide cross-linkages in the completed peptidoglycan.

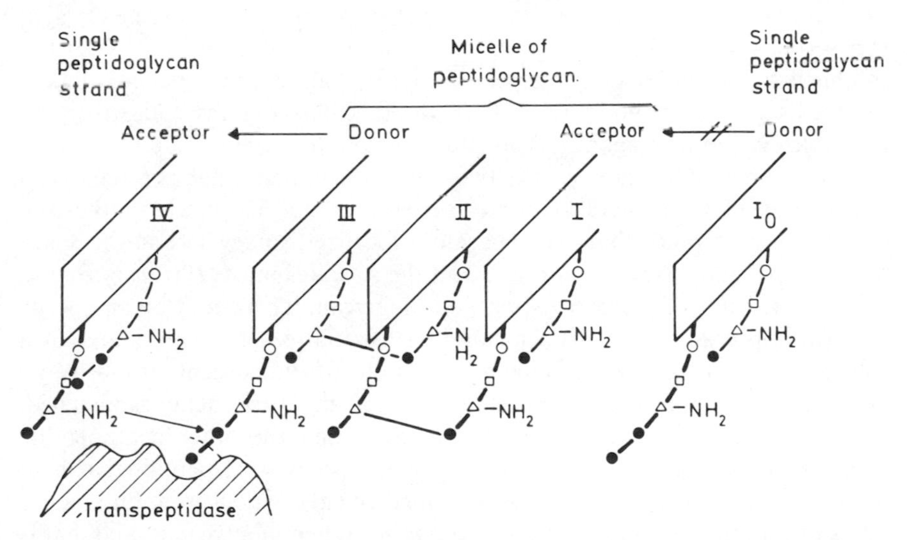

Figure 11 Proposed mechanism of incorporation of newly synthesized peptidoglycan into the preformed *G. hommari* cell wall peptidoglycan by means of transpeptidation. Peptidoglycan strands I, II, and III represent the cell wall peptidoglycan; I_0 and IV represent the newly synthesized peptidoglycan. The amino acids of peptide subunits are: (○) L-alanine, (□) D-glutamic acid, (△) L-lysine, (●) D-alanine. Reprinted from Hammes and Kandler (1976) with permission of the Federation of European Biochemical Societies.

4.3 Cross-linkage of Peptidoglycan in *E. coli*

As mentioned above, the synthesis of cross-linked peptidoglycan and all the reactions involved in the bridge formation of the *E. coli* cell wall were identified and studied to a considerable extent with a particulate membrane preparation (Izaki et al., 1968), as well as with isolated enzymes and model compounds (Nguyen-Distèche et al., 1974a, b; Pollock et al., 1974). However, the peptidoglycan obtained by the *E. coli* particulate membrane preparations was synthesized under conditions in which the natural controls of the enzymatic reactions were partially lost and the product formed was not identical to that made by intact cells (Izaki et al., 1968). Furthermore, the enzymes that were solubilized and isolated from *E. coli* membranes and studied in the artificial DD-carboxypeptidase–transpeptidase system are now believed not to be physiologically important and their activities are apparently not relevant to the biosynthesis of the peptidoglycan (Nguyen-Distèche et al., 1974a; Suzuki et al., 1978).

In a novel system that was recently introduced (Mirelman et al., 1976), *E. coli* cells, made permeable to nucleotide precursors by ether treatment, catalyze peptidoglycan synthesis and its subsequent attachment to the preexisting peptidoglycan of the sacculus by a mechanism that is similar to the mechanism found in intact cells. Approximately half of the newly synthesized peptidoglycan obtained was material that was insoluble in TCA but soluble in hot 4% SDS. Only the peptidoglycan that became covalently linked to preexisting wall peptidoglycan (i.e., the sacculus, Figures 7 and 10), was insoluble in both reagents. The transfer of the SDS-soluble, nascent oligomeric peptidoglycan onto a two- or three-dimensional peptidoglycan matrix most likely involves a complex series of events. The newly synthesized peptidoglycan strands that are apparently localized somewhere in the space between the cytoplasmic membrane and the sacculus must be transported to the close vicinity of the preexisting peptidoglycan. This could be an energy requiring process or a hydrogen bonding association that culminates in a covalent point attachment ("anchor" reaction) of the nascent strand to the preexisting one by transpeptidation. As soon as this first anchor is fastened, the new strand is no longer soluble in SDS and the other pentapeptide residues of the chain gradually approach the preexisting peptidoglycan to produce additional cross-links without further energy consumption (Figure 10). In addition, it is quite possible that a nascent peptidoglycan strand that is already anchored (or cross-linked) to the preexisting sacculus can continue at the same time to elongate its glycan chain by the translocation of activated disaccharide-pentapeptide units at its reducing end. It is possible that two transpeptidases are involved in the process of peptidoglycan attachment. The first one begins cross-linking the nascent peptidoglycan strands, perhaps at the stage of their chain elongation within the membrane and this step is apparently not very sensitive to most β-lactam antibiotics. A second transpeptidase may be involved in the "anchoring" reaction between the nascent peptidoglycan and the preexisting one of the sacculus. Apparently this reaction is very sensitive to a large number of β-lactam antibiotics.

The presence of two types of transpeptidases has not yet been found in *E. coli* but has been observed by Ghuysen and his co-workers in *Streptomyces* R_{61}. The first transpeptidase is an enzyme that is strongly associated with the membrane, is quite difficult to extract, and is almost insensitive to β-lactams. The second transpeptidase, which is only weakly associated to the membrane, is very sensitive to β-lactams (Leyh-Bouille et al., 1977; Dusart et al., 1977).

The degree of peptidoglycan cross-linkages obtained with the *in vitro* system in *E. coli* was very similar to those values found for intact cells (Weidel and Pelzer, 1964; Mirelman et al., 1976, 1978). Covalent attachment of new peptidoglycan to the sacculus as well as the release of labeled *C*-terminal D-alanine residues by both the transpeptidase and DD-carboxypeptidase was completely inhibited by high concentrations of ampicillin (50 μg/ml). Low doses of ampicillin (0.5 μg/ml) that were known to inhibit the DD-carboxypeptidase (Nguyen-Distèche et al., 1974b) caused a small stimulation of incorporation and of peptide cross-linkage formation without affecting transpeptidation. Of great interest was the finding that the amount of pentapeptide units in the SDS-soluble peptidoglycan intermediate was under the control of a DD-carboxypeptidase and this enzymatic activity actually determines the degree of peptide cross-linkage, as well as the amount of newly synthesized peptidoglycan that will incorporate (and become insolubilized) into the sacculus (Mirelman et al., 1976, 1977, 1978).

4.4 Peptide Cross-linkage and Septum Formation

The division of the rod-shaped *E. coli* cell involves an ingrowth of the cytoplasmic membrane and peptidoglycan layers to form the septum, which subsequently become the new polar caps of the daughter cells (Rogers, 1970; Slater and Schaechter, 1974; Burdett and Murray, 1974b). For this process to occur, a considerable number of synthetic and hydrolytic reactions that modify the morphology of the peptidoglycan layer must take place with very precise topological regulation. Cell division of *E. coli* normally occurs only after completion of a round of DNA replication (Zusman et al., 1972; Pardee et al., 1973) and has been shown to be inhibited by a great variety of chemical and physical methods (Slater and Schaechter, 1974). A protein X of 40,000 daltons has been shown to be produced upon inhibition of DNA replication by several drugs, including nalidixic acid. This protein, which is the product of the *recA* gene (McEntee et al., 1976 and Gudas and Mount, 1977) binds to single stranded DNA regions and this prevents their subsequent attachment to specific sites on the membrane. Satta and Pardee (1978) propose, therefore, that protein X may function as an inhibitor of septation.

Low concentrations of β-lactams such as ampicillin (0.2–1.0 μg/ml) or cephalexin (1–50 μg/ml) are known to elicit a typical morphological response, that is, inhibition of cell septation and cell division and the subsequent growth of long, nonseptated filamentous cells (Greenwood and O'Grady, 1973a, b). Furthermore, a number of thermosensitive division mutants that are incapable of synthesizing the septum region at the nonpermissive temperature have been isolated (Hirota et al., 1968; Spratt, 1975, 1977). We have studied the synthesis and extent of peptidoglycan cross-linkage in two thermosensitive division mutants, *E. coli* PAT 84 and SP-63, as well as in two types of *E. coli* cells that were induced to grow as nonseptated filaments by adding to their growth medium either cephalexin (25 μg/ml) or nalidixic acid (10 μg/ml). All four forms of filamentous cells were found to have higher levels of peptide side chain cross-linkages in their newly synthesized peptidoglycan as compared with those found in normally dividing rodlike cells of the same strain (Table 2). Furthermore, in contrast to normal rod-shaped cells, all filaments had a lower level of DD-carboxypeptidase activity as compared to that of the transpeptidase.

To find out whether the levels of DD-carboxypeptidase and transpeptidase of *E. coli* are uniform during a cell division cycle, peptidoglycan biosynthesis was studied in various synchronously growing cells at different stages (Mirelman et al., 1978). Both the extent of peptide side chain cross-linkage and the ratios between the transpeptidase and DD-carboxypeptidase activity were found to fluctuate during a cycle. Maximal transpeptidation and cross-linking were observed in newly divided cells, whereas maximal DD-carboxypeptidase activity and lower degrees of cross-linkage were detected in the nascent peptidoglycan of cells before division, at a time when cells form their septum. This new evidence gives further support to our theory that a dynamic balance between the relative activities of transpeptidase and DD-carboxypeptidase is the important factor in septum formation and synthesis of cell polar caps. A model showing the possible regulatory role of DD-carboxypeptidase in peptidoglycan expansion during growth and division of the cell is shown in Figure 12.

According to this model, the *E. coli* nascent peptidoglycan with its intact pentapeptides can act as the carboxyl donor for transpeptidation reactions with two types of acceptors. In the first pathway they can react with free ϵ-amino groups that are abundantly present on the preexisting peptidoglycan of the lateral wall (sacculus) and this will lead to extension and elongation of the

Table 2 Enzymatic Activities of DD-Carboxypeptidase and Transpeptidase as Calculated From the Release of *C*-Terminal [^{14}C]Alanine by Ether-Treated Rodlike or Filamentous Cells Incubated with UDP-MurNAc-L-Ala-D Glu-*meso*-DAP-D-Ala[^{14}C-D]Ala. The Degree of Peptide Side Chain Cross Linkages in the Newly Synthesized Peptidoglycan of the Different Cells is Also Shown[a]

E. coli Strain	Growth Conditions[b]	Cell Shape	Transpeptidase Activity[c]	DD-Carboxypeptidase Activity[d]		Transpeptidase/ Carboxypeptidase Ratio	Ratio of[e] Peptide Side Chain Cross Linkages $\frac{C_3 + C_4}{C_3+C_4/C_5+C_6}$
				Ether-Treated Cells	Sonicated Cells		
PAT 84	30°C	Normal rods	260	380	910	0.70	1.01
PAT 84	30°C + cephalexin (25 μg/ml), 60 min	Filaments	320	205	290	1.50	1.23
PAT 84	30°C + nalidixic acid (10 μg/ml), 120 min	Filaments	240	210	690	1.15	1.45
PAT 84	42°C, 60 min	Filaments	290	210	780	1.40	1.39
SP-63	30°C	Normal rods	310	420		0.74	0.93
SP-63	42°C, 60 min	Filaments	340	205		1.65	1.28

[a] The thermosensitive division mutants used were *E. coli* PAT 84 (Hirota et al., 1968) and *E. coli* SP-63 (Spratt, 1975).

[b] For details on experimental conditions see Mirelman et al. (1977). Results are given in picomoles of [^{14}C]alanine released per milligram of cell protein.

[c] Transpeptidase activity was calculated from the difference in amounts of [^{14}C]alanine released by reactions carried out in the presence or absence of UDP-GlcNAc and 0.5 μg of ampicillin per milliliter.

[d] The ampicillin-sensitive D-alanine-carboxypeptidase activity was determined from the difference in amounts of [^{14}C]alanine released by reactions carried out with UDP-GlcNAc in the absence or presence of ampicillin (0.5 μg/ml). A control was also done with the same ether-treated cells which were sonically disrupted before incubation for 1 min at maximal amplitude.

[e] Ratios of cross-linked ($C_3 + C_4$) to un-cross-linked ($C_5 + C_6$) peptidoglycan fragments were obtained after chromatography of lysozyme digests. The amounts of cross-linked and un-cross-linked fragments in the digests were computed from the amount of radioactivity in each of the respective peptidoglycan fragments. Differences of less than 15% were obtained in two independent experiments.

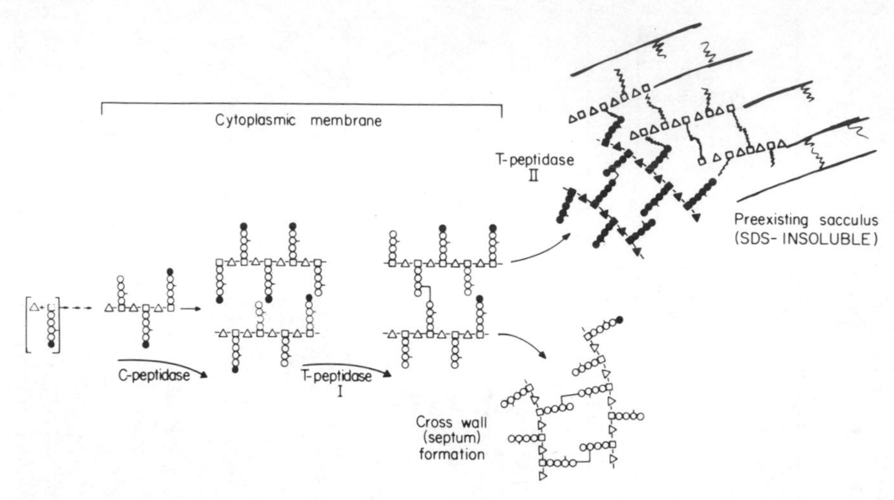

Figure 12 Schematic representation of the regulatory role of the D-alanine carboxypeptidase (C-peptidase) activity in the incorporation and attachment of newly synthesized strands by transpeptidation (T-peptidase II) to the preexisting peptidoglycan in *E. Coli* cell walls. Crosslinkage of nascent peptidoglycan strands begins in the cytoplasmic membrane and is catalyzed by T-peptidase I. Peptidoglycan strands from which most of their *C*-terminal D-alanine moieties have been cleaved off by a D-alanine carboxypeptidase are converted into acceptors to which donor peptides can cross-link. Formation of the new cross wall is then achieved by transpeptidation between the newly synthesized donor and acceptor strands. The acceptor strands apparently begin accumulating before cell division at their site of synthesis due to an increase in the activity of the DD-carboxypeptidase (Mirelman et al (1977).

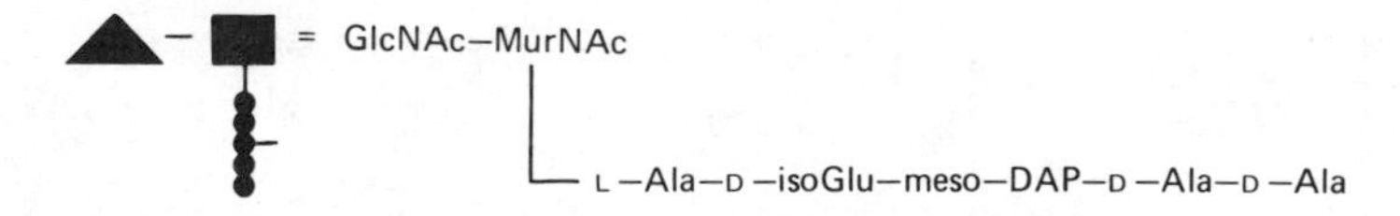

wall. In the second pathway, the intact pentapeptide donors can react with free ε-amino groups that are present on the nascent peptidoglycan strands that were previously deprived of their own D-alanine carboxyl donor residues by the action of a DD-carboxypeptidase. The removal of *C*-terminal D-alanine residues by the DD-carboxypeptidase, especially before cell division (Mirelman et al., 1978), leads to the accumulation at the equator of the cell (i.e., at their site of synthesis) of an increasing amount of "acceptor" substrate. Nascent "donor" substrate then has the option of transpeptidating either to the nascent or to the preexisting acceptor and this probably depends on the relative local concentration of each of the acceptor substrates. At the time of cell division there is apparently more attachment to the nascent acceptor and this leads to the septum synthesis and formation of a new wall across the cell. The specific inhibition (by cephalexin) or the lack of DD-carboxypeptidase activity (in the thermosensitive mutants) apparently upsets the delicate balance between donor and acceptor substrates and the net result is that more donors are available for transpeptidation. These donors therefore continue to become attached to acceptors on the preexisting peptidoglycan of the lateral wall, causing it to elongate without synthesizing the septum region.

A major difficulty in the interpretation of these findings, especially with regard to the specific role prescribed for the DD-carboxypeptidase in septum formation and cell division, is due to our lack of knowledge about the real identity of the DD-carboxypeptidase(s) that operate(s) in our *in vitro* systems. As is mentioned earlier, the enzyme(s) can operate with at least two different specificities, as a DD-carboxypeptidase and as an endopeptidase (peptide cross-link splitting enzyme). Unfortunately, we have been unable to determine the main type of enzymatic activity that functions in the intact cell. If the enzyme, the activity of which we determine *in vitro* from the release of *C*-terminal D-alanine, acts in the intact cell mainly as an endopeptidase, it may function under topological control at the cell equators to hydrolyze peptide cross-linkages for the morphological conversion of the wall cylinder into two hemispherical ends. The liberated ϵ-amino DAP groups could then serve as new acceptors for nascent peptidoglycan donors and this could lead to the formation of the septum. Inhibition of such a topological enzymatic process could interfere with the cell's ability to divide and cause the growth of nonseptated filaments.

A number of *E. coli* mutants that lack most of their D-alanine carboxypeptidase I activity (Tamura *et al*., 1976a) yet grow and divide as normal cells have recently been isolated (Matsuhashi et al., 1977, 1978a; Iwaya and Strominger, 1977; Suzuki et al., 1978). The finding of these mutants implies that the activities of DD-carboxypeptidase I do not play a significant physiological role in regulating peptidoglycan insertion. The discrepancy between these observations and the role ascribed by us for a DD-carboxypeptidase activity in septum formation (Mirelman et al., 1976, 1977) may be explained by the different *in vitro* systems used for determining the enzymatic activities. Whereas in the ether-treated cell system, only a limited activity of DD-carboxypeptidase can be detected (Mirelman et al., 1976), the mechanically disrupted membrane preparations always have very high levels of DD-carboxypeptidase I activity (Matsuhashi et al., 1977, 1978a, b). Furthermore, the mutants were not completely devoid of DD-carboxypeptidase and contained 10–20% of the activity present in wild-type *E. coli* cells. Thus it may be that in the ether-treated cells, as apparently happens in intact cells also, most of the potential DD-carboxypeptidase activity of the cell (arising mainly from DD-carboxypeptidases Ia and Ib) is not expressed and the release of *C*-terminal D-alanine that is detected in this system is due to another DD-carboxypeptidase or endopeptidase. This enzyme could be the one that catalyzes the low level of DD-carboxypeptidase activity still detected in the mutants that are lacking DD-carboxypeptidase Ia and Ib (Matsuhashi et al., 1977, 1978a; Suzuki et al., 1978).

Another possibility that should be borne in mind is that the DD-carboxypeptidase–transpeptidase–endopeptidase enzymes of the *E. coli* cross-linking system may have two different active forms, one of them sensitive to low levels of ampicillin (and perhaps involved in cell septation) and the other sensitive to only much higher levels of ampicillin and perhaps involved in cell elongation (Pollock et al., 1974; Nguyen-Distèche et al., 1974a, b). Preferential inhibition (at low levels of ampicillin) of one set of enzymes could lead to filamentation.

The specific inhibition of DD-carboxypeptidase activity in *B. subtilis* and *B. stearothermophilus* by low concentrations of penicillin does not cause detectable damage to the cell (Blumberg and Strominger, 1971; Yocum et al., 1974), suggesting that in these organisms the transpeptidase is the physiologically important enzyme.

In contrast to these observations, with *G. hommari* the specific inhibition of the DD-carboxypeptidase (by 0.25 μg/ml penicillin) is sufficient to prevent incorporation of nascent peptidoglycan and to cause cessation of cell growth (Hammes and Kandler, 1976; Hammes, 1976). Thus, in some organisms, the role of the DD-carboxypeptidase (or its activity in relation to that of the transpeptidase) may be crucial for the correct insertion of nascent peptidoglycan and for the expansion of the sacculus.

5 INTERACTION BETWEEN β-LACTAMS AND ENZYMES OF THE CROSS-LINKAGE SYSTEM

The interaction between the isolated membrane transpeptidase and β-lactams fits the general scheme (Frère et al., 1974; 1975a)

$$E + I \underset{}{\overset{k}{\rightleftharpoons}} EI \xrightarrow{k_3} EI^* \xrightarrow{k_4} E + \text{degradation products}$$

Enzyme and β-lactam (inhibitor) first react to form a complex (EI), which in turn isomerizes into a stable complex (EI*). This complex has been shown to break down under a variety of conditions and its stability depends very much on the nature of the β-lactam and enzyme used. During breakdown, the enzyme is reactivated and recovers its ability to react with β-lactams. With some enzymes, penicillin is released in the form of penicilloic acid, whereas in other cases, other breakdown products of penicillin have been found (Figure 13) (Frère et al., 1975a, b, 1976a; Hammerström and

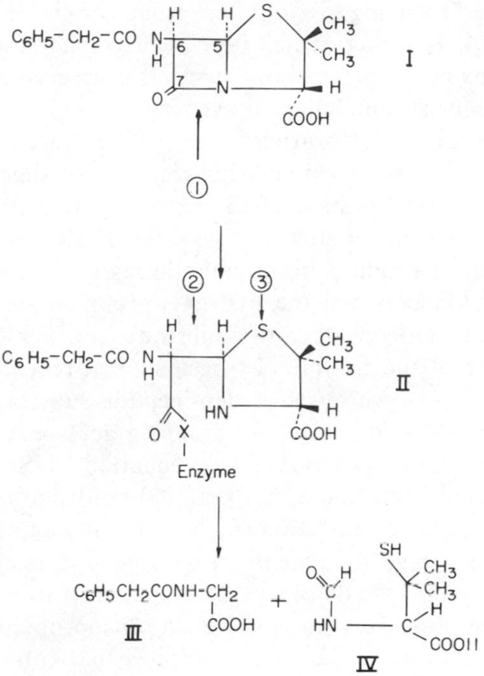

Figure 13 Hypothetical mechanism for the fragmentation of benzylpenicillin catalyzed by the DD-carboxypeptidase. (*I*) Benzylpenicillin; (*II*) benzylpenicilloic acid; (*III*) phenyl acetyl glycine; (*IV*) *N*-formyl-D-penicillanine. For further details on this reaction see Ghuysen (1978) and Hammarström and Strominger (1975).

Strominger, 1975). The β-lactams are therefore, in a sense, substrates for the enzymes and only by immobilizing the inactive complexes (EI*), at least for some time, do they behave also as inhibitors. The finding that some penicillin-sensitive enzymes have a weak β-lactamase activity is of particular interest in view of the frequently expressed suggestion that β-lactamases may have evolved from the penicillin-sensitive cross-linkage enzymes (Ghuysen, 1977a, b, c).

5.1 The Penicilloyl–Enzyme Linkage

According to the well-known model of penicillin as a structural analogue of D-Ala-D-Ala (Tipper and Strominger, 1965), it is the active CO—N bond of the β-lactam ring that reacts with the enzyme. Early work showing that bound penicillin could be removed from a purified D-alanine carboxypeptidase of *B. subtilis* with hydroxylamine or thiol derivatives suggested that the bond is a thioester involving the sulfydryl groups of cysteine (Lawrence and Strominger, 1970). Recently, it has been realized that the cleavage of the penicilloyl–enzyme bond by the above reagents is an enzyme catalyzed reaction that is somehow stimulated by those reagents and does not occur when the enzyme is denatured (Kozarich et al., 1977; Curtis and Strominger, 1978). Determination of the amino acid sequence of the peptide fragment obtained by proteolytic digestion of the penicilloylated transpeptidase–carboxypeptidase enzyme from *Streptomyces* R_{61} showed that the penicillin was bound to a serine residue, presumably through an ester linkage between carbon 7 of the β-lactam and the hydroxyl group of serine (Frère et al., 1976b). The question of whether the substrate and penicillin bind at the same catalytic site of the enzyme has been studied recently by Rasmussen and Strominger (1978), who used a depsipeptide substrate analogue (diacetyl-L-lysyl-D-alanyl-D-lactic acid) for trapping acyl–enzyme intermediates (EP) of D-alanine carboxypeptidase (see equation in Section 3.4.2). The lactyl ester substrate had a lower K_m than did the usual D-alanyl substrate and a much higher rate of acylation (k_2) and upon denaturation and lowering of the pH to 3.0 gave good yields of a stable acyl–enzyme complex. As in the case of penicillin, the depsipeptide appeared to form an ester linkage with the enzyme that was insensitive to diisopropylphosphofluoridate (Georgopapadakou et al., 1977). This indicates that DD-carboxypeptidases may belong to a particular type of serine proteases (Blow, 1971).

5.2 Binding of β-Lactams to Enzymes of the Cross-linkage System

A novel approach for studying the constituents of the peptidoglycan cross-linking system and their interrelationship has involved the investigation of membrane-bound proteins responsible for penicillin binding (Blumberg and Strominger, 1974). With few exceptions, the enzyme activities performed by the peptidoglycan cross-linking complex are located in the plasma membrane of bacteria. Isolated membranes bind about 50–150 pmole of [^{14}C]penicillin G per milligram protein and the working hypothesis was that penicillin-sensitive enzymes (PSEs) and penicillin-binding proteins (PBPs) are synonymous.

Penicillin-binding proteins were detected by autoradiography of membrane proteins separated by SDS–polyacrylamide gel electrophoresis and were numbered in order of decreasing molecular weights. This technique has been applied to *E. coli* (Spratt, 1975, 1978), *B. megaterium* (Chase et al., 1977), *B. subtilis* (Blumberg and Strominger, 1972; Buchanan and Strominger, 1976), *B. stearothermophilus* (Yocum et al., 1974), *Salmonella typhi* (Shepherd et al., 1977), *P. aeruginosa* (Noguchi et al., 1978), *S. aureus*

(Kozarich and Strominger, 1978), and *S. faecalis* (Coyette et al., 1978). These taxonomically different bacteria represent a whole range of peptidoglycans of different primary structures, and the profiles of their PBPs differ significantly. No attempt is made here to review all of this work, except for the evidence pertaining to *E. coli*, and the reader is referred for further details to a number of recent reviews (Spratt, 1978; Ghuysen, 1977a, b, c). At least six main penicillin-binding components have been detected in *E. coli*. Two complementary approaches have been taken. The first one has been to try to relate one or several of the PBPs to the various morphological responses that β-lactams elicit in this organism. The other approach has been to try to isolate the individual PBPs and to characterize the particular enzymatic activities of the different penicillin-binding components. To distinguish between the physiological roles that each of the PBPs play in the expansion of peptidoglycan, a whole variety of genetically defined thermosensitive mutants exhibiting morphological abnormalities have been isolated and studied. In addition, a number of β-lactams that bind almost exclusively to a single PBP have been detected and the relative affinities of several β-lactams for the various PBPs has been determined (Spratt, 1977; Spratt et al., 1977a, b; Matsuhashi et al., 1977, 1978a, b; Tamaki et al., 1977; Iwaya and Strominger, 1977).

A summary of the properties, enzymatic activities and presumed roles of the different PBPs of *E. coli* is given in Table 1. Proteins 4, 5, and 6 are probably not required for normal peptidoglycan synthesis and appear not to be involved in the effects on cell morphology (Matsuhashi et al., 1977, 1978a). Proteins 5 and 6 have been shown to be identical with the isolated D-alanine carboxypeptidase Ia (Tamura et al., 1976a; Spratt and Strominger, 1976). Mutants lacking carboxypeptidase Ia activity, which maps at 13.3 min, grow without any detectable deficiencies. The mutation, however, was not associated with disappearance of PBP 5 and 6 (Suzuki et al., 1978; Matsuhashi et al., 1978a). Both penicillin and substrate were found to acylate the active site of the mutant's enzyme and the mutation is apparently in the inability of the DD-carboxypeptidase to catalyze the deacylation step (K_3) (J. L. Strominger, personal communication).

Mutants lacking carboxypeptidase Ib (which maps at 68 min) have no discernible changes of phenotype but lack PBP 4 (Matsuhashi et al., 1977; Iwaya and Strominger, 1977). However, the identity between D-alanine carboxypeptidase Ib and PBP 4 is still not completely clear, since Tamura et al. (1976a) failed to bind [^{14}C]penicillin to the isolated and purified D-alanine carboxypeptidase Ib. This may be due to modified binding properties of the isolated enzyme or to a very short half-life of the penicilloyl–enzyme linkage.

Cephalexin binds almost exclusively to protein 3, the product of a gene located at 2 min, and this binding causes inhibition of cell division and filamentation (Spratt, 1975). A number of thermosensitive mutants that do not bind penicillin to protein 3 at the nonpermissive temperature and that grow as filaments have been isolated (Spratt, 1975; Matsuhashi et al., 1977). As is described in Section 4.3, these mutants have a hyper-cross-linked newly synthesized peptidoglycan and a higher transpeptidase versus DD-carboxypeptidase activity ratio (Table 2), suggesting that PBP 3 may be a cephalexin-sensitive DD-carboxypeptidase–endopeptidase, which is required for septum formation (Mirelman et al., 1976, 1977, 1978).

Mecillinam (amidinopenicillanic acid, FL 1060) binds to protein 2 and induces the growth of ovoid cells (Spratt, 1975). Thienamycin and clavulanic acid also have affinity for PBP 2 (Spratt et al., 1977b). *E. coli* mutants that grow as spheres, such as *rodX* (maps at 14 min) and *rodY* (70 min) (Matsuzawa et al., 1973; Westling-Häggström and Normark, 1975; Spratt, 1977; Iwaya et al., 1978), were found to lack PBP 2. No defect was detected in the peptidoglycan synthesized by ether-treated cells of two of the ovoid mutants, SP-6 and SP-137, as compared to the rod-shaped, wild-type organism (D. Mirelman and B. G. Spratt, unpublished observations).

Protein 1 was recently resolved into several components (PBP 1A and three PBPs 1B) by the use of electrophoresis on loosely cross-linked polyacrylamide gels (Spratt et al., 1977a). PBPs 1B, probably the products of a single gene located at 3.3 min, appear to be the targets to which β-lactams bind, causing cessation of cell growth and lysis. These PBPs are therefore supposed to be the "main" transpeptidase(s) responsible for cell elongation (Spratt, 1975; Spratt et al., 1977a, b; Matsuhashi et al., 1978a; Noguchi et al., 1978; Spratt, 1978). Mutants lacking PBPs 1B have been shown to be hypersensitive to cephalosporins and penicillins, and their loss has been coupled with lack of enzyme activity of the membrane fraction for *in vitro* synthesis of peptidoglycan (Tamaki et al., 1977; Suzuki et al., 1978). Protein 1A, which maps at 73.5 min, has a very high affinity for cephalosporins (Noguchi et al., 1978; Suzuki et al., 1978). Mutants lacking in PBP 1A grow normally and synthesize a cross-linked peptidoglycan (Matsuhashi et al., 1978a; Suzuki et al., 1978). A very interesting connection seems to exist between PBP 1A and the three PBPs 1B in that a compensating activity by PBP 1A takes over whenever the PBPs 1B are missing.

Based on the large number of penicillin-binding proteins and their activities, it may be pertinent to assume that in *E. coli* there are duplicate or triplicate enzymatic systems for performing the same reaction, that is, cross-linking of peptidoglycan, and that if one system is blocked another will compensate for its lack of activity. It is still difficult to see, however, how so many copies of enzymes, exhibiting compensating functions, but without apparent important physiological roles could be synthesized by the cell.

Although the assay for penicillin-binding proteins has introduced an exciting new approach, their full significance cannot be evaluated until a complete correlation between the multiple penicillin-binding enzymes and their role in peptidoglycan metabolism has been established. One of the difficulties in attaining this goal may be the relative stability of the penicillin–protein linkages. In spite of all the available evidence, there is still some chance that inhibition of cell elongation and cessation of growth may be caused by the inhibition of a penicillin-sensitive enzyme that has not yet been detected as a penicillin-binding protein. Furthermore, there is also a possibility that the PBP that serves as the lethal target for β-lactams is devoid of any enzymatic activity but functions as a trigger for uncontrolled autolytic activity (Tomasz and Waks, 1975). The results of studies to determine the effect of penicillin on bacterial mutants defective in peptidoglycan hydrolases indicate that the hydrolases are essential for penicillin induced lysis and possibly even for killing of the bacteria. The attachment of penicillin to the binding proteins and the inhibition of enzymes of the cross-linking system cause only inhibition of bacterial growth, while the irreversible antibacterial effects seem to require additional cellular factors, such as peptidoglycan hydrolases (Tomasz and Höltje, 1977).

The β-lactam group of antibiotics, because of their specific inhibitory action on the peptide cross-linkage reaction which is unique to the bacterial world, are likely to remain of major medical importance for many years. A staggering number (> 10,000) of different derivatives of penicillins, cephalosporins, and monocyclic β-lactams have been synthesized and studied, but only a very limited number of these are widely used as antibiotics. Screening of antibiotics is usually done by measuring their inhibitory effects on the growth of various bacterial strains, but the design of such derivatives as effective antibiotics is usually a "hit and miss" affair. More work is required to determine which features of the β-lactam structure are essential to make the drug (1) a poor substrate for β-lactamases, (2) a good affinity binder to the physiologically important enzyme complex that cross-links peptidoglycan, (3) permeable across bacterial outer membranes, and (4) nontoxic to the host.

There is no doubt that recent advances in our understanding of the biosynthesis of peptidoglycan and its assembly process, the effect and mode of binding of β-lactam antibiotics to them, and the development of sensitive assays for the detection of PBPs and membrane permeability parameters (Zimmerman and Rooselet, 1977; Nikaido et al., 1977a, b) will contribute significantly to the rational design strategy of antibiotics.

6 INHIBITION OF CELL WALL ASSEMBLY

6.1 Changes in Cell Morphology Induced by β-Lactams

Complete inhibition of peptidoglycan synthesis in growing bacterial cultures usually results in osmotically fragile forms that lyse in the absence of an osmotic stabilizer. A wide variety of different types of morphological effects, dependent on the β-lactam and the concentration used, have been observed in *E. coli* (Greenwood and O'Grady, 1973a, b). Low concentrations of penicillin G (30–50 units/ml) inhibited cell division, but the cells continued to elongate and bulges developed at the sites where cells normally would have divided (Schwarz et al., 1969). On continued incubation, cells were observed to burst at the bulges. Burdett and Murray (1974a, b) confirmed these observations and noted that the bulges develop and appear in the region of new septum synthesis mostly just before cells divide. Since the septum region has been found to be the main zone of the cell where new synthesis of peptidoglycan occurs (Ryter et al., 1973; Schwarz et al., 1975), the new wall in this area may be very labile and β-lactams may upset the delicate balance of the peptidoglycan metabolizing machinery so that hydrolytic activity takes over, producing bulges of weakened wall that eventually bursts.

Ovoid cells of *E. coli* induced by growth with mecillinam were found to incorporate less (50%) peptidoglycan, and the lipoprotein content of the rigid peptidoglycan layer was twice that of a normal rod cell (Braun, 1975). The binding of mecillinam to PBP 2 had no direct effect on DNA synthesis but affected the appearance and disappearance, respectively, of two outer membrane proteins, protein D (mol. wt. 80,000) and protein G (mol. wt. 15,000), which are assumed to be involved in the regulation of cell elongation (James et al., 1975; Gudas et al., 1976; Churchward and Holland, 1976a, b). No conclusive evidence is available, however, on the role of these or other outer membrane proteins in controlling cell morphology. The failure to detect any inhibition of DD-carboxypeptidase, transpeptidase, or endopeptidase activity by mecillinam (Park and Burnam, 1973; Matsuhashi et al., 1974) is still compatible with the idea that PBP 2 is a minor, yet important, species of the penicillin-sensitive enzymes. Nothing is known about the function of PBP 2 in peptidoglycan metabolism, and purification of this protein will hopefully lead to a better understanding of its role.

6.2 The Role of Peptidoglycan in Cell Morphogenesis

There appears to be general acceptance of the concept that peptidoglycan has a definite role in maintaining bacterial rod shape. The mechanism through which this occurs, however, is still not clear. In many cases, cell shape is maintained in the absence of peptidoglycan in the envelope. For example, several extreme halophiles (such as *Halobacterium salinarium*) grow as perfect rods and are devoid of muramic acid (Brown, 1964; Brown and Cho, 1970). Other cell envelope components, such as several of the outer membrane proteins, have been suggested as having some role in shape maintenance (Henning *et al.*, 1973a, b). However, the recent isolation of a variety of outer membrane protein-deficient mutants, which have perfect rod morphology, has weakened this earlier argument (Henning, 1975; Henning and Haller, 1975). It seems that the mechanism for shape determination is a most complex one and it probably involves the interaction and participation of a number of envelope and membrane components, all of which operate in unison in the biogenesis of the rod-shaped cell.

Protoplasts and autoplasts (produced by the action of the native autolytic enzymes) of a number of organisms have been shown to produce and excrete fibrils of the soluble nascent peptidoglycan (Elliott et al., 1975a, b; Fitz-James, 1974; Rosenthal and Shockman, 1975; Rosenthal et al., 1975). When incubated on the surface of a medium containing 2.5% agar, however, protoplasts of *B. lichenformis* can successfully revert to normal bacilli. Hence insolubilization of the nascent peptidoglycan by peptide cross-linking can be achieved by cells that have no detectable preexisting walls by immobilization of the protoplasts on a medium of low fluidity that gives it structural support. Study of the sequence of events occurring during reversion of protoplast to bacilli showed that at early stages the glycan chains were very short and the peptides were poorly cross-linked. At these stages the cells still rounded up when they were removed from the agar and suspended in iso-osmolal solutions. As glycan chain length and cross-linking increased, reversion became complete and the wall material became rod shaped.

E. coli spherical cells (obtained either by treatment with mecillinam or by a thermosensitive mutation in PBP 2) apparently retain a mechanically intact peptidoglycan sacculus. When these cells are allowed to reshape themselves into rods (by changing to proper growth conditions), the areas of the cell envelope that had been the end poles of the original rods again become the poles of the newly reverted rods (Goodell and Schwarz, 1975). On the other hand, spheroplasts of *E. coli* completely lacking the rigid layer were unable to revert to rods even though they were able to synthesize a new, but spherical, peptidoglycan layer. Thus the complete absence of the original peptidoglycan layer may prevent cells of certain organisms from reverting to their original shape.

The peptidoglycan synthesis in a number of bacteria that can shift from sphere to rod shape have been studied. Unfortunately, no conclusive evidence has been obtained that would pinpoint an alteration in the peptidoglycan biosynthetic machinery as the mechanism responsible for the change in morphology. In a number of cases the rate of incorporation and content of peptidoglycan in the wall of the spherical bacteria was found to be lower than that found for the rod shape. For example, in *E. coli* grown with mecillinam (Braun, 1975), and in a pH-dependent, morphologically conditional mutant of *Klebsiella pneumonia* (Satta and Fontana, 1974 and Satta et al., 1979), the incorporation of nascent peptidoglycan into the walls of the spherical cells was almost half of that of the rod-shaped organism, whereas rates of DNA, RNA, and protein synthesis were comparable. In *K. pneumonia*, the cocci-shaped cells (grown at pH 7.0) were found to be more sensitive to cephalexin (sevenfold) than the rods grown at pH 5.8. On the other hand, mecillinam had an effect only on the rods (converting them to cocci). While these results clearly suggest that the cocci-shaped mutants have lost the capability to synthesize the cell elongation portion (Spratt, 1975, 1978), this does not necessarily explain why cocci cells should have less peptidoglycan. On the other hand, the fact that cocci incorporate less peptidoglycan may result in a partial loss of rigidity of the peptidoglycan layer, and osmotic swelling of the cells would cause their reshaping into spheres. This shape is obtained because it is the most physically favored, osmotically stable structure. A possible, but unsubstantiated, reason for the formation of these round cells under certain conditions can be derived also from the following observations. *Arthrobacter crystallopoietes* has been shown to grow as a rod or a sphere depending on nutritional conditions (Krulwitch et al., 1967a, b). An increased level of *N*-acetylmuramidase activity was found in the spherical form (Krulwitch and Ensign, 1968). Similarly, the abnormally shaped *E. coli* MAD-1 mutants obtained by Lazdunski and Shapiro (1972), as well as the spherical cells of the E. coli K-12 lss-12 thermosensitive-rod mutant of Henning et al. (1972), were observed to have accelerated rates of lysis or "lysis from within." These increased rates of autolysis, perhaps due to an increased concentration or topological activation of one of the autolytic enzymes of the cell, may cause the hydrolysis of certain linkages in the peptidoglycan and gradually give rise to a weakened matrix that is unable to maintain the original rod shape in the face of the osmotic pressure.

As much as these preliminary observations and casual conclusions can perhaps suggest a mechanism by which rods sometimes convert to spheres, no rational explanation is available to explain how and why the reverse process occurs. An experimental model that attempts to explain how the synthesis of the peptidoglycan that forms the cell septum (or cell poles) and the cylindrical portions are controlled, and under certain conditions inhibited, is described in Section 4.3 (Figure 12). The inability to form the cell-dividing septum was suggested to be connected with a lower activity of one of the hydrolytic enzymes, DD-carboxypeptidase-endopeptidase (Mirelman et al., 1976). Thus we may conclude again that "normal" peptidoglycan synthesis in a rod-shaped organism is a dynamic process and is regulated by an unknown mechanism that controls the delicate balance among the numerous synthetic and hydrolytic reactions (Figure 14). This equilibrium probably has considerable backup and alternative systems that allow it some degree of freedom without affecting the shape of the cell. However, any major interference with these reactions, either artificially by the use of β-lactams that preferentially inhibit (or partially inhibit) one, or several of them, or by a mutation in one of the metabolically important functions will disturb the balance between the rates of synthesis and hydrolysis and may cause the formation of an abnormally shaped organism.

6.3 Regulation of Peptidoglycan-Metabolizing Enzymes

As mentioned earlier, the autolytic enzymes of the cells are actively involved and play an important role in the regulation of cell surface growth and division. Their expression is, however, apparently very well regulated and a "barrier" of unknown nature is believed to exist between these enzymes and their substrates in the intact cell (Hakenbeck et al., 1974).

The peptidoglycan of *E. coli* cells, either from exponential or from stationary growth phase, is not hydrolyzed by the various types of membrane-bound hydrolases if cells are maintained in buffer at 37°C (Hartmann et al., 1974). Triggering of hydrolase activity can be achieved only after breakage of the cells or by pretreatments with reagents such as 5% TCA, 1 *M* NaCl, 20% sucrose, chelating agents, and certain β-lactam antibiotics.

Localized hydrolase action appears to be triggered at a given stage of the cell life cycle and the activities of a number of peptidoglycan-metabolizing

GROWTH AND SHAPE DETERMINATION OF *E. coli* CELL WALLS

(Penicillin sensitivity, - , +)

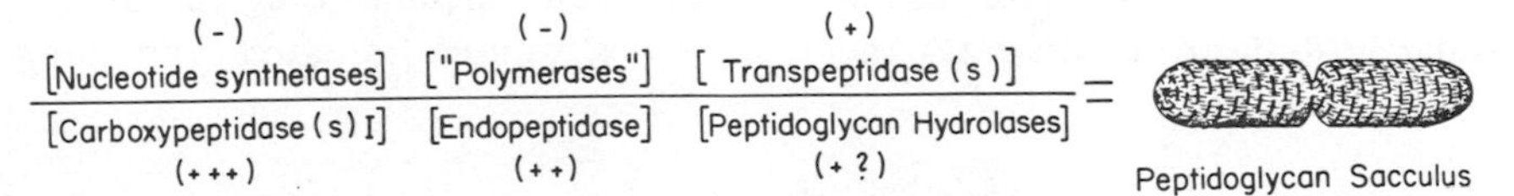

Figure 14 Hypothetical equation that attempts to emphasize the delicate balance existing between the synthetic and hydrolytic enzymatic reactions that participate in the construction of the cell wall peptidoglycan. Enzymatic activities that are very sensitive to penicillin are indicated by (+++).

enzymes in synchronously dividing cells have been found to oscillate. Evidence obtained with various strains of *E. coli* suggest that towards the end of the cell cycle (D period, Daneo-Moore and Shockman, 1977), the rate of incorporation of pulse-labeled precursors into the peptidoglycan increases (Hoffmann et al., 1972). Similar results were observed by testing the activity of the peptidoglycan biosynthetic machinery in synchronized cells (Mirelman et al., 1976, 1978). Furthermore, the activity of membrane-bound peptidoglycan hydrolases, as well as the activity of carboxypeptidase II (the LD-carboxypeptidase), increased as the cells approached the time of division (Hakenbeck and Messer, 1977a, b; Beck and Park, 1976).

It is still not clear whether the apparent inactivity of various peptidoglycan-metabolizing enzymes during certain periods of the cell cycle, or in mutants such as the thermosensitive division mutant, *E. coli* PAT 84, is due to a lack of enzyme synthesis or to an inhibition of the activation mechanism. Various endogenous inhibitors of autolytic activity, such as lipoteichoic acids and modifier proteins, have been found in a number of gram-positive organisms (Cleveland et al., 1975; Tomasz and Höltje, 1977; Daneo-Moore and Shockman, 1977). It is very possible that in gram-negative cells, an analogous controlling system exists in the membranes that provides an appropriately organized arrangement for the rapid, precise, and efficient topological control of autolysin action.

An interesting hypothesis on an additional possible mechanism of regulation of peptidoglycan synthesis was recently proposed by Ishiguro and Ramey (Ishiguro and Ramey, 1976; Ramey and Ishiguro, 1978). These authors found that the tetra- and pentaphosphate derivatives of guanosine (ppGpp and pppGpp), which are known to accumulate during amino acid starvation of stringently controlled *E. coli* strains, not only control the synthesis of proteins and fatty acids of cells, but apparently are also capable of inhibiting peptidoglycan biosynthesis. Analysis of incorporation of labeled DAP into cells revealed an eightfold accumulation of lipid intermediates in amino acid-starved cells. This indicates that the control is apparently at the level of the utilization of lipid intermediates.

Bacteria therefore have a number of possible mechanisms by which the synthesis of peptidoglycan can be controlled at different levels (see also Section 3.2). Much more information is needed, however, before we can understand how the cell integrates them all into a well-coordinated regulatory system.

7 CONCLUSIONS

During the past decade, and as a result of collaboration as well as competition among many laboratories, we have seen tremendous progress in the elucidation of the mechanism by which the cells assemble the extremely complex peptidoglycan macromolecule. An attempt has been made in this chapter to asses the present state of progress in the field of assembly of peptidoglycan, mainly as it concerns gram-negative bacteria. Some of the most important conclusions obtained are given below:

1. There is a main membrane growth zone, most likely at the equator of the cell, that synthesizes nascent peptidoglycan strands by a sequential chain elongation mechanism.
2. The insertion and insolubilization of nascent peptidoglycan into the preformed wall is catalyzed by a transpeptidation reaction while the glycan chain is probably still elongating.
3. Inhibition of transpeptidation (or in some cases of DD-carboxypeptidation) by β-lactam antibiotics causes an arrest of peptidoglycan insertion and expansion.
4. DD-Carboxypeptidase catalyzed hydrolysis and transpeptidations are bimolecular reactions, and β-lactams inhibit the enzymes by temporarily acylating a serine residue, presumably located at the active site.
5. Numerous enzymes that markedly differ in their abilities to bind substrates and in their sensitivities and binding capabilities to the various penicillins and other β-lactam antibiotics are apparently involved in the process of peptide cross-linkage and attachment of nascent peptidoglycan.
6. Cell lysis and death are most likely the result of uncontrolled autolytic activity that is somehow triggered by the action of certain β-lactams that block synthesis.

Perhaps to a reader who is not actively involved in this field, it may appear as if most of the relevant questions and problems regarding peptidoglycan have been more or less resolved. This may be partially true; however, as has happened in many fields, our newly acquired knowledge has only led us to new problems and questions that need to be answered. Although the structure of peptidoglycan is now well established, more work is still needed to understand the mode of its expansion and growth and its relation and interaction with other components of the wall, which together contribute to the unique properties of the bacterial envelope.

There is no doubt that to understand, for example, the mechanism of cell shape expression and determination we will have to know much more about the exact functioning, reciprocal control, and coordination of the multi-enzyme assembly centers that synthesis the various wall polymers in the microenvironment of the plasma membrane. We still know very little about the exact functioning, regulation, and triggering of the synthetic and autolytic systems in a specific cell area for a defined period during the cell life cycle. Some of these enzymes are known to be targets of important antibiotics and other enzymes may be the potential targets for new antimicrobial agents. Studies of these enzymes with respect to their primary, secondary, and tertiary structures will surely lead us to a more rational approach in the designing of better antibiotics.

One must not forget that a growing number of crucial developments in the understanding of the synthesis and assembly of mammalian cell surface components have originated from studies of the bacterial cell envelope. I believe therefore that elucidation of some of the problems discussed above will be of great importance not only for the understanding of the assembly of bacterial envelope, but for application to other fields as well.

ABBREVIATIONS

DAP	Diaminopimelic acid
SDS	Sodium dodecyl sulfate
TCA	Trichloroacetic acid
PBPs	Penicillin-binding proteins

REFERENCES

Anderson, J. S., Matsuhashi, M., Haskin, M. A., and Strominger, J. L. (1965), *Proc. Natl. Acad. Sci. U.S.*, **53,** 881–889.

Anderson, J. S., Meadow, P. M., Haskin, M. A., and Strominger, J. L. (1966), *Arch. Biochem. Biophys.*, **116,** 487–515.

Beck, B. D., and Park, J. T. (1976), *J. Bacteriol.*, **126,** 1250–1260.

Berberich, R., Kaback, M., and Freese, E. (1968), *J. Biol. Chem.*, **243,** 1006–1011.

Bettinger, G. E., and Young, F. E. (1975), *Biochem. Biophys. Res. Commun.*, **67,** 16–21.

Blow, D. M. (1971), in *The Enzymes* (P. D. Boyer, ed.), Vol. 3, 3rd ed., Academic Press, New York, pp. 185–212.

Blumberg, P. M., and Strominger, J. L. (1971), *Proc. Natl. Acad. Sci. U.S.*, **68,** 2814–2817.

Blumberg, P. M., and Strominger, J. L. (1972), *J. Biol. Chem.*, **247,** 8107–8113.

Blumberg, P. M., and Strominger, J. L. (1974), *Bacteriol. Rev.*, **38,** 291–335.

Boy de la Tour, E., Bolle, A., and Kellenberg, E., (1965), *Pathol. Microbiol.*, **28,** 229–234.

Braun, V. (1975), *Biochim. Biophys. Acta*, **415,** 335–377.

Brown, A. D. (1964), *Bacteriol. Rev.*, **28,** 296–329.

Brown, A. D., and Cho, K. Y. (1970), *J. Gen. Microbiol.*, **62,** 262–270.

Buchanan, C. E., and Strominger, J. L. (1976), *Proc. Natl. Acad. Sci. U.S.*, **73,** 1816–1820.

Burdett, I. D. J., and Murray, R. G. E. (1974a), *J. Bacteriol.*, **119,** 303–324.

Burdett, I. D. J., and Murray, R. G. E. (1974b), *J. Bacteriol.*, **119,** 1039–1056.

Chase, H. A., Shepherd, S. T., and Reynolds, P. E. (1977), *FEBS Lett.*, **76,** 199–203.

Chipman, D. M., Pollock, J. J., and Sharon, N. (1968), *J. Biol. Chem.*, **243,** 487–496.

Churchward, G. G., and Holland, I. B. (1976a), *FEBS Lett.*, **62,** 347–350.

Churchward, G. G., and Holland, I. B. (1976b), *J. Mol. Biol.*, **105,** 245–261.

Cleveland, E. F., and Gilvarg, C. (1976), *Proc. Natl. Acad. Sci. U.S.*, **73,** 4200–4204.

Cleveland, R. F., Höltje, J. V., Wicken, A. J., Tomasz, A., Daneo-Moore, L., and Shockman, G. D. (1975), *Biochem. Biophys. Res. Commun.*, **67,** 1128–1135.

Coyette, J., Ghuysen, J. M., Binot, F., Adrianes, P., Meesschaert, B., and Vanderhaeghe, H. (1977), *Eur. J. Biochem.*, **75,** 231–239.

Coyette, J., Ghuysen, J. M., and Fontana, R. (1978), *Eur. J. Biochem.*, **88,** 297–305.

Curtis, S. J., and Strominger, J. L. (1978), *J. Biol. Chem.*, **253,** 2584–2588.

Daneo-Moore, L., and Shockman, G. D. (1977), in *Cell Surface Reviews*, Vol. IV, (G. Poste and G. L. Nicolson, eds.), North Holland, Amsterdam, pp. 596–795.

Diedrich, D. L., and Schnaitman, C. A. (1978), *Proc. Natl. Acad. Sci. U.S.*, **75,** 3708–3712.

Dusart, J., Marquet, A., Ghuysen, J. M., and Perkins, H. R. (1975), *Eur. J. Biochem.*, **56,** 57–65.

Dusart, J., Leyh-Bouille, M., and Ghuysen, J. M. (1977), *Eur. J. Biochem.*, **81,** 33–44.

Elliott, T. S. J., Ward, J. B., and Rogers, H. J. (1975a), *J. Bacteriol.*, **124,** 623–632.

Elliott, T. S. J., Ward, J. B., Wyrick, P. B., and Rogers, H. J. (1975b), *J. Bacteriol.*, **124.** 905–917.

Fiedler, F., and Glaser, L. (1973), *Biochim. Biophys. Acta*, **300,** 467–485.

Fitz James, P. (1974), *Ann. N.Y. Acad. Sci.*, **235,** 345–346.

Formanek, H. (1978), *Biophys. Struct. Mech.*, **4,** 1–14.

Formanek, H., Formanek, S., and Wawra, H. (1974), *Eur. J. Biochem.*, **46,** 279–294.

Frère, J. M., Leyh-Bouille, M., Ghuyesen, J. M., and Perkins, H. R. (1974), *Eur. J. Biochem.*, **50,** 203–214.

Frère, J. M., Ghuyesen, J. M., and Iwatsubo, M. (1975a), *Eur. J. Biochem.*, **57,** 345–351.

Frère, J. M., Ghuysen, J. M., Degelaen, J., Loffet, A., and Perkins, H. R. (1975b), *Nature*, **258,** 168–170.

Frère, J. M., Ghuysen, J. M., Vanderhaeghe, H., Adrianes, P., Degelaen, J., and De Graeve, J. (1976a), *Nature*, **260,** 451–454.

Frère, J. M., Duez, C., Ghuysen, J. M., and Vandekerkhove, J. (1976b), *FEBS Lett.*, **70,** 257–260.

Georgopapadakou, N., Hammarström, S., and Strominger, J. L. (1977), *Proc. Natl. Acad. Sci. U.S.*, **74,** 1009–1012.

Ghuysen, J. M. (1968), *Bacteriol. Rev.*, **32,** 425–464.

Ghuysen, J. M. (1977a), in *Cell Surface Reviews*, Vol. IV, (G. Poste and G. L. Nicolson, eds.), North Holland, Amsterdam, pp. 463–595.

Ghuysen, J. M. (1977b), in *The Bacterial* DD-*Carboxypeptidase-Transpeptidase Enzyme System*, E. R. Squibb Lectures on Chemistry of Microbial Products Products (W. E. Brown, ed.), University of Tokyo Press, 162 pp.

Ghuysen, J. M. (1977c), *J. Gen. Microbiol.*, **101,** 13–33.

Ghuysen, J. M. (1979), *Annu. Rev. Biochem.*, **48,** 73–101.

Ghuysen, J. M., and Shockman, G. D. (1973), in *Bacterial Membranes and Walls*, Vol. I, (L. Leive, ed.), Dekker, New York, pp. 37–130.

Glaser, L. (1973), *Annu. Rev. Biochem.*, **42,** 91–112.

Goodell, E. W., and Schwarz, U. (1975), *J. Gen. Microbiol.*, **86,** 201–209.

Greenwood, D., and O'Grady, F. (1973a), *J. Clin. Pathol.*, **26,** 1–6.

Greenwood, D., and O'Grady, F. (1973b), *J. Infect. Dis.*, **128,** 791–794.

Gudas, L. J., and Mount, D. W. (1977), *Proc. Natl. Acad. Sci. U.S.*, **74,** 5280–5284.

Gudas, L. J., James, R., and Pardee, A. B. (1976), *J. Biol. Chem.*, **251,** 3470–3479.

Hakenbeck, R., and Messer, W. (1977a), *J. Bacteriol.*, **129,** 1234–1238.

Hakenbeck, R., and Messer, W. (1977b), *J. Bacteriol.*, **129,** 1239–1244.

Hakenbeck, R., Goodell, E. W., and Schwarz, U. (1974), *FEBS Lett.*, **40,** 261–264.

Hammerström, S., and Strominger, J. L. (1975), *Proc. Natl. Acad. Sci. U.S.*, **72,** 2463–3467.

Hammes, W. P. (1976), *Eur. J. Biochem.*, **70,** 107–113.

Hammes, W. P., and Kandler, O. (1976), *Eur. J. Biochem.*, **70,** 97–106.

Hammes, W. P., and Neuhaus, F. C. (1974), *J. Bacteriol.*, **120,** 210–218.

Hartmann, R., Höltje, J. V., and Schwarz, U. (1972), *Nature*, **235,** 426–429.

Hartmann, R., Bock-Henning, S. B., and Schwarz, U. (1974), *Eur. J. Biochem.*, **41,** 203–208.

Hasegawa, Y., Yamada, H., and Mizushima, S. (1976), *J. Biochem.*, **80,** 1401–1409.

Henning, U. (1975), *Annu. Rev. Microbiol.*, **29,** 45–60.

Henning, U., and Haller, I. (1975), *FEBS Lett.*, **55,** 161–164.

Henning, U., Rehn, K., Braun, V., Höhn, B., and Schwarz, U. (1972), *Eur. J. Biochem.*, **26,** 570–586.

Henning, U., Höhn, B., and Sonntag, I. (1973a), *Eur. J. Biochem.*, **39,** 27–36.

Henning, U., Rehn, K., and Höhn, B. (1973b), *Proc. Natl. Acad. Sci. U.S.*, **70,** 2033–2036.

Higashi, Y., Strominger, J. L., and Sweeley, C. C. (1967), *Proc. Natl. Acad. Sci. U.S.*, **57,** 1878–1884.

Higashi, Y., Siewert, G., and Strominger, J. L. (1970), *J. Biol. Chem.*, **245,** 3683–3690.

Hirashima, A., Wu, H. C., Venkateswaran, P. S., and Inouye, M. (1973), *J. Biol. Chem.*, **248,** 5654–5659.

Hirota, Y., Ryter, A., and Jacob, F. (1968), *Cold Spring Harbor Symp. Quant. Biol.*, **33,** 677–693.

Hirota, Y., Suzuki, Y., Nishimura, Y., and Yasuda, Y. (1977), *Proc. Natl. Acad. Sci. U.S.*, **74,** 1417–1420.

Hoffman, B., Messer, W., and Schwarz, U. (1972), *J. Supramol. Struct.*, **1,** 29–37.

Höltje, J. V., Mirelman, D., Sharon, N., and Schwarz, U. (1975), *J. Bacteriol.*, **124,** 1067–1076.

Ishiguro, E. E., and Ramey, W. D. (1976), *J. Bacteriol.*, **127,** 1119–1126.

Iwaya, M., and Strominger, J. L. (1977), *Proc. Natl. Acad. Sci. U.S.*, **74,** 2980–2984.

Iwaya, M., Jones, C. W., Khorana, J., and Strominger, J. L. (1978), *J. Bacteriol.*, **133,** 196–202.

Izaki, K., and Strominger, J. L. (1968), *J. Biol. Chem.*, **243,** 3193–3201.

Izaki, K., Matsuhashi, M., and Strominger, J. L. (1966), *Proc. Natl. Acad. Sci. U.S.*, **55,** 656–663.

Izaki, K., Matsuhashi, M., and Strominger, J. L. (1968), *J. Biol. Chem.*, **243,** 3180–3192.

James, R., Haga, J. Y., and Pardee, A. B. (1975), *J. Bacteriol.*, **122,** 1283–1292.

Johnston, L. S., and Neuhaus, F. C. (1975), *Biochemistry*, **14,** 2754–2760.

Johnston, L. S., and Neuhaus, F. C. (1977), *Biochemistry*, **16,** 1251–1257.

Johnston, R. B., Scholz, J. J., Diven, W. F., and Shepard, S. (1966), in *Enzymes and Model Systems* (E. E. Snell, A. E. Braunstein, E. S. Severin, and Y. M. Torshinsky, eds.), Wiley-Interscience, New York, pp. 537–547.

Kahan, F. M., Kahan, J. S., Cassidy, P. J., and Kropp, H. (1974), *Ann. N.Y. Acad. Sci.*, **235,** 364–385.

Keglevic, D., Ladesic, B., Hadzija, J., Tomasic, Z., Valinger, M., Pokorny, M., and Naumski, R. (1974), *Eur. J. Biochem.*, **42,** 389–400.

Kelemen, M. V., and Rogers, H. J. (1971), *Proc. Natl. Acad. Sci. U.S.*, **68,** 992–996.

Kozarich, J. W. and Strominger, J. L. (1978), *J. Biol. Chem.*, **253,** 1272–1278.

Kozarich, J. W., Nishino, T., Willoughby, E., and Strominger, J. L. (1977), *J. Biol. Chem.*, **252,** 7525–7529.

Krulwitch, T. A., and Ensign, J. C. (1968), *J. Bacteriol.*, **96,** 857–859.

Krulwitch, T. A., Ensign, J. C., Tipper, D. J., and Strominger, J. L. (1967a), *J. Bacteriol.*, **94,** 734–740.

Krulwitch, T. A., Ensign, J. C., Tipper, D. J., and Strominger, J. L. (1967b), *J. Bacteriol.*, **94,** 741–750.

Lawrence, P. J., and Strominger, J. L. (1970), *J. Biol. Chem.*, **245,** 3653–3659.

Lazdunski, G. and Shapiro, B. M. (1972), *J. Bacteriol.*, **111,** 499–509.

Leduc, M., and Van Heijenoort, J. (1975), 10th FEBS Meeting, Paris, Abstr. 1055.

Lee, N., and Inouye, M. (1974), *FEBS Lett.*, **39,** 167–170.

Leyh-Bouille, M., Dusart, J., Nguyen-Disteche, M., Ghuysen, J. M., Reynolds, P. E., and Perkins, H. R. (1977), *Eur. J. Biochem.*, **81,** 19–28.

Lugtenberg, E. J. J. (1972), *J. Bacteriol.*, **110,** 26–34.

Lugtenberg, E. J. J., and Van Dam, A. V. S. (1972a), *J. Bacteriol.*, **110,** 35–40.

Lugtenberg, E. J. J., and Van Dam, A. V. S. (1972b), *J. Bacteriol.*, **110,** 41–46.

Lugtenberg, E. J. J., Haas-Menger, L., and Ruyters, W. H. M. (1972), *J. Bacteriol.*, **109,** 326–335.

Lugtenberg, E. J. J., Bronstein, H., Van Selm, N., and Peters, R. (1977), *Biochim. Biophys. Acta*, **465,** 571–578.

McEntee, K., Hesse, J. E., and Epstein, W. (1976), *Proc. Natl. Acad. Sci. U.S.*, **73,** 3979–3983.

Mardarowicz, C. (1966), *Z. Naturforsch.*, **21B,** 1006–1007.

Martin, H. H. (1964), *J. Gen. Microbiol.*, **36,** 441–450.

Martin, H. H., Schiff, W., and Maskos, C. (1976), *Eur. J. Biochem.*, **71,** 585–593.

Matsuhashi, M., Kamiyro, T., Blumberg, P. M., Linnett, P., Willoughby, E. and Strominger, J. L. (1974), *J. Bacteriol.*, **117,** 578–587.

Matsuhashi, M., Takagaki, Y., Maruyama, J. N., Tamaki, S., Nishimura, Y., Suzuki, H., Ogino, U., and Hirota, Y. (1977), *Proc. Natl. Acad. Sci. U.S.*, **74,** 2976–2979.

Matsuhashi, M., Maruyama, F. N., Takagaki, Y., Tamaki, S., Nishimura, Y., and Hirota, Y. (1978a), *Proc. Natl. Acad. Sci. U.S.*, **75,** 2631–2635.

Matsuhashi, M., Tamaki, S., Nakajima, S., Nakagawa, J., Tomioka, S. and Takagaki, Y. (1978b), in *Microbial Drug Resistance and Related Plasmid* (S. Mitsuhashi, ed.) University of Tokyo Press, in press.

Matsuzawa, H., Matsuhashi, M., Oka, A., and Seguno, Y. (1969), *Biochem. Biophys. Res. Commun.*, **36,** 682–689.

Matsuzawa, H., Hayakawa, K., Sato, T., and Imahori, K. (1973), *J. Bacteriol.*, **115,** 436–442.

Mirelman, D. and Nuchamowitz, Y. (1979a), *Eur. J. Biochem.*, **94,** 541–548.

Mirelman, D. and Nuchamowitz, Y. (1979b), *Eur. J. Biochem.*, **94,** 549–556.

Mirelman, D., and Sharon, N. (1972), *Biochem. Biophys. Res. Commun.*, **46,** 1909–1917.

Mirelman, D. and Siegel, R. C. (1979), *J. Biol. Chem.*, **254,** 571–574.

Mirelman, D., Bracha, R., and Sharon, N. (1972), *Proc. Natl. Acad. Sci. U.S.*, **69,** 3355–3359.

Mirelman, D., Bracha, R., and Sharon, N. (1974a), *Ann. N.Y. Acad. Sci.*, **235,** 326–347.

Mirelman, D., Bracha, R., and Sharon, N. (1974b), *Biochemistry*, **13,** 5045–5053.

Mirelman, D., Yashouv-Gan, Y., and Schwarz, U. (1976), *Biochemistry*, **15,** 1781–1790.

Mirelman, D., Yashouv-Gan, Y., and Schwarz, U. (1977), *J. Bacteriol.*, **129,** 1593–1600.

Mirelman, D., Yashouv-Gan, Y. Nuchamowitz, Y., Rozenhak, S., and Ron, E. Z. (1978), *J. Bacteriol.*, **134,** 453–461.

Miyakawa, T., Matsuzawa, H., Matsuhashi, M., and Sugino, Y. (1972), *J. Bacteriol.*, **112,** 950–958.

Nakae, T., and Nikaido, H. (1976), *J. Biol. Chem.*, **250,** 7359–7365.

Neuhaus, F. C., Carpenter, C. V., Lambert, M. P., and Wargel, R. J. (1972), in *Molecular Mechanisms of Antibiotic Action on Protein Biosynthesis and Membranes* (E. Munoz, F. Ferrandiz, and D. Vazquez, eds.), Elsevier, Amsterdam, pp. 339–362.

Nguyen-Distèche, M., Ghuysen, J. M., Pollack, J. J., Reynolds, P. E., Perkins, H. R., Coyette, J., and Salton, M. R. J. (1974a), *Eur. J. Biochem.*, **41,** 447–455.

Nguyen-Distèche, M., Pollock, J. J., Ghuysen, J. M., Puig, J., Reynolds, P. E., Perkins, H. R., Coyette, J., and Salton, M. R. J. (1974b), *Eur. J. Biochem.*, **41,** 457–463.

Nikaido, H., Bavoil, P., and Hirota, Y. (1977a), *J. Bacteriol.*, **132,** 1045–1047.

Nikaido, H., Song, S. A., Shaltiel, L., and Nurminen, M. (1977b), *Biochem. Biophys. Res. Commun.*, **76,** 324–330.

Nishino, T., Kozarich, T. W., and Strominger, J. L. (1977), *J. Biol. Chem.*, **252,** 2934–2939.

Noguchi, T. H., Matsuhashi, M., Takaoka, M., and Mitsuhashi, S. (1978), *Antimicrob. Agents Chemother.* **14,** 617–624.

Oldmixon, E. H., Glauser, S., and Higgins, M. L. (1974), *Biopolymers*, **13,** 2037–2060.

Osborn, M. J., Gander, J. E., Parisi, E., and Carson, J. (1972), *J. Biol. Chem.*, **247,** 3962–3972.

Pardee, A. B., Wu, P. C., and Zusman, D. R. (1973), in *Bacterial Membranes and Walls* (L. Leive, ed.), Dekker, New York, pp. 357–312.

Park, J. T. (1952), *J. Biol. Chem.*, **194,** 897–904.

Park, J. T., and Burnam, L. (1973), *Biochem. Biophys. Res. Commun.*, **51,** 863–868.

Perkins, H. R., and Nieto, M. (1972), in *Mechanisms of Antibiotic Action on Protein Biosynthesis and Membranes* (E. Muñoz, F. Ferrandiz, and D. Vazquez, eds.), Elsevier, Amsterdam, pp. 363–387.

Pollock, J. J., Nguyen-Distèche, M., Ghuysen, J. M., Coyette, J., Linder, R., Salton, M. R. J., Kim, K. S. Perkins, H. R., and Reynolds, P. E. (1974), *Eur. J. Biochem.*, **41,** 439–446.

Previc, E. P. (1970), *J. Theor. Biol.*, **27,** 471–497.

Ramey, W. D., and Ishiguro, E. E. (1978), *J. Bacteriol.*, **135,** 71–77.

Rasmussen, J. R., and Strominger, J. L. (1978), *Proc. Natl. Acad. Sci. U.S.*, **75,** 84–88.

Rogers, H. J. (1970), *Bacteriol. Rev.*, **34,** 194–214.

Rogers, H. J. (1974), *Ann. N.Y. Acad. Sci.*, **235,** 29–51.

Rosenberger, R. F., Grover, N. B., Zaritsky, A., and Woldringh, C. L. (1978), *Nature*, **271,** 244–245.

Rosenbusch, J. P. (1974), *J. Biol. Chem.*, **249,** 8019–8029.

Rosenthal, R. S., and Shockman, G. D. (1975), *J. Bacteriol.*, **124,** 419–423.

Rosenthal, R. S., Jungkind, D., Daneo-Moore, L., and Shockman, G. D. (1975), *J. Bacteriol.*, **124,** 398–409.

Ryter, A., Hirota, Y., and Schwarz, U. (1973), *J. Mol. Biol.*, **78,** 185–195.

Salton, M. R. J. (1953), *Biochim. Biophys. Acta*, **10,** 512–523.

Satta, G., and Fontana, R. (1974), *J. Gen. Microbiol.*, **80,** 51–75.

Satta, G., and Pardee, A. B. (1978), *J. Bacteriol.*, **133,** 1492–1500.

Satta, G., Fontana, R. Canepari, P., and Botta, G. (1979), *J. Bacteriol.*, **137,** 727–734.

Schindler, M., Mirelman, D., and Schwarz, U. (1976), *Eur. J. Biochem.*, **71,** 131–134.

Schleifer, K. H., and Kandler, O. (1972), *Bacteriol. Rev.*, **36,** 407–477.

Schwarz, U., Asmus, A., and Frank, H. (1969), *J. Mol. Biol.*, **41,** 419–429.

Schwarz, U., Ryter, A., Rambach, A., Hellio, R., and Hirota, Y. (1975), *J. Mol. Biol.*, **98,** 749–759.

Sharon, N., Osawa, T., Flowers, H. M., and Jeanloz, R. W. (1966), *J. Biol. Chem.*, **241,** 223–230.

Shepherd, S. T., Chase, H. A., and Reynolds, P. E. (1977), *Eur. J. Biochem.*, **78,** 521–532.

Siewert, G., and Strominger, J. L. (1967), *Proc. Natl. Acad. Sci. U.S.*, **57,** 767–773.

Slater, M., and Schaechter, M. (1974), *Bacteriol. Rev.*, **38,** 199–221.

Spratt, B. G. (1975), *Proc. Natl. Acad. Sci. U.S.*, **72,** 2999–3003.

Spratt, B. G. (1977), *J. Bacteriol.*, **131,** 293–305.

Spratt, B. G. (1978), *Sci. Prog. Oxf.*, **65,** 101–128.

Spratt, B. G., and Strominger, J. L. (1976), *J. Bacteriol.*, **127,** 660–663.

Spratt, B. G., Jobanputra, V., and Schwarz, U. (1977a), *FEBS Lett.*, **79,** 374–378.

Spratt, B. G., Jobanputra, V., and Zimmerman, W. (1977b), *Antimicrob. Agents. Chemother.*, **12,** 406–409.

Storm, D. R. (1974), *Ann. N.Y. Acad. Sci.*, **235,** 387–398.

Strominger, J. L. (1970), *The Harvey Lectures*, **64,** 171–213.

Strominger, J. L., Izaki, K., Matsuhashi, M., and Tipper, D. J. (1967), *Fed. Proc.*, **26,** 9–22.

Suzuki, H., Nishimura, Y., and Hirota, Y. (1978), *Proc. Natl. Acad. Sci. U.S.*, **75,** 664–668.

Takatsuki, A., Kawamura, K., Okina, M., Kodama, Y., Ito, T., and Tamura, G. (1977), *Agric. Biol. Chem.*, **41,** (11), 2307–2309.

Tamaki, S., Nakajima, S., and Matsuhashi, M. (1977), *Proc. Natl. Acad. Sci. U.S.*, **74,** 5472–5476.

Tamura, T., Imae, Y., and Strominger, J. L. (1976a), *J. Biol. Chem.*, **251,** 414–423.

Tamura, G., Sasaki, Y., Matsuhashi, M., Takatsuki, A., and Yamasaki, M. (1976b), *Agric. Biol. Chem.*, **40** (2), 447–449.

Taylor, A., Das, B. C., and Van Heijenoort, J. (1975), *Eur. J. Biochem.*, **53,** 47–54.

Tipper, D. J., and Strominger, J. L. (1965), *Proc. Natl. Acad. Sci. U.S.*, **54,** 1133–1141.

Tomasz, A., and Höltje, J. V. (1977), in *Microbiology* (D. Schlessinger, ed.), American Society for Microbiology, pp. 209–215.

Tomasz, A., and Waks, S. (1975), *Proc. Natl. Acad. Sci. U.S.*, **72,** 4162–4166.

Tynecka, Z., and Ward, J. B. (1975), *Biochem. J.* **146,** 253–267.

Van Alphen, W. V., Nelke, V. S., and Lugtenberg, B. (1978), *Mol. Gen Genet.*, **159,** 75–83.

Van Heijenoort, Y., Derrien, M., and Van Heijenoort, J. (1978), *FEBS Lett.*, **89,** 141–144.

Ward, J. B. (1973), *Biochem. J.*, **133,** 395–398.

Ward, J. B., and Perkins, H. R. (1973), *Biochem. J.*, **135,** 721–728.

Ward, J. B., and Perkins, H. R. (1974a), *Biochem. J.*, **139,** 781–784.

Ward, J. B., and Perkins, H. R. (1974b), *Biochem. J.*, **141,** 227–241.

Weidel, W., and Pelzer, H. (1964), *Adv. Enzymol.*, **26,** 193–232.

Weigand, R. A., Vinci, K. D., and Rothfield, L. I. (1976), *Proc. Natl. Acad. Sci. U.S.*, **73,** 1882–1886.

Weppner, W. A., and Neuhaus, F. C. (1977), *J. Biol. Chem.*, **252,** 2296–2303.

Weppner, W. A., and Neuhaus, F. C. (1978), *J. Biol. Chem.*, **253,** 472–478.

Westling-Häggström, B., and Normark, S. (1975), *J. Bacteriol.*, **123,** 75–82.

Weston, A., Ward, J. B., and Perkins, H. R. (1977), *J. Gen. Microbiol.*, **99,** 171–181.

Wickus, G. G., and Strominger, J. L. (1972), *J. Biol. Chem.*, **247,** 5307–5311.

Wijsman, H. J. W. (1972), *Genet. Res.*, **20,** 65–74.

Wise, E. M., Jr., and Park, J. T. (1965), *Proc. Natl. Acad. Sci. U.S.*, **54,** 75–81.

Wright, A., Dankert, M., Fennessey, P., and Robbins, P. W. (1967), *Proc. Natl. Acad. Sci. U.S.*, **57,** 1798–1803.

Yocum, R. R., Blumberg, P. M., and Strominger, J. L. (1974), *J. Biol. Chem.*, **249,** 4863–4871.

Zaritsky, A., Woldringh, C. L. and Mirelman, D. (1979), *FEBS Lett.*, **98,** 29–32.

Zimmerman, W., and Rooselet, A. (1977), *Antimicrob. Agents. Chemother.*, **12,** 368–372.

Zusman, D. R., Inouye, M., and Pardee, A. B. (1972), *J. Mol. Biol.*, **69,** 119–136.

Chapter 6 The Fusion Sites Between Outer Membrane and Cytoplasmic Membrane of Bacteria: Their Role in Membrane Assembly and Virus Infection

MANFRED E. BAYER

The Institute for Cancer Research, Fox Chase Cancer Center, Philadelphia, Pennsylvania

1 INTRODUCTION

In this chapter, we discuss the structural and functional organization of specialized domains of the bacterial envelope that play a central role in the life of the microorganism. These "adhesion sites" or sites of fusion between inner and outer membrane have been found in the envelope of growing rod-shaped organisms such as gram-negative enterobacteria and fresh water bacteria. Their main feature is an intimate contact between inner and outer membrane of the cell's envelope, becoming apparent only when the two membranes are slightly separated by plasmolysis (Bayer, 1968a, b, 1975). Adhesion sites represent growing zones at which newly synthesized lipopolysaccharides, capsule polysaccharides, and proteins are inserted into the cell wall; furthermore, at these sites, a great variety of bacteriophages inject their nucleic acid. The adhesion areas not only serve as sites for export and import, but also provide sites of adsorption and uptake for nutrients and vitamins of higher molecular weight; in addition, they seem to be involved in complex membrane functions such as motility and, possibly, in the sensory responses of the organism.

Studies of eukaryotic cells and of tissues reveal that membrane contact and membrane fusion are not a unique design principle of bacterial envelopes, but are rather widely employed in the organization of cells and tissues in general: we briefly discuss here the functions of membrane contact or fusion sites of eukaryotic cells and compare them with the functional capability of the adhesion sites of the bacterial membrane systems.

Contact domains of membranes in eukaryotic cells and tissues represent frequent structural elements and can often be identified as being of rather simple organization, whereas other types comprise highly differentiated organelles. Tight junctions of eukaryotic cells, at which two neighboring plasma membranes are interlinked by rows of integral membrane proteins, create a seal impermeable to fluids from extracellular compartments (Hull and Staehelin, 1976). The various types of desmosomes, at which neighboring epithelial cells are held together, might be compared to spot welds and involve peripheral, as well as integral, membrane proteins; they are connected to intracellular fibrillar tonofilaments and provide the anchoring sites for the tensile structures of the interior skeleton of the cells. A third kind of membrane "adhesive" area is the complex gap junction (Caspar et al., 1977) at which a functional continuity with the neighboring cells is established by special protein molecules that aggregate to form cylindrical structures containing a central pore; these pores are organized in the membrane to form ducts between adjacent plasma membranes and endow the cells with communication capabilities (Bennett, 1973). The gap junction integrates the individual cells of an organ to a functional unit by providing transfer of chemical and electrical signals to the adjacent cell (Loewenstein, 1964; Pitts and Simms, 1977; Revel and Karnovsky, 1976); such signals seem to play a major role in processes of cell–cell recognition and in the coordination of development of tissues and organisms.

The fusion of membranes as an organizational principle opens up a number of mechanisms of transfer of signals and of information-carrying molecules between adjacent cells and membranes. Fusion has been observed in nucleated cells, in mitochondria, and in bacteria. It has been employed in vesicle studies and to form hybrids of cells and cytoplasts. (Maggio et al 1978); these fusions are mediated either by coats of viruses or by a variety of "fusogenic" substances. Fusion has also been achieved in bacterial protoplasts (Schaeffer and Hotchkiss, 1978) and appears to be a rather slow process. However, faster means of molecular transfer among eukaryotic membrane components have been observed: a transient cell contact of a few minutes was sufficient to achieve transfer of lipid label and of lipid material containing nonlipid label (Collard et al., 1978). Recent studies also revealed that whole cells take up lipid components from vesicle membranes (Huang and Pagano, 1975; Dunnick et al., 1976) and substances trapped within vesicles (Weinstein et al., 1977). In these instances, not only a fusion with the cytoplasmic membrane seems to occur, but also endocytotic processes (Poste and Papahadjopoulos, 1976) are involved by which proteins, virus particles, and nucleic acid are taken up, in analogy to phagocytosis of larger objects.

A serial sequence of membrane contact and fusion emerges as a rather common phenomenon in highly specialized cells; during excretion of proteins these cells employ sequentially membrane contact, fusion, and permeation (Palade, 1959, 1975); proteins are transferred from membranes of the rough endoplasmic reticulum to the golgi zone and storage organelles; subsequently, vesicle fusion with the plasma membrane and expulsion of the vesicle contents takes place. Interestingly, the membranes involved in the process of synthesis and excretion have a half-life of approximately 2 orders of magnitude longer than the excretion product, so that the membrane material might be reutilized by employing either endocytosis or degradation (Meldolesi, 1974). Furthermore, the membrane components (lipids, proteins) of fused vesicles and cell membranes appear to remain unmixed (Meldolesi et al., 1971; Bergeron et al., 1973). Fusion processes have also been described for the "import" of lipids and lipopolysaccharides in bacteria. Lipopolysaccharide label of extracellular artificial vesicle suspensions was found to be taken up by the outer membrane of the intact cell, whereas phospholipids were incorporated also into the inner membrane (Jones and Osborn, 1977). It is not known whether this phenomenon might be analogous to that described above for the transfer of lipids during short-time interaction of eukaryotic cells, or whether it might involve a slower membrane fusion process.

In summary, the various types of membrane contact zones and fusion sites in eukaryotic cells are employed (*a*) in mechanical stabilization ("spot weldings") between membrane systems and organelles and as anchors for tensile intracellular structures; (*b*) in synthesis, export, and import of membrane lipids and secretory proteins; and (*c*) in the control of metabolic and electrical coupling of the cells in organs and tissues. We will show in the subsequent text that the membrane adhesion sites of the bacterial cell also may serve in many analogous functions.

The functional significance of membranes in association with protein excretion processes is underscored by the discovery that excretory proteins are synthesized with an additional "signal" sequence (Blobel and Dobberstein 1975), to be cleaved upon release of the final protein. In this model the synthesizing ribosomes are attached to the membrane by the nascent polypeptide chain. The formation of such a ribosome/membrane complex has also been proposed for the binding of cytoplasmic ribosomes to the external membrane of mitochondria of yeast cells; such a complex would allow for the transfer of the translation products across the mitochondria membrane as shown by Butow et al., 1975. These authors provided electron micrographs in which complexes of cytoplasmic ribosomes can be seen as they are preferentially attached to the mitochondrial surface at areas of fusion between the inner and outer membrane. Butow et al. (1975) hypothesize that at these sites, the transfer of nascent cytoplasmic protein might occur from the outside of the organelle into its inner compartment.

The model of membrane-associated production sites of proteins appears to be applicable to bacterial membranes. The "flow" of nascent cell surface products in bacterial cells, however, would have to be in the opposite direction to explain the numerous "export" phenomena taking place from the inside of the cell toward the cell surface. The clustering of ribosomes at export sites of membrane proteins has recently been suggested by DeLeij et al. (1978, 1979). These authors propose that the messenger RNA is moving past a stationary set of ribosomes that are bound to the membrane by the transmembrane polypeptide chain.

The bacterial cell surface exhibits to the environment characteristic, specific antigens (Kaufmann, 1966), including organelles such as pili and flagella (Brinton, 1965; Lawn et al., 1967). A multitude of functions seem to be concentrated at the membrane adhesion sites and a delicate balance must exist for maintenance, production, and shedding of the variety of cell surface components. It appears from recent experiments that at the adhesion sites rapid membrane changes originate in response to virus infections. These signals appear to be indistinguishable from those generated by attractants, metabolic poisons, and bacteriocins. Although the bacterial surface is apparently equipped with several hundred adhesion sites, they comprise only a relatively small area ($\sim$5%) of the total surface of growing wild-type *E. coli* or *Salmonella* species; most of the cell surface is covered by more "inert" material that does not seem to respond qualitatively or quantitatively to fast environmental changes. This inert cell coat is nevertheless of great importance to the physiological performance and the survival of the cell (Brown and Corner, 1977).

The integrated properties of the outer surface determine the general surface qualities of the bacterium; for example, long polymer molecules of capsular polysaccharides may cover the entire cell. This extremely hydrated coat endows the organism with varying degrees of protection against most viruses and drugs, as well as against phagocytosis by macrophages. In contrast to this extremely extended and pliable surface layer of capsulated *E. coli* strains, a cell coat may consist of rather compact, cross-linked cell wall material, typical for the design of gram-positive bacteria. Furthermore, additional protein aceous layers may form highly ordered subunit arrangements (Sleytr et al. 1974; Beveridge and Murray, 1974; Buckmire and Murray 1973).

It has been demonstrated that not only the chemical, macromolecular composition of the surface is responsible for characteristic properties of the cell, but that the density of charged surface molecules influences electric charge and adhesiveness to neighboring cells or inorganic surfaces; these parameters in turn, affect uptake or exclusion of nutrients, macromolecules, and drugs (Jones, 1978; Magnusson and Johannsson 1977; Stendahl et al. 1977).

The predominant feature in the structural organization of the gram-negative envelope is its composition of two separate membrane systems, the inner (or protoplasmic) membrane, and the outer membrane, to which in many species the peptidoglycan layer is attached. Since the subunit molecules for the outer membrane are assembled at the inner membrane (Kanegasaki and Wright, 1973; Osborn et al., 1972) a transfer of these molecules to the outer membrane must be postulated; the membrane adhesion sites have been shown to serve in this function (see below). One must also postulate an organizational control that integrates the action of individual growing sites into a balanced overall maintenance of the cell's characteristic shape and dimension. These equilibria respond to environmental, nutritional, and physical parameters (see Leive, 1973). Specialized structures such as pili and flagellae comprise an additional challenge to the understanding of the functional composition of the total cell surface, since the synthesis and operation of these organelles may play major roles in chemotaxis, adherence, and promotion of disease. Here, too, sites with the structural features of membrane adhesion zones seem to be involved in the biosynthesis of these organelles (Bayer, 1975).

The discovery of functionally active domains indicated a mosaicistic assembly of the growing cell surface, a model that received most of its support from the observation that several of the molecular species of the cell envelope are being "exported" to the cell surface at areas of adhesion. As the cell surface grows, the freshly exported material is integrated in the existing outer membrane. However, once these molecules (for example, lipopolysaccharides) are incorporated into the surface, there seems to be a separation of some of their functions; this has been demonstrated in the case of virus receptor activity. The general surface behaves differently from the specialized injection site for bacteriophages: while the cell surface maintains phage binding activity or "receptor" activity, the infection of the cell takes place over the adhesion sites. To reach these sites, a macromolecule either must desorb and readsorb, or it must be able to execute a surface walk (Wong et al., 1978). Particles that do not possess this capability (colicins) might be trapped within the LPS layer, and only a "direct hit" of a sensitive area might cause a specific effect. With regard to bacteriophages, the signal for release of infectious nucleic acid is generated at the adhesion sites, possibly in association with exposure of the virion (*a*) to additional receptors such as membrane proteins, and/or (*b*) to unique conformational arrangements of the macromolecules at the adhesion site.

2 MEMBRANE ADHESION SITE

The fact that the envelope of rod-shaped bacteria can be separated into two membranes has been known for about 80 years, when the effects of plasmolysis were studied in plants and bacteria (Fischer, 1903), whereas the chemical and functional description of cell walls and protoplasmic membranes had to wait until the work of Salton (1952) and Weibull (1953) opened up the possibility of enzymatically separating the cell membranes from cell walls. Although preparation of membrane vesicles of whole envelopes provides suitable systems for transport studies (Kaback, 1973), the two membranes of the envelop of gram-negative bacteria such as *E. coli* and *Salmonella* species may be separated by gradient centrifugation after spheroplast formation, exposure of the cells to media of relatively high osmotic pressure, and subsequent disruption of the cells (Miura and Mizushima, 1969; Schnaitman, 1970; Osborn et al., 1972). Several modifications of this separation scheme are employed in the current membrane isolation procedures. A temporary separation of the membranes is used in the pretreatment of whole cells by osmotic shock for the release of periplasmic proteins (Heppel, 1969, Rosen and Heppel, 1973). Plasmolysis is also required for visualization of the adhesion sites. The significance of the adhesion sites was only slowly recognized, probably because their preservation requires careful adherence to a sequence of procedures involving fixation in aldehydes, control and maintenance of the concentration of plasmolyzing agents, and a proper timing of postfixation. Whenever these parameters were met, the ultrathin sections of plasmolyzed *E. coli* and *Salmonella* species revealed the adhesion sites. The lack of an earlier acceptance of the sites as a structural reality may have been caused by two factors: (1) their fragility to osmotically incompatible fixations such as the "standard" primary osmiumtetroxide fixation; thus, considerable time elapsed before confirming reports from various laboratories became available, and (2) the existence at that time of any correlation between these structures and cell functions was lacking. Although speculations existed that the sites might represent zones of growth (Bayer, 1968a), the adhesion sites were a purely morphological concept without an obvious role in the life of the bacterium; thus investigators unfamiliar with, or skeptical of, the evaluation of electron microscope data had to wait for supporting results from virological and biochemical studies. Membrane adhesion sites had originally been demonstrated in *E. coli* strains (B, C1, K-12) (Bayer, 1967, 1968a; Bayer and Starkey, 1972); they were subsequently also found in *Salmonella typhimurium* (Mühlradt et al., 1973; Crowlesmith et al., 1978) and *Salmonella anatum* (Bayer, 1974, 1975), in capsulated strains of *E. coli* (K-29, K-26) (Bayer and Thurow, 1977), in *Caulobacter* (M. E. Bayer and N. Agabian, unpublished observations), as well as in *Shigella* and a variety of other bacterial species. It seems reasonable to expect that other rod-shaped microorganisms also possess membrane adhesion sites.

3 STRUCTURE OF THE ADHESION SITE

The size of the individual membrane adhesion site is too small to be recognizable under the light microscope. Of the current methods of electron microscopy, only ultrathin sections of plasmolyzed cells reveal them. Normally, the turgor within the cytoplasm of the living unplasmolyzed cell seems to keep the inner membrane attached to the outer membrane. In order to describe the fine structure of the adhesion site, a few of the main features of the cell envelope are outlined here. More detailed analyses of the structure of the envelope of gram-negative bacteria have been the subject of previous reports and reviews (Brauet et al., 1976; Costerton, 1977; Costerton et al., 1974; Salton, 1971). Ultrathin sections show a double contour in the contrast distribution of outer membrane, an image predominantly attributed to the scattering properties of the heavy metal of the fixing and staining solution. The metals are bound to the lipid portion of the membrane, its phospholipids, and the lipid A of the lipopolysaccharides (LPS), and possibly also to the lipoprotein (Hirashima et al; 1973; Braun

1975); in contrast, the integral and transmembrane proteins of the membrane, as well as the polysaccharide of the LPS facing the outside environment of the cell (Mühlradt et al., 1973; 1974; Smit et al., 1975), retain comparatively little, if any, stain and therefore are not visible in the electron microscope; special staining procedures must be employed to make carbohydrate portions of LPS and capsules recognizable (Shands, 1971; Bayer and Thurow, 1977). The contrast distribution of the double-track contour of the outer membrane is asymmetric due to the inner "track" being considerably thicker than the outer track. This added thickness is most likely caused by the peptidoglycan and by its associated proteins (Murray et al., 1965). In contrast, the cross-sectioned profile of the inner membrane (i.e., protoplasmic membrane) is that of the typical lipid bilayer and reveals in ultrathin sections no obvious asymmetry; in spite of this aspect, the inner membrane has been shown to be highly asymmetric when membrane markers are used, such as transport of calcium, proteins (Rosen and McClees, 1974), ATPase activity, and NADH ferricyanide reductase activity (Futai, 1974). However, membrane fractions and vesicles may pose problems because they may form in varying proportions inversed (inside-out) vesicles (Kaback, 1973), and a redistribution of the LPS from an asymmetric distribution in the living cell to a more random localization at both faces of the membrane may occur (Mühlradt and Golecki, 1975).

Measurement of membrane antigenicity and enzyme activity of whole cells and protoplasts indicated that ATPase is located at the inner surface of the protoplasmic membrane (Owen and Salton, 1975). Furthermore, freeze-etching and freeze-fracturing replicas of the membranes of whole cells and of vesicle preparations expose an asymmetric arrangement of particulate elements in the hydrophobic fracture faces of the outer membrane. These particles most likely represent integral and transmembrane proteins, possibly complexed with membrane lipids (Smit et al., 1975; Bayer et al., 1977; Verkleij et al., 1977). Obviously, these investigations point to an intriguing complexity of the envelope. Although a great number of functional markers are present at the membranes, one must remember that most biochemical data were obtained with methods that do not discern individual sites but rather collect the information in a randomized fashion from the total material of a particular type of membrane. To discern localized domains of altered structural and functional composition, microscopy is needed to inspect individual cells and membrane domains and a search for suitable "markers becomes essential." Since the contribution of the membrane adhesion site to the total cell surface is small and does not comprise a major antigenic component, electron microscopy is presently the only method to reveal them unambiguously, and stable markers for the isolation of these sites seem to be rather difficult to isolate. For example, Crowlesmith et al. (1978) found that phage P22, adsorbed to the cells before vesicle preparation, cannot be safely employed as membrane marker during purification by density gradient centrifugation. However, recent work on the isolation of adhesion sites seems to show encouraging progress (see below, also Bayer and Bayer, 1979).

The typical structure of the adhesion site has been described earlier (Bayer, 1968a) and is shown in the micrographs in Figures 1, 2, 3 and 4; Figure 5 depicts its general features. At the site of a membrane adhesion, the outer membrane (OM) is apparently continuous and often seems to be pulled in by the inner membrane (IM) in the direction of the shrinking protoplast. The contact sites are usually rather small and measure 200–300 Å across, with a few larger contact zones occasionally also being present. At the contact area of the inner membrane with the inside of the outer membrane, the relatively thick inner track of the outer membrane seems to be thinner, suggesting to us that some of its constituents are possibly missing at the inner membrane insertion. The exact definition of the insertion or fusion site is difficult to ascertain because of an accumulation of granular amorphous material at these areas. In cross sections of adhesion sites, the extended parts of the inner membrane are more or less circular, with a clear open central portion, suggesting a structure of a conical tube (Bayer, 1968a; Figures 2 and 3). Since the outer membrane appears to be continuous, the adhesion area seems to be structurally "sealed" and does not exhibit an

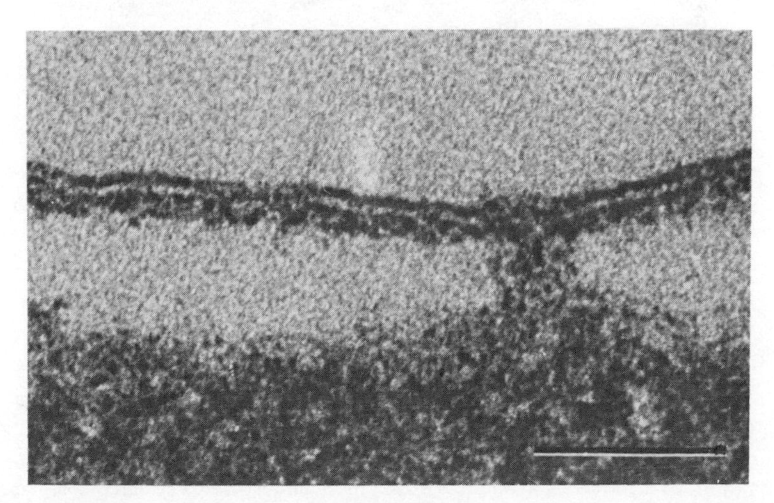

Figure 1 Ultrathin cross section through the outer membrane and cytoplasmic (inner membrane, including a membrane adhesion site. The cell (*Escherichia coli*) was fixed and embedded in the plasmolyzed state. The bar in this and the subsequent micrographs represents 1000 Å = 100 nm.

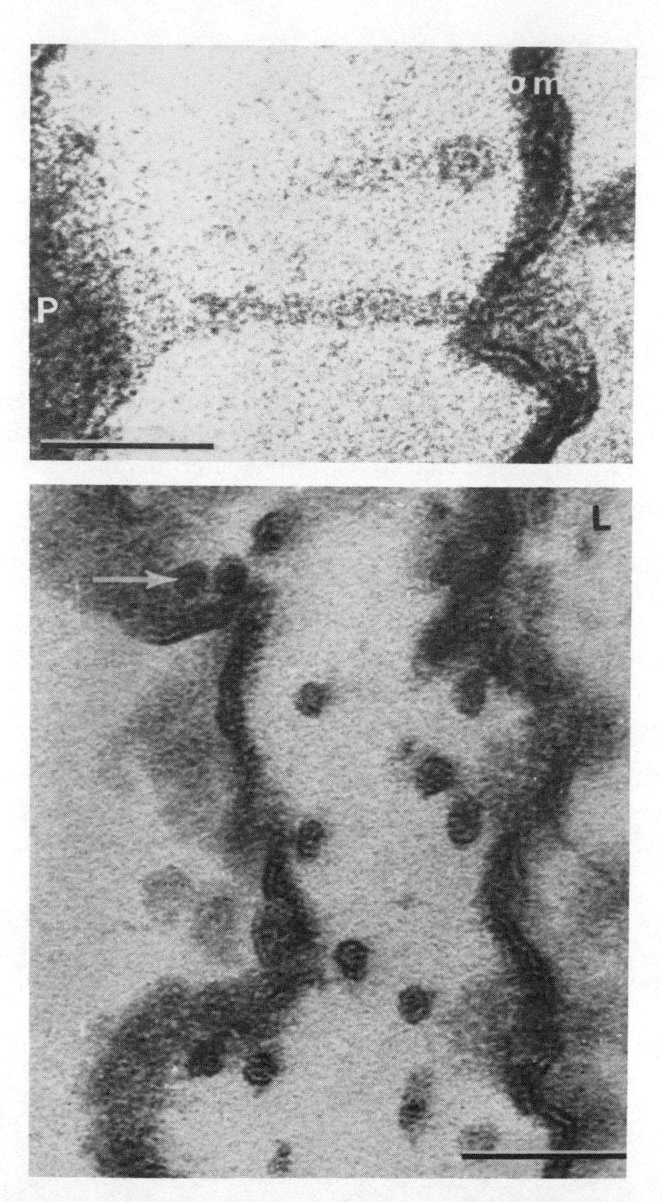

Figure 2 In this micrograph the plane of the ultrathin section is oblique relative to the axis of the inner membrane's tubular extension. These extensions are caused by the pull of the inner membrane, which is retracting at all areas except the adhesion sites. The outer membrane (om) faces the right side of the micrograph; the protoplasmic contents (P) face the left side. The cut through the upper extension reveals a beaded circumference.

Figure 3 Cross section showing a number of inner membrane extensions as they bridge the gap between inner and outer membrane.

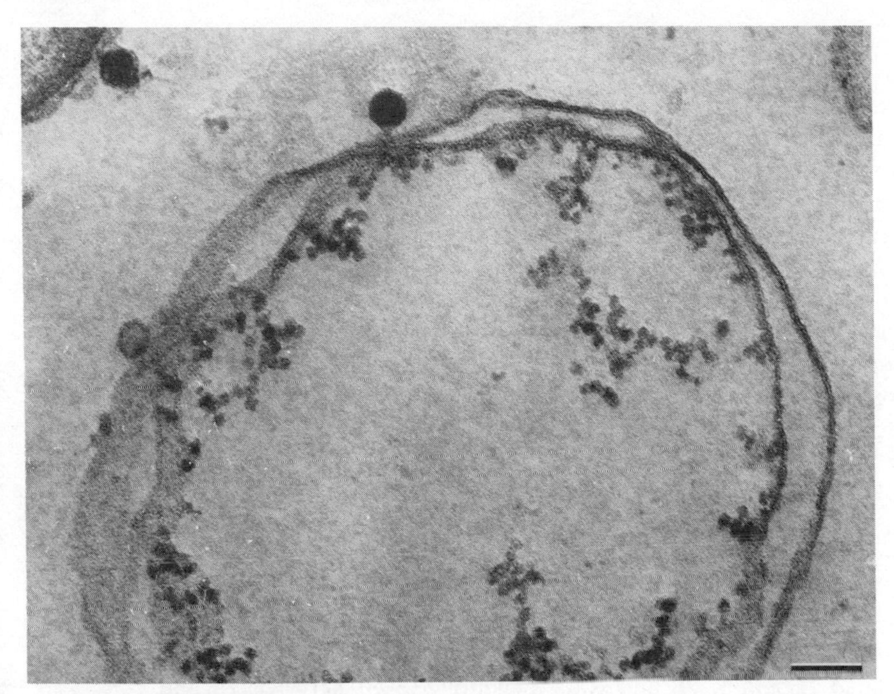

Figure 4 Ultrathin section of membrane adhesion zones in the envelope of a lysed *Caulobacter* cell; two phage particles are adsorbed to the cell surface. Ribosomal clusters remain attached to the inner membrane.

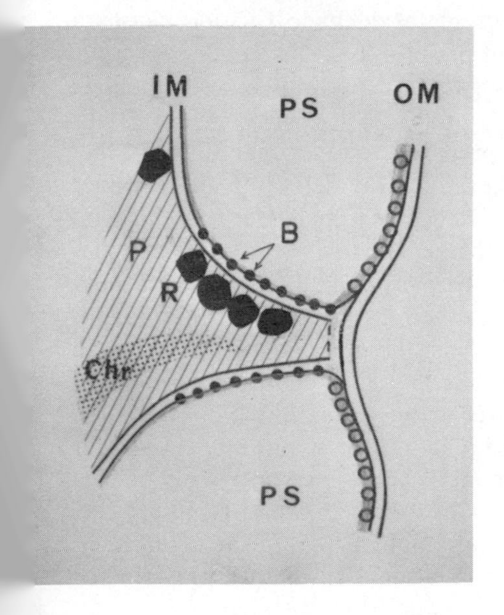

Figure 5 Diagram of an adhesion site. Om, outer membrane; IM, inner membrane; PS, periplasmic space; B, beaded appearance of IM; P, protoplasmic contents; R, ribosomes adhering to inner membrane; Chr, portion of chromosome.

obvious opening in the envelope. Freeze-etching and freeze-fracture studies have given no indication for a discontinuity (a "hole") in the outer membrane; however, such a discontinuity could be obscured by "nonetchable" material filling the hole. Estimates from ultrathin sections revealed a total of 200–400 adhesion sites per growing *E. coli* (Bayer, 1968a). The surface area taken up by the insertion domains was estimated to be about 5% of the total surface of the growing cell. In stationary cells (*E. coli* B, Cla), adhesion sites are not visible. Instead, rather extended areas of contact between inner and outer membrane have been observed (Bayer, unpublished observations). Since a certain degree of plasmolysis is necessary to visualize adhesion sites, one may speculate that the presence of plasmolyzing sugars could exert an effect on the structure of the cell envelope. So far, we have not found any evidence that this is the case. On the other hand, plasmolysis can be too forceful, so that rupture of the junctions occurs (Bayer and Starkey, 1972). The fine structure of the adhesion sites seems to be unaffected by the composition of the plasmolyzing sugars such as sucrose, mannitol, and lactose.

The extent of plasmolysis produced by a given concentration of plasmolyzing agent may differ significantly among the various bacterial strains and their mutants. *E. coli* B requires about 10% (w/v) sucrose, capsulated *E. coli* K-26 and K-29 plasmolize well between 15 and 20%, and *Salmonella anatum* needs at least 20% sucrose. An increased plasmolysis is observed when the cells (*E. coli* B) are placed in KCN, either before or during plasmolysis. It should also be mentioned here that fixation affects the degree of plasmolysis: formaldehyde or glutaraldehyde fixation causes a relatively small, but recognizable, loss of plasmolysis, whereas, for example, osmium tetroxide or potassium permanganate deplasmolyze cells quickly, with partial lysis ensuing.

In conclusion: (1) Plasmolysis is the prerequisite for recognition of adhesion sites. (2) At the contact area, the structural definition of the inner membrane in general is not clear enough to decide (*a*) whether true "fusion" of the membranes has occurred or (*b*) whether the two membranes are continuous and are "welded" together by their proximal surfaces, with both double tracks running parallel to each other. Many of our micrographs show the arrangment described in *a*, in which the inner membrane seems to be discontinuous at the adhesion zone while the outer membrane retains its double track appearance. (3) A specialized arrangement of membrane particles comparable to the rather organized junctions of eukaryotic cells so far has not been observed, with the exception of some insertion areas of bacterial flagellae, which are described later in this chapter. In lysed cells we have often observed an attachment of ribosomal clusters to the inner membrane, including adhesion sites. Determination of whether this finding depicts the ribosome/membrane association described above, awaits further data.

4 FUNCTION OF THE ADHESION SITE

The existence of adhesion sites suggests that the cell surface is composed of a mosaic of structurally unique domains. The first indication for a particular function associated with the adhesion areas was the discovery that a large variety of different species of bacteriophages infected the cells at these sites (Bayer, 1968b; Bayer and Starkey, 1972; Bayer, 1975; Bayer et al., 1979). Subsequently, other functions of the adhesion sites were found namely, the production (1) of O-antigens (Bayer, 1973, 1974, 1975, 1977; Mühlradt et al., 1973; Mühlradt and Golecki, 1975), (2) of specialized proteinaceous organelles (F pili) (Bayer, 1973, 1975) and (3) of outer membrane proteins (Smit and Nikaido, 1978). These data are discussed in the following sequence: (*a*) synthesis (export) of cell surface components: polysaccharides, proteins; (*b*) basis for insertion of specialized organelles: pili, flagellae; (*c*) virus receptors, virus injection sites, associated functions of receptors and; (*d*) multifunctionality of the adhesion site, anchoring site for chromosomes, site of origin of membrane signals.

4.1 Insertion of Lipopolysaccharides and Capsular Polysaccharides

The synthesis of components of the cell envelope has been the topic of many detailed investigations and reviews (Sutherland, 1977; Kanegasaki and Wright, 1973; Osborn et al., 1972; Lüderitz et al. 1971; Braun et al., 1976; Machtiger and Fox, 1973; DiRienzo et al., 1978; Costerton et al., 1974; Costerton, 1977). Since lipopolysaccharides are assembled in association with the inner membrane, they have to be translocated to the outer membrane, where they provide the areas of newly synthesized surface material. As soon as the new LPS is available as antigen and as specific receptor for viruses, the material can be labeled and observed with either fluorescence microscopy or electron microscopy; other methods such as immune precipitation and virus adsorption provide indirect, but critical, supporting data for the morphological results. Fluorescence microscopy, although rather low in resolution, provided the first clues to the insertion mode, as well as to the timing of surface antigen production. The data suggested that lipopolysaccharide insertion is a multiple site intercalation process (Beachey and Cole, 1966). Such an intercalation process is not unique for gram-negative rod-shaped organisms but has also been shown to be one of the mechanisms by which cell wall enlargement takes place in the cylindrical portion of *Bacillus* species (see, for a tabulation of these data, Doyle et al., 1977). The criticism that the marker antigen may be rapidly diffusing over the surface and therefore a clustering of fluorescence might be an artifact, turned out to be largely unjustified. The work of Mühlradt et al. (1974), who studied *Salmonella* species, and the work of our own group using *E. coli* B and *Salmonella anatum* (Bayer, 1974) and *E. coli* Cla (Bayer and Leive, 1977), as well as the results of Smit and Nikaido (1978) with *Salmonella typhimurium* clearly showed that the mobility of LPS molecules within the outer membrane is extremely slow and therefore negligible as long as surface labeling is executed at low temperatures. Mühlradt et al. (1974) calculated a diffusion constant (D) for a newly synthesized LPS molecule of $D = 2.9 \times 10^{-13}$ cm^2/sec. It is remarkable that the value is several orders of magnitude smaller than that for diffusion of markers in animal cell surfaces (Frye and Edidin, 1970). Mühlradt et al. (1974) suggested that the low phospholipid content (13–18%) of the outer membrane (Osborn et al., 1972) might be responsible for the slow diffusion. Additional factors such as LPS–LPS interactions and relative protein concentration seem also to be significant in determining the liquidity of the outer membrane (Bayer et al., 1975; Nikaido et al., 1977; Rottem and Leive, 1977). The localization of the production sites of specific surface antigens was achieved by employing electron microscopy of ferritin-labeled antibody. Ultrathin sections and freeze-etching replicas (Bayer, 1973, 1974, 1975; Mühlradt et al., 1973, 1974) of the outer membrane revealed clearly small patches at which new antigen (such as LPS) was labeled with ferritin conjugate. These labeled patches were preferentially positioned over membrane adhesion sites. The density distribution by which new LPS was introduced into the cell surface varied between the experiments of Mühlradt et al. (1973, 1974) and those of our laboratory (Bayer, 1974, 1975). The former group used a *Salmonella typhimurium* mutant that was lacking UDP-galactose 4-epimerase; wild-type LPS was producted when the growth medium was substituted with galactose (Nikaido, 1960). Wild-type LPS appeared on the cell surface within 30 sec after transfer into galactose-containing medium at about 220 patches/cell (Mühlradt et al., 1974). Our laboratory employed a different approach: *Salmonella anatum* cultures were infected with phage ϵ15, which "converts" the newly produced LPS from O-antigens type 3,10 (E_1) to O-antigen 3,15 (E_2) (Uetake et al., 1958). This change is controlled by three phage genes and involves synthesis of a β-polymerase that causes a switch in the synthesis of the O-antigen from an α-galactosidic linkage of the subunits to a β-galactosidic linkage; in addition, an inhibitor is produced blocking the "old" α-polymerase, and an *O*-acetyl side group in the converted LPS is also omitted (Robbins and Uchida, 1962). In the conversion system, other characteristics of the LPS, including its chain length, seem to remain unaltered, a fact that makes the system ideal for studies of the cell surface viscosity. Phage ϵ15 is a lysogenic phage; however, the conversion genes of the virus do not require for their expression an incorporation into the bacterial chromosome. This makes possible a fast response of the cells to the infection event. Employing this system, the shortest time for antigen production (i.e., conversion) was obtained in fast-growing cells within about 1 min; this time span included the mixing of virus and cell culture, virus adsorption, infection, synthesis of the β-polymerase, and the export of the new LPS to the cell surface (Bayer, 1975). Certain parameters make the conversion system ideally suited for investigations of the synthesis of LPS, as well as for studies of mechanisms of virus adsorption; for example, (1) the onset of conversion of a culture can be regulated by cell density, since the time taken up by virus diffusion through the medium is reduced in cultures of higher densities, (2) we found that the onset of conversion is also dependent on the number of phages adsorbing per host cell, possibly as a result of shorter time spent for surface "walk" of the virus before infection (M. E. Bayer et al., in preparation), (3) the time span between addition of virus and conversion can be controlled by the growth conditions of the cultures (Bayer, M. H. and Bayer M. E., 1979).

Employing immunoelectron microscopy, we found that within 10 min after conversion, LPS is exported to the cell surface at 20–40 patches, whereas most of the remaining sites reveal no such activity. The LPS production sites are randomly distributed over the entire cell surface, including cell poles and areas around the septum. The new LPS is extruded over membrane adhesion sites. At later times, beyond 10 min growth at 37° counts can no longer, be performed since the growing patches of new LPS start to make contact with each other. Since introduction of LPS can be stopped by KCN and deoxyglucose, as well as by low temperatures, it is also possible to measure the diffusional expansion of the newly exported LPS patches: in poisoned cells at 37°C, it takes at least 10–15 min to recognize a small amount of expansion of the ferritin label. This slow rate is in agreement with related data of Nikaido et al. (1977), Mühlradt et al. (1974), and Rottem and Leive (1977).

The striking difference between the conversion system and the systems using mutants lacking galactose-epimerase is the number of LPS production sites: the conversion system involves about 10% of the total number of potentially available adhesion sites, whereas in the mutants lacking epimerase, apparently most or all of the adhesion sites are producing LPS.

For an estimate of the speed of conversion, the following considerations are offered. Calculations of the time required for a phage to adsorb (Wong et al., 1978), as well as recent fluorescence methods measuring the time required for injection of virus DNA (Bayer and Bayer, 1978), suggest that the production of the conversion LPS seems to be as fast, or even faster, than the 30 sec measured for LPS export in epimerase-less mutants (Mühlradt et al., 1973). Conversion executed under optimal conditions requires about 70 sec, when immunological and virological methods are used. Phage ϵ15 adsorbs with diffusion-limited speed; the phage induces a fluorescence signal (a membrane response) of the cell after 40–50 sec (Bayer and Bayer, 1978), leaving 20–30 sec for the conversion antigens and receptors to manifest themselves on the cell surface. Thus the response of the molecular machinery producing LPS seems to be surprisingly fast.

In addition to lipopolysaccharides, enterobacteria exhibit *capsular antigens* (Lüderitz et al., 1971). A capsular strain (K-29) of *E. coli*, whose antigenic structure has been determined (Choy et al., 1975; Fehmel et al., 1975), and several of its temperature-sensitive mutants were studied to find the sites of synthesis of the capsule polysaccharides; employment of immunoprecipitation, virus adsorption, and electron microscopy revealed that polysaccharide strands occur at a limited number of cell surface sites; they are exported at some of the membrane adhesion sites (Bayer and Thurow, 1977). In contrast to LPS production, these mutants require a considerably longer time, namely, 15 min, for production of measurable quantities of antigen. The slow induction might be linked to the slow growth rate of these cell strains (Bayer and Thurow, 1977).

In summary (see also Figure 13) we conclude that lipopolysaccharides and capsule polysaccharides are produced over membrane adhesion sites. The number of production sites per cell varies among the systems employed from apparently all adhesion sites being involved in the *Salmonella* species lacking galactose-epimerase to only 10% of the available adhesion sites exporting LPS in the conversion system of *Salmonella anatum*. A similarly small number of production sites was found in the capsular mutants of *E. coli* K-29. It seems that adhesion sites are involved in different functions, or they might undergo different functional states, in contrast to their relatively uniform morphological appearance in the electron microscope.

4.2 Insertion of Pili and Flagella in the Outer Membrane

In addition to lipopolysaccharides, capsule polysaccharides, and phospholipids, proteins constitute a large fraction of the outer membrane. The proteins of the outer membrane are grouped into a relatively small number of "major proteins" and a larger number of "minor proteins." Their characterization, mainly by polyacrylamide gel electrophoresis, has led to the use of several systems of nomenclature (Dirienzo et al., 1978; see also Chapter 1). The major proteins are present in comparatively large amounts (7×10^5 copies/cell), whereas minor membrane proteins may occur in quantities of up to 10^2 copies/cell. Morphologically, most obvious are those proteins that are organized in specialized structures, such as the various types of pili (Brinton, 1965; Lawn et al., 1967; Manning and Achtmann, see Chapter 12) and flagellae (DePamphilis and Adler, 1971). The unambiguous assignment of the inserted segments of these organelles to either the outer or the inner membrane is difficult because F pili, as well as flagellae, have their origin at a "fusion site" between the two membrane systems (Bayer, 1975) and Fig 5.; obviously, these adhesion areas are clearly distinguished by their assembly product from the "typical" adhesion site. Synthesis of these proteinaceous organelles occurs in different assembly steps: from the pilin, composition of the F pili, assembly of the pilus and extension through the outer membrane. The site of the deposition of pilin, either at the inner or outer membrane is discussed in Chapter 12 by Achtmann. The flexible organelle serves as virus receptor and also functions as recognition element in the initiation of conjugation (Manning and Achtmann, see Chapter 12). The reverse reaction, namely, a retraction of the pilus, has been reported to occur in the presence of KCN (Novotny and Fives-Taylor, 1974). The suggested capability for the cell to either assemble ("grow") or retract (disassemble?) the pilus would make the insertion area of the organelle a highly complex structure (Figure 6).

Bacterial flagellae, on the other hand, exhibit a different mechanism of growth. Their newly synthesized subunit proteins are attached to the distal end of the flagellum, so that this organelle grows from its tip. The proximal end of the flagellum, where it is attached to the envelope (Coulton and Murray, 1978), has been the topic of detailed studies. A "motor" responsible for the action of the flagellum is anchored at a fusion site of the membranes. Berg (1975) proposed a "rotational engine" turning the flagellum in either of two directions (Berg and Brown, 1972; Silverman and Simon, 1974) in response to gradients of attractants or repellants. In a morphological study of flagellar bases in *Spirillum serpens*, Coulton and Murray (1978) reported that the fine structure of the areas of flagellar insertion consists of a complex structure, revealing concentric rings around the flagellar shaft. These rings have been assigned to the inner leaflet of the outer membrane. It is surprising that the insertion area of flagella in *E. coli* B and in *Salmonella* strains does not display such an extended structure. Instead, only the area directly around the inserting flagellar shaft seems to be organized in a single circle of spherical elements (Coulton and Murray, 1978). In spite of an extensive search in our own laboratory for typical structural markers in freeze-etched membranes of *E. coli* and *Salmonella* strains, there has been a scarcity of identifiable insertions, even in preparations of multiflagellated (peritrichous) cells; only in a few instances have we found structures such as the one shown in Figure 7. This micrograph of the inner membrane's outer

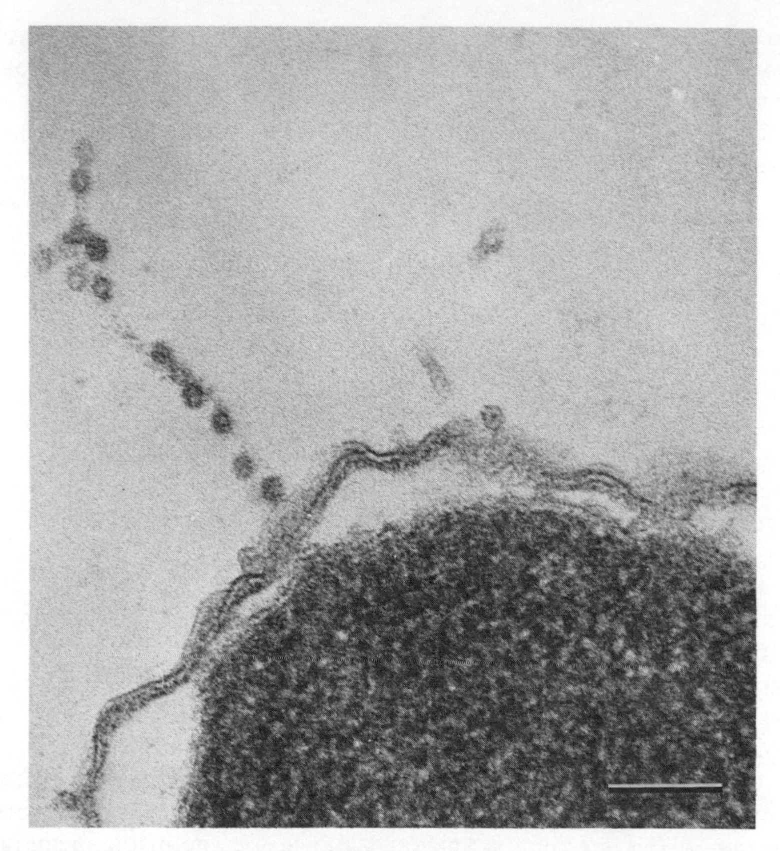

Figure 6 Ultrathin section of an *E. coli* HfrH showing the attachment area of an F pilus; the pilus is labeled with bacteriophage MS2.

leaflet possibly depicts a cluster of flagella (Bayer, unpublished work). In ultrathin sections, the insertion site represents itself as a membrane adhesion site, with a smooth membrane profile (Bayer, 1975).

The study of the basis of flagella derives its increasing importance from the rapid progress in the field of chemotaxis (Berg, 1975), with implications of membrane signal transfer and its linkage to the action of the flagellar "motor." Ridgeway et al. (1977) described the localization (within the envelope) of proteins controlling mobility and chemotaxis and the essential role of methyl-accepting chemotaxis proteins in rapid excitation (Silverman and Simon, 1977; Springer et al., 1977), as well as in the slower process of adaptation (Goy et al., 1977). Both processes govern the transient signal and the response to a stimulus and show features in common with mechanisms

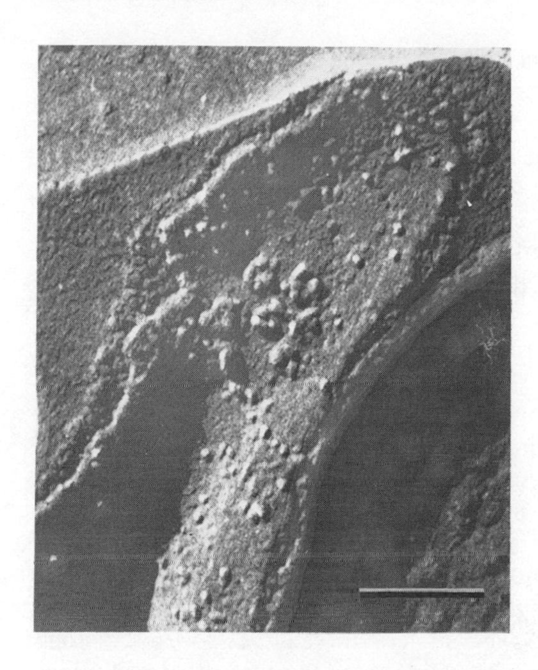

Figure 7 Freeze-cleaved inner membrane (concave plane) of *E. coli* AW 405 showing insertion area of flagella.

of sensory adaptation (Delbrück and Reichardt, 1956). Furthermore, the studies on the flagellar insertion encompass the relationship of membrane potentials, cell motility, and chemotaxis (Miller and Koshland, 1977). It appears that the "sensory activities" of the bacterial membrane system significantly contribute to the regulation of the general physiology of the microorganisms (Harold, 1977). Attachment of bacteriophages to adhesion sites seems to have a severe and rapid effect on the cell membrane potential and on several membrane functions; it is interesting that flagellotropic phage chi, which infects at the base of the flagellum, seems to be indistinguishable in this respect from other lytic bacteriophages (M. E. Bayer et al. in preparation).

4.3 Outer Membrane Proteins and Adhesion Sites

Insertion of proteins into the outer membrane is discussed above only for specialized organelles, such as pili and flagella. Other proteinaceous organelles of the cell surface (Brinton, 1965) are often abundant and seem to be scattered over the entire cell surface. Pilus-like structures such as antigen K88 (L) and K99 (L) are striking proteinaceous elements that play important roles in the virulence of the cells (Stirm et al., 1967; Orskov et al., 1975). However, their localization and production sites are not known.

Major outer membrane proteins comprise more than 50% of the mass of all other membrane proteins. There are close enough to be crosslinked with each other (Haller, and Henning, 1974). Until recently, their sites of synthesis and incorporation into the membrane were not known except for a few indirect observations relating to cell growth and peptidoglycan synthesis (Doyle et al., 1977). Insertion of specific labels into the oligopeptide of the peptidoglycan was followed by autoradiography, and a diffuse intercalation was reported, as well as an insertion at the septum areas (Schwartz et al., 1975). Multiple randomly positioned zones of penicillin-sensitive sites in the peptidoglycan-associated structures of the envelope have been described earlier (see for example Bayer, 1974) and were interpreted as expressions of disturbances in the distribution of the major membrane proteins. However, an involvement of the adhesion sites in peptidoglycan synthesis has not been demonstrated. A recent study of Smit and Nikaido (1978) provided evidence for the involvement of adhesion sites in the insertion of major outer membrane proteins of *Salmonella typhimurium*. These proteins, the "porins" (Nakae, 1976a, b), correspond to the matrix proteins in *E. coli* (Rosenbusch, 1974). Nearest neighbor analysis suggests a dimeric protein (Palva and Randall 1976). They have a molecular weight of 34–36 k daltons and have been shown to be largely responsible for the formation of "transmembrane" channels. By using nonpermeable label, Kamio and Nikaido (1977) showed that matrix protein is exposed on the cell surface of *Salmonella* and *E. coli* B (see also Chapter 11 by Nikaido).

The porins are present in various relative amounts depending on the cell strain and culture condition (medium, temperature, pH). The production of the proteins can be repressed and derepressed in accordance with the salt concentration of the medium. Employing this approach, and using ferritin-labeled anti-porin IgG, Smit and Nikaido (1978) demonstrated in mutants of *Salmonella typhimurium* that porins are integrated in the outer membrane at numerous discrete sites. These authors found that the insertion sites are distributed randomly over the entire cell surface. A preferential labeling of the cell poles or septum regions was not seen. With increasing time or porin production, the ferritin patches did not increase in size, but the number of production sites increased rapidly. The porins first became demonstrable at around 10 min after onset of synthesis. This time span is in agreement with the onset of synthesis of capsular polysaccharides (Bayer and Thurow, 1977), but it is relatively slow in comparison to the 1.5–2 min needed for LPS export. Smit and Nikaido demonstrated further that the label for the newly synthesized outer membrane proteins was located over membrane adhesion sites, adding one more function to these cell surface domains.

4.4 Multifunctionality of Adhesion Sites

From the published data, one can postulate that the membrane adhesion site is involved in a rather complex set of functions in the cell envelope: (1) export of lipopolysaccharides, (2) export of capsular polysaccharides, (3) export of major membrane proteins, (4) export of F pili, and (5) anchoring of flagellae. Obviously, in the last two cases, the membrane fusion site is only related by structural similarity to the more common adhesion site; this fact does not exclude the presence of similar organizational principles possibly common to all adhesion sites. Since the proteins, as well as the LPS, are implicated in the permeability and hydrophobicity of the outer membrane (Inouye, 1974), their role in providing a barrier to macromolecules and antibiotics (Ichihara and Mizushima, 1978; Costerton, Ingram, and Cheng, 1974) becomes an important factor to the survival of the cell. (6) there is also morphological evidence that the bacterial chromosome is attached to some of these sites (Bayer, 1968a). Although we have not studied this aspect in greater detail, we find the chromosomal associations frequently enough to exclude their adventitious association with the inner membrane at adhesion sites (Figure 8). (7) the disturbance of the adhesion site, for example, by phages binding firmly to the envelope, seems to cause a membrane signal that spreads rapidly over the cell surface. The quality of the signal is not known but seems to implicate changes in membrane potential (M. E. Bayer et al., in preparation).

4.5 Bacteriophage Infection and Adhesion Sites

The study of the adsorption and infection mechanisms of virus particles further emphasized the multifunctionality of adhesion sites. It appears that, in general, all types of surface elements of the bacterial cell may potentially serve as receptors for bacterial viruses. However, the individual virus strains are highly specialized with regard to the range of cell surface receptors; genetic studies have shown that the specificity of the adsorption apparatus, rather than the morphological type of virus, determines the unique virus–receptor relationship. The adsorption organelle of a phage is composed of several proteins; these proteins may comprise tail fibers of various length and composition, and base plates of varying complexities. Other phages exhibit rather simple "spikes" that form hexagonal arrays at one (Figure 9*a*) or all vertices of the virus capsid (see Ackermann, 1973). The major function of these organelles is the recognition and attachment to the specific cell surface receptor, in most cases a polysaccharide and/or a protein. In an increasing number of phages, adsorption organelles with recep-

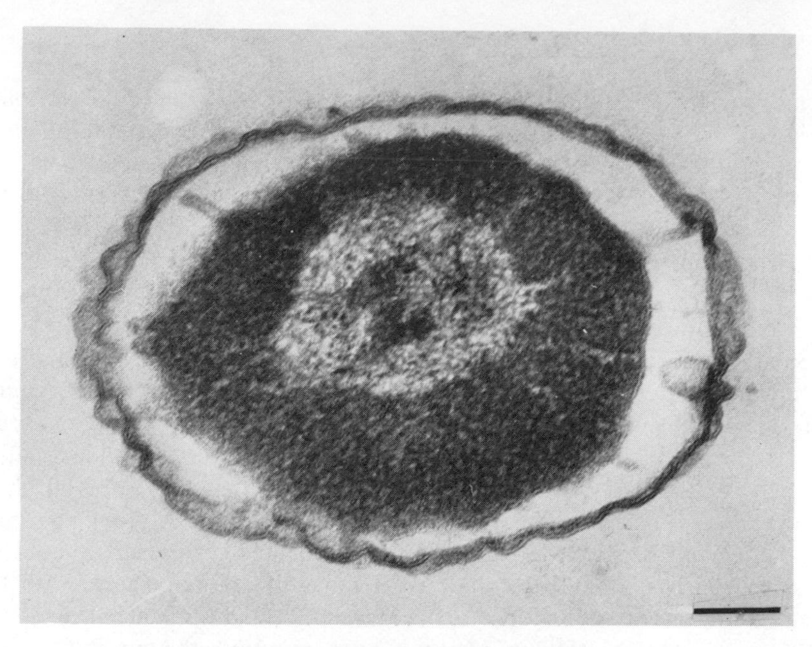

Figure 8 In this plasmolyzed *E. coli* B, the chromosome is visible as a relatively light area in the center of the cell. Delicate protrusions of the chromosomal material can be seen that seem to connect to some of the membrane adhesion sites.

tor-destroying capability have been found (Lindberg, 1977). Such hydrolysis of the receptor enables the phage to penetrate enzymatically the relatively thick polysaccharide layer of the host cell (Figure 9*b*). For a number of phages, the proteins of the adsorption organelles have been well identified: Phage P22 requires for adsorption the product of gene 9, which controls the proteins for the six spikes of the baseplate of the virion (King et al., 1973); LPS hydrolysis is not essential for adsorption; successful infection requires the presence of three more proteins (P6, P16, P20) (Susskind and Botstein, 1978). Israel (1977) has obtained evidence that these three proteins and another protein (P26) are ejected from the virion after adsorption. There might be considerable similarities to phage P22 in the arrangement, as well as in the function, of the adsorption organelles of phages that are structurally related to P22 (such as phages ϵ15 and ϵ34). The wide range of receptors of the bacterial surface is paralleled by the variety of virus structures and specificities (Lindberg, 1973, 1977; Braun and Hantke, 1978; Bartell 1977). We discuss here mainly those virus–cell systems for which an involvement of the adhesion sites in the infection mechanism has been demonstrated. However, it is our belief that the cases studied are not exceptions, but rather, that the involvement of the adhesion site in bacteriophage infection is a general feature.

After sufficient time of stable growth, the receptors such as LPS or capsule polysaccharide are evenly distributed over the entire cell surface. Early steps of interaction of virus and cell surface consist of collision and adsorption. In later steps, the virus is firmly bound to the cell and virus nucleic acid is released. To initiate the latter event, additional signals are required, which appear to be provided in living cells by the adhesion sites.

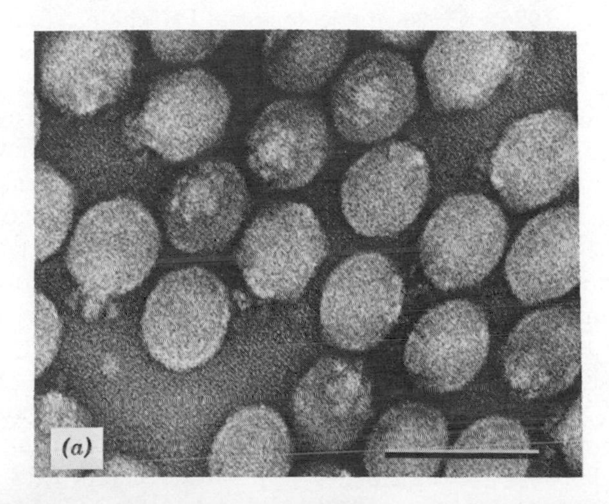

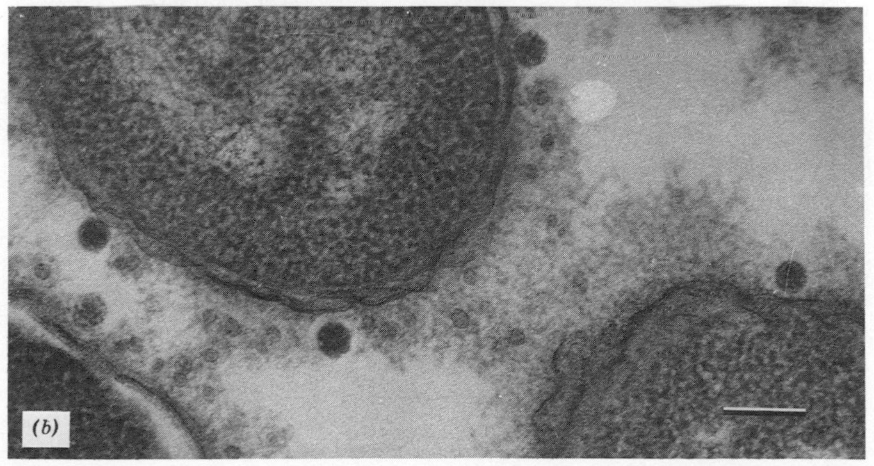

Figure 9 Bacteriophage ε15 (*a*) negatively stained in uranylacetate (note the adsorption apparatus with its sixfold symmetry and its center "trunk") and (*b*) after adsorption to the host cell, *Salmonella anatum*; the cells were subsequently exposed to O-specific antiserum. The O-antigenic coat of the cells is destroyed in the vicinity of the virion.

Virus receptors have been classified as (*a*) lipopolysaccharides (rough and smooth types), capsule polysaccharides, and (*b*) proteins.

4.5.1 Lipopolysaccharides and Capsule Polysaccharide Receptors Viruses requiring the short ("rough") LPS phenotype for adsorption include T3, T4, T7, and $\phi\chi$174. Bacteriophage T4 attaches to a glucose disaccharide linked to the heptose of the core region of the LPS. This core region is common to a variety of species. Thus phage T4 covers a wide host range.* Removal of the terminal glucose leads to a 10–50% reduction of receptor activity (Prehm et al., 1976). Considerable longer LPS chains, found in typical smooth cell strains or capsulated cells, do not permit rough-specific phages to reach the deeper portion of the LPS; these fail to adsorb. Furthermore rough-specific phages do not seem to have receptor-destroying enzyme activity. In contrast, the *Salmonella* phages, such as ϵ15, ϵ34, and g341, have a requirement for LPS of a specific smooth type O-antigen. The phages hydrolyze their receptor enzymatically. Without this activity, the phage would not be able to engage in close contact with the deeper region of the outer membrane and with the membrane adhesion site (Figure 9*b*). Phage ϵ15, for example, adsorbs to the rhamnose–galactose linkage (Uetake and Hagiwara, 1969) in the O-antigen repeat unit only when an α-linkage between repeat units is present. Phage ϵ34, on the other hand, adsorbs to the β-linkage between the *Salmonella* O-antigen subunits only when the galactose of the subunits is not substituted. Bacteriophage P22 adsorbs to *Salmonella* O-antigen 12 (Zinder, 1958); its base plates destroy the receptor LPS (Israel et al., 1972; Erikson and Lindberg, 1977) at the rhamnose galactose α-linkage (Lindberg, 1973; Iwashita and Kanegasaki, 1973). Substitution of the galactose or alteration of the subunit linkages renders the cells P22 resistant. A similarly narrow host range is found in the capsule-specific phages: Out of 82 heterologous exopolysaccharides, only one, that of *Klebsiella* K31, was hydrolyzed by the phage K29 endoglucosidase (Thurow et al., 1975). Phage K29 hydrolyzes the glucose $1\rightarrow3$ glucoronic acid linkage of the hexasaccharide subunit of the K29 polysaccharide (Fehmel et al., 1975). With regard to the second step in the virus–cell interaction, the irreversible attachment, all the LPS-specific phages, as well as the capsule-specific phages K29 and K26, were found to inject their DNA preferentially at membrane adhesion sites Figure 10 (Bayer, 1968a, 1975; Bayer and Starkey, 1972; Crowlesmith et al., 1978; Bayer et al., 1979).

4.5.2 Membrane Proteins as Phage Receptors For these studies, the binding and loss of infectivity (eclipse) of the phage is measured after

* It is competitively inhibited from adsorption by a number of di- and trisaccharides (Dawes, 1975).

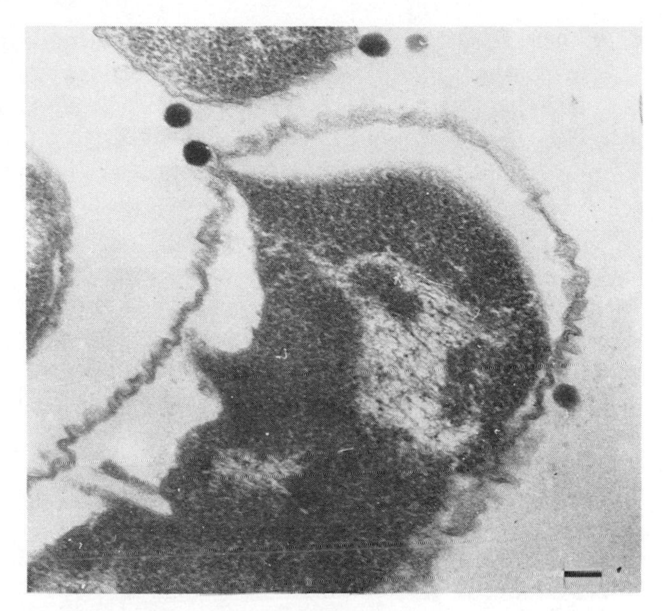

Figure 10 Bacteriophage K29 adsorbed to plasmolyzed *E. coli* K29. The capsule has not been contrasted in this preparation.

interaction with isolated membrane proteins. In an increasing number of instances, the combination of the protein with the LPS results in a higher yield of phage inactivation (Datta et al., 1977). The major proteins of the outer membrane, such as matrix protein, or protein Ia, Ib, and ompA protein, comprise > 10^5 molecules per cell; these proteins not only provide phage receptors (Schnaitman et al., 1975; van Alphen et al., 1977; Verhoef et al. 1977), but also have a number of other functions (Dirienzo et al., 1978), such as provision of transmembrane "pores" (NaKae, 1976a,b) and mating pair formation (Skurray et al., 1974; Manning et al, 1976). Bacteriophage T2, which requires a major outer membrane protein for adsorption and injection (Michael, 1968; DePamphilis, 1971), adsorbs to the host cells with adsorption rates approaching the diffusion limit but it releases its DNA over adhesion sites (Bayer, 1975). The distribution of the injection sites of this phage is random (Bayer, 1968). The infection mechanism of this phage is sensitive to membrane changes such as those associated with superinfection: superinfecting virions do not recognize the adhesion sites as injection sites (Bayer, 1975). Recently, Hantke provided evidence that matrix protein Ia serves as receptor for phage T2 (Hantke, 1978).

Other membranes, which would be classified as minor, also serve as receptors for a variety of bacteriophages. The outstanding features of protein receptors, with their apparent multifunctionality, is especially obvious in these minor proteins, which are of vital importance for cell growth; it appears as if they have been parasitized by the bacteriophages (Braun et al., 1976). The adsorption sites of a number of these phages have been localized with the electron microscope, and their identity with membrane adhesion sites has been established (Bayer, 1968). The electron microscopic and virologic data point to the multifunctionality of the sites:

1 The receptor protein for phage BF23 has been found to be identical to the outer membrane protein required for uptake of vitamin B_{12} (White et al., 1973), and is identical to the receptor for colicins of type E. Normally, 200–250 receptor molecules are present per cell (Bradbeer et al., 1976), and the different macromolecules compete for their receptor. Similarly, *ton A* mutants, resistant to phages T1, T5 and ϕ80, are impaired in ferrichrome-iron uptake. The tonA protein receptor is also shared with colicin M (Wayne and Neilands, 1975; Hantke and Braun, 1975) and is present in an inactive form in colicin M-resistant and T5-resistant cells. LPS does not seem to be required for its receptor activity. While tonA is sufficient for T5 infection, a further protein, tonB, is needed for the uptake of ferrichrome and for the infection of phages T1 and ϕ80, as well as for colicin M killing (Hancock and Braun, 1976). These processes are energy dependent. We found previously that phages T1, T5, and BF23 release their DNA in contact with the membrane adhesion sites (Bayer, 1968b, 1975) (Figures 11 and 12).
2 Like the phages mentioned above, phage lambda also infects the host cell over adhesion sites (M. E. Bayer, unpublished observations); the phage adsorbs to the (inducible) receptor protein (lamB), which is required for maltose transport (Szmeleman and Hofnung, 1975). The receptor protein is incorporated into the cell envelope only during a relatively short period in the cell cycle. In contrast to the other membrane proteins described above, the incorporation of the lambda receptor is not random but takes place at the cell division site (Ryter et al., 1975).
3 Another protein (gene *tsx*) serves as receptor for colicin K and bacteriophage T6; the protein is also involved in uptake of thymidine, uridine, adenosine, and deoxyadenosine (Hantke, 1976). The insertion of this membrane protein has been described in various mutants by Begg and Donachie (1977), who found an uneven distribution of the receptor. However, in wild-type *E. coli* B, the distribution of injection sites for phage T6 was found to be random (Bayer, 1968b, 1975).

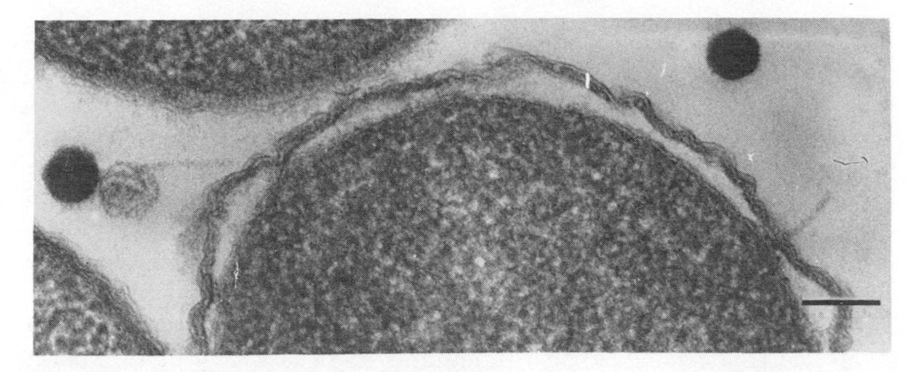

Figure 11 *E. coli* B to which phage BF23 was adsorbed. At the left side of the micrograph, the full capsid of the virion is seen with its long tail attached. On the right side of the micrograph, the phage tail is visible only, and the head is absent.

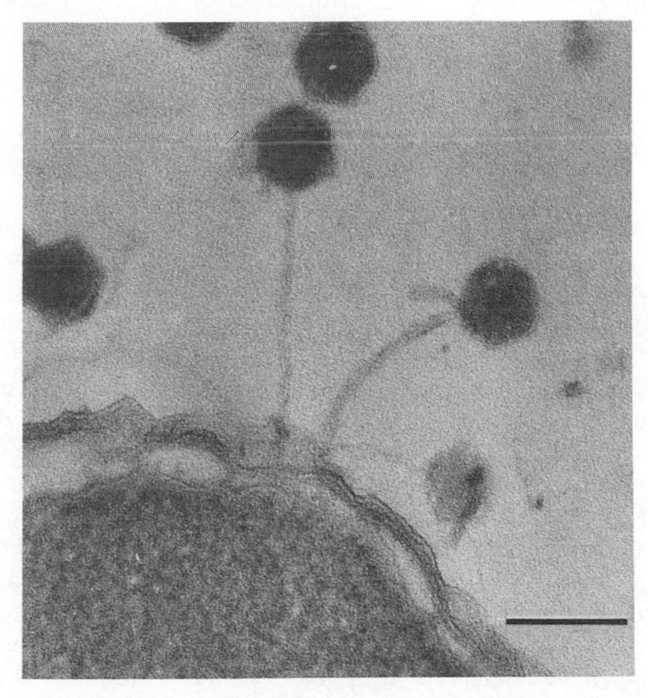

Figure 12 Bacteriophage T5 adsorbed to *E. coli* B at high multiplicity (500 phages/cell). More than one phage may adsorb to a (relatively broad) area of membrane adhesion.

Recently, the data of Begg and Donachie were reevaluated by Smit and Nikaido (1978).

4.6 Interaction of the Receptor(s) With the Virion

There is increasing evidence that the outer membrane receptors to which virus particles attach without DNA release "prepare" the virion for DNA injection (Yamamoto and Uchida, 1975). For example, it has been demonstrated that the requirement for the outer membrane as phage receptor may be circumvented by infecting protoplasts with urea-treated T4 (Wais and Goldberg, 1969) or with mutants of T4 lacking gene 12 product (Benz and Goldberg, 1969). These phages have in common an altered (*contracted*) tail structure exposing the core or needle through which DNA is injected (Dawes and Goldberg, 1973a, b). Without the alteration in the virus particle, the infection of the protoplasts cannot take place. The concept of a changing capsid upon contact with the LPS receptor, without causing release of DNA, is in agreement with data on capsule-penetrating viruses (Bayer et al., 1979; Fehmel et al., 1975), with data for LPS-specific ϵ-phages (McConnel et al., 1979) and with inactivation data on $\phi\chi174$ (Jazwinski et al., 1975). In the case of the ϵ-phages, a distinctive action was observed for different portions of the LPS molecule: In the presence of lipopolysaccharides, the phages were inactivated, whereas they remained infectious when exposed to lipid-free polysaccharides, lacking lipid A (Takeda and Uetake, 1973). Furthermore, the interaction of the *Salmonella* phages $\epsilon 15$ with several LPS mutant cell strains (Kanegasaki and Tomita, 1976) showed that the LPS of the mutants had adsorption and inactivation characteristics differing substantially between two phages, the ϵ-phages and phage Felix O. This prompted the authors to suggest that more than the receptor LPS is involved in phage infection or DNA release. A further aspect in the distinction of injection site of phage DNA came from the work of Tomita et al. (1976). Employing sucrose density gradients for separation of membrane vesicles of infected cells, these authors described a temperature-dependent migration of the phage label from the outer membrane fraction to a fraction at the "denser shoulder" of the inner membrane fraction; the virus particles found at the inner membrane fraction seemed to have released their nucleic acid. The migration of the labeled virus protein was not found in injection-deficient mutants (Tomita et al., 1976). This work supports the model of the adhesion site as the site of DNA release. However, the use of virus particles such as phage P22 as markers in the isolation procedures of adhesion sites has been discouraged by Crowlesmith et al. (1978); these authors observed detachment and possibly redistribution of phage particles and proteins during preparation of membrane vesicles from *Salmonella typhimurium* envelopes. It is obvious from this that much further work is necessary to purify and chemically characterize the adhesion site. Ongoing work in our laboratory suggests that membrane fractions with phage receptor activity and antigenicity specific for adhesion sites can be isolated (Bayer, M. E. and Bayer, M. H. 1979). One has to expect a complex physicochemical composition of the adhesion site in order to explain its full spectrum of functions. The data of Datta et al. (1977) and of Mutoh et al. (1978), who showed enhanced protein receptor activity in the presence of LPS, point toward a complex formation with lipid.

5 SUMMARY STATEMENT AND FURTHER EVIDENCE FOR A LOCALIZED HETEROGENEITY OF THE OUTER MEMBRANE

Studies on growth and division of bacteria indicate a heterogeneous surface in gram-positive, as well as gram-negative, cells. With the discovery of the adhesion sites, the cell envelope of gram-negative bacteria became a rather complex mosaic in which small discrete areas were interspersed with the bulk membrane. However, the evidence for the existence of adhesion sites was often questioned (also by our laboratory), since the cells had to be plasmolyzed and chemically fixed. However, the functions of the sites as production and export zones for polysaccharides, lipopolysaccharides, and membrane proteins served as strong supportive evidence for the reality of the adhesion zone. The capability of bacteriophages to "find" these zones and the release of the virus nucleic acid over these sites initiated further search for markers of a possible microheterogeneity in the cell envelope at the adhesion sites. The work dealing with conversion of LPS after infection with phage ϵ15 revealed that not all the 200–400 adhesion sites, but only 10% of them, participated in the export of new LPS. Thus the number of sites possibly showing some structural differences was small enough to be distinguishable with ultrastructural methods, such as freeze-fracture replication. This technique requires neither chemical fixation nor any cryoprotection with sucrose or glycerol. It has been established that the fracture of the envelope of gram-negative bacteria takes place preferentially in the inner membrane; fractures occur with preference in membranes with relatively high lipid content. Employment of a certain fracture technique revealed the existence of discrete domains in *E. coli* envelopes at which the fracture had switched from the preferred plane of the inner membrane to small areas of the outer membrane (Bayer and Leive, 1977), as if, at these sites only, the outer membrane had become enriched in lipid relative to the surrounding domains. These plateaus are seen neither in stationary cells, nor in cells that have been metabolically poisoned for about 20 min at 37°C; within this time period, diffusion seems to average the physicochemical differences between the plateaus and their neighboring membrane portions. Plateaus of identical structure have been observed in various *E. coli* and *Salmonella* strains. Since number, size, and distribution of the plateaus closely resembles that of the LPS-producing sites observed after ϵ15 conversion, we hypothesized that the plateaus might pinpoint the location of active adhesion sites. Studies on the effect of EDTA extraction of *E. coli* envelopes seem to support this suggestion (Bayer and Leive, 1977).

The adhesion sites are domains at which inner and outer membranes of gram-negative organisms are attached to each other. There seem to be various structural and functional features common to specialized membrane fusion sites found in mitochondria and membranous organelles of eukaryotic cells. The adhesion sites of the bacteria are involved in the growth of the cell. They represent the areas of export of surface components such as lipopolysaccharides, capsule polysaccharides, and major and minor proteins of the outer membrane. They cause the firm binding of phages and provide the signal for injection of the virus nucleic acid. Flagellae and F pili are anchored in the cell envelope at domains that are structurally indistinguishable from the "common" adhesion site. Chromosomes are often seen to be attached to areas of adhesion. There seem to exist either different adhesion sites, or at least different stages of activity, since LPS synthesis (after phage conversion) takes place at only 10% of them; in other instances (in epimerase-less mutants), most of the sites seem to be involved in LPS export. For some phages, the number of available injection sites seems to be restricted to about 20 or 40 per cell; others seem to be able to infect at most or all of the sites. Major proteins, such as *Salmonella* porins, are exported at an increasing number of sites. Figure 13 summarizes these data.

The discovery of adhesion sites provided considerable input for the understanding of virus infection mechanisms, and a model of the stepwise events during adsorption and infection provided the base for the experimental design and proof of some of the stages. As a consequence, the term virus "receptor" is not identical with site of infection. The virion colliding with the host cell recognizes the receptor molecule, which is abundantly available on the cell surface, and a loose, but specific, binding permits the virus particle to move on, or into, the cell surface coat. The receptor also seems to prepare the capsid of the virion for the injection event. A surface walk ends at the adhesion site and the energy-dependent infection process commences. These latter events cause, within a few seconds, the entire cell envelope to change its biophysical character, such as its state of energization. We have recently observed that the membrane fluorescence of *E. coli* responds to phage T4 infection at the same time (10–12 sec) as the electron microscope reveals tail contraction of the adsorbed virion (Bayer and Bayer,

Figure 13 Composite diagram of various functional domains of the surface of *E. coli* and *Salmonella* strains. The right portion of the diagram shows a capsule (C), which is penetrated by a virion with capsule-specific depolymerase. Various sites of synthesis and macromolecular assembly are indicated as they are positioned over the adhesion or fusion sites (WmF) between wall and inner membrane. O, export site for LPS (= O-antigen); similarly, major membrane proteins are exported over these sites; K, capsular antigen (K-antigen) production, either visible in fibers or in clusters; F, F-pilus insertion; the pilus is specifically labeled with bacteriophage MS2; T, virion of the T-even group with contracted tail sheath; ϕX, bacteriophage ϕX174; W, cell wall; R, the peptidoglycan plus major protein (R layer); PM, protoplasmic membrane; CHR, bacterial chromosome.

1979). The cellular response to this event originates from the adhesion site, which seems to serve as a center for signal propagation either by diffusion of substances through the cytoplasm of the cell or by a response traveling along the cell membrane. Bacteriophage receptor proteins often have been found to be identical with the receptor proteins involved in passive and active uptake of nutrients. The limited number and area occupied by adhesion sites per cell surface and the multitude of their activities requires a sharing of functions at the sites; The established competition among the various "substrates" for their receptor also favors the model of a clustering of activities. The adhesion sites appear to be relatively stable structures in mechanical terms; their isolation seems possible. However, their lifetime is not known; the possibility of their movement "underneath" the outer membrane has been raised but has not been proven; some indirect evidence seems to make a fast migration unlikely. In the absence of firm data relating to the origin and lifetime of adhesion sites, we hypothesize that new membrane attachments and fusion sites are generated as the cell grows, possibly leaving only a fraction of the several hundred sites highly active and others in less active, even inactive, states.

ACKNOWLEDGMENT

This work was supported by National Institutes of Health grants AI-10414, CA-06927 and RR-05539; by National Science Foundation grant PCM 78-13637; and by an appropriation from the Commonwealth of Pennsylvania.

REFERENCES

Achtmann, M. (1973) see this book, chapter 12.

Ackermann, H. W. (1973), in *Handbook of Microbiology* (A. I. Laskin and H. A. Lechevalier, eds.), Vol. 1, CRC Press, Cleveland, Ohio, p. 599.

Bartell, P. F. (1977), in *Microbiology—1977* (D. Schlesinger, ed.), American Society of Microbiology, Washington, D.C., pp. 134–137.

Bayer, M. E. (1967), *J. Gen. Microbiol.*, **46,** 237–246.

Bayer, M. E. (1968a), *J. Gen. Microbiol.*, **53,** 395–404.

Bayer, M. E. (1968b), *J. Virol.*, **2,** 346–356.

Bayer, M. E. (1973), in, 1st Int. Congr. Bacteriol., *Jerusalem*.

Bayer, M. E. (1974), *Ann. N. Y. Acad. Sci.*, **235,** 6–28.

Bayer, M. E. (1975), in *Membrane Biogenesis* (A. Tzagaloff, ed.), Plenum Press, New York, pp. 393–427.

Bayer, M. E., and Bayer, M. H. (1978), Abstract S51, *Annual Meeting of the American Society of Microbiology*.

Bayer, M. E. and Bayer, M. H. (1979), Abstract J1, *Annual Meeting of the American Society of Microbiology*.

Bayer, M. H., and Bayer, M. E. (1979), in *Proc. Am. Soc. Microbiol., Los Angeles* (abstr. S247)

Bayer, M. E., and Leive, L. (1977), *J Bacteriol.*, **130,** 1364–1381.

Bayer, M. E., and Starkey, T. (1972), *Virology*, **49,** 236–256.

Bayer, M. E., and Thurow, H. (1977), *J. Bacteriol.*, **130,** 911–936.

Bayer, M. E., Koplow, J., and Goldfine, H. (1975), *Proc. Natl. Acad. Sci. U.S.*, **72,** 5145–5149.

Bayer, M. E., Dolack, M., and Houser, E. (1977), *J. Bacteriol.*, **129,** 1563–1573.

Bayer, M. E., Thurow, H., and Bayer, M. H. (1979), *Virology*, **94,** 95–118.

Beachey, E. H., and Cole, R. M. (1966), *J. Bacteriol.*, **92,** 1245–1251.

Begg, K. J. (1978), *J. Bacteriol.*, **135,** 307–310.

Begg, K. J. and Donachie, W. D. (1977), *J. Bacteriol.*, **129,** 1524–1536.

Bennett, Michael, V. L. (1973), *Fed. Proc.*, **32,** 65–75.

Benz, W. C., and Goldberg, E. B. (1973), *Virology*, **53,** 225–235.

Berg, H. C. (1975), *Annu. Rev. Biophys. and Bioeng.*, **4,** 119–136.

Berg, H. C., and Brown, D. A. (1972), *Nature*, **239,** 500–504.

Bergeron, J. J. M., J. H. Ehrenreich, P. Siekevitz, and G. E. Palade (1973), *J. Cell Biol.*, **59,** 73–89.

Beveridge, T. J., and Murray, R. G. E. (1974), *J. Bacteriol.*, **119,** 1019–1038.

Blobel, G., and Dobberstein, B. (1975), *J. Cell Biol.*, **67,** 835–851.

Bradbeer, C., Woodrow, M. L., and Khalifah, L. I. (1976), *J. Bacteriol.*, **125,** 1032–1039.

Braun, V. (1975), *Biochim. Biophys. Acta*, **415,** 335–377.

Braun, V., and Hantke, K. (1978), in *Receptors and Recognition* series B, Vol. 3, (J. L. Reissig, ed.) Wiley, New York, pp. 101–137.

Braun, V., Hancock, R. E. W., Hantke, K., and Hartmann, A. (1976), *J. Supramol. Struct.*, **5,** 37–58.

Brinton, C. C. (1965), *Trans. N.Y. Acad. Sci.*, **27,** 1003–1054.

Brown, R. J., and Corner, T. R. (1977), in *Microbiology—1977* (D. Schlesinger, ed.), American Society of Microbiology, Washington, D.C., pp. 93–99.

Buckmire, F. L. A., and Murray, R. G. E. (1973), *Can. J. Microbiol.*, **19,** 59–66.

Butow, R. A., Bennett, W. F., Finkelstein, D. B., and Kellems, R. E. (1975), in *Membrane Biogenesis* (A. Tragaloff, ed.), Plenum Press, New York, pp. 155–191.

Caspar, D. L. D., Goodenough, D. A., Makowski, Lee, and Phillips, W. C. (1977), *J. Cell Biol.*, **74,** 605–628.

Choy, Y. M., Fehmel, F., Frank, N., and Stirm, S. (1975), *J. Virol.*, **16,** 581–590.

Collard, J. G., De Wildt, A., and Inbar, M. (1978), *FEBS Lett.*, **90,** 24–28.

Costerton, J. W. (1977), in *Microbiology—1977* (D. Schlesinger, ed.), American Society of Microbiology, Washington, D.C., pp. 151–157.

Costerton, J. W., Ingram, J. M., and Cheng, K. J. (1974), *Bacteriol. Rev.*, **38,** 87–110.

Coulton, J. W., and Murray, R. G. E. (1978), *J. Bacteriol.*, **136,** 1037–1049.

Crowlesmith, I., Schindler, M., and Osborn, M. J. (1978), *J. Bacteriol.*, **135,** 259–269.

Datta, D. B., Arden, B., and Henning, U. (1977), *J. Bacteriol.*, **131,** 821–829.

Dawes, J. (1975), *Nature*, **256,** 127–128.

Dawes, J., and Goldberg, E. B. (1973a), *Virology*, **55,** 380–390.

Dawes, J., and Goldberg, E. B. (1973b), *Virology*, **55,** 391–396.

Delbrück, M., and Reichardt, W. (1956), in *Cellular Mechanisms in Differentiation and Growth* (D. Rudnick, ed.), Princeton University Press, Princeton, New Jersey, pp. 3–44.

DeLeij, L., Kingma, J., and Witholt, B. (1978), *Biochim. Biophys. Acta*, **512,** 365–376.

DePamphilis, M. L. (1971), *J. Virol.*, **7,** 683–686.

DePamphilis, M. L., and Adler, J. (1971), *J. Bacteriol.*, **105,** 384–395.

Dirienzo, J. M., Nakamura, K., and Inouye, M. (1978), *Annu. Rev. Biochem.*, **47,** 481–532.

Doyle, R. J., Streips, U. N., and Helman, J. R. (1977), in *Microbiology—1977* (D. Schlesinger, ed.), American Society of Microbiology, Washington, D.C., pp. 44–49.

Dunnick, J. K., Rooke, J. D., Aragon, S., and Kriss, J. P. (1976), *Cancer Res.*, **36,** 2385–2389.

Erikson, U., and Lindberg, A. A. (1977), *J. Gen. Virol.*, **34,** 207–221.

Fehmel, F., Feige, V., Niemann, H., and Stirm, S. (1975), *J. Virol.*, **16,** 591–601.

Fischer, A. (1903), *Vorlesunger über Bakterien*, 2nd ed., Fischer, Jena.

Frye, L. D., and Edidin, M. J. (1970), *J. Cell Sci.*, **7,** 319–324.

Futai, M. (1974), *J. Membrane Biol.*, **15,** 15–28.

Goy, M. F., Springer, M. S., and Adler, J. (1977), *Proc. Natl. Acad. Sci. U.S.*, **74,** 4964–4968.

Haller, I., and Henning, U. (1974), *Proc. Natl. Acad. Sci. U.S.*, **71,** 2018–2021.

Hancock, R. E. W., and Braun, V. (1976), *J. Bacteriol.*, **125,** 409–415.

Hantke, K. (1976), *FEBS Lett.*, **70,** 109–112.

Hantke, K. (1978), *Mol. Gen. Genet.*, **164,** 131–135.

Hantke, K., and Braun, V. (1975), *FEBS Lett.*, **49,** 301–305.

Harold, F. M. (1977), *Annu. Rev. Microbiol.*, **31,** 181–203.

Heppel, L. A. (1969), *J. Gen. Physiol.*, **54,** 95s–113s.

Hirashima, A., Wu, H. C., Venkateswaran, P. S., and Inouye, M. (1973), *J. Biol. Chem.*, **248,** 5654–5659.

Huang, L., and Pagano, R. E. (1975), *J. Cell Biol.*, **67,** 38–48.

Hull, B. E., and Staehelin, L. A. (1976), *J. Cell Biol.*, **68,** 688–704.

Ichihara, S., and Mizushima, S. (1978), *J. Biochem.*, **83,** 137–140.

Inouye, M. (1974), *Proc. Natl. Acad. Sci. U.S.*, **71,** 2396–2400.

Israel, J. V. (1977), *J. Virol.*, **23,** 91–97.

Israel, J. V., Rosen, H., and Levine, M. (1972), *J. Virol.*, **10,** 1152–1158.

Iwashita, S., and Kanegasaki, S. (1973) *Biochem. Biophys. Res. Commun.*, **55,** 403–409.

Jazwinski, S. M., Lindberg, A. A., and Kornberg, A. (1975), *Virology*, **66,** 268–282.

Jones, G. W. (1978), in *Receptors and Recognition* Series B, Vol. 3, (J. L. Reissig, ed.), Wiley, New York pp. 141–176.

Jones, N. C., and Osborn, M. J. (1977), *J. Biol. Chem.*, **252,** 7405–7412.

Kaback, H. R. (1973), in *Bacterial Membranes and Walls* (L. Leive, ed.), Dekker, New York pp. 241–292.

Kamio, Y., and Nikaido, H. (1977), *Biochim. Biophys. Acta*, **464,** 589–601.

Kanegasaki, S., and Tomita, T. (1976), *J. Bacteriol.*, **127,** 7–13.

Kanegasaki, S., and Wright, A. (1973), *Virology*, **52,** 160–173.

Kauffmann, F. (1966), *The Bacteriology of Enterobacteriaceae*, Williams & Wilkins Co., Baltimore, Maryland.

King, J., Lenk, E. V., and Botstein, D. (1973), *J. Mol. Biol.*, **80,** 697–731.

Lawn, A. M., Meynell, G. G., Meynell, E., and Datta, N. (1967), *Nature* (*Lond.*), **216,** 343–346.

Leive, L. (1973), *Bacterial Membranes and Walls*, Dekker, New York.

Lindberg, A. A. (1973), *Annu. Rev. Microbiol.*, **27,** 205–241.

Lindberg, A. A. (1977), in *Surface Carbohydrates of the Prokaryotic Cell* (I. W. Sutherland, ed.), Academic Press, London and New York, pp. 289–356.

Loewenstein, W. R., and Kanno, Y. (1964), *J. Cell Biol.*, **22,** 565–586.

Lüderitz, O., Westphal, O., Staub, A., and Nikaido, H. (1971), in *Microbiol Toxins* (G. Weinbaum, S. Kadis, and S. J. Ajl, eds.), Vol. 4, Academic Press, New York and London, pp. 145–233.

Machtiger, N. A., and Fox, C. F. (1973), *Annu. Rev. Biochem.*, **42,** 575–600.

Maggio, B., Cumar, F. A., and Caputto, R. (1978), *FEBS Lett.*, **90,** 149–152.

Magnusson, K.-E., and Johansson, J. (1977), *FEMS Microbiol. Lett.*, **2,** 225–228.

Manning, P. A., Puspurs, A., and Reeves, P. (1976), *J. Bacteriol.*, **127,** 1080–1084.

McConnell, M., A. Reznick and A. Wright (1979), *Virology*, **94,** 10–23.

Meldolesi, J. (1974), *Phil. Trans. R. Soc. Lond. B*, **268,** 39–53.

Meldolesi, J., J. D. Jamieson, and G. E. Palade (1971), *J. Cell Biol.*, **49,** 109–120.

Michael, J. G. (1968), *Proc. Soc. Exp. Biol. Med.*, **128,** 434–438.

Miller, J. B., and Koshland, Jr., D. E. (1977), *Proc. Natl. Acad. Sci. U.S.*, **74,** 4752–4756.

Miura, T., and Mizushima, S. (1969), *Biochim. Biophys. Acta*, **193,** 268–273.

Mühlradt, P., and Golecki, J. (1975), *Eur. J. Biochem.*, **51,** 343–352.

Mühlradt, P., Menzel, J., Golecki, J., and Speth, V. (1973), *Eur. J. Biochem.*, **35,** 471–481.

Mühlradt, P., Menzel, J., Golecki, J., and Speth, F. (1974), *Eur. J. Biochem.*, **43,** 533–539.

Murray, R. G. E., Steed, P., and Elson, H. E. (1965), *Can. J. Microbiol.*, **11,** 547–560.

Mutoh, N., Furukawa, H., and Mizushima, S. (1978), *J. Bacteriol.*, **136,** 693–701.

Nakae, T. (1976a), *J. Biol. Chem.*, **251,** 2176–2178.

Nakae, T. (1976b), *Biochem. Biophys. Res. Commun.*, **71,** 877–884.

Nikaido, H. (1960), *Biochim. Biophys. Acta*, **48,** 460–469.

Nikaido, H., Takeuchi, Y., Ohnishi, S., and Nakae, T. (1977), *Biochim. Biophys. Acta*, **465,** 152–164.

Ørskov, I., Ørskov, P., Smith, H. W., and Soyka, W. J. (1975), *Acta Pathol. Microbiol. Scand. Sect. B*, **83,** 31–36.

Osborn, M. J., Gander, J. E., Parisi, E., and Carson, J. (1972), *J. Biol. Chem.*, **247,** 3962–2972.

Owen, P., and Salton, M. R. J. (1975), *Biochim. Biophys. Acta*, **406,** 214–226.

Palade, G. E. (1959), in *Subcellular Particles* (T. Hayashi, ed.), Ronald Press, New York, pp. 64–72.

Pitts, J. D., and Simms, J. W. (1977), *Exp. Cell Res.*, **104,** 153–163.

Poste, G., and Papahadjopoulos, D. (1976) *Proc. Natl. Acad. Sci. U.S.*, 73, 1603–1607.

Prehm, P., Jann, B., Jann, K., Schmidt, G., and Stirm, S. (1976) *J. Mol. Biol.* 101, 277–281.

Revel, J. P., and Karnovsky, M. J. (1976), *J. Cell Biol.*, **33,** C7–C12.

Ridgeway, H. F., Silverman, M., and Simon, M. I. (1977), *J. Bacteriol.*, **132,** 657–665.

Robbins, P. W., and Uchida, T. (1962), *Biochemistry* **1,** 323.

Rosen, B. P., and Heppel, L. A. (1973), in *Bacterial Membranes and Walls* (L, Leive, ed.), Dekker, New York, pp. 209–239.

Rosenbusch, J. P. (1974), *J. Biol. Chem.*, **249,** 8019–8029.

Rottem, S., and Leive, L. (1977), *J. Biol. Chem.*, **252,** 2077–2081.

Ryter, A., Shuman, H., and Schwartz, M. (1975), *J. Bacteriol.*, **122,** 295–301.

Salton, M. R. J. (1952), *Nature* (*Lond.*), **170,** 746–747.

Salton, M. R. J. (1971) *CRC Crit. Rev. Microbiol.*, **1,** 161–197.

Schaeffer, P., and Hotchkiss, R. D. (1978), in *Methods in Cell Biology* (D. M. Prescott, ed.), Vol. 20, Academic Press, New York, pp. 149–158.

Schnaitman, C. A. (1970), *J. Bacteriol.*, **104,** 890–901.

Schnaitman, C. A., Smith, D., and Forn de Salsas, M. (1975), *J Virol.*, **15,** 1121–1130.

Schwartz, U., Ryter, A., Rambach, A., Hellio, R., and Hirota, Y. (1975), *J. Mol. Biol.*, **98,** 743–759.

Shands, Jr., J. W. (1971), in *Microbiol Toxins* (G. Weinbaum, S. Kadig, and S. J. Ajl, eds.), Academic Press, New York and London, pp. 127–144.

Silverman, M., and Simon, M. (1974), *Nature*, **249,** 73–74.

Silverman, M., and Simon, M. (1977), *Proc. Natl. Acad. Sci. U.S.*, **74,** 3317–3321.

Skurray, R. A., Hancock, R. E. W., and Reeves, P. (1974), *J. Bacteriol.*, **119,** 726–735.

Sleytr, U. B., Thornley, M. J., and Glauert, A. M. (1974), *J. Bacteriol.*, **118,** 693–707.

Smit, J., and Nikaido, H. (1978), *J. Bacteriol.*, **135,** 687–702.

Smit, J., Kamio, Y., and Nikaido, H. (1975), *J. Bacteriol.*, **124,** 942–958.

Springer, M. S., Goy, M. F., and Adler, J. (1977), *Proc. Natl. Acad. Sci.*, **74,** 3312–3316.

Stendahl, O., Edebo, L., Magnusson, K.-E., Tagesson, C., and Hjerten, S. (1977), *Acta Pathol. Microbiol. Scand. Sect. B*, **85,** 334–340.

Stirm, S., Ørskov, F., Ørskov, I., and Mansa, B. (1967), *J. Bacteriol.*, **93,** 731–739.

Susskind, M. M., and Botstein, D. (1978), *Microbiol. Rev.*, **42,** 385–413.

Sutherland, I. W. (1977), in *Surface Carbohydrates of the Prokaryotic Cell* (I. W. Sutherland, ed.), Academic Press, New York and London, pp. 27–96.

Szmeleman, S., and Hofnung, M. (1975), *J. Bacteriol.*, **124,** 112–118.

Takeda, K., and Uetake, H. (1973), *Virology*, **52,** 148–159.

Thurow, H., Choy, Y.-M., Frank, N., Niemann, H., and Stirm, S. (1975), *Carbohydr. Res.*, **41,** 241–255.

Tomita, T., Iwashita, S., and Kanegasaki, S. (1976), *Biochem. Biophys. Res. Commun.*, **73,** 807–813.

Uetake, H., and Hagiwara, S. (1969), *Virology*, **37,** 8–14.

Uetake, H., Luria, S. E., and Burrows, J. W. (1958), *Virology*, **5,** 68–91.

Van Alphen, L., Havekes, L., and Lugtenberg, B. (1977), *FEBS Lett.*, **75,** 285–290.

Verhoef, C., DeGraaff, P. J., and Lugtenberg, P. J. J. (1977), *Mol. Gen Genet.*, **150,** 103–105.

Verkleij, A., van Alphen, L. V., Bijvelt, J., and Lugtenberg, B. (1977), *Biochim. Biophys. Acta*, **466,** 269–282.

Wais, A. C., and Goldberg, E. B. (1969), *Virology*, **39,** 153–161.

Wayne, R., and Neilands, J. B. (1975), *J. Bacteriol.*, **121,** 497–503.

Weibull, C. (1953), *J. Bacteriol.*, **66,** 688–702.

Weinstein, J. N., Hoshikami, S., Henkhart, P., Blumenthal, R., and Hagins, W. A. (1977), *Science*, 195, 483–491.

White, J. C., DiGirolamo, P. M., Fu, M. L., Preston, Y. A., and Bradbeer, C. (1973), *J. Biol. Chem.*, **248,** 3978–3986.

Wong, M., Bayer, M. E., and Litwin, S. (1978), *FEBS Lett.*, **95,** 26–30.

Yamamoto, M., and Uchida, H. (1975), *J. Mol. Biol.*, **92,** 207–233.

Zinder, N. D. (1958), *Virology*, **5,** 291–326.

Chapter 7 A Genetic Approach to the Study of Protein Localization in *Escherichia coli*

THOMAS J. SILHAVY, PHILIP J. BASSFORD, JR., AND JONATHAN R. BECKWITH

Department of Microbiology and Molecular Genetics, Harvard Medical School, Boston, Massachusetts

1 INTRODUCTION

The proteins of *E. coli* are found in at least four different compartments: the cytoplasm, the cytoplasmic or inner membrane, the outer membrane, and the periplasmic space between the two membranes. The bulk of the protein (approximately 75%) is in the cytoplasm. However, there are a large number of proteins in the inner membrane, including those involved in transport, electron transport, and lipid biosynthesis. The outer membrane incorporates far fewer proteins, although in total amount it surpasses the inner membrane. Many of these proteins act as receptors for bacteriophage or colicins, and more recently, certain of the same proteins have been shown to allow transport or diffusion of certain molecules across this membrane. The periplasmic space is comprised of at least two classes of proteins: the binding proteins involved in transport of various molecules and degradative enzymes. There is little evidence for the secretion into the medium of any *E. coli* proteins. It seems likely that the outer membrane acts as a barrier to such transfer. In nearly every case where it has been studied, the location of the particular protein is restricted to one of these four compartments. A major question confronting those interested in bacterial envelope structure is how each of the noncytoplasmic (envelope) proteins gets exported and directed to its ultimate location.

Within the last few years, considerable evidence has accumulated on the mechanism of protein export in *E. coli*. Most of the proteins that have been studied are those localized either in the periplasm or in the outer membrane. In addition, the study of a bacteriophage protein initially located in the cytoplasmic membrane has suggested a model for localization of proteins in this compartment. The results of this work indicate that there are certain *common* features of the mechanisms involved in the export to these three *different* cellular sites. In particular, many aspects of the export process for these proteins in bacteria conform to predictions of a model, the *signal hypothesis*, presented to account for the secretion of proteins and peptide hormones in eukaryotic systems. In this chapter, we discuss the biochemical evidence that supports a signal hypothesis model for the initial steps in the export of *E. coli* proteins. In addition, we discuss the possible variations that must be superimposed upon this mechanism to explain the variety of envelope locations for *E. coli* proteins. Finally, we describe recent genetic evidence that provides further support for the signal hypothesis and that opens up the prospect of a detailed analysis of the mechanisms of protein export.

1.1 The Signal Hypothesis

In a model initially presented by Blobel and Sabatini (1971) and subsequently elaborated on by Blobel and Dobberstein (1975), a number of steps are proposed for secretion of proteins into the lumen of the rough endoplasmic reticulum (RER). First, initiation of translation of a messenger RNA coding for a protein destined to be secreted takes place in the same fashion as with any other messenger RNA, that is, ribosomes from the cytoplasmic pool attach, initiate, and begin to translate the messenger RNA. The divergence between messenger RNAs coding for proteins destined to remain in the cytoplasm and those destined to be secreted comes with the synthesis and/or appearance in the cytoplasm of the initial amino acid sequence.

As proposed, the signal hypothesis predicts that for secreted proteins, the first 15–30 amino acids possess an unusually hydrophobic nature and probably some other features that "signal" the beginning of the secretion process. The polysomes that initially are soluble are attracted to the RER membrane by this *signal sequence.* Ribosomes carrying the nascent peptide then attach to special sites on the membrane and stimulate pores that form a channel to the RER lumen. Then, by a mechanism not well defined, the nascent polypeptide chain is discharged into the lumen as the polypeptide chain is being extended during protein synthesis. After the entire signal amino acid sequence has passed through the membrane of the RER, this sequence is cleaved from the polypeptide chain. The polypeptide chain continues to be discharged into the RER lumen during its synthesis, ultimately yielding the mature form of the particular protein or peptide hormone. This means that secreted proteins are made initially as precursors.

This model has been extended, again in eukaryotic systems, to explain the synthesis of certain transmembrane proteins. In such cases, most aspects of the model pertain, except it is suggested that an amino acid sequence later in the polypeptide chain causes the process of discharge into the lumen to stop, resulting in the protein remaining embedded in the membrane (Katz et al., 1977; Rothman and Lenard, 1977; Sabatini and Kreibich, 1976).

This model is based on data from a number of systems. Since its proposal, a much greater number of examples have confirmed many predictions of the model. In addition, specific binding proteins, termed ribophorins, that bind ribosomes tightly have been found in the RER membrane (Kreibich et al., 1978a, b). It is suggested that these proteins are the site of attachment of polysomes engaged in the synthesis of secreted proteins. However, results that implicate specific ribosome–membrane interaction are still somewhat controversial, with contradictory results being obtained in different systems (see e.g. Lodish and Froshauer, 1977). One other feature of the model might seem to be contradicted by the finding that the molecular weight of ovalbumin, a protein secreted in the same way as other secreted proteins, is identical with the initial translation product of its messenger RNA (Palmiter et al., 1978). Yet the secretion of this protein involves the same cotranslational transfer process (Lingappa et al., 1978) as the secreted proteins. This result suggests either that there is no signal sequence or that no processing takes place. However, since it is not known whether processing is necessary for secretion, or, in fact, what role it plays, this result does not necessarily bring into question major features of the model.

1.2 Biochemical Evidence for the Signal Hypothesis for Protein Export in *E. coli*

1.2.1 Precursors of Envelope Proteins As essential element of the signal hypothesis is that the initial translation product of the messenger RNA for a secreted protein is a larger precursor molecule that has a sequence of amino acids, unusually rich in hydrophobic amino acids, at the aminoterminus of the polypeptide chain. In most cases, this extra sequence is ultimately cleaved off. In the last few years, evidence has accumulated that many envelope proteins of *E. coli* are made initially in precursor form (Table 1). We describe here the evidence for precursors of proteins in each of the three extracytoplasmic compartments. In general, it appears that precursors of envelope proteins are short lived *in vivo*, being rapidly processed to the mature form. This makes their detection in growing bacteria quite difficult. As a result, many of the studies that have revealed such precursors have utilized *in vitro* translation systems or specially treated cells.

In the case of periplasmic proteins, alkaline phosphatase was the first to be shown to be made initially as a longer precursor polypeptide chain. The enzyme can be synthesized *in vitro* using the Zubay system for protein syn-

Table 1 Evidence for Precursors of Exported Proteins in *E. coli*[a]

Protein	System
Alkaline phosphates	(1) *In vitro* synthesis using DNA template (2) Synthesis on membrane-bound polysomes (3) Accumulation in cells grown at low temperature
Three outer membrane proteins	Synthesis in toluene-treated cells
λ receptor, maltose- and arabinose-binding proteins	Synthesis on membrane-bound polysomes
β-Lactamase	Deduced from DNA sequence
Glycerol-3-phosphate transport protein	*In vitro* synthesis using DNA template
fl, fd, and M13 coat proteins	(1) *In vitro* synthesis using phage DNA template (2) Deduced from messenger RNA sequence

[a] See text for references and discussion.

thesis (Zubay, 1973) and bacteriophage ϕ80DNA templates carrying the *phoA* gene (structural gene for the enzyme) (H. Inouye et al., 1977). Precipitation with anti-alkaline phosphatase antibody revealed that the major product of the *phoA* gene *in vitro* is a polypeptide chain several thousand daltons larger than the monomeric subunit of the enzyme, as obtained from the periplasm of *E. coli* (Inouye and Beckwith, 1977).* Further substantiation of these results comes from studies using a membrane-bound polysome preparation for protein synthesis (for further discussion of this system see subsequent section). Employing this approach, Smith et al., 1977 have also shown synthesis of a larger polypeptide chain immunologically related to alkaline phosphatase. In this case, the polysomes were extracted from cells constitutive for the synthesis of the enzyme. In neither case was it determined whether the extra amino acid sequence was amino terminal or carboxy terminal. However, affinity for a Decyl Agarose column of the precursor suggested that the extra amino acid sequence was highly hydrophobic (Inouye and Beckwith, 1977). Finally, Lazdunski and coworkers (Pagès et al., 1978; C. Lazdunski, personal communication) have evidence that the precursor of alkaline phosphatase accumulates in the cell membrane fraction when bacteria are grown at very low temperature. It is proposed that the reduction in membrane fluidity at low temperatures may be responsible for this effect. Other studies on temperature effects and the effects of altering the fatty acid composition of the membrane suggest that membrane fluidity may be essential for the secretion process (Ito et al., 1977; Kimura and Izui, 1976).

Randall and co-workers (Randall and Hardy, 1978) employed a membrane-bound polysome system to demonstrate synthesis of larger molecular weight polypeptide chains related to the periplasmic arabinose- and maltose-binding proteins. These proteins were shown to be related to the mature forms by both immunological means and analysis of peptides produced by limited proteolysis. Further, the extra amino acid sequence of the arabinose-binding protein was shown to be located at the amino terminus of the proteins (Randall et al., 1978; Hardy and Randall, 1978). In addition, the precursor of the arabinose-binding protein was detected *in vivo* after pulse labeling of proteins. Some evidence has been presented for the existence of a precursor of another periplasmic protein involved in carbon source transport, the product of the *glpT* gene (Schumacher and Bussmann,

* In this case, the larger molecular weight was inferred from the slow mobility on SDS-polyacrylamide gels. However, it has been shown that other factors can affect this mobility (Tanford and Reynolds, 1976). As a result, the conclusions concerning molecular weight differences stated here must be somewhat qualified and await confirmation by more direct sequencing techniques. Nevertheless, the consistent finding of related proteins that *appear* to have higher molecular weights in all these cases is more than suggestive.

1978). The product is necessary for glycerol-3-phosphate transport. The protein can be synthesized *in vitro* using the Zubay system and a colEl-*glpT* DNA preparation. A major band has been observed on SDS–polyacrylamide gels, indicating an apparent molecular weight 1000–2000 higher than the corresponding protein found *in vivo*. However, in this case, no direct evidence was presented, indicating that this protein was related to the glpT protein.

Finally, evidence for a precursor of a β-lactamase has been obtained by comparing the amino acid sequence of the protein with the DNA sequence of the gene that codes for it. The gene in question is carried by a plasmid, pBR322, and confers ampicillin resistance on its host. Preliminary evidence suggests that the β-lactamase is a periplasmic protein (J. Knowles, personal communication). The DNA coding sequence of the gene that corresponds to the amino acid sequence of the mature enzyme begins 69 bp (base pairs) after the probable translation initiation site (Sutcliffe, 1978). Thus the DNA sequence predicts a longer polypeptide chain as the initial translation product.

Several outer membrane proteins have been shown to be made initially as longer precursor molecules. M. Inouye and co-workers found that when cells were treated with toluene, they began to accumulate precursors of the lipoprotein, the matrix protein, and the protein product of the *ompA* gene (S. Inouye et al., 1977; Sekizawa et al., 1977). This work is described in detail elsewhere in this volume. Randall and Hardy (1978), using the membrane-bound polysome system, have shown *in vitro* synthesis of a precursor of bacteriophage λ receptor, another outer membrane protein. Again, the protein was shown to be related to λ receptor by limited proteolysis.

While no native *E. coli* inner membrane proteins have been analyzed for signal sequences, it might be expected that at least those that span the cytoplasmic membrane would be synthesized in a fashion similar to that of other exported proteins. In fact, the study of certain *E. coli* phage capsid proteins that, during the process of phage infection, span the inner membrane as integral proteins suggests that the signal hypothesis can apply for such proteins also. Three DNA phages, fl, fd, and M13, code for identical capsid proteins. When fl or M13 DNA is used to prime protein synthesis *in vitro* (Gold and Schweiger, 1971), the capsid protein is made as a precursor (Konings et al., 1975; Chang et al., 1978). With fl, a 23 amino acid amino-terminal extension is found. In addition, the sequencing of the messenger RNA for the analogous protein of phage fd reveals a sequence of codons corresponding to an amino-terminal sequence of 23 amino acids not found in the mature product (Sugimoto et al., 1977). The two sequences are identical.

While these results suggest certain cytoplasmic membrane proteins may be inserted according to the signal hypothesis, it seems quite possible that at least some of the membrane proteins that are not transmembranal may be incorporated by other mechanisms. In some cases, the insertion may be dependent only on the structure of the protein itself. Further, those signal sequences that have already been determined in *E. coli* do not appear to be as hydrophobic as many of their eukaryotic counterparts. Thus it is not clear whether the hydrophobic nature of this region is important (see Chapter 4).

1.2.2 Membrane-Bound Polysomes According to the signal hypothesis and the evidence supporting it from eukaryotic systems, the synthesis of secreted proteins takes place on membrane-bound polysomes. A number of years ago, Cancedda and Schlessinger (1974) attempted to determine whether the polysomes for alkaline phosphatase (a periplasmic protein) were membrane bound. While there were some positive indications, a control experiment suggested that the messenger RNA for a cytoplasmic enzyme (β-galactosidase) was also, to some extent, membrane bound. Subsequently, evidence accumulated for a difference in properties of those polysomes involved in the secretion of certain noncytoplasmic proteins. The synthesis of the murein lipoprotein has different properties in terms of resistance to antibiotics. For instance, it is unusually resistant to puromycin, as is the synthesis of other membrane proteins (Halegoua et al., 1976; Hirashima et al., 1973). The messenger RNAs for these proteins also appear to be more stable than those for cytoplasmic proteins (Hirashima et al., 1973). More recently, Randall and co-workers have presented some evidence that much of outer membrane protein synthesis, as well as the synthesis of a specific periplasmic protein, the maltose-binding protein, takes place preferentially on membrane-bound polysomes (Randall and Hardy, 1977). In addition, they have obtained synthesis of the arabinose-binding protein (periplasmic) and bacteriophage λ receptor (outer membrane) on membrane-bound polysomes, although they did not show preferential synthesis on such polysomes (Randall and Hardy, 1978). Varenne et al. (1978) have shown preferential synthesis of alkaline phosphatase on membrane-bound polysomes compared to total protein synthesized. Similarly, Smith et al. 1977 have shown that the synthesis of alkaline phosphatase can be promoted by membrane-bound polysomes preferentially. In contrast a cytoplasmic protein (EFG) is synthesized predominantly by soluble polysomes.

There is not enough evidence at present to determine whether the ribosomes or translation factors that are components of these membrane-bound polysomes are different from those found in soluble polysomes. One attempt to detect differences revealed none (Randall and Hardy, 1975). The evidence from eukaryotic systems is consistent with the hypothesis that the polysomes involved in the synthesis of secreted proteins are initially cytoplasmic and presumably the signal sequence itself determines their ultimate membrane localization. However, the results still do not rule out the possibility that there are special ribosomal or translation factors for this synthesis. If such factors exist, it is possible that genetic studies in bacteria, which are discussed below, will reveal them.

In addition, there is no evidence similar to that from eukaryotic systems to indicate that there is a special interaction between the ribosome and the cytoplasmic membrane in such membrane-bound polysomes. Smith et al. (1978a) have reported that puromycin releases from the membrane those polysomes engaged in the synthesis of secreted proteins. While these authors have proposed that the results suggest that there is nothing special about the attachment of the ribosomes to the membrane and that the signal sequence alone is responsible for that attachment, other effects of puromycin or of the termination of protein synthesis could alter any ribosome–membrane interaction.

1.2.3 Cotranslational Transfer of Proteins Across the Membrane The requirement for membrane-bound polysomes in the synthesis of secreted proteins is connected with a basic tenet of the signal hypothesis: the transfer of secreted proteins through membranes takes place during translation. This component of the model has been verified directly in *E. coli* (Smith et al., 1977; 1978a) and in *B. subtilis* (Smith et al., 1978b) by *extracellular* labeling of *growing* polypeptide chains as they traversed the bacterial membrane. Specifically, *E. coli* alkaline phosphatase (along with other unidentified proteins) was labeled in this way.

However, it is known that proteins can probably enter and at least span membranes in the absence of translation. Such proteins include colicins in *E. coli*, toxins in eukaryotic cells, and certain chloroplast and mitochondrial proteins (see Silverstein, 1978, for review articles on these subjects). In the last case, these proteins are synthesized on cytoplasmic polysomes in precursor form and are subsequently incorporated into chloroplast membranes and processed. In possibly an analogous situation, the precursor of the phage M13 coat protein can be incorporated into the membranes of vesicles in the proper orientation *after* completion of the polypeptide chain (Wickner et al., 1978). It is not clear, in this case, that the protein is imbedded in the membrane in exactly the same way it is *in vivo*. It may well be that *in vivo* the protein is incorporated into the membrane during synthesis but that *in vitro* insertion is possible only because of the nature of this particular protein. Nevertheless, the results from this system and from the chloroplast studies indicate that the existence of a presumed signal sequence in a precursor of an exported protein does not necessarily indicate cotranslational transfer.

1.2.4 Processing Another step in the secretion process, according to the signal hypothesis, is the proteolytic cleavage that removes the signal sequence. Clearly, this event must occur with the bacterial protein precursors also. Some evidence for a processing activity comes from the studies on alkaline phosphatase. An outer membrane activity can convert the precursor made in the Zubay system to a molecular weight close to that of the mature form of the enzyme subunit (Inouye and Beckwith, 1977). This finding was subsequently confirmed with the precursor made from membrane-bound polysomes (W. P. Smith, personal communication). In contrast, an activity that processes the fl coat protein precursor was found in the inner membrane after treatment with detergent (Chang et al., 1978).

Further, this latter activity only appears to be active during translation. These two activities have recently been compared using alkaline phosphatase precursor as substrate. Both activities reduce the apparent molecular weight of the precursor to one that is close to native alkaline phosphatase monomer size, but it appears that the molecular weights in the two cases are different (N. C. Chang, H. Inouye, and P. Model, personal communication). Randall and co-workers (1978) have observed that addition of detergent (Triton X-100) to a membrane-bound polysome system making arabinose- or maltose-binding proteins results in processing of the two precursors.

The detection of a "processing" activity *in vivo* does not necessarily mean that the functional enzyme has been discovered. It is quite possible that signal sequences did not acquire the ability to fold properly since they do not ordinarily serve as part of the structure of the functional protein. If this is the case, signal sequences may be susceptible to a variety of different proteases. Then a cellular proteolytic activity that cleaves a precursor to approximately the approriate size may not be the true *in vivo* processing enzyme.

Several lines of evidence suggest that cleavage of the signal sequence is not a necessary component of the export process. First, there are examples of exported proteins that do not differ in molecular weight from the initial translation product of the gene in question. The example of ovalbumin has been cited. In *E. coli*, preliminary evidence indicates that certain outer membrane proteins coded for by the sex factor, F, are not processed (Kennedy et al., 1977). Again, it is not clear in this example whether the results indicate an alternative mechanism of export or merely a variation on the signal hypothesis. Second, a mutant has been isolated in which the signal sequence of the *E. coli* lipoprotein is not removed and yet there is substantial export of the protein to the outer membrane. This mutation results in a change of amino acid 14 of the signal sequence from glycine to aspartic acid (Lin et al., 1978). While there appear to be some defects in proper localization, the finding of the precursor intact in the outer membrane indicates that processing is not necessary for export.

We suggest that a further genetic analysis of this sort along with isolation of mutants lacking processing activity is necessary to shed light on the following questions: Which are the true processing activities? Is there more than one such activity? What role does processing play in export and proper localization of envelope proteins?

1.3 Determinants of Ultimate Location of Envelope Proteins

In eukaryotic systems, proteins transferred across the RER membrane are destined for several different locations. Some become integral membrane proteins, others are incorporated into cellular organelles such as lysosomes and peroxisomes, and others are secreted entirely from the cell (for reviews, see Silverstein, 1978 and Palade, 1975). While the initial steps in the secretion process are beginning to be worked out in great detail, subsequent steps are still obscure. Similarly, in *E. coli*, while common features are beginning to emerge for the intracellular and membrane-associated steps in the export process of many proteins, the mechanisms that result in assortment of different proteins into the inner membrane, the outer membrane, and the periplasm have not been clarified.

However, studies on certain periplasmic and outer membrane proteins have revealed differences in the process of export that may be important for determining ultimate locations. By use of reagents that recognize specific outer membrane proteins, it has been possible to observe with the electron microscope the sites of appearance of newly synthesized protein. In the case of bacteriophage λ receptor, adsorption of λ was used to show that this protein appears to be incorporated initially only at the cell septum (Ryter et al., 1975). In contrast, in *Salmonella typhimurium* the outer membrane matrix protein and certain other outer membrane proteins appear only at the sites of adhesion between the inner and outer membranes known as Bayer patches (Smit and Nikaido, 1978). In the case of the galactose- and maltose-binding proteins located in the periplasm, evidence suggests that they are only synthesized during periods of cell division (Shen and Boos, 1973; Dietzel et al., 1978). In contrast, alkaline phosphatase is synthesized throughout the cell division cycle (Shen and Boos, 1973). The apparent temporal regulation of synthesis of these proteins raises the possibility that transcriptional or translational control may be required in addition to the signal sequence and any other protein structural features for proper localization.

Studies on genetic fusions with envelope proteins suggest a model that explains some features of the differential compartmentalization of these proteins in *E. coli*. This model is presented and discussed in a subsequent section.

Thus the bulk of the evidence accumulated on *E. coli* envelope proteins conforms remarkably to predictions of the signal hypothesis. However, these results do not of themselves confirm the model. What is necessary for further confirmation and elaboration of new features is a combined genetic and biochemical analysis. As an example cited earlier, the lipoprotein signal sequence mutant provides direct evidence that processing is not necessary for export. Other kinds of mutants that will be important in such studies are those that alter export by changing the signal sequence and possibly mutants in membrane proteins, the ribosomes, or processing activities, and, in general, mutants that alter the export of proteins can be used as a handle for identifying those factors essential to the process.

2 GENE FUSIONS IN THE STUDY OF ENVELOPE PROTEIN LOCALIZATION

The isolation of certain of the classes of mutants affecting protein localization has been made possible by a new genetic approach to this problem, the technique of gene fusion (Bassford et al., 1978; Franklin, 1978). We are interested in determining which components of the genes for envelope proteins are important in the export process. One way of studying this is by isolating certain of the classes of mutants described above. Another approach is to tag *fragments* of the envelope protein under study with a label that can be followed. Then, by determining the location of the label, it should be possible to conclude whether or not the particular fragment can promote export. Specifically, we have developed a technique that allows us to tag amino-terminal fragments of envelope proteins (of varying lengths) with the easily detectable enzyme β-galactosidase. This was done using gene fusion techniques.

The finding that the amino-terminal portion of β-galactosidase can be removed and replaced by the amino-terminal sequence of another protein made this technique feasible. First, it was shown that the inactive β-galactosidase coded for by a *lacZ* deletion mutation, M15 (or XA21), which removes amino acids 11–41 from the protein, can be restored to activity in the presence of antibody to β-galactosidase (Accola and Celada, 1976). The enzymatic activity can also be restored to *lacZ* gene products missing an amino-terminal sequence by fusing them to the amino-terminal sequence of many other proteins. This was first shown by Müller-Hill and Kania (1974) in the following way. A strain was constructed carrying a *lacI* mutation (I^Q) that caused overproduction of repressor protein, and a *lacZ* chain terminating mutation (U118) (corresponding to amino acid 17 in the protein) that resulted in termination of protein synthesis early in the gene (Figure 1). Selection for Lac^+ revertants of this strain yielded back-mutations, suppressor mutations, and deletions *that fused the* lacZ *gene to the* lacI *gene.* Such deletions removed the protein chain termination signal at the end of *lacI*, the chain initiation signal at the beginning of *lacZ*, and the *lacZ* U118 termination signal. The substitution of varying extents of the amino-terminal sequence of the *lac* repressor for the missing amino-terminal sequence of β-galactosidase was apparently sufficient to allow β-galactosidase activity. In one case, the hybrid protein had both repressor and β-galactosidase activity.

A number of other cases have been described in which two genes that ordinarily code for *two* independent polypeptide chains were genetically fused so that they code for only *one* polypeptide chain comprised of an amino-terminal fragment of one and a carboxyl-terminal fragment of the other. A major step in permitting such an approach to be generalized was the development of methods for taking a gene from its normal position on the chromosome and causing it to be inserted close to a gene elsewhere on the chromosome. Such a step permits, under the the appropriate conditions, selections for fusion of the two genes. What follows is a description of a new approach to isolating such fusions and the logic behind it (Casadaban, 1976).

Let us consider a hypothetical operon that includes gene X, to which we wish to fuse the *lacZ* and *lacY* genes. One possible first step would be to transpose the *lac* region to a site close to or within the gene X. (Ordinarily, this can not be conveniently done for most regions of the chromosome by previously described techniques.) If there were genetic homology between the two regions, then it would be possible to select for recombination events that generate the appropriate transposition. Since in most cases, there will not be any extensive homology between any two regions chosen at random

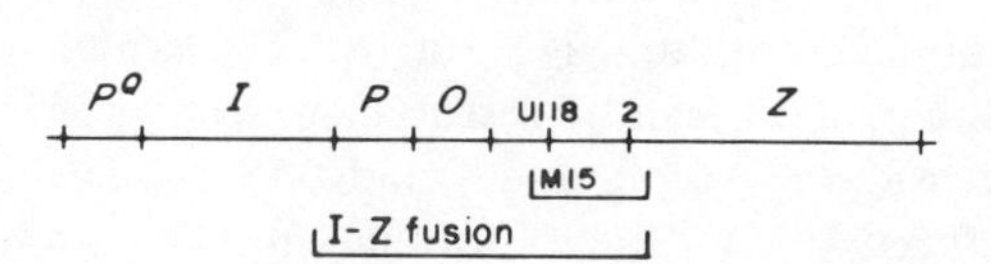

Figure 1 Fusions of the *lacZ* and *lacI* genes. The U118 mutation has been shown to alter the codon for amino acid 17 in β galactosidase. Reprinted with permission from P. Bassford et al., in *Molecular Aspects of Operon Control*, J. H. Miller and W. S. Reznikoff, eds. Copyright 1978 by the Cold Spring Harbor Laboratory.

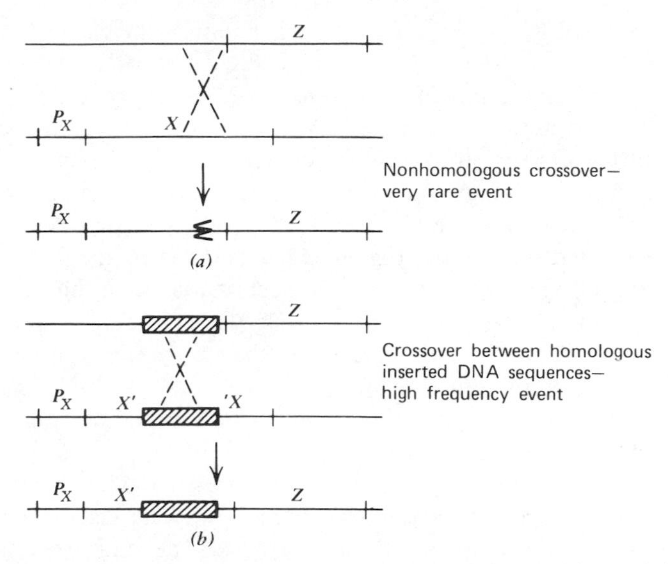

Figure 2 Mechanisms for transposition of the *lac operon*. Reprinted with permission from P. Bassford et al., in *Molecular Aspects of Operon Control*, J. H. Miller and W. S. Reznikoff, eds. Copyright 1978 by the Cold Spring Harbor Laboratory.

(Figure 2*a*), these events will be very rare. Therefore, it is necessary to generate that homology. This can be accomplished by the insertion of identical segments of DNA into the two regions, near or in gene X and in the *lac* operon (Figure 2*b*). Given genetic homology between the two regions, recombination events may occur at a reasonable frequency.

Several methods exist for inserting DNA sequences in the chromosome. These include isolation of insertion sequence mutations, translocation of antibiotic resistance markers, insertion of the bacteriophage Mu-1 chromosome, and insertation of the bacteriophage λ genome at different positions on the chromosome (all reviewed in Bukhari et al., 1977). Casadaban chose to use bacteriophage Mu-1 as a tool for simplifying gene fusion technology. What follows is a description of this approach (Casadaban, 1976).

Bacteriophage Mu-1 is a temperate phage that upon lysogenization inserts its genome (28×10^6 daltons) at random around the bacterial chromosome. A significant percentage of Mu-1 lysogens are auxotrophic because of insertion of the Mu-1 genome into a gene for a biosynthetic pathway (Taylor, 1963). Thus it is possible using appropriate selection or screening procedures to detect Mu-1 insertions in or near a gene of interest (e.g., gene X and *lac*). In addition, the approaches exist for detecting Mu-1 insertions in essential genes (Nomura and Engbaek, 1972). Once strains have been isolated that carry the Mu-1 genome in or near gene X and in or near *lac*, a genetic cross can be performed (Figure 3*b*) that will result in the production of recombinants in which *lac* has been transposed close to gene X. To facilitate crosses of this type, a λ transducing phage was constructed that carried the *lac* region and one end of the Mu-1 genome (Figure 3*a*). The only genetic homology that exists between such a λ phage (missing its *att* site) and the chromosome of strains that carry the Mu-1 genome in gene X and are deleted for the *lac* region (Figure 3*b*) is in the gene X region. Selection of λ lysogens after infection of such strains with this special phage yields strains with *lac* integrated near gene X (Figure 3*c*). To facilitate selection for fusions, the *lac* operon carried on the λ genome is missing its promoter region and is, therefore, inactive.

The steps described above bring *lac* and gene X close together but do not fuse them. What prevents transcription initiated at P_X from proceeding into the *lac* operon is the long sequence of Mu-1 DNA that must contain at least one barrier to continued transcription (Figure 3*c*). The last step in generating the fusions is the removal of such barriers. The initial use of a Mu-1 phage that is thermoinducible makes this step possible. Strains carrying the structure indicated in Figure 3*c* can be used to select for Lac^+ revertants that are also thermoresistant. A single mutational event that can generate such revertants is a deletion that removes those genes of Mu-1 lethal to the cell and, at the same time, removes all barriers to transcription in the Mu-1 genome. It is this approach that has been used successfully to fuse the *lac* operon to a number of different bacterial genes (Bassford et al., 1978). In each case, the synthesis of β-galactosidase exhibits the regulation expected for the promoter to which it has been fused. For instance, fusions of *lacZ* to the *ara* operon result in stains in which β-galactosidase synthesis is induced by the presence of arabinose in the growth media.

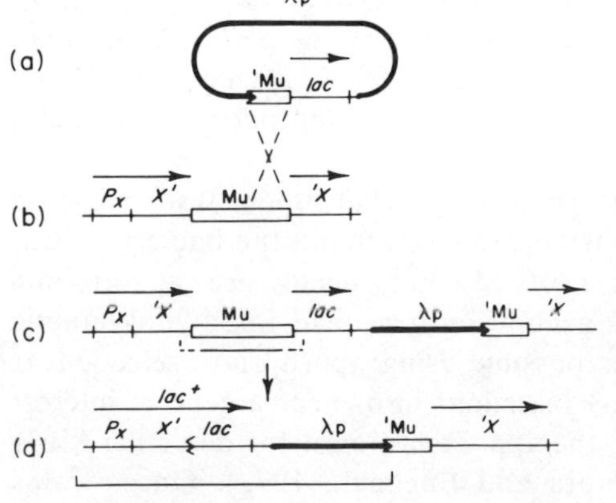

Figure 3 Steps leading to fusion of the *lac* operon to other bacterial genes. The horizontal arrows indicate the directions of transcription. Reprinted with permission from P. Bassford et al., in *Molecular Aspects of Operon Control*, J. H. Miller and W. S. Reznikoff, eds. Copyright 1978 by the Cold Spring Harbor Laboratory.

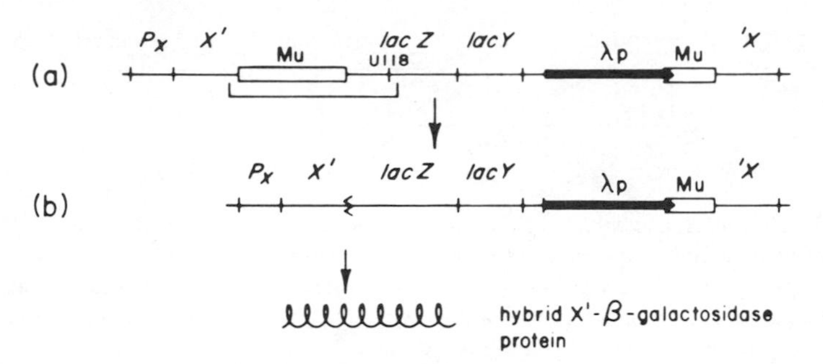

Figure 4 Steps leading to the formation of hybrid proteins between β-galactosidase and other bacterial proteins. Reprinted with permission from P. Bassford et al., in *Molecular Aspects of Operon Control*, J. H. Miller and W. S. Reznikoff, eds. Copyright 1978 by the Cold Spring Harbor Laboratory.

In our laboratory, such fusions can be isolated within 1 month using genes in which the detection of Mu-1 insertions is easily achieved.

Employing this approach, fusions have been isolated in which an intact *lacZ* structural gene is brought under the control of a new promoter. The β-galactosidase protein that is produced by such fusion strains is identical with that produced by a wild-type *lac*$^+$ strain. We call such fusion strains *operon fusions*. In contrast, we can use the same general approach to isolate strains that produce β-galactosidase molecules whose amino-terminal sequence has been replaced by the amino-terminal sequence of another protein. This can be done by incorporating the *lacZ* U118 mutation onto the λ*plac* Mu phage before initiating the sequence of steps leading to selection for fusions (Figure 4). Using this phage derivative, the ultimate selection for Lac$^+$ colonies at 42°C yields strains in which the early portion of the *lacZ* gene (including the U118 site) has been removed and replaced by a portion of a nearby gene as with the *lacI–lacZ* fusion of Muller-Hill and Kania. Such strains produce hybrid β-galactosidase molecules and are termed *protein fusions*.

3 ENVELOPE PROTEINS INVOLVED IN MALTOSE TRANSPORT

We have extended the protein fusion approach to study envelope proteins of *E. coli*. As a convenient system for the study of protein export in *E. coli*, we have chosen the set of genes determining the transport of maltose and maltodextrins. These genes are clustered in one locus (*malB*) of the chromosome and comprise two operons transcribed in opposite directions (Hofnung, 1974; Figure 5). The products of two of the genes are well char-

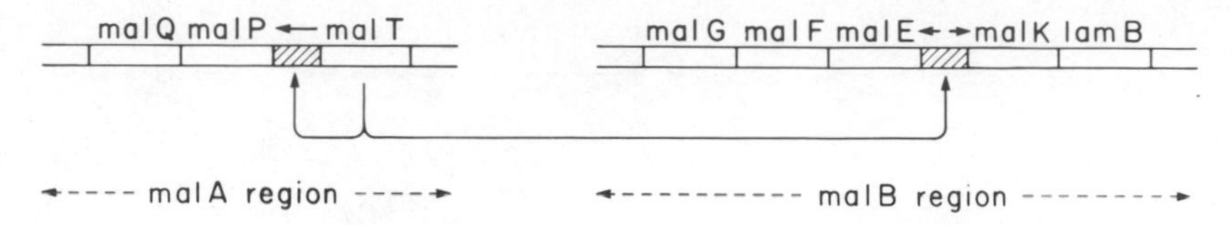

Figure 5 The *malA* and *malB* regions of *Escherichia coli*. Genes *malP* and *malQ* are the structural genes encoding the enzymes maltodextrin phosphorylase and amylomaltase, respectively. All the genes in the *malB* region code for proteins involved in maltose transport (see text). The product of the *malT* gene is a positive control factor that regulates transcription at the promoters for the three *mal* operons, as indicated.

acterized; the *malE* gene encodes a periplasmic maltose binding protein (Kellermann and Szmelcman, 1974) and the *lamB* gene codes for an outer membrane protein. This latter protein serves as the cell surface receptor for bacteriophage λ (Randall-Hazelbauer and Schwartz, 1973), is essential for penetration of maltodextrins into the cell, and facilitates transport of maltose when the sugar is present at low concentrations in the medium (Szmelcman et al., 1976). Recent evidence referred to later in this article indicates that the *malF* gene product is a protein located in the cytoplasmic membrane. The location of the *malK* and *malG* gene products has not been determined.

The synthesis of these proteins is inducible by maltose, the regulation being determined by the *malT* locus, which is located elsewhere on the chromosome. The *malT* gene codes for a positive control factor; inactivation of the *malT* gene by mutation results in a Mal$^-$ phenotype (Schwartz, 1967; Debarbouille et al., 1978).

Since the regulation of these genes is well understood and since their products are found in the three extracytoplasmic cellular compartments, they appeared to us to be ideally suited as a system for genetic analysis of protein export.

4 ISOLATION AND PROPERTIES OF VARIOUS *malB-lacZ* FUSIONS

Using techniques described above, a number of fusions between each of the genes in the *malB* region and the *lacZ* gene have been constructed. Results that we have obtained to date show a remarkable consistency and a common pattern is emerging. Fusions constructed so as to contain very little of the particular *malB* gene invariably produce a hybrid protein that is located in the cytoplasm. This is true for all five *malB* genes. If, however, the hybrid gene contains a substantial amount of a particular *malB* gene, the cellular location of the hybrid protein is altered in at least four of the five cases.

These results indicate that information directing a protein to a noncytoplasmic location is contained within the gene that codes for that protein.

They also suggest, in agreement with what is known concerning "*in vitro*" precursors (see above), that at least part of this information is located at the amino terminus of the protein. Finally, in certain cases (in particular, *lamB–lacZ*) these results suggest that any protein, regardless of its properties, can be directed to a noncytoplasmic location provided the correct genetic sequence is present.

In this section we summarize the properties of *lacZ* fusions to the following three *malB* genes: *malF*, *malE*, and *lamB*. We have chosen these three genes because each codes for a protein located in a different cellular compartment. The products of these three genes are found, respectively, in the inner membrane, the periplasm, and the outer membrane.

4.1 Characterization of Fusion Strains

Potential *malB–lacZ* fusions isolated as described above are all characterized in the following manner:

1 The existence of the fusion is verified by demonstrating that the β-galactosidase activity produced by these strains is inducible by maltose. In the case of *lamB–lacZ* fusions this can be done simply by showing that maltose added to the growth media increases the amount of β-galactosidase activity produced. The *lamB* gene product, the phage λ receptor, is not required for growth on high concentrations of maltose. In the case of *malF* or *malE–lacZ* fusions, adding maltose to the growth media has no effect, since, in such fusions strains, essential maltose transport genes have been destroyed. To demonstrate that the β-galactosidase activity in these strains is maltose inducible, a wild-type *malB* region must be introduced to enable entry of external inducer. This is done by constructing merodiploids with either an episome or a λp*malB* transducing phage.

2 Deletions that generate a hybrid gene are exceedingly rare. This is not surprising, as the deletion event required must be very specific (see above). Accordingly, it is not unusual for us to leave selection plates in the warm room at 37°C for 40–50 days. We have learned from experience that fusions continue to appear for as long as the plates are incubated. Presumably, slow background growth on the selection media is responsible for this phenomenon. In any event, before any studies can be performed it is necessary to demonstrate that the event that generated the Lac^+ phenotype was the predicted deletion that generated the hybrid gene. Occasionally, we find that the Lac^+ phenotype is not the result of the formation of a hybrid gene but, in fact, the result of a double mutational event.

One common double mutational event leading to a Lac^+ phenotype is a reversion of the *lacZ* U118 mutation and a deletion that removes the Mu (*c* ts) prophage but does not enter the *lacZ* gene. The β-galactosidase produced by such a strain is identical with wild-type β-galactosidase. Such a fusion is an operon fusion rather than a protein fusion.

Furthermore, as is described in more detail below, we have also found genetically unlinked mutations that were responsible for generating a Lac$^+$ phenotype in several fusion strains. The existence of these mutations was demonstrated by first isolating the λp transducing phages for the various fusions and then transducing the fusion into a wild-type strain. In some cases it was found that these transducing phages did not transduce a Lac$^+$ phenotype even though they carry the *malB–lacZ* fusion and *lacY*. The reason for this must be that the initial fusion strain contains a secondary unlinked mutation that renders the fusion Lac$^+$.

To verify the existence of a hybrid gene it is necessary to demonstrate that the deletion event that generated the fusion extended into the *lacZ* gene beyond the site of the U118 mutation on the one side. Also, it is important to show that the deletion event extended beyond the Mu (*c* ts) prophage on the other side into the *malB* gene in question (see Figure 4). Mapping of the fusion joint in the *lacZ* portion of the hybrid gene is done by mating a series of episomes containing various *lacZ* point mutations. Recombination on the episome to Lac$^+$ indicates that the corresponding *lacZ* DNA in the fusion was not removed by the deletion event that generated the fusion (Brickman et al., in press). It should be noted at this point that so far all fusions examined contain a similar, but not identical, portion of the *lacZ* gene. In four fusions examined so far; 18, 19, 20 and 25 amino acids were found to have been removed from the NH_2-terminal end of β-galactosidase (A. Fowler and M. Casadaban, personal communication). Genetic studies show that in no case are more than 40 amino acids removed from β-galactosidase (Brickman et al., in press). The fusion joint in the *malB* portion of the hybrid gene is determined by deletion mapping (Silhavy et al., 1977; T. Silhavy et al., in press).

3 Once genetic evidence for the existence of the hybrid gene is obtained, fusion strains are then further characterized biochemically. Whole cell extracts are prepared and run on sodium dodecyl sulfate (SDS) polyacrylamide gels. The monomer subunit of wild-type β-galactosidase has a molecular weight of 116,000. Since the deletion events generating hybrid proteins remove a relatively small number of amino acids from the enzyme, all these proteins have molecular weights of approximately 116,000 or higher. *E. coli* produces very few proteins in the molecular weight range of these hybrid proteins and none that are inducible by maltose (Figure 6). Accordingly, we can easily identify and approximate the molecular weight of the hybrid protein produced. Since most fusions contain a similar length of the *lacZ* gene, the

molecular weight of the hybrid protein provides a good approximation of the amount of *malB* DNA present in the hybrid gene. This is not absolutely reliable since it is apparent that proteolysis occurs in certain strains, reducing the hybrid proteins to a smaller size (see below). For this reason, nearly all experiments are performed on cells grown at 30°C, since low temperature is known to stablize certain mutant proteins (Lin and Zabin, 1972).

4 Strains that produce a β-galactosidase activity that is located outside the cytoplasmic membrane are capable of growth on lactose even in the absence of a lactose transport system. Presumably, the external β-

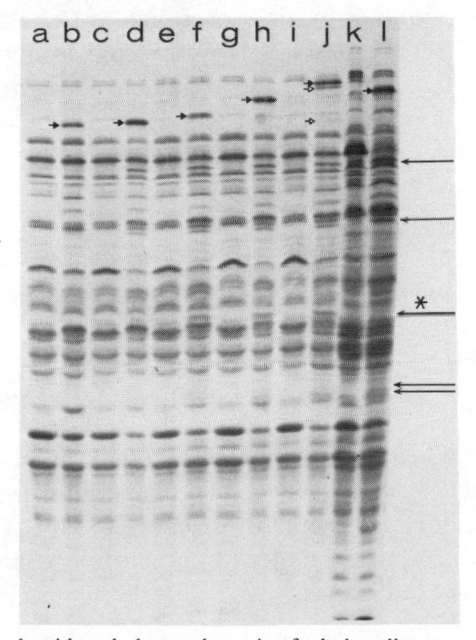

Figure 6 SDS–polyacrylamide gel electrophoresis of whole cell extracts of *malE–lacZ* protein fusion strains. Whole cell extracts were prepared and electrophoresis was performed as described (Silhavy et al., 1977). (*a*) Class I fusion, uninduced; (*b*) class I fusion, induced; (*c*) class II fusion, uninduced; (*d*) class II fusion, induced; (*e*) class III fusion, uninduced; (*f*) class III fusion, induced; (*g*) class IV fusion, uninduced; (*h*) class IV fusion, induced; (*i*) class V fusion, uninduced; (*j*) class V fusion, induced; (*k*) class III *lamB–lacZ* fusion strain pop3186, uninduced; (*l*) strain pop3186, induced. The hybrid proteins are easily identified as maltose-inducible proteins of high molecular weight (small, filled arrows). In addition, the class V *malE–lacZ* fusion strain exhibits several additional maltose-inducible protein bands that are apparently breakdown products of the hybrid protein (small, open arrows, see text). A number of new protein bands are discerned when maltose-sensitive fusion strains are induced with maltose (large arrows, on right). These apparently correspond to precursors of normal exported proteins that accumulate under these conditions (see Section 5). The protein band whose position is marked with an asterisk corresponds to the precursor of the phage λ receptor. This protein band is observed in all the maltose-sensitive strains except pop3186, which does not have an intact *lamB* gene (see text).

galactosidase hydrolyzes the lactose into glucose and galactose that can then be transported into the cell by means of other transport systems. Fusion strains are made *lacY* by P1 transduction to chloramphenicol resistance with a lysate prepared on a strain that contains the chloramphenicol resistance-conferring translocon (TN9) inserted in the *lacY* gene. Acquisition of the translocon confers the *lacY* phenotype. If such transductants are capable of growth on lactose, then the β-galactosidase activity is probably extracytoplasmic. Erroneous results may be obtained, however, when the fusion strain produces very large amounts of β-galactosidase activity.

5 Finally, cell fractionation techniques are performed. Membranes are isolated and separated by the procedure of Osborn et al., (1972); the periplasmic fraction is isolated by the cold osmotic shock procedure of Neu and Heppel (1965), and β-galactosidase activity is assayed to determine the location of the hybrid protein. The activity produced by a particular fusion is said to be cytoplasmic if this activity is soluble and not present in the periplasmic fraction, and if the *lacY* derivative of the fusion cannot grow on lactose.

4.2 *malF–lacZ* Fusions

So far we have constructed three different classes of *malF–lacZ* fusions. The relevant data with respect to these three classes of fusions is summarized in Table 2. These classes of fusions differ only in the amount of *malF* DNA present in the hybrid gene. The fusion, *malF–lacZ* hybrid 6-3, contains only a very small amount of *malF* DNA. The fusion joint for this fusion in fact maps in the first (amino-terminal) deletion group. The hybrid protein produced by this fusion has a molecular weight very close to that of wild-type β-galactosidase and appears to be localized exclusively in the cytoplasm.

The *malF–lacZ* fusion 11-1 was the first *malF* fusion to be characterized (Silhavy et al., 1976). The fusion, and several others that are apparently identical, contains a substantial amount of *malF* DNA. The fusion joint maps in the ninth deletion group from the portion of the gene corresponding to the amino terminus of the protein. The hybrid protein produced has a molecular weight of approximately 146,00. Localization studies indicate that this hybrid protein is localized almost entirely in the cytoplasmic membrane. On the basis of the *lacY* experiments explained in the previous section, we believe that the β-galactosidase protion of this hybrid protein is localized to the inner face (cytoplasmic side) of this membrane. In other words, the β-galactosidase activity, although membrane bound, is not external to the cell's permeability barrier, the cytoplasmic membrane.

Table 2 Characterization of the Hybrid Proteins Produced by Various *malF–lacZ* Fusions Strains

Strain	Fusion	Induced β-Galactosidase Activity (units)	Cellular Hybrid Protein Localization (%)				Approximate Mol. Wt.	Estimate of Amount of *malF* DNA Present in the Hybrid Gene[a]
			Inner Membrane	Outer Membrane	Periplasm	Cytoplasm		
pop3124	6-3	1600	NT	NT	<1	92	116,000	1/11
MC4416	11-1	70	80	<5	<1	14	146,000	9/11
MC4419	53-1	130	80	<5	<1	—	155,000	10/11

[a] The *malF* gene has been divided into 11 segments by deletion mapping. The fraction listed represents the number of segments present in the hybrid gene. It should be noted that this is a genetic result, not a physical result, and accordingly, these data are only an estimate. If deletion end points are nonrandom, it is possible that these estimates are substantially incorrect.

[b] Exact fractionation data with fusion 53-1 is complicated by the existence of proteolytic degradation products of the hybrid protein. See text for details.

The final class of *malF–lacZ* fusions, represented by hybrid 53-1, is the largest yet isolated by genetic tests and appears to contain nearly all the *malF* gene. The fusion joint maps in the penultimate deletion group. The hybrid protein produced has a molecular weight of approximately 155,000 and appears to be localized in the cytoplasmic membrane in a manner exactly analogous to the previous fusion 11-1. However, this determination is complicated for the following reason. If the fusion stain is fractionated according to our standard procedure, we find about 50% of the activity to be membrane bound while the remainder is soluble. This result was at first quite puzzling. More careful analysis revealed that the soluble hybrid protein has an apparent molecular weight of only 116,000. It is the membrane-bound hybrid protein that has a molecular weight of 155,000. We believe that the hybrid protein produced by this fusion is in fact the species with a molecular weight of 155,000. This species is localized entirely in the cytoplasmic membrane. Possibly by virtue of the abnormal nature of this hybrid protein, it is somewhat protease sensitive. As is seen later, many other hybrid proteins also appear to be protease sensitive and invariably the result of degradation is a species with an apparent molecular weight identical to normal β-galactosidase. We suspect that these hybrid proteins are molecules with two different domains: one domain containing the β-galactosidase portion of the hybrid and the other containg the portion coded for by *malB* DNA. The two domains would be connected by a somewhat exposed and thus a proteolytically sensitive portion of the polypeptide backbone. This has in fact been demonstrated with certain *lacI–lacZ* fusions (Heidecker and Müller-Hill, 1977). This region may correspond to the fusion joint. Hence proteolytic attack would result in the release of a relatively stable domain of the hybrid protein corresponding to β-galactoidase. The large species of the *malF–lacZ* hybrid 53-1 is located in the membrane with the β-galactosidase portion facing the cytoplasm. Proteolytic attack by a cytoplasmic protease releases the β-galactosidase portion (domain) of the hybrid molecule, thus effectively solubilizing 50% of the β-galactosidase activity.

Until recently, very little was known about the product of the *malF* gene. It has been generally assumed that the product is a protein located in the cytoplasmic membrane. In view of its function in the active transport of maltose, it has been proposed that this protein may be a transmembrane carrier for maltose and other transport substrates. Alternatively, it may play a role in energy coupling (Hofnung, 1974). Since none of these hybrid proteins exhibits any *malF* activity, they shed no light on the possible function of this essential transport component.

The existence of these fusions has made possible a biochemical study aimed at identifying the *malF* gene product. The *malF–lacZ* hybrid protein 11-1 has been purified and antisera directed against the hybrid protein have been prepared. It is expected that antibodies directed against the hybrid will cross-react with both wild-type β-galactosidase and the wild-type *malF* gene product. Using this antibody, attempts are being made to identify the *malF* gene product in a strain that is *mal*$^+$Δ*lac*. Preliminary results indicate that the *malF* gene product is a protein of approximately 50,000 daltons localized in the cytoplasmic membrane (H. Shuman, personal communication).

Identification of the *malF* gene product will enable a careful determination of the protein's configuration in the cytoplasmic membrane. In particular, we would like to know if the malF protein (*a*) is simply stuck to the inner face of the membrane, (*b*) is simply stuck to the outer face of the inner membrane or (*c*) actually passes through the membrane bilayer with portions of the molecule exposed on both the inner and outer face of the bilayer. As discussed by Rothman and Lenard (1977), since little or no hydrophilic portions of the protein project beyond the extracytoplasmic surface, proteins such as *a* (called endoproteins by these authors) do not require any special mechanism to be correctly localized in the bacterial membrane. Presumably, the molecule could by synthesized on "free" ribosomes and could then become associated with the cytoplasmic membrane simply by virtue of a hydrophobic segment of the molecule. On the other hand, proteins of classes *b* and *c* have substantial hydrophilic mass projecting beyond the extracytoplasmic surface (ectoproteins) and, as such, proteins of this type may require specialized mechanisms for correct localization. Rothman and Lenard (1977) have proposed a variation of the signal hypothesis to account for this localization (see Section 7).

Results obtained so far using the fusion technique do not distinguish among these possibilities, as the β-galactosidase moiety of the hybrid molecules is always located on the inner face of the cytoplasmic membrane. They do, however, indicate that information for localizing the *malF* gene product is located in the *malF* gene itself. This must be the case, since simply varying the amount of *malF* DNA present in an otherwise isogenic hybrid gene alters cellular location of the hybrid protein. If the malF protein is an "endoprotein," then it follows that the portion of the molecule responsible for membrane attachment is located somewhere in the amino-terminal portion of the protein. Presumably, the isolation of a number of *malF–lacZ* fusions could better define this region. If, on the other hand, the malF protein is an ectoprotein, then the situation becomes more complicated, yet also more interesting. Here again, the isolation of more fusions could serve to define the signal for localization.

4.3 *lamB–lacZ* Fusions

From the beginning, the isolation of *lamB–lacZ* protein fusions was found to be more difficult than the isolation of *malF–lacZ* protein fusions. The frequency of the former is several orders of magnitude less than that of the latter. In order to obtain any *lamB–lacZ* fusions at all, selection plates had to be left in the warm room for as long as 50 days. There are numerous possible explanations for this requirement. We believe though, that the most likely explanation involves the respective cellular locations of the *malF* and *lamB* gene products. Unlike the malF protein, the *lamB* gene product, the phage λ receptor, is an extracytoplasmic protein located in the outer membrane. It may well be that isolating viable fusions between β-galactosidase and an extracytoplasmic protein is inherently difficult. It also is not surprising that many of the *lamB–lacZ* fusions that were isolated after such long incubations are double mutants. As is seen below, the nature of certain of the double mutants is still unresolved.

At the time of this writing, five classes of *lamB–lacZ* fusions have been isolated. The relevant properties of these fusions are summarized in Table 3.

There exist a large number of fusions belonging to class I. Their properties are all very similar. The example shown in the table, fusion 61-4, is the most thoroughly studied (Silhavy et al., 1977). This fusion contains very small amounts of *lamB*. In fact, the fusion joint for this fusion maps earlier in *lamB* (corresponds to the extreme amino terminus of the λ receptor) than any known mutation in the gene. However, genetic evidence shows that this fusion is indeed a *lamB–lacZ* fusion rather than a *malK–lacZ* fusion. Nonpolar nonsense mutations in the promoter proximal gene *malK*, when incorporated into the fusion strain, do not affect hybrid protein synthesis. The hybrid protein produced by this fusion strain has an apparent molecular weight slightly smaller than that of wild-type β-galactosidase (M. Schwartz, personal communication), and it is localized exclusively in the cytoplasm. The strain in which this fusion was isolated contains no detectable secondary mutations, and the phenotype is, as expected, Lac^+Mal^+. The strain is Mal^+ because the λ receptor is not required for growth on high concentrations of maltose. The one class II fusion strain that exists is identical to the fusions of class I in all respects except that it contains slightly more *lamB* DNA by genetic mapping, and accordingly, the hybrid protein produced is slightly larger (T. J. Silhavy, M. Hall and M. Schwartz, manuscript in preparation).

Several fusions belonging to class III now exist. For a number of reasons that become clear later, these fusions have proven to be the most interesting and the most useful. Again, fusion 42-1 is the one that has been most thoroughly studied (Silhavy et al., 1977). Genetic mapping suggests that fusion 42-1 contains a substantial portion of the *lamB* gene (Table 3). The hybrid protein produced has an apparent molecular weight of 137,000. Using three different methods we have obtained evidence that a fraction of the hybrid protein produced is extracytoplasmic and probably in the outer membrane:

Table 3 Relevant Properties of Various *lamB–lacZ* Fusion Strains and Hybrid Proteins

Fusion Class	Example	Phenotype	Induced β-Galactosidase Activity (units)	Cellular Hybrid Protein Localization (%)				Approximate Hybrid Protein Molecular Weight	Approximate Amount of *lamB* DNA Present in Hybrid Gene[a]	Evidence for Unlinked Mutations
				Inner Membrane	Outer Membrane	Periplasm	Cytoplasm			
I	61-4	Mal^+Lac^+	900–1100	<10	<1	<1	90	116,000	<1/9	No
II	52-4	Mal^+Lac^+	2000–3000	<15	<1	<1	85	116,000	1/9	No
III	42-1	Mal^SLac^+	1000–1300	~25	~30	2	43	137,000	5/9	No
IV	42-18	Mal^+Lac^+	100–200	<10	>80	NT	<10[b]	141,000	6/9	Yes, unknown
V	42-12	$Mal^-Lac^\pm$	100–200	<10	>80	NT	<10[b]	141,000	6/9	Yes, *malQ*

[a] The *lamB* gene has been divided into nine segments by deletion mapping. The value listed represents the fraction of segments that are present in the hybrid gene and that were not removed during construction (M. Schwartz, personal communication; see Table 2).

[b] Exact fractionation data with these strains are difficult to give because of the extremely low specific activity of the hybrid protein. These values represent estimates from gels of the various cellular fractions.

1. By appropriate genetic strain constructions, we have shown that the fusion strain can grow on lactose even in the absence of a lactose transport system. This result indicates the presence of β-galactosidase activity outside the cytoplasmic membrane barrier.
2. Biochemically, using sucrose density gradients, we have demonstrated the presence of a significant amount of β-galactosidase activity in an outer membrane fraction.
3. Finally, using immunofluorescence, we have obtained evidence for the presence of some β-galactosidase cross-reacting material at the cell surface.

The strain in which the fusion 42-1 was isolated does not contain any detectable secondary mutations. If the fusion is moved nonselectively into the parental strain background, the percent β-galactosidase activity located in the outer membrane fraction remains unchanged. However, strains that carry the *lamB–lacZ* fusion 42-1 or any other class III fusion do exhibit a very characteristic and unusual phenotype. These strains are sensitive to the presence of the inducer maltose. This Mal^s phenotype is described in more detail in the following section. For the present though, it is important to know that evidence exists that the Mal^S phenotype is the result of the cell's difficulty in exporting the hybrid molecule.

Fusions of classes IV and V are alike in many ways. Genetic mapping indicates that both contain a substantial portion of the *lamB* gene. Both result in the production of a hybrid protein with a molecular weight of approximately 141,000. Also, the activities produced by fusions of class IV and V are substantially lower than other *lamB–lacZ* fusions. Fractionation studies reveal that nearly all the hybrid protein produced by fusions of classes IV and V is located in the outer membrane. We do not know if the hybrid protein produced by these fusions is active or not. If it has any activity, it must certainly be very low. Both class IV and class V fusions exhibit a Lac^- phenotype in the parental genetic background; a secondary mutation is necessary to uncover a Lac^+ phenotype. These secondary mutations, however, do *not* affect localization of the hybrid proteins in any detectable manner. Fusion proteins of classes IV or V are present almost exclusively in the outer membrane regardless of the presence or absence of the secondary mutation (T. J. Silhavy, M. Hall, and M. Schwartz, manuscript in preparation). Accordingly, we believe they have little bearing on the present discussion.

Surprisingly, fusions of classes IV and V do not exhibit the maltose-sensitive phenotype discussed above. Since both classes contain more *lamB* DNA then class III and since both classes export the hybrid protein to the outer membrane with very high efficiency, we expected these fusions to be extremely Mal^S. This, however, is not the case. Fusions of class IV and V are nearly normal Mal^+ in the parental strain background.

Why are strains carrying fusions of classes IV or V *not* Mal^S? If export of the hybrid protein produced by class III fusions causes inducer sensitivity, then why doesn't export of the larger hybrid proteins produced by class IV and V fusion strains? The fact is that hybrid proteins produced by class IV and class V fusion strains are exported to the outer membrane in higher absolute amounts with much greater efficiency than in any class III fusion strain. We believe that the answer to these questions will shed some light on the mechanisms of protein localization. In the Section 7 we present some possible explanations for this somewhat surprising result.

4.4 *malE–lacZ* Fusions

The *lacZ* gene has also been fused to the *malE* gene encoding the periplasmic maltose-binding protein (P. Bassford, T. Silhavy, and J. Beckwith, in press). Five classes of *malE–lacZ* protein fusions have been characterized (Table 4). These different classes are recognized on the basis of (*a*) the location of the fusion joint between the *malE* gene and the *lacZ* gene as determined by genetic mapping; and (*b*) the molecular weight of the corresponding hybrid proteins as determined by their migration on SDS–polyacrylamide gels. As is mentioned previously, the end point within the *lacZ* gene is very nearly the same for all the protein fusions described in this chapter. Thus the molecular weight differences discerned between different *malE–lacZ* hybrid proteins reflect differences in the contribution of the maltose-binding protein to the hybrid product. A SDS–polyacrylamide gel of whole cell extracts of strains representative of each of the five classes of *malE–lacZ* protein fusions is presented in Figure 6. For each class, the hybrid protein is readily identified as a maltose-inducible protein of unusually high molecular weight. In some cases, this has been verified by specifically precipitating these proteins with antisera prepared against wild-type β-galactosidase.

As was the case for both the *malF* and the *lamB* fusions, the major interest in isolating these *malE–lacZ* protein fusion strains was determining the cellular location of the hybrid protein. *Although the maltose-binding protein is a protein normally exported to the periplasm, essentially no β-galactosidase activity was released from cells of any of the fusion strains by several procedures known to release periplasmic constituents*. However, for the different classes of fusion strains, the relative amount of the *malE* product in the protein fusion apparently does determine where the hybrid protein is localized. In the case of class I fusions, the hybrid gene includes only a small portion of the *malE* gene. Hence the hybrid protein synthesized

Table 4 Relevant Properties of Various *malE–lacZ* Fusion Strains and Hybrid Proteins

Strain	Induced β-Galactosidase Activity (units)	Cellular Hybrid Protein Localization (%)				Approximate Mol. Wt.	Approximate Amount of *malE* DNA Present in Hybrid Gene[a]
		Inner Membrane	Outer Membrane	Periplasm	Cytoplasm		
PB4-1	3370	NT	NT	1	88	116,000	2/13
PB41-4	909	32	5	1	64	117,000	3/13
PB62-32	676	65	5	1	32[b]	120,000	4/13
PB72-47	960	73	5	1	24[b]	131,000	8/13
PB179-3	475	60	5	1	39[b]	150,000	12/13

[a] The *malE* gene has been divided into 13 segments by deletion mapping. The value listed represents the fraction of segments that are present in the hybrid gene and that were not removed during construction. It should be noted that the genetic map is not a physical map, and thus the data given are only estimates (see Table 2).

[b] The soluble β-galactosidase activity in these strains probably results from the proteolytic breakdown of the hybrid protein that releases an enzymatically active portion into the cytoplasm (see text).

by these strains has a molecular weight virtually indistinguishable from that of *E. coli* β-galactosidase (mol. wt. 116,000) and is located in the cytoplasm, as are *malF–lacZ* and *lamB–lacZ* hybrid proteins of comparable size. (Preliminary amino-terminal sequencing of one class I hybrid protein reveals that this protein is only two amino acids smaller than native β-galactosidase. The first 15 amino acid residues are derived from the amino-terminal end of what is presumably the maltose binding protein precursor. The sixteenth amino acid of the hybrid protein is amino acid residue number 18 of β-galactosidase [A. Fowler, personal communication]).

The hybrid proteins synthesized by the other four classes of *malE–lacZ* fusion strains have approximate molecular weights that range from 117,000 (class II) to 150,000 (class V). In fact, a genetic analysis of these fusions indicates that the hybrid gene in class V fusions includes nearly the entire *malE* gene, and the others contain proportionately less. For each of these classes, a significant portion of the β-galactosidase activity is membrane bound. Furthermore, most of this membrane-bound activity appears to fractionate on sucrose gradients with the cytoplasmic membrane. Initially, this would appear to be similar to that class of *malF–lacZ* protein fusion strains in which the hybrid proteins are also associated with the cytoplasmic membrane. However, we feel that localization of the *malE–lacZ* hybrid proteins to this particular extracytoplasmic region requires an entirely different explanation. As it turns out, these fusion strains (like class III *lamB–lacZ* fusion strains and unlike any *malF–lacZ* fusion strain) have the unusual property of being maltose sensitive. For reasons discussed in the following section, we feel that the maltose-sensitive phenotype, as well as the localization of these hybrid proteins in the cytoplasmic membrane, results from the cell's attempt to export these hybrid proteins to the periplasm.

Finally, a word about the stability of these hybrid proteins is required. As is also the case with protein fusions to other *malB* genes, extensive degradation of the larger hybrid proteins is frequently observed. For example, it is obvious that in the gel pictured in Figure 6, there are several maltose-inducible protein bands visible in the region of the hybrid protein for the class V fusion strain. All these bands are apparently derived from the hybrid protein synthesized by this strain. Since these hybrid proteins are to some extent abnormal proteins, at least as far as their amino termini are concerned, they must be somewhat susceptible to proteolytic action. The proteolytic breakdown is seen most dramatically when these cells are grown at 37°C, while at 30°C, the breakdown is less marked. (It is known that growth at lower temperatures tends to stabilize nonsense fragments of β-galactosidase.) Our results indicate that for *malE–lacZ* fusion classes II–V, the soluble β-galactosidase activity determined in these strains probably results from the proteolytic breakdown of the amino-terminal ends of the hybrid protein that releases an enzymatically active fragment from the cytoplasmic membrane. A mutation that is known to affect the stability of β-galactosidase nonsense fragments *in vivo* (*lon*, or *deg* reviewed by Goldberg and St. John, 1976) has no effect on the stability of the hybrid proteins in any of the fusion strains described in this chapter.

5 THE "MALTOSE-SENSITIVE" PHENOTYPE

As is mentioned in the preceding section, certain *lamB–lacZ* and *malE–lacZ* protein fusion strains exhibit an unusual phenotype. The growth of these strains is markedly inhibited in the presence of maltose. This "maltose-sensitive" phenotype is illustrated by the experiment presented in Figure 7. Strain PB72-47 synthesizes a hybrid protein in which a substantial portion of the amino-terminal end of the maltose binding protein is attached to β-galactosidase. When this strain is induced for the synthesis of hybrid protein, cell division is inhibited concomitantly with the appearance of β-galactosidase enzyme activity. The cells at this point can be observed by phase contrast microscopy to elongate to several times their normal length and, shortly thereafter, begin to lyse. (These cells, it should be noted, are maltose sensitive as opposed to being maltose negative. The cells are quite sensitive to maltose in the presence of metabolizable carbon sources such as glycerol and succinate. Indeed, since strain PB72-47 has an intact *malE* gene, it synthesizes all of the machinery required to transport and metabolize maltose.)

Among the different classes of *lamB–lacZ* and *malE–lacZ* protein fusion strains that have been described, there are qualitative differences in their sensitivity to maltose. Generally, the sensitivity increases with the amount of the amino-terminal portion of the *lamB* or *malE* product, respectively, that is incorporated into the hybrid protein. Hence the smallest fusions, in which all of the resultant β-galactosidase activity is found in the cytoplasm, are maltose insensitive. The larger fusions, where the major portion of the enzyme activity is membrane bound, are exquisitely maltose sensitive. (This is not true for those class IV and class V *lamB–lacZ* protein fusion strains in which essentially 100% of the hybrid protein produced is localized in the outer membrane. These strains grow normally on maltose. The significance of this is discussed in a later section).

We have concluded that this growth inhibition by maltose is a direct consequence of the induction of synthesis of the hybrid proteins in these strains. In addition to the strong correlation between the appearance of β-galactosidase activity and growth inhibition, we have shown that eliminating the synthesis of the hybrid protein can restore normal growth properties to these strains. For example, when a series of known nonsense (amber and ochre)

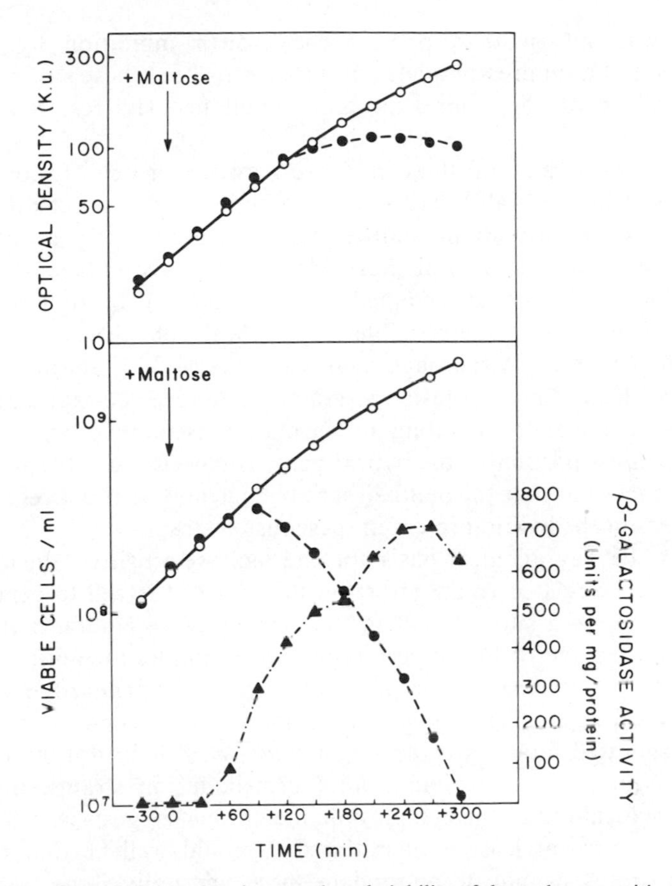

Figure 7 The effect of maltose on the growth and viability of the maltose-sensitive *malE–lacZ* protein fusion strain PB72-47. Cells growing logarithmically in glycerol minimal medium were induced for expression of *mal* genes by addition of 0.2% maltose at 0 min. (Upper panel) the effect of maltose on optical density of culture (expressed in Klett units); (lower panel) the effect of maltose on cell viability (○, ●) and the induction of β-galactosidase enzyme activity ▲. (○), Control culture; (●, ▲) maltose-induced culture.

mutations in the *lacZ* gene were recombined into the *malE–lacZ* hybrid gene encoding the hybrid protein in strain PB72-47 (the resultant strains lacked β-galactosidase activity and became Lac⁻), the response to maltose differed depending on the location of the chain-terminating mutation. When the nonsense mutation was fairly late in the gene, there was virtually no effect on the maltose-sensitive phenotype. However, when the mutation was known to map early in the *lacZ* gene, the strain became maltose insensitive. Mutations mapping in the middle of the *lacZ* gene had an intermediate effect. Thus the early termination of synthesis of the hybrid protein in strain

PB72-47 was sufficient to prevent the growth inhibition by maltose. Introduction of a suppressor allele (su^+) that permitted these strains to complete synthesis of the hybrid protein simultaneously restored maltose sensitivity.

Beginning with any of these maltose-sensitive protein fusion strains, "maltose-resistant" (Mal^R) derivatives can be isolated without difficulty as strains that grow normally on maltose when it is used as the carbon source. The great majority (>99%) of these Mal^R strains simultaneously became Lac^- (a few Mal^R isolated remained Lac^+; see below). In all cases investigated, the genetic lesion responsible for the Mal^R phenotype maps within the hybrid gene itself. A thorough analysis of the Mal^R mutations obtained with the *lamB–lacZ* protein fusion strain pop3186 revealed that many of the mutations were chain-terminating amber mutations, some of which mapped within the *lamB* portion of the hybrid gene. Again, these experiments support the conclusion that the maltose-sensitive phenotype is a direct result of the synthesis of the hybrid protein in these fusion strains.

What is the physiological basis for the maltose-sensitive phenotype? It may, in fact, be related to the effort on the part of the cell to export these hybrid proteins to a site external to the cytoplasmic membrane, that is, to the outer membrane or to the periplasm. For example, the amino-terminal extension of the maltose-binding protein thought to be primarily responsible for its localization to the periplasmic compartment of the cell should have been incorporated intact into the larger *malE–lacZ* hybrid molecules. The observation that, in certain *malE–lacZ* protein fusion strains, the hybrid proteins are localized to the cytoplasmic membrane suggests that the export process may have at least been initiated. It could well be that the cell's export apparatus cannot accommodate these unusually large polypeptide chains. Alternatively, there may be some feature of the β-galactosidase moiety of these hybrid proteins that precludes export. These hybrid molecules may be literally "stuck" in the cytoplasmic membrane, anchored there by a somewhat hydrophobic signal sequence at their amino terminus. Support for this reasoning is partly derived from the isolation of Mal^R strains that are still Lac^+. In these strains, the hybrid protein is still synthesized, but it is localized primarily in the cytoplasm. As is discussed in detail in the next section, these Mal^R strains result from an alteration in the amino-terminal end of the hybrid protein such that it is no longer recognized by the cell for export to the periplasm.

Still, how does the aborted attempt to export the hybrid protein to the periplasm adversely affect cell division and, ultimately, result in cell lysis? Preliminary evidence in this laboratory suggests that the synthesis of these large *malE–lacZ* hybrid proteins interferes with the export of at least certain normal periplasmic and outer membrane proteins. When maltose-sensitive *malE–lacZ* protein fusion strains are cultured at 30°C and induced for hybrid protein synthesis by growth in the presence of maltose for several hours, a number of new protein bands can be visualized when whole cells are prepared and run on a SDS–polyacrylamide gel* (Figure 6). One of the proteins visualized is the precursor form of the phage λ receptor (see legend to Figure 6). Our interpretation of this result is that the hybrid protein has "plugged" the site of export for a number of proteins, including the outer membrane λ receptor protein. As a consequence, these same proteins are accumulated in their unprocessed precursor states. (This experiment also suggests that the periplasmic maltose binding protein and the outer membrane λ receptor may be exported at a common site).

Similar results are obtained with the maltose-sensitive *lamB–lacZ* protein fusion strain pop3186. Although, in this case, some of the hybrid protein is localized to the outer membrane, it is accomplished with rather poor efficiency. The major portion is found in the cytoplasmic membrane and cytoplasm. Thus the hybrid protein synthesized by this strain also tends to interfere with the export of certain protein species. The same new protein bands observed in the *malE–lacZ* protein fusion strains are observed here, except for the pre λ receptor, which this strain does not synthesize since it lacks an intact *lamB* gene. It seems likely that the failure to export one or more of these affected proteins ultimately manifests itself in the maltose-sensitive phenotype described above.

The observation that maltose-sensitive cells accumulate a number of exported proteins in their precursor form when induced with maltose may prove to be a useful phenomenon. Recent evidence (discussed in Section 1) suggests that among those proteins exported to the periplasm and to the outer membrane there may be two general classes with regard to their mechanism of export. One class includes those proteins whose synthesis is temporal and that seem to be exported in the region of the cell septum. The second class represents those proteins whose synthesis is cell cycle independent and that are exported at many different sites on the cell surface, most probably by way of the zones of adhesion between the cytoplasmic and outer membranes, the so-called Bayer's patches. If, in fact, there is a real distinction between these two classes of proteins and the mechanism governing their export, it may well be that only the export of the first class of proteins will be adversely affected when maltose-sensitive fusion strains are cultured on maltose. It seems reasonable that those proteins exported at sites distinct from the export of the λ receptor and the maltose-binding protein will not be affected under these same circumstances. Preliminary results in

* These proteins are not observed when the cells are cultured and induced at 37°C. Apparently, these proteins are rapidly degraded by a protease activity that is more active at the higher temperature.

this laboratory suggest that this may indeed be the case (P. Bassford, unpublished work).

6 THE USE OF GENE FUSIONS FOR GENETIC STUDIES OF PROTEIN LOCALIZATION

A number of different models for the localization of proteins in the cell have been proposed. Most of the models proposed in recent years have relied heavily on the basic aspects of the signal hypothesis. Although this hypothesis may well be correct in most, if not all, of its details, many aspects of the actual mechanism remain to be elucidated, and others are still controversial (see above). Using *E. coli* as a model system, we hope to gain insight into the mechanism(s) of protein localization by isolating mutants that are defective in different stages of the localization process. A variety of such mutations could be envisioned. For example, a mutation in the signal sequence of a particular gene, or an unlinked mutation that causes a defect in the export process, could be expected to affect protein localization. Until recently, however, the search for such mutants has been complicated because there was no way, outside of brute force techniques, to distinguish between mutations affecting localization and mutations that either prevent synthesis or destroy a protein's normal function.

What is needed is a selection or screening procedure that can be used to detect directly those mutants specifically affecting export of a protein. The evidence we now review shows that certain of the *malE–lacZ* and *lamB–lacZ* fusion strains provide just such a selection. Since we believe that the maltose-sensitive phenotype of these strains was due to the attempt to export the hybrid proteins, we predicted that mutants defective in the export process would lose their sensitivity to maltose and could be selected as maltose-resistant mutants (Mal^R). Unlike mutants that are resistant to maltose because they are defective in the *synthesis* of the hybrid proteins, mutants that interfere with *localization* should retain β-galactosidase activity. Our results obtained to date demonstrate that this reasoning was correct. This characteristic and unexpected Mal^S phenotype therefore has provided us with a simple and direct selection for mutants that fail to export these noncytoplasmic proteins to their correct cellular locations.

6.1 Isolation of *lamB* Mutants That Alter the Cellular Localization of the λ Receptor

Starting with a strain (pop3186) that carries the Mal^S Lac^+ *lamB–lacZ* fusion 42-1, we have isolated a series of mutant strains that are resistant to maltose but still produce the hybrid protein in the same amounts. Two such mutants, SE1068 and SE1069, have been characterized in some detail (Emr et al., 1978). As is mentioned earlier, the hybrid protein produced by the *lamB–lacZ* fusion 42-1 is localized, at least in part, in the outer membrane. The results given in Table 5 show that the hybrid protein produced by the two mutant strains is localized almost exclusively in the cytoplasm.

The mutations responsible for the alteration in hybrid protein localization were found to lie very early in the *lamB* portion of the hybrid gene, at a position corresponding to the extreme amino terminus of the protein. To determine the effect of these mutations on λ receptor localization itself, as opposed to the hybrid protein, these mutations were recombined into an otherwise wild-type *lamB* gene (Figure 8). Strains containing the mutations *lamB*S68 (SE2068) and *lamB*S69 exhibit a Dex^{-}λ^{R} phenotype. We have shown that in the case of at least one mutant (*lamB*S69), this phenotype is due to a failure to export the λ receptor and its retention in the cytoplasm.

We have shown that one of these mutations (*lamB*S68) is a small deletion internal to the *lamB* gene. When this mutation is present in an otherwise wild-type *lamB* gene, the protein produced is slightly smaller than wild-type λ receptor. The other mutation (*lamB*S69) behaves as a point mutation. When this mutation is present in an otherwise wild-type *lamB* gene, the protein produced is larger than the wild type gene product by approximately 2000 daltons (Figure 9). The position of this band approximately corresponds to the molecular weight of the λ receptor precursor as identified by Randall et al. (1978).

The simplest explanation for the properties of these mutations is that they are mutations in the portion of the *lamB* gene coding for the signal sequence such that the protein can no longer be exported (Figure 10). Thus these mutations, when present in the *lamB–lacZ* fusion, prevent secretion of the hybrid protein that is now found in the cytoplasm. Similarly, these muta-

Table 5 Characterization of the Hybrid Proteins Produced by Strain pop3186 and Its Mal^{R} Derivatives[a]

Strain	Phenotype	Induced β-Galactosidase Activity (units)	Cellular Hybrid Protein Localization (%)			Molecular Weight
			Cytoplasm	Membrane	Periplasm	
pop3186	Mal^{S}Lac^{+}	1000–1300	43	55	2	145,000
SE1068	Mal^{R}Lac^{+}	1200–1500	92	6	2	145,000
SE1069	Mal^{R}Lac^{+}	1200–1500	87	11	2	142,000

[a] Enzyme activities are expressed in percentage of the activity present in the sonicate after the removal of remaining intact cells.

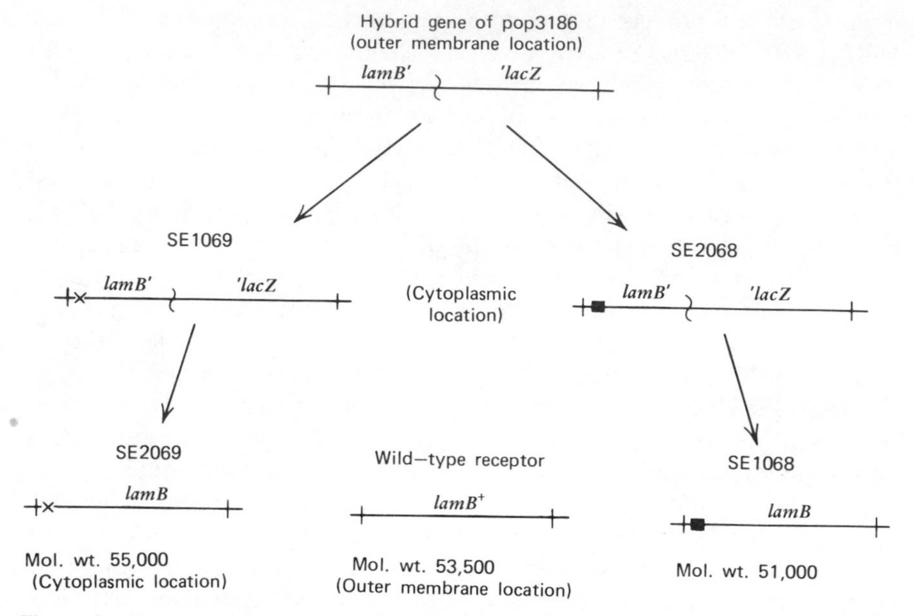

Figure 8 The construction of *lamB* signal sequence mutant strains. The mutation in strain SE2069 (×) is a point mutation; that in strain SE2068 (■) is a small deletion (see text).

tions, when present in a wild-type *lamB* gene, prevent secretion of the receptor. This was directly shown with mutation *lamB*S69.

If this interpretation is correct, the characterization of such mutants should yield much information concerning the secretion process. First, it would provide direct evidence that the signal sequence is *essential* for transport of the proteins. Until now, this mechanism was suggested by the correlation between signal sequences and exported proteins. However, there was no direct evidence to show that the signal sequence did not play a role in a step subsequent to transport. In addition, since one of the mutants behaves as a point mutation, we suggest that a single amino acid change is sufficient to alter cellular location of the λ receptor. Thus sequencing studies with this and other similar mutants should reveal the essential elements of this important sequence.

6.2 Mutations Affecting the Localization of the Maltose-Binding Protein

An approach identical to the one described above was employed to isolate mutations affecting the periplasmic localization of the *malE* product, the maltose-binding protein (Bassford and Beckwith, 1979). Maltose-resistant derivatives of the *malE–lacZ* protein fusion strain PB72-47 were obtained that were still Lac^{+}. In each case, the genetic lesion responsible for the Mal phenotype was located in the *malE* portion of the hybrid gene. Some were point mutations; others were apparently small deletions. The major portion of the β-galactosidase activity produced by these strains was soluble and was located in the cytoplasm. Presumably, these strains were maltose insensitive because they no longer attempted to export the hybrid proteins. Thus, as was also the case for the Mal *lamB-lacZ* fusions, it appears that we obtained mutations that either altered or deleted an essential portion of the amino-terminal end of the hybrid protein required to initiate the export process.

These mutations were recombined into the wild-type *malE* gene to determine their effect on the localization of the maltose-binding protein itself. The resultant recombinant strains score as Mal^{-} on indicator plates, indicating that their ability to utilize maltose as a carbon source has been impaired. However, in the case of several point mutations (presumably missense mutations that revert to wild type), these strains grow slowly on maltose minimal medium. Thus *malB* encoded proteins in these mutant strains can be induced. Whole cell extracts of one of these strains (PB1101) was prepared from a culture that had been grown on glycerol (uninduced) or maltose (induced) and examined on an SDS–polyacrylamide gel (Figure 11). The maltose-inducible band corresponding to wild-type maltose binding

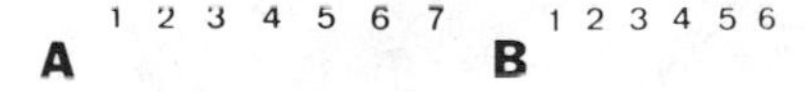

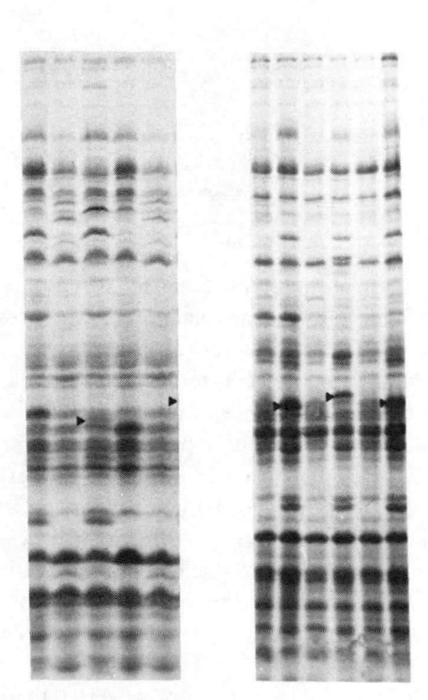

Figure 9 SDS–polyacrylamide gel electrophoresis of radioactively labeled whole cell extracts of strains SE2068 and SE2069. Approximately 2 to 3 $\times$ 10^{8} growing cells were pulse labeled at 30°C with 1 μCi of ^{14}C uniformly labeled amino acids for 7 min. Whole cell extracts were then prepared and electrophoresis and subsequent autoradiography were performed as described (Emr et al., 1978). Gel *A*: (*1*) succinate-grown MC4100, (*2*) maltose-grown MC4100, (*3*) succinate-grown SE2068, (*4*) maltose-grown SE2068, (*5*) glycerol-grown SE2069, (*6*) succinate-grown SE2069, (*7*) maltose-grown SE2069. Gel *B*: (*1*) succinate-grown MC4100, (*2*) maltose-grown MC4100, (*3*) succinate-grown SE2069, (*4*) maltose-grown SE2069, (*5*) succinate-grown revertant of SE2069, (*6*) maltose-grown revertant of SE2069. Wild-type λ receptor and the mutant λ receptor proteins are indicated by arrows. Gel length is 28 cm.

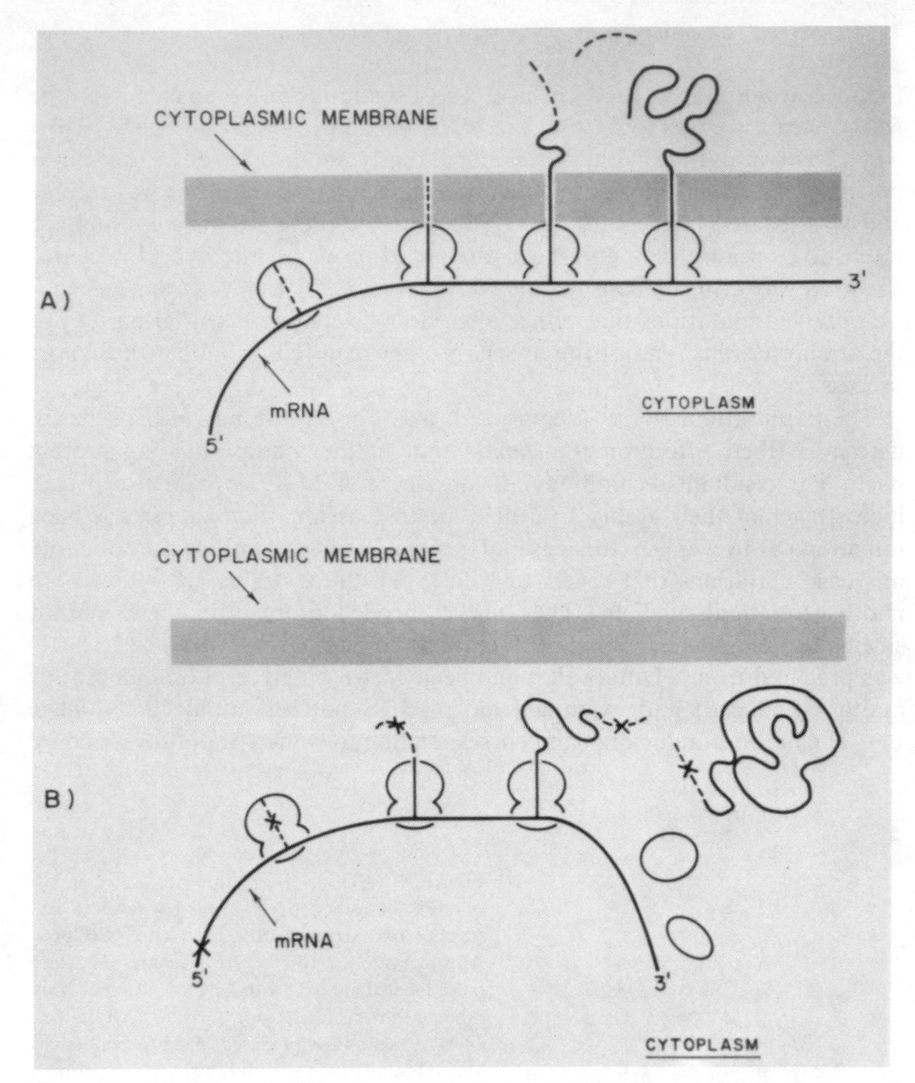

Figure 10 (*A*) Schematic illustration of the first step in the export of the receptor to its normal outer membrane location. The mechanism by which the nascent chain traverses the cytoplasmic membrane and subsequent steps for λ-receptor incorporation into the outer membrane is not shown. The signal sequence (dotted line) at the NH_2 terminus of the nascent polypeptide chain emerges from the ribosome and initiates attachment of the polysome to the cytoplasmic membrane. As translation proceeds, the nascent chain is transferred vectorially across the membrane. This schematic diagram is analogous to the signal hypothesis as proposed by Blobel and Dobberstein (1975). It is proposed that proteolytic processing of the signal sequence from the λ-receptor precursor occurs outside the cytoplasm. Such processing may or may not occur before synthesis of the protein is complete. (*B*) Schematic illustration of the effect of the presumed signal sequence mutation S69 on the export of the λ receptor. The mutant signal sequence mutation is presumed to prevent proper interaction between the sequence and the cytoplasmic membrane. As a result, protein synthesis continues in the absence of a vectorial transfer of the polypeptide chain across the membrane. After the completion of synthesis the mutant λ receptor remains in the cytoplasm. Since processing is proposed to occur outside the cytoplasm, the mutant protein cannot be processed and it is found in precursor form.

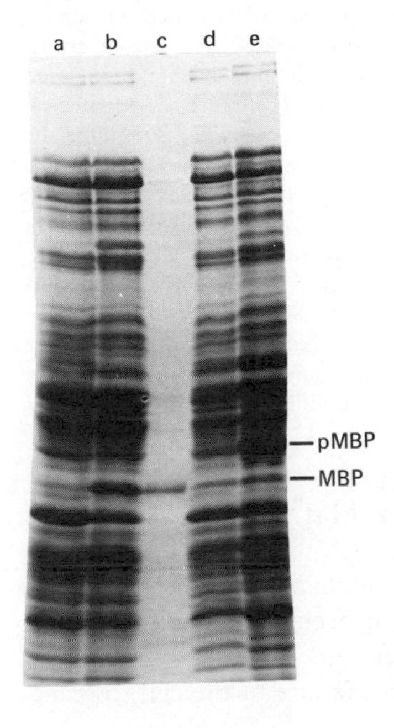

Figure 11 SDS–polyacrylamide gel electrophoresis of whole cell extracts of the Mal$^+$ strain MC4100 and the *malE* signal sequence mutant strain PB1101. Whole cell extracts were prepared and electrophoresis was performed as described (P. Bassford and J. Beckwith, submitted for publication). (*a*) Strain MC4100, uninduced; (*b*) strain MC4100, induced; (*c*) purified maltose binding protein, 2.5 μg; (*d*) strain PB1101, uninduced; (*e*) strain PB1101, induced. The position of native maltose-binding protein (MBP) and its presumed precursor (pMBP) are indicated.

protein is not present for the mutant strain, which most probably accounts for its defect in maltose utilization. However, a new maltose-inducible band is discerned that migrates with a molecular weight that is 2000–3000 higher than that of native maltose-binding protein. This protein was considered a likely candidate for a maltose binding-protein precursor.

Different cell fractions of strain PB1101 were run on SDS–polyacrylamide gels to determine the cellular location of this new, maltose-inducible protein. Somewhat surprisingly, this protein could be found in neither the soluble nor membrane fractions. However, a new protein band intermediate in size between native maltose-binding protein and its putative precursor could be observed in the soluble fraction of the mutant. This protein is not released from the cells by osmotic shock and, therefore, is presumably localized in the cytoplasm. Antisera prepared against native maltose-binding protein specifically precipitated this protein from the soluble fraction of a culture of PB1101 (Figure 12). In addition, it precipitated several minor proteins in the same molecular weight range. All these proteins were shown to be antigenically related to maltose binding protein. One of the minor proteins precipitated had the same molecular weight as the putative precursor observed in whole cells. Apparently, this precusor protein is subject to some degradation when the cells are disrupted, and very little remains intact even though all manipulations were performed at 4°C. Similar results have been

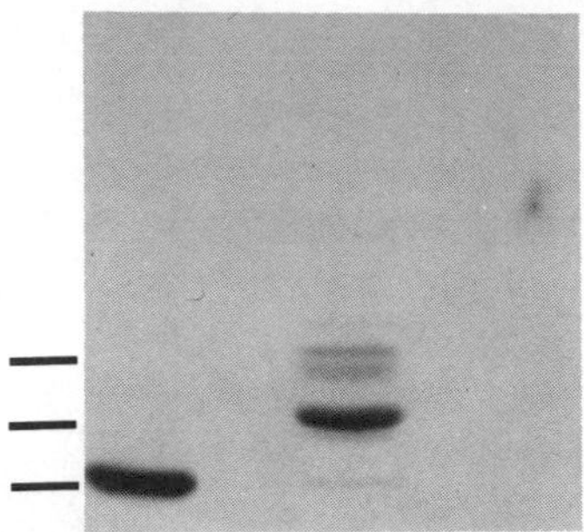

Figure 12 Immune precipitation of radioactively labeled maltose binding protein from soluble extracts of strains MC4100 and PB1101. The immune precipitation, electrophoresis, and autoradiography were performed as described (P. Bassford and J. Beckwith, submitted for publication). (*a* and *b*) Strain MC4100; (*c* and *d*) strain PB1101. For *b* and *d* the anti-MBP serum was preincubated for 30 min with 10 μg of unlabeled native *E. coli* MBP. The three lines along the left side of the gel indicate the corresponding positions where the following three proteins are observed to run: (top line) the MBP precursor observed in whole cells of strain PB1101; (center line) the maltose inducible protein observed in soluble cell extracts of strain PB1101 (see text); and (bottom line) purified, native *E. coli* MBP.

obtained with other Mal$^-$ strains derived from Mal^R derivatives of strain PB72-47.

Despite the problems encountered with protein degradation in these strains, it is apparent that mutant strains have been obtained that fail to export the maltose-binding protein to the periplasm. Instead, it is accumulated in its prccursor form in the cytoplasm. This failure of the cell to recognize and export the *malE* product most probably results from an alteration in the amino acid sequence at the amino-terminal end of the precursor protein. Preliminary mapping studies indicate that the mutations map very early in the *malE* gene and, based on the reversion analysis, are probably point mutations leading to single amino acid substitutions. Elucidation of the exact nature of the alteration in these mutant precursor proteins should help to define the essential feature at the amino-terminus of the maltose binding-protein precursor, which is essential for its correct localization to the periplasm.

6.3 Identification of Unlinked Genetic Loci Involved in the Localization of Envelope Proteins

An important goal of a genetic analysis of protein export is to help identify those cellular components (e.g., of the ribosome or membrane) that are involved in the export process. Mutations that alter the localization of a protein (or a group of proteins) but are not in the structural gene for the protein itself should prove key to such genetic studies. It is possible that some Mal^R mutations of the various Mal^S fusions described above, which change the location of *malB* coded proteins, will be genetically unlinked to the *malB* genes in question. For example, if ribosome–membrane interactions are important, a ribosomal protein may prevent attachment of the polysome to the membrane and hence prevent protein export. Another possibility is that mutants may be found that alter a pore or channel through which proteins pass during the export process or that block the proteolytic precursor processing activity. In any event, other genetic loci are likely to be involved in the rather complex process of protein localization. Characterization of unlinked mutants could provide us with insight as to the nature of these other components and possibly the mechanism of the export process.

However, it is also possible that mutations that affect the localization of either the λ receptor or the maltose-binding protein would affect the process of protein localization in general and as such, would prove to be lethal. So far, all the Mal^R Lac^+ mutants of the various Mal^S fusions characterized are the result of linked mutations, that is, presumably signal sequence mutations. Accordingly, other genetic techniques may be required to uncover new genetic loci involved in the export process. In particular, approaches involving conditional lethal screening procedures may prove valuable in detecting mutant strains in which export is defective. This approach involves seeking mutants that are export *defective*. Another approach is to devise selections for mutants in which an internalized protein is exported. The mutations we have isolated in which the precursor of the *lamB* or *malE* gene product is found in the cytoplasm provide such a selection.

When the *lamBS69* mutation, a point mutation, is present in an otherwise wild-type *lamB* gene, reversion of the mutation can be detected by selecting for return of the wild-type Dex^+ phenotype. The mutation responsible for reversion must restore some degree of export of λ receptor to the outer membrane. Many such revertants appear to be "true revertants," in that the reversion mutation was found to map at the same place as the original mutation. Such true revertants produce apparently normal amounts of λ receptor protein. As expected, such revertants exhibit wild-type sensitivity to phage λ. We have also detected a second class of revertants. However, unlike true revertants, these revertants produce barely detectable levels of a protein that correspond to the λ receptor in the outer membrane. Genetic mapping studies revealed that this class of reversion mutations is not linked to the *lamB* gene. In fact, these mutations are found to be 98% cotransducible by bacteriophage Pl with the mutation conferring resistance to the antibiotic spectinomycin (S. Emr, S. Way, and T. J. Silhavy, unpublished results). The latter mutation, *rpsE*, maps at approximately 72 min on the current *E. coli* linkage map in a region of the chromosome commonly referred to as the "ribosomal cluster." In this tightly linked cluster are approximately 30 genes that code for proteins that are a part of both the large and small ribosomal subunits. In light of these mapping results, we suspect that these reversion mutations (presently termed *emr* for *e*xport of *m*utant *r*eceptor) lie within a gene coding for a ribosomal protein. The possibility that *emr* is a mutation causing ambiguity in the translation process does not seem likely as the phenotype exhibited by *emr* mutants is quite different from that of known ribosomal ambiguity mutants. We favor the idea that *emr* is a mutation altering some aspect of the localization process. This contention would appear to be strengthened by the observation that *emr* also phenotypically suppresses several of the *malE* signal sequence mutations (P. Bassford, unpublished observations).

As is mentioned in Section 1, there exists some controversy about whether or not special ribosomal factors exist that function in the export process. Furthermore, there is a similar controversy about whether or not there is a ribosome–membrane interaction during export. If biochemical evidence supporting our contention that *emr* is a ribosomal mutation can be obtained, then studies utilizing this mutant and other similar mutants may provide insight into the mechanism of protein export.

7 DISCUSSION

This chapter summarizes our results obtained with fusions between *lacZ* and three genes that code for proteins involved in maltose transport: *malF*, *malE*, and *lamB*. These three genes were chosen because each codes for a protein destined to be localized in a different noncytoplasmic location: the inner membrane, the periplasm, or the outer membrane, respectively.

Results that we have accumulated to date clearly indicate the usefulness of gene fusions in the study of protein localization. Gene fusions between *malF* and *lacZ* constructed so as to contain a substantial portion of *malF*, produce a hybrid protein that is tightly bound to the cytoplasmic membrane. However, *malF–lacZ* fusions constructed so as to contain only a small portion of *malF*, produce a hybrid protein that is localized in the cytoplasm. Similarly, *lamB–lacZ* fusions containing a substantial portion of the *lamB* gene produce a hybrid protein that is localized in the outer membrane. Again, *lamB–lacZ* fusions containing only a small portion of the *lamB* gene produce a hybrid protein that is localized in the cytoplasm. Finally, *malE–lacZ* fusions that contain a substantial portion of the *malE* gene also produce hybrid proteins that are found in a noncytoplasmic location. Such hybrid proteins appear not to be localized in the periplasm, however. Rather, they are found to be associated with the inner membrane. Although this appears at first glance to be an anomaly, we believe, for reasons discussed below, that it is not. What is important for the present is that such fusions do alter β-galactosidase localization. To complete the summary, *malE–lacZ* fusions that contain only a small portion of the *malE* gene produce a hybrid protein that, as expected, is localized in the cytoplasm.

Clearly, a pattern is emerging. These results indicate that information directing a protein to a noncytoplasmic location is contained within the gene that codes for that protein. They also suggest, in agreement with what is known concerning "*in vitro*" precursors and the signal hypothesis, that this information is located at the amino-terminal end of that protein. Possibly more important (with the notable exception of the *malE* gene and the periplasmic location), these results indicate that, whatever the mechanism, the sequence present in the gene that is responsible for localization is capable of directing even a very large macromolecule such as β-galactosidase to the prescribed noncytoplasmic location. In other words, our results suggest that any protein can be directed to certain noncytoplasmic locations if the correct genetic sequence is present.

The application of the technique of gene fusion to the study of protein localization has also provided us with some unexpected benefits. One of the most useful results that we have found in these studies is that fusions of *lacZ* to either *malE* or *lamB*, provided that they contain the sequence responsible for localization, result in a very characteristic phenotype. Such strains are inducer sensitive. They grow normally on a variety of media; however, their growth is strongly inhibited by the addition of the inducer, maltose. When maltose is added to growing cultures, after a short time we observe that the cells do not divide properly and often elongate to four or five times their normal length. They appear to form septa but do not separate and, after more than 3 hr at 37°C, cell lysis becomes apparent.

For a variety of reasons discussed above, it seems likely that the Mal^{S} phenotype of such fusion strains is due to the cell's inability to export these hybrid proteins. Perhaps the strongest evidence for this conclusion comes from our studies of mutants of the various Mal^{S} *malE–lacZ* or *lamB–lacZ* fusions that are no longer sensitive to the inducer maltose. If we exclude mutants that, for one reason or another, prevent synthesis of the hybrid protein, we found, in all cases examined to date, that mutations that confer Mal^{R} simultaneously prevent noncytoplasmic localization of the hybrid protein. Furthermore, if these mutations are recombined into an otherwise wild-type *malE* or *lamB* gene, they prevent correct localization of the wild-type gene product, resulting in a phenotype that can be scored. By all criteria provided us by the signal hypothesis, such mutants behave as signal sequence mutants. Such mutations map extremely early in their respective genes, at a position that is quite likely to correspond to a signal sequence. In addition, such mutations prevent correct localization of the gene product and result in the accumulation of a "precursor" product that has a molecular weight corresponding to that of the precursor detected *in vitro*.

We believe that the existence of these mutations can provide concrete evidence that signal sequences are intimately involved in the cellular localization of their respective gene products. Furthermore, since some of the mutations that have been characterized so far behave as point mutations, we believe that there is an inherent specificity associated with each such region. Thus the detailed study of various gene fusions has uncovered a direct selection for mutants that are blocked in the localization process. We believe that the study of a series of such mutants, at the level of protein and/or DNA sequencing, will reveal the essential elements of this important sequence and perhaps shed light on the important first step in protein localization.

Given the availability of what appear to be signal sequence mutations present in an otherwise wild-type gene, we hope to extend our knowledge of the mechanism of protein localization using standard genetic approaches. If a thorough study is made of revertants of such mutations, it is possible that unlinked "suppressor-type" mutations may be uncovered. Using the analogy of nonsense or polarity suppressor mutations, it is possible that such a study may reveal other important components of the localization machinery. Such studies have to date revealed what appears to be a mutation that alters a ribosomal protein. This mutation, which maps in the midst of the largest ribosomal gene cluster and which suppresses both *lamB* and *malE* signal sequence mutations, may provide information as to the exact nature of ribosomal function in the localization process. Without further data though, it is difficult to ascribe any potential meaning to this unlinked suppressor-type mutation.

By this point, it should be obvious to the reader that there are numerous similarities between maltose-binding protein (*malE*) and λ receptor (*lamB*) export. First, both *malE–lacZ* and *lamB–lacZ* fusions that contain the appropriate signal sequence region exhibit the same characteristic Mal^S phenotype. Second, mutations of either class of fusions that relieve the Mal^S phenotype (Mal^R) simultaneously prevent hybrid protein localization, and these mutations, when present in an otherwise wild-type gene, also prevent wild-type gene product localization, resulting in a negative phenotype. third, unlinked suppressor mutations, which relieve the defect caused by a mutation in the signal sequence of the *lamB* gene, also relieve the defect caused by a mutation in the signal sequence of the *malE* gene. Finally, we have presented evidence that suggests the export of the *malE* gene product may interfere directly with the export of the *lamB* gene product. If a Mal^S *malE–lacZ* fusion is induced by the addition of maltose to the growth media, it can be demonstrated by SDS–gel electrophoresis that precursor λ receptor is accumulated (Figure 6). Accordingly, it seems possible that the initial step, that is, the signal sequence-mediated step, in both λ receptor and maltose-binding protein export utilizes the same localization machinery.

It is equally obvious, however, that λ receptor and maltose-binding protein export must diverge at some point. After all, the former is destined to be found in the outer membrane, while the latter is destined to be localized in the soluble periplasmic compartment. Since the initial steps in the localization of each appear similar, it follows that differentiation may be made at a later step in the export process. Two striking differences between *malE–lacZ* and *lamB–lacZ* fusions exist that may be related to this divergence of export mechanisms.

First of all, none of the *malE–lacZ* fusions isolated to date produce a hybrid protein that is localized in the periplasm, the cellular localization of the wild-type *malE* product. All *malE–lacZ* fusions, except those that contain only a small portion of the *malE* gene, produce a hybrid protein that is localized to the cytoplasmic membrane. This is true even if a very substantial portion of the *malE* gene is present in the hybrid gene. This is clearly not the case with *lamB–lacZ* fusions, where fusions that contain a substantial portion of the *lamB* gene incorporate the hybrid protein into the outer membrane with nearly 100% efficiency. Thus, at this juncture, we have been successful in incorporating hybrid proteins into the outer membrane, but a similar strategy employed with a periplasmic protein has not yielded comparable results.

The second important difference relates to the Mal^S phenotype exhibited by a number of these fusions. In the case of *malE–lacZ* fusions, the Mal^S phenotype is proportional to the amount of *malE* DNA present in the hybrid gene. If very little *malE* DNA is present, the fusion causes no detectable Mal^S phenotype. Larger amounts of *malE* DNA present in the hybrid gene, however, result in an increasingly apparent inducer sensitivity. As is discussed above, the Mal^S phenotype is due to the cell's inability to export the hybrid protein. It is as though the *malE–lacZ* hybrid protein has started its journey to the periplasm but for some reason the localization process is aborted, leaving the hybrid protein stuck in the cytoplasmic membrane. Apparently, this hitch in hybrid protein export ties up essential elements of the export machinery. Thus, as the cell attempts to export large amounts of this hybrid protein, export of other essential proteins is compromised, resulting in a Mal^S phenotype. This would explain why precursor λ receptor, for example, can be detected during periods when a *malE–lacZ* fusion strain is trying to export large amounts of hybrid protein. While it is true that certain (class III) *lamB–lacZ* fusions behave in a manner analogous to the Mal^S *malE–lacZ* fusions, it is also true that other *lamB–lacZ* fusions (classes IV and V) do not exhibit a Mal^S phenotype, even though they contain a larger portion of the *lamB* gene and produce a larger hybrid protein. Thus, in the case of *lamB–lacZ* fusions, the Mal^S phenotype is not proportional to the amount of *lamB* DNA present in the hybrid gene.

To clarify some of the concepts presented in this chapter, we have prepared a schematic (Figure 13) that expresses our current working hypothesis on the mechanism of cellular localization of the various noncytoplasmic proteins coded for by the *malB* locus. While this model may well be a vast oversimplification of what actually occurs *in vivo*, it is at least consistent with all the data obtained to date using the technique of gene fusion.

We propose that the initial steps in the cellular localization of the *malE* and *lamB* gene products are identical and are analogous to the steps depicted in the signal hypothesis as described by Blobel and Dobberstein (1975). Both the maltose-binding protein and the λ receptor contain a signal sequence at their amino-terminus. As protein synthesis proceeds and this signal sequence emerges from the larger ribosomal subunit, this signal causes attachment of the polysome to the cytoplasmic membrane. As translation proceeds, the nascent chain is discharged vectorially across the cytoplasmic membrane. Proteolytic processing of the signal sequence probably occurs before synthesis of the protein is complete.

In the case of the periplasmic protein, the maltose-binding protein, translation simply proceeds, ultimately discharging the completed molecule into the periplasm. In the case of the outer membrane protein, the λ receptor, the process of cellular localization diverges at this point. We propose the existence of a second localization signal within the *lamB* gene. We have termed this signal the "dissociation sequence." Translation of this region and subsequent appearance of the corresponding amino acid sequence outside the large ribosomal subunit causes a dissociation of the polysome from the membrane. This dissociation prevents further vectorial discharge of the nascent chain across the membrane and subsequent translation completes the carboxy-terminal portion of the molecule in the cytoplasm. At the completion of synthesis, the λ receptor is found tightly embedded in the inner membrane with its amino terminus facing the periplasm and the carboxy terminus facing the cytoplasm.

That part of the model up to and including the dissociation sequence is analogous to the previously proposed explanation of cellular localization of cytoplasmic membrane proteins in eukaryotic cells (Katz et al., 1977; Rothman and Leonard, 1977; Sabatini and Kreibich, 1976). Indeed, if the *malF* gene product is found to exist in a transmembranal fashion in the cytoplasmic membrane, then a mechanism such as this could explain how this protein reaches its cellular location.

To return to the model, several alternative mechanisms could be envisioned that would transport the λ receptor embedded in the inner membrane to its final outer membrane location. For example, vesicles containing the

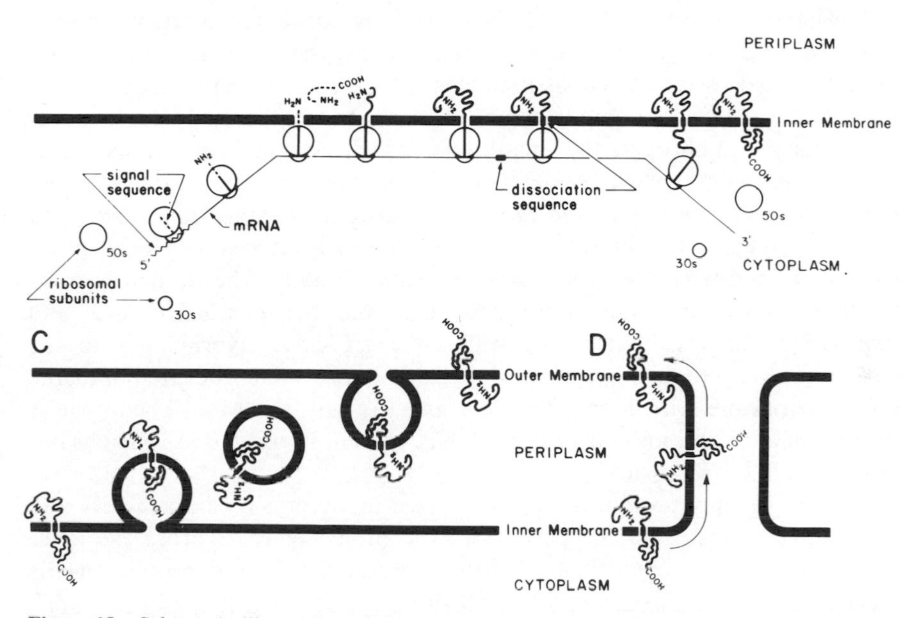

Figure 13 Schematic illustration of the export of newly synthesized λ receptor to its normal outer membrane location. (*A*) The signal sequence, represented by a jagged line at the 5′ end of the mRNA and by a dotted line in the nascent polypeptide chain, emerges from the ribosome and initiates attachment of the polysome to the cytoplasmic membrane. As translation proceeds, the nascent chain is transferred vectorially across the inner membrane. Proteolytic processing of the signal sequence from the λ-receptor precursor occurs outside the cytoplasm. Such processing probably occurs before synthesis of the protein is complete. This section of the schematic is analogous to Figure 10*A* and to the signal hypothesis as proposed by Blobel and Dobberstein (1975). We propose that the *lamB* gene, which codes for the outer membrane protein λ receptor has an additional information sequence relative to protein localization. We have termed the sequence the dissociation sequence, represented here as a heavy dark line in both the mRNA and the λ-receptor protein. (*B*) Translation of this information sequence and subsequent emergence of the dissociation sequence region of the polypeptide chain from the ribosome triggers dissociation of the polysome from the membrane. Subsequent translation of λ receptor mRNA completes the COOH-terminal end of the protein in the cytoplasm, leaving the protein tightly embedded in the inner membrane with its NH_2 terminus facing the periplasm and its COOH terminus facing the cytoplasm. Two possible mechanisms by which λ receptor, embedded in the inner membrane immediately after synthesis is complete, is subsequently transported to its final outer membrane location are depicted: (*C*) vesicles blebbing off the inner membrane and subsequently fusing with the outer membrane, and (*D*) λ-receptor protein diffusing to the outer membrane through sites of inner membrane–outer membrane fusion. Both hypothetical mechanisms result in λ-receptor incorporation into the outer membrane with its NH_2 terminus facing the periplasm and its COOH terminus exposed on the surface of the cell.

protein may bleb off the inner membrane and fuse with the outer membrane. Alternatively, the protein may diffuse into the outer membrane through sites of inner membrane–outer membrane fusion.

We have proposed this model because it is consistent with our results obtained with various fusions. First of all, the signal sequence mutants we have obtained so far behave as predicted by the model that is, they prevent association of the polysome to the membrane, consequently preventing vectorial discharge through the membrane, which results in precursor accumulation in the cytoplasm (see Figure 10). Furthermore, the model offers an explanation of how *malE* and *lamB* gene product localizations can be so similar in their initial steps, yet, by diverging at a later stage, result in localization in different cellular compartments. Finally, the model explains the two striking differences we have observed between *malE–lacZ* and *lamB–lacZ* fusions, that is (*a*) why no *malE–lacZ* hybrids get to the periplasm even though *lamB–lacZ* hybrids can be incorporated into the outer membrane with high efficiency, and (*b*) why the Mal^{S} phenotype is proportional to the amount of *malE* DNA present in *malE–lacZ* fusions but not in *lamB–lacZ* hybrids.

Clearly, for any protein to be synthesized by cytoplasmic machinery and yet wind up located in the periplasm, the molecule must cross the inner membrane. Since no *malE–lacZ* fusions produce a hybrid protein that is located in the periplasm (even though in one fusion nearly all of the *malE* gene is present in the hybrid gene), we believe that it may not be possible for the export machinery to vectorially discharge molecules with a β-galactosidase moiety. In fact, we believe that the Mal^{S} phenotype is due to the jamming of the export machinery with β-galactosidase sequences.

This model we propose offers an explanation of how it would be possible to get a large portion of β-galactosidase into the outer membrane without the molecule even passing through the inner membrane. Provided that the β-galactosidase was fused to the *lamB* gene at a point past the dissociation sequence (class IV or V fusions), the β-galactosidase portion of the hybrid would not even enter the cytoplasmic membrane. The model predicts that only the amino-terminal portion of the hybrid molecule, that is, λ receptor sequences, have to pass through the membrane.

Two predictions can be readily made. First, the *lamB–lacZ* hybrid proteins should be orientated in the outer membrane with the carboxy terminus (β-galactosidase portion) on the cell surface. Since *lamB–lacZ* hybrid strains appear to have material at the cell surface that immunologically cross-reacts with β-galactosidase (Silhavy et al., 1977), this would appear to be the case. Second, the model predicts that *lamB–lacZ* fusions that contain the *lamB* signal sequence but not the dissociation sequence (class III fusions) would behave in a manner analogous to comparable *malE–lacZ* fusions in that β-galactosidase sequences would jam the export machinery resulting in a Mal^S phenotype. This is what we observe. The *lamB–lacZ* fusions that contain a very substantial portion of the *lamB* gene, including the dissociation sequence (class IV or V), incorporate the hybrid protein into the outer membrane with high efficiency and are not Mal^S. However, fusions containing only slightly less *lamB* DNA are very Mal^S, and only a portion of the hybrid protein reaches the outer membrane. An approximately equal amount remains stuck in the inner membrane in exactly the same manner as with *malE–lacZ* fusions.

This model, however, even if substantially correct, still leaves a number of very important unanswered questions. Signal sequences are involved in protein localization. Presumably they function by causing the polysome to associate with the membrane. Even this, though, has not been conclusively demonstrated. Sequence studies with a few known bacterial signal sequences reveal no substantial homology among them and little insight into their mechanistic role. Other important questions remain: How do the nascent polypeptide chains cross the cytoplasmic membrane? Is this process mediated by a transient pore, or does the nascent chain pass directly through the bilayer? Arguments favoring either contention can be, at least theoretically, supported. What provides the energy for pushing the nascent chain vectorially across the bilayer? Presumably the energy of translation itself could be sufficient, but this requires that the ribosome be tightly bound to the cytoplasmic membrane. This point, particularly in bacterial systems, is controversial. There may, in fact, be no such interaction, and other mechanisms may exist that drive the nascent chain across the bilayer. The steps in our model, from the dissociation sequence on, are at this point only speculative. At present we have too little data that bear on these steps in the localization process to comment further. Finally, a word must be said about the potential generality of our proposed model. Different proteins appear in the outer membrane at different places. There is evidence that the synthesis of certain periplasmic proteins may be cell cycle-dependent while others may not be (see Section 1). Keeping this in mind, it is certainly possible that a variety of mechanisms may exist that accomplish cellular compartmentalization even in the relatively simple prokaryotic cell.

Despite these numerous gaps in the current understanding of the localization process, the recent progress in this field, both with eukaryotic and prokaryotic systems, has been unusually rapid. There is a striking concordance of predictions of the signal hypothesis with many aspects of the export process in both systems. Until now, nearly all the features of this model have been derived from work on eukaryotic systems. However, we believe that many further details require a genetic analysis for which bacteria, particularly *E. coli*, are clearly the organisms of choice. The results from recent genetic studies described in this review indicate the potential of this approach. To our mind, the field of protein export is one of the few in which there is a concrete two-way interaction among researchers working on a range of organisms, from bacteria to bumble bees (Suchanek et al., 1978) to humans.

ACKNOWLEDGMENTS

We wish to thank the large number of investigators who kindly provided reprints and/or preprints of their work. The contributions of our colleagues, in the laboratory, M. Casadaban, H. Shuman, M. Schwartz, S. Emr, M. Hall, H. Inouye, and E. Brickman, are greatly appreciated. We thank R. MacGillivray for excellent technical assistance and A. McIntosh for typing (and retyping) this manuscript. Our own research was supported by grants from the American Cancer Society and the National Science Foundation (to J. Beckwith), and by grants from the Medical Foundation, Inc., Boston, Massachusetts, the National Institute of General Medical Sciences, and the American Cancer Society, Massachusetts Division, Inc. (to T. Silhavy). P. Bassford is a fellow of The Helen Hay Whitney Foundation. T. Silhavy is a Medical Foundation Research Fellow.

REFERENCES

Accolla, R. S., and Celada, F. (1976), *Fed. Eur. Biochem. Soc.*, **67,** 299–303.

Bassford, P., Beckwith, J., Berman, M., Brickman, E., Casadaban, M., Guarente, L., Saint-Girons, I., Sarthy, A., Schwartz, M., Shuman, H., and Silhavy, T. (1978), in *Molecular Aspects of Operon Control*, (J. H. Miller, and W. S. Reznikoff, eds.) Cold Spring Harbor Laboratory, New York.

Brickman, E., Silhavy, T. J., Bassford, P. J., Jr., Shuman, H. A., and Beckwith, J. R., *J. Bacteriol.*, in press.

Blobel, G., and Dobberstein, B. (1975), *J. Cell. Biol.*, **67,** 835–851.

Blobel, G., and D. Sabatini (1971), in *Biomembranes*, Vol. 2, (L. A. Manson, ed.) Plenum Press, New York, pp. 193–196.

Bukhari, A. I., Shapiro, J. A., and Adhya, S. L., eds., 1977, *DNA Insertion Elements, Plasmids and Episomes*, Cold Spring Harbor Laboratory, New York.

Cancedda, R., and Schlessinger, M. J. (1974), *J. Bacteriol.*, **117,** 290–301.

Casadaban, M. (1976), *J. Mol. Biol.*, **104,** 541–555.

Chang, N. C., Blobel, G., and Model, P. (1978), *Proc. Natl. Acad. Sci. U.S.*, **75,** 361–365.

Débarbouillé, M., Shuman, H. A., Silhavy, T. J., and Schwartz, M. (1978), *J. Mol. Biol.*, **124,** 359–371.

Dietzel, I., Kolb, U., and Boos, W. (1978), *Arch. Microbiol.*, **118,** 207–218.

Emr, S. D., Schwartz, M., and Silhavy, T. J. (1978), *Proc. Natl. Acad. Sci. U.S.*, **75,** 5802–5806.

Franklin, N. C. (1978), *Annu. Rev. Genet.*, **12,** 193–221.

Gold, L. M., and Schweiger, M. (1971), in *Methods in Enzymology*, (K. Moldave and L. Grossman, eds.), Academic Press, New York, pp. 535–542.

Goldberg, A. L., and St.-John, A. C. (1976), *Annu. Rev. Biochem*, **45,** 747–803.

Halegoua, S., Hirashima, A., and Inouye, M. (1976), *J. Bacteriol.*, **126,** 183–191.

Hardy, S. J. S., and Randall, L. L. (1978), *J. Bacteriol.*, **135,** 291–293.

Heidecker, G., and Müller-Hill, B. (1977), *Mol. Gen. Genet.*, **155,** 301–307.

Hirashima, A., Childs, G., and Inouye, M. (1973), *J. Mol. Biol.*, **79,** 373–389.

Hofnung, M. (1974), *Genetics*, **76,** 169–184.

Inouye, H., and Beckwith, J. (1977), *Proc. Natl. Acad. Sci. U.S.*, **74,** 1440–1444.

Inouye, H., Pratt, C., Beckwith, J., and Torriani, A. M. (1977), *J. Mol. Biol.*, **110,** 75–87.

Inouye, S., Wang, S., Sekizawa, J., Halegoua, S., and Inouye, M. (1977), *Proc. Natl. Acad. Sci. U.S.*, **74,** 1004–1008.

Ito, K., Sato, T., and Yura, T. (1977), *Cell*, **11,** 551–559.

Katz, F. N., Rothman, J. E., Lingappa, U. R., Blobel, G., and Lodish, J. F. (1977), *Proc. Natl. Acad. Sci. U.S.*, **74,** 3278–3282.

Kellermann, O, and Szmelcman, S. (1974), *Eur. J. Biochem.*, **47,** 139–149.

Kennedy, N., Beutin, L., Achtman, M., Skurray, R., Rahmsdorf, U., and Herrlich, P. (1977), *Nature*, **270,** 580–585.

Kimura, K., and Izui, K. (1976), *Biochem. Biophys. Res. Commun.*, **70,** 900–906.

Konings, R. N. H., Hulsebos, T., and van den Hondel, L. A. (1975), *J. Virol*, **15,** 570–584.

Kreibich, G., Freienstein, C. M., Pereyra, P. N., Ulrich, B. C., and Sabatini, D. D. (1978a), *J. Cell Biol.*, **77,** 488–505.

Kreibich, G., Ulrich, B. C., and Sabatini, D. D. (1978b), *J. Cell Biol.*, **77,** 464–487.

Lin, J. J. C., Kanazawa, H., Ozols, J., and Wu, H. C. (1978), *Proc. Natl. Acad. Sci. U.S.*, **75,** 4891–4895.

Lin, S., and Zabin, I. (1972), *J. Biol. Chem.*, **247,** 2205–2211.

Lingappa, V. R.; Shields, D.; Woo, S. L. C. and Blobel, B. (1978), *J. Cell Biol.*, **79,** 567–572.

Lodish, H. F., and Froshauer, S. (1977), *J. Cell Biol.*, **74,** 358–364.

Müller-Hill, B., and Kania, J. (1974), *Nature*, **249,** 561–563.

Neu, H. C., and Heppel, L. A. (1965), *J. Biol. Chem.*, **240,** 3685–3692.

Nomura, M., and Engbaek, F. (1972), *Proc. Natl. Acad. Sci. U.S.*, **69,** 1526–1530.

Osborn, M. J., Gander, J. E., Parisi, E., and Carson, J. (1972), *J. Biol. Chem.*, **247,** 3962–3972.

Pagès, J. M., Piovant, M., Varenne, S., and Lazdunski, C. (1978), *Eur. J. Biochem.*, **86,** 589–602.

Palade, G. E. (1975), *Science*, **189,** 347–358.

Palmiter, R. D., Gagnon, J., and Walsh, K. A. (1978), *Proc. Natl. Acad. Sci.*, **75,** 94–98.

Randall-Hazelbauer, L., and Schwartz, M. (1973), *J. Bacteriol.*, **116,** 1436–1446.

Randall, L. L., and Hardy, S. J. S. (1975), *Mol. Gen. Genet.*, **137,** 151–160.

Randall, L. L., and Hardy, S. J. S. (1977), *Eur. J. Biochem.*, **75,** 43–53.

Randall, L. L., and Hardy, S. J. S. (1978), *Proc. Natl. Acad. Sci. U.S.*, **75**, 1209–1212.

Randall, L. L., Josefsson, L.-G., and Hardy, S. J. S. (1978), *Eur. J. Biochem.*, **92**, 411–415.

Rothman, J. E., and Lenard, J. (1977), *Science*, **195**, 743–755.

Ryter, A., Shuman, H., and Schwartz, M. (1975), *J. Bacteriol.*, **122**, 295–301.

Sabatini, D. D., and Kreibich, G. (1976), in *The Enzymes of Biological Membranes*, Vol. 2 (A. Martonosi, ed.), Plenum Press, New York, p. 531–579.

Schumacher, G., and Bussmann, K. (1978), *J. Bacteriol.*, **135**, 239–250.

Schwartz, M. (1976), *Ann. Inst. Pasteur Paris*, **113**, 685–704.

Sekizawa, J., Inouye, S., Halegoua, S., and Inouye, M. (1977), *Biochem. Biophys. Res. Commun.*, **77**, 1126–1133.

Shen, B. H. P., and Boos, W. (1973), *Proc. Natl. Acad. Sci.*, **70**, 1481–1485.

Silhavy, T., Casadaban, M. J., Shuman, H. A., and Beckwith, J. R. (1976), *Proc. Natl. Acad. Sci. U.S.*, **73**, 3423–3427.

Silhavy, T., Shuman, H., Beckwith, J., and Schwartz, M. (1977), *Proc. Natl. Acad. Sci. U.S.*, **74**, 814–817.

Silhavy, T., Brickman, E., Bassford, P. J., Jr., Casadaban, M. J., Shuman, H. A., Schwartz, V., Guarente, L., Schwartz, M., and Beckwith, J. R., *Molec. Gen. Genet.*, in press.

Silverstein, S., ed. (1978), *Transport of Macromolecules in Cellular Systems*, Dahlem-konferenzen, Berlin.

Smit, J., and Nikaido, H. (1978), *J. Bacteriol.*, **135**, 687–702.

Smith, W. P., Tai, P.-C., Thompson, R. C., and Davis, B. D. (1977), *Proc. Natl. Acad. Sci. U.S.*, **74**, 2830–2834.

Smith, W. P., Tai, P.-C., and Davis, B. D. (1978a), *Proc. Natl. Acad. Sci. U.S.*, **75**, 814–817.

Smith, W. P., Tai, P.-C., and Davis, B. D. (1978b), *Biochemistry*, **18**, 198–202.

Suchanek, G., Kreil, G. and Hermodson, M. A. (1978), *Proc. Natl. Acad. Sci. U.S.*, **75**, 701–704.

Sugimoto, K., Sugisceki, H., Okamoto, T., and Takanami, M. (1977), *J. Mol. Biol.*, **111**, 487–507.

Sutcliffe, J. G. (1978), *Proc. Natl. Acad. Sci. U.S.*, **75**, 3737–3741.

Szmelcman, S., Schwartz, M., Silhavy, T. J., and Boos, W. (1976), *Eur. J. Biochem.*, **65**, 13–19.

Tanford, C., and Reynolds, J. A. (1976), *Biochem. Biophys. Acta*, **457**, 133–170.

Taylor, A. L. (1963), *Proc. Natl. Acad. Sci. U.S.*, **50**, 1043–1051.

Varenne, S., Piovant, M., Pagès, J. M., and Lazdunski, C. (1978), *Eur. J. Biochem.*, **86**, 603–606.

Wickner, W., Madel, G., Zwizinski, C., Bates, M., and Killick, T. (1978), *Proc. Natl. Acad. Sci. U.S.*, **75**, 1754–1758.

Zabin, I., Fowler, A. V., and Beckwith, J. R. (1977), *J. Bacteriol.*, **133**, 437–438.

Zubay, G. (1973), *Annu. Rev. Genet.*, **7**, 267–287.

Chapter 8 The Genetics of Outer Membrane Proteins

PETER REEVES

Department of Microbiology and Immunology, University of Adelaide, Adelaide, South Australia

1 INTRODUCTION

Before reviewing our present understanding of the genetics of outer membrane proteins, it is as well to ask what we can hope to learn from genetic analysis. First, each of the outer membrane proteins, or to be more precise each polypeptide, will have a structural gene that encodes its sequence. For several outer membrane proteins a structural gene can be identified with reasonable certainty. Mutants in the structural gene may lead to complete absence of functional protein: a full deletion mutation clearly lacks any functional protein, as do many nonsense mutations and insertion mutations. Even some missense mutations effectively lack protein if the altered product is unstable. However, the majority of missense mutants and also some small deletion mutants will have an altered protein, as will some nonsense mutants terminating towards the —COOH end of the polypeptide.

Mutant strains have been used in many of the studies on function of outer membrane proteins. A good understanding of the fine structure of a structural gene would enable mutant bacteria of defined type to be used in such studies, whether *in vivo* or *in vitro*, but as yet our understanding of outer membrane protein genetics is at that stage only for the *lamB* gene. In general, it is not yet possible to be sure that physiological studies carried out on mutants involve deletion mutations with absolutely no gene product present. However, for several structural genes, mutation can give rise to different phenotypes, indicating that the protein can be altered by mutation in specific ways, and as the genetic analysis proceeds it should help unravel the structure–function relationships of these proteins.

In addition to mutations in the structural genes for the outer membrane proteins, we can expect to find mutations that affect the expression of these proteins. As we see later, mutations are known that affect control of matrix proteins Ia and Ib, or of certain so-called new membrane proteins. Analysis of these mutants should lead to an understanding of the control of expression of outer membrane protein genes, and perhaps also of the insertion of proteins into the outer membrane. As yet, however, the analysis of these control mutants is at an early stage and indeed it is not yet absolutely clear which genes are control genes and which are structural genes for matrix proteins Ia and Ib.

1.1 Isolation of Mutants

Genetic analysis can only proceed effectively when mutant strains are available. Most of the mutants affecting outer membrane proteins were obtained using colicins or bacteriophages. Mutants resistant to bacteriophages are generally found to lack the surface receptor to which the phage first attaches, and for phage that use outer membrane proteins as a receptor this offers a very easy specific method for isolation of mutants lacking a given protein. Several phages specific for particular *Escherichia coli* outer membrane proteins have been recognized in recent years and are discussed in the sections on appropriate genes; many are also given in Table 1. In the case of *Salmonella*, although phages using lipopolysaccharide as receptor have been instrumental in genetic and chemical studies of lipopolysaccharides (see Stocker and Makela, 1978, for review), it is only recently that phages using specific protein receptors have been recognized (Siitonen et al., 1977; Graham and Stocker, 1977; M. Nurminen and P. H. Makela, personal communication). Some of the data of the Helsinki group is given in Table 2 and individual examples are referred to with the appropriate gene.

Isolation of the desired phage-resistant mutants is frequently hampered, as mucoid mutants are resistant to most phages, presumably because the capsular polysaccharide prevents phages from approaching the cell surface. These mucoid mutants frequently outnumber and overgrow the desired specific mutants. For *E. coli* K-12 this problem can be overcome by use of *non* mutants (Radke and Siegel, 1971; Hancock and Reeves, 1975) that are blocked in synthesis of the capsular polysaccharide such that the addition of a mutation in the control gene *capR* cannot lead to overproduction of the polysaccharide. The readily isolated colicin-resistant mutants also frequently lack outer membrane proteins: in some cases the protein acts as a receptor for the colicin, but this is not always the case.

Table 1 *E. coli* **Genes (Other Than *nmp*) Involved With Outer Membrane Proteins**

Gene	Map Position	Proteins Affected	Agents for Selection[a]		
			Phages	Colicins	Other
ompA	21	3A	K3, TuII	L	Recipient deficiency
ompB	74	1a, 1b	As for *ompC*, *ompF*	K, L, A, S4, N E2, E3	As for *ompC*, *ompF*
ompC	47	1b	PA2, Tu1b, Me1 434		
ompF	21	1a	Tu1a, T2	K, L, A, S4, N	Chloramphenicol, Ag^{3+} Cu^{2+}
tsx	9	tsx	T6	K	
lpp	36	Lipoprotein	—	—	
lamB	90	lamB	—	—	
btuB	88	btuB	BF23	A, E1, E2, E3	B_{12} utilization
fepA	13	fepA	—	B, D	
tonA	3	tonA	T1, T5, ϕ80	M	Albomycin
cir	44	cir	—	I, V	

[a] Not all the phage and colicins work well for all genes listed.

In this chapter we look at each of the genes that has been identified as having a major effect on outer membrane proteins of *E. coli* and *Salmonella*. Tables 1 and 2 give the genes, their synonyms, and the proteins affected by mutation.

1.2 Genetic Nomenclature

The system of nomenclature used in *E. coli* and *Salmonella* is based on the recommendations of Demerec et al., (1966), and the latest published linkage maps (Figure 1) (Bachmann et al. 1976; Sanderson and Hartman, 1978) are

Table 2 ***Salmonella Typhimurium*** **Genes Involved With Outer Membrane Proteins**

Gene	Map Position	Proteins Affected	Phages	Colicins	Other Selection
ompA	21	33K		4–59	
ompB	74	35K, 36K	As for *ompC*, *ompD*		
ompC	46	36K	PM31, PM51, PM41, PM42		
ompD	28	34K	PM105, P221		
ompF	21	35K			
sid	4		ES18		Albomycin
bfe (*btu*)	89		BF23		

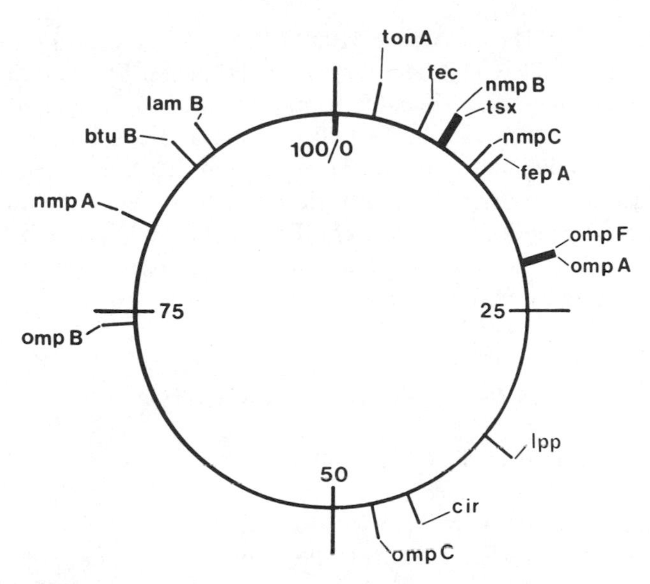

Figure 1 Genetic map of *E. coli* K-12, based on map of Bachman et al. (1976), showing genes coding for or controlling outer membrane proteins. Genetic symbols as used in the text.

generally taken as authoritative with regard both to gene usage and gene symbols. In this review the *E. coli* nomenclature used is that to be used in the forthcoming *E. coli* linkage map (B. J. Bachmann, personal communication), as this will resolve current confusion due to use of several synonyms for some genes. In many cases the *E. coli* and *Salmonella* gene symbols are the same, but this is not always the case. With regard to the current situation, a quotation from Bachmann et al. (1976) may be relevant. They "strongly urge that persons assigning new gene symbols attempt to . . . identify and use symbols already assigned to the loci in question unless these are clearly unsuitable." One might add that even unsuitable terminology is best altered on the occasion of the publication of the linkage map or after discussion with the clearing houses for *E. coli* and *Salmonella*, as the plethera of names used for some genes confuses those closely involved and discourages others from trying to understand us.

The system for designation of allele numbers does not seem to have been widely published and perhaps should be mentioned here. To quote from Sanderson and Hartman (1978): "the clearing houses for gene symbols and allele numbers will attempt to continue co-ordination as they have in the past: *Salmonella* users should consult Kenneth F. Sanderson (Dept. of Biology, University of Calgary, Calgary, Alberta, Canada T2N, 1N4) and *E. coli* users should consult Barbara J. Bachmann (Yale University School of Medicine, New Haven, CT 06510, U.S.A.)." The clearing houses will allocate blocks of allele numbers for particular genes. For example, tsx-200 through tsx-249 have been allocated to my laboratory. Careful adherence to the assignment of allele numbers to all new mutations will make the literature easier to follow, and also, as I learned to my cost, help keep ones own strain collection in order. It is appropriate to acknowledge the indebtedness of all bacterial geneticists to Barbara Bachmann and Kenneth Sanderson for their contributions to maintenance of law and order among us. Without them the confusion would be worse.

2 THE *ompA* GENE (SYNONYMS *tolG, con, tut*)

2.1 Characterization of the Gene Product

Mutants lacking a major protein with a molecular weight of about 30,000 have been isolated in several laboratories and all have been found to map at 21.5 min at a locus now properly known as *ompA* (Foulds, 1974; Manning and Reeves, 1975c; Manning and Reeves, 1976b; Henning et al. 1976) but previously described as *tolG*, *con*, and *tut*. The gene product is designated as ompA protein.

ompA mutants are pleiotropic and are altered in phage sensitivity, colicin sensitivity, recipient ability and other ways (Foulds and Barrett, 1973; Skurray et al., 1974; Foulds, 1974; Davies and Reeves, 1975b; Manning et al., 1977). They were first isolated by Foulds and Barrett (1973) as colicin L-tolerant (*tolG*) mutants and were later shown to lack ompA protein (Chai and Foulds, 1974). They were independently isolated by Skurray et al. (1974), Davies and Reeves (1975b, c and Hancock and Reeves (1975) as phage-resistant, colicin-tolerant mutants with a conjugation defect (*con*) and lacking ompA protein.

Phage K3, originally a *Shigella* typing phage (Krzwy et al., 1972) was used during a survey of coliphages (Hancock and Reeves, 1975) and has been extensively used for isolation of *ompA* mutants, as has phage TuII*. This latter phage was isolated from sewage (Henning and Haller, 1975) as unable to propagate on an *ompA* mutant. Both of these phages and several others (Hancock and Reeves, 1975) are probably independent isolates of essentially the same phage.

2.2 Isolation of *ompA* Mutants

Isolation of *ompA* mutants using phages has proved very satisfactory, as all nonmucoid mutants resistant to phage K3 or related phages are found to map at the *ompA* locus (Manning et al. 1976; Henning et al. 1976). Substantial resistance to K3-like phage is conferred by some mutations affecting lipopolysaccharide (Hancock and Reeves, 1975), but mutants of this type have not been isolated after selection with K3-like phages.

Selection of *ompA* mutants using colicin L is straightforward (Foulds and Barrett, 1973; Davies and Reeves, 1975b); mutants of other colicin-tolerant classes are also selected, but these can be readily distinguished on the basis of other properties.

Havekes and Hoekstra (1976) and Havekes (1978) have also isolated *ompA* mutants using a selection for loss of recipient functions.

2.3 Genetic Analysis of *ompA* Mutants

As yet no complementation studies have been reported for *ompA* mutants; thus it is not established that all are in the one cistron or gene, but there is no reason to propose more than one *omp* gene in this region.

Several *ompA* mutants have been described in which ompA protein has an altered electrophoretic mobility (Henning et al. 1976; Achtman et al., 1978) or altered phage receptor activity after purification (P. A. Manning and P. Reeves, unpublished data). The *ompA* region of the *E. coli* K-12 chromosome has also been transferred to *Salmonella typhimurium* and *Proteus mirabilis* using an F′ plasmid derived from E. coli K-12. The presence of this F′ in these species leads to the presence in the outer membrane of a heat modifiable protein with electrophoretic mobility similar to that of *E. coli* ompA protein, in addition to the normal proteins of *S. typhimurium* or *P. mirabilis*, respectively (Datta et al., 1976). In the course of these experiments it was established that there is no gene dosage effect for *ompA*, as diploids produce only the normal amount of ompA protein. There is thus very good evidence for *ompA* being the structural gene for ompA protein.

Generally, *ompA* mutants selected on the basis of bacteriophage resistance appear to lack ompA protein, with only 1 out of 48 in one study (Manning et al. 1976) and 8 of 150 in another study (Datta et al., 1976) still producing the protein in readily detectable amounts. However, about half of the mutants isolated as resistant to bacteriophage K3 and lacking detectable ompA protein were able to plaque host range mutants of phage K3, selected as able to plaque on such *ompA* bacteria.

Those *ompA* mutants unable to plaque host range mutants of phage K3 are fully resistant to colicin L and their recipient ability in crosses with F′ and Hfr donors is reduced by a thousandfold (Manning et al., 1976). Mutants of this type are frequently amber mutations (A. Puspurs and P. Reeves, unpublished data; U. Henning, personal communication) and presumably totally lack functional ompA protein.

The only simple explanation for *ompA* mutants still able to plaque phage host range mutants is that ompA protein is present but in amounts too low to detect by acrylamide gel electrophoresis, and that host range mutants of the phage are less exacting with regard to the density of ompA protein required for infection (Manning et al., 1976). The alternative explanation, that these mutants completely lack ompA protein but that phage K3 host range mutants can absorb in its absence, poses the question of the nature of the other large group on which no host range phage mutants can plaque. In support of the hypothesis that host range sensitive mutants had residual amounts of ompA protein is the fact that they are frequently sensitive to colicin L and their recipient ability with respect to wild type ranges from 0.001 (as for an amber mutant) to 0.06.

Mutants selected as colicin L-resistant generally lack detectable ompA protein [30 out of 30 lacked protein (J. Foulds, personal communication)].

In contrast to *ompA* mutants selected as phage or colicin resistant, three out of five mutants selected on the basis of conjugation deficiency had high levels of ompA protein in the outer membrane and remained sensitive to bacteriophage K3; another mutant had reduced levels of ompA protein and was resistant to phage K3, while the fifth lacked detectable ompA protein and was also resistant to phage K3 (Havekes and Hoestra, 1976; Havekes, 1978; Achtman et al., 1978). It seems clear from these data that colicin L action, phage K3 adsorbtion, and recipient function all require ompA protein, and that any mutants totally lacking functional protein will be isolable by selection based on any of these three properties. We can then use as a bases for comparing these three sets of mutants, the number of mutants lacking any trace of ompA protein on acrylamide gels and totally resistant to phage K3 and its host range mutants. The analysis of mutants isolated as K3 resistant tells us that there arise a number, approximately equal to the base number, of mutants with ompA protein present in such low amounts as to be detectable only because of their ability to plaque K3 host range mutants, but that there are relatively very few mutants with normal levels of protein but lacking phage receptor function. Analysis of the mutants isolated as defective in recipient functions indicates that *ompA* mutants with normal levels of a presumably altered ompA protein that are partially defective as recipients arise at several times the base level of totally defective mutants. These data are summarized in Table 3.

We are faced then with two questions:

1 Why do so few phage- or colicin-resistant mutants have detectable levels of protein in the outer membrane?
2 How large is the number of mutants that have low undetected levels of ompA protein?

Table 3 Types of *ompA* Mutants; Frequency With Which Different *ompA* Mutant Types Occur Relative to Type 1 Mutants

Type	Approximate Relative Frequency	Sensitivity			Recipient Ability	ompA Protein	Description
		Phage K3	K3h	Colicin L			
Parent	—	S	S	S	1	+	—
1	1	R	R	R	0.001	–	Full loss of *ompA* function (mostly nonsense mutation)
2	1	R	S	Variable	0.001–0.06	Very low level	Low protein level missense?)
3	0.05	R	S[a]	S	1	+	Altered protein affecting phage receptor function (missense)
4	3	S	N.D.	N.D.[b]	0.02	+	Altered protein affecting recipient ability (missense)

[a] Only one tested (Manning et al., 1976).
[b] Probably sensitive, as 30/30 colicin-resistant mutants lacked protein (Foulds, personal communication).

The answer to the first question probably lies in the nature of phage receptor function. If receptor function were not very exacting, then relatively few amino acid substitutions would affect this property and most would go undetected during selection based on phage resistance. This explanation is speculative but may be tested by fine structure analysis of the gene now under way in our laboratory. With regard to colicin L, the 30 mutants selected by Foulds (personal communication) resemble those selected by phage in that no protein is detected by PAGE. We may assume that most missense mutations giving altered protein remain sensitive to colicin L. We can speculate as we did for phage K3, that it is not simple to modify ompA protein in such a way that it is still able to locate in the outer membrane but is not functional in colicin L action. At present we have little knowledge of the role of ompA protein in colicin action other than that it is probably involved in transport of colicin across the outer membrane (Foulds and Chai, 1978a).

The answer to the second question is not so easy, but there must be a large class of mutant proteins that are not readily put into the outer membrane. It appears that the efficient insertion of ompA protein into the outer membrane is very sensitive to minor changes in the protein. Fine structure analysis will tell us if these mutations are clustered within the gene.

2.4 The *ompA* Gene of *Salmonella*

The 33K protein of *Salmonella typhimurium* is thought to be equivalent to ompA protein of *E. coli* (see Chapter 11). It has been found that mutants resistant to bacteriocin 4-59 lack the 33K protein, map at 21 min and are 30% cotransducible with *pyrD*. It seems that the mutations are in the *S. typhimurium ompA* gene (M. Nurminen and P. H. Makela, personal communication).

3 GENETIC CONTROL OF PORINS

3.1 Characterization of Porins

Matrix proteins Ia and Ib of *E. coli* K-12 and the 34, 35, and 36K proteins of *Salmonella* have been shown to have general porin function: three membrane proteins produced in *E. coli* K-12 only after mutation, and known as new membrane proteins, and protein 2, produced by some *E. coli* strains other than K-12, also act as porins (see Chapter 11).

The genetics of matrix proteins Ia and Ib are more complex than that of the other proteins and there is not yet agreement on which is the structural gene (or genes). Mutations at three loci have been studied in some detail; *ompB* mutants generally lack both matrix protein Ia and Ib, while *ompF* and *ompC* mutants lack matrix protein Ia *or* Ib, respectively. There are two fundamentally different hypotheses to explain this phenomenon. Matrix proteins Ia and Ib may be forms of the same peptide (Schmitges and Henning, 1976; Bassford et al., 1977) and, if so, neither *ompF* nor *ompC* could be structural genes for these proteins, but must affect modification, while *ompB* could be the structural gene for the unmodified matrix protein I (Bassford et al., 1977; Sarma and Reeves, 1977). The alternative hypothesis proposed recently is that matrix proteins Ia and Ib do not derive from a common peptide chain despite their apparent similarities; *ompF* and *ompC* are then considered to be strong candidates for the structural genes for matrix proteins Ia and Ib, respectively (Ichihara and Mizushima, 1978; Sato et al., 1977; van Alphen et al., 1979; Verhoef et al., 1979). Matrix proteins Ia and Ib are clearly similar, although neither the published information on their chemistry nor that on the genetics is yet sufficient to unequivocally determine which of the above hypotheses is true; the data at present favor the hypothesis that *ompC* and *ompF* are structural genes for different polypeptides. Matrix proteins Ia and Ib resolve on PAGE, but this could be due to either differences in amino acid sequence (two structural genes hypothesis) or differences in modification of one polypeptide (one structural gene hypothesis); the two proteins give similar, but different, cyanogen bromide cleavage peptide patterns, but again this could be easily accommodated by either hypothesis, and finally two differences have been reported in the sequence of the first 14 amino acids at the *N*-terminal end.

This last observation is considered by the authors (Ichihara and Mizushima, 1978) to be proof that the two proteins derive from different structural genes. However, even these data must still be treated with caution as the amino acids do not appear to have been rigorously identified. The conventional methods for amino acid identification in sequence analysis are designed specifically to distinguish quickly among the "natural" amino acids and are not designed to identify modified amino acids. Posttranslational modification could conceivably give amino acid derivatives that after release by the Edman degradation and analysis by conventional thin layer chromatography are indistinguishable from one of the other natural amino acids. To rigorously exclude the hypothesis of posttranslational modification, it is in this instance, necessary to characterize the released amino acids in several systems. In conclusion, it appears that matrix protein Ia and Ib are encoded by different structural genes, but we cannot yet fully discount the alternative hypothesis.

3.2 The *ompF* Gene (Synonyms *tolF, cmlB, coa*)

A well-defined class of *E. coli* mutants lacking matrix protein Ia has been mapped at 21 min at a locus known as *ompF* (previously known as *tolF*, *cmlB*, and *coa*). There were probably *ompF* mutants among those isolated in early analysis of colicin resistance, as those described as TolI(Nomura and Witten, 1967) have an appropriate phenotype. TolI mutants were later divided into two phenotypic classes. Tol1a and Tol1b (Davies and Reeves, 1975b), and it now appears likely that these classes (and also others) are different classes of *ompF* mutants. The Ktw2 and Ktwl types of phage-resistant mutants from our laboratory (Hancock and Reeves, 1975) are identical to Tol1a and Tol1b types (Hancock et al. 1976b). The locus was first mapped as *cmlB* by Reeve and Doherty (1968), and again by Foulds and Barrett (1973) as *tolF* at 21 min. It was also mapped as *coa* by Lavoie and Mathieu (1975). In the forthcoming edition of the *E. coli* linkage map (B. J. Bachmann, personal communication), *tolF*, which is identical with *cmlB* (Foulds, 1976) and with *coa* (J. Foulds, personal communication), is to be named *ompF*. The deficiency in protein 1 was observed by Davies (1974) (who at that time did not distinguish matrix proteins Ia and Ib) and by Foulds who observed the absence of protein Ia in *tolF* mutants (Chai and Foulds, 1977).

3.3 Isolation of *ompF* Mutants in *E. coli*

Strains mutant in *ompF* are readily isolated as colicin tolerant and one would expect phage Tu1a to be useful in isolation since *ompF* mutants are resistant (Foulds and Chai, 1978b), although Pugsley and Schnaitman (1978a) report only partial resistance. For whatever reason, attempts to isolate *tolF* mutants using phage Tu1a have been unsuccessful (Pugsley and Schnaitman, 1978a; Verhoef et al., 1979).

Studies show that *ompF* mutants are tolerant to colicins K, L, A, S4, and N and partially tolerant to colicins E2 and E3 (Foulds and Barrett, 1973; Davies and Reeves, 1975b; Foulds, 1976; Chai and Foulds, 1977); their tolerance depends in part on genetic background (V. Sarma, unpublished data). Colicin L has been the preferred colicin for mutant isolation (Foulds and Barrett, 1973; J. K. Davies, V. Sarma and P. Reeves, unpublished data). (Note that the colicin L produced by *Serratia marcescens* JF246 was for a while referred to as a marcescin, Foulds, 1976), but colicin A (J. Foulds, personal communication; Lavoie and Mathieu; 1975) and colicin K (J. K. Davies, personal communication) can be used. Isolation of colicin-tolerant mutants yields several other mutant types in addition to *ompF* mutants, but these are easily distinguished by both colicin-tolerance pattern and phage-resistance patterns, as well as by map position and protein composition.

Some *ompF* mutants were isolated as chloramphenicol-resistant (Reeve and Doherty 1968; Foulds, 1976) and some of the mutants isolated as cryptic for periplasmic enzymes (Beachman et al., 1973, 1977) have been shown to be *ompF* mutants (Verhoef et al., 1979).

3.4 Complementation Studies with *ompF* Mutations

Foulds (1976) used an F′ carrying *ompF* to demonstrate that mutants isolated as colicin tolerant or chloramphenicol resistant are all in the same complementation group or gene. Thus there is only one known gene in this region affecting outer membrane proteins, although in that complementation study, the variants known as Tolla (Ktw2) and Tollb (Ktw1) were not distinguished.

3.5 Is *ompF* the Structural Gene for Protein Ia?

Mutants with an altered protein Ia have not been isolated (Verhoef et al., 1978) and so cannot be used to identify the structural gene. However, Sato et al. (1977 and personal communication) transferred the 21 min region of the chromosome of *S. typhimurium* to *E. coli* by transduction and found that many of the transductants (presumably those carrying the *Salmonella ompF* gene) produced two outer membrane proteins not found in *E. coli* in place of Matrix protein Ia; both new proteins were considered to be similar to a *S. typhimurium* protein on the basis of tryptic peptide patterns. They interpret their results to mean that the structural gene for matrix protein Ia lies in this region and that it was replaced by the structural gene for the *S. typhimurium* protein. They do not have a full explanation for the proposed *S. typhimurium* structural gene giving *two* proteins in *E. coli*, but this may reflect possible differences in structural modification of the nascent protein molecules synthesized in the heterologous (*E. coli*) and homologous (*S. typhimurium*) cells. Although the *Salmonella* protein involved was first reported as being 36K, it is probably the 35K *Salmonella* protein that is produced in the transductants (T. Sato and T. Yura, personal communication), in agreement with the homology proposed elsewhere for *E. coli* and *Salmonella ompF* genes (see below). Sato and Yura (personal communication) also transferred an *E. coli* F′ plasmid carrying the *ompF* gene to *Salmonella*: the merozygotes produced a protein with an isoelectric point identical to that of matrix protein Ia, in addition to the *Salmonella* proteins. Sato and Yura interpret their data as evidence for *ompF* being the structural gene for matrix protein Ia, but their data could, with difficuly, be interpreted as favoring the hypothesis that *ompF* is a gene for a modification system, by arguing that *E. coli* and *Salmonella* have different modification systems that do not interact.

3.6 The *ompC* Gene (Synonyms *par, meoA*)

Mutants lacking only matrix protein Ib have been mapped at about 48 min at the *ompC* locus, also referred to as *par* and *meoA* (Bassford et al., 1977; Verhoef et al., 1977, 1979; van Alphen et al., 1979). Strains mutant in *ompC* have no known colicin resistance (Bassford et al., 1977) and are isolated using bacteriophage. the first phages to be used had the host range of phage PA2, a lambdoid phage isolated by Schnaitman; virulent mutants of PA2 or hybrids with phage λ were used (Schnaitman et al., 1975; Bassford et al., 1977), and other laboratories have isolated other phages: phage Tulb (originally known as phage Tul) (Schmitges and Henning, 1976; Henning et al. 1977; Datta et al., 1977), phage Mel (Verhoef et al., 1977, 1979; van Alphen et al., 1979), and phage A8 (V. Sarma, unpublished data), all select *ompC* mutants, and phage 434 probably could also be used (Hantke, 1978).

Mapping of mutations in this region is incomplete, as *nalA* and *glpT* are difficult markers to use and the precise gene order is still not certain (Verhoef et al., 1979). There is thus no reason as yet to propose more than one locus for outer membrane proteins near 48 min, but no complementation studies have been reported. Two classes of *ompC* mutants, *ompC I* and *ompC II*, have been distinguished (the third class of PA2-resistant mutants described have normal outer membrane and do not map near 48 min). While *ompC I* mutants are fully resistant to phage PA2 *ompC II* mutants are slightly sensitive, and whereas *ompC I* mutants produce protein 2 (in a suitable lysogen), *ompC II* mutants do not.

3.7 Is *ompC* the Structural Gene for Protein Ib?

Two mutations that map at the *ompC* locus lead to altered matrix protein Ib (van Alphen et al., 1979). One of the mutants (CE1081), which arose fortuitously produces a protein that has a slightly lower electrophoretic mobility than normal matrix protein Ib. It was identified as an altered matrix protein Ib because it is absent in phage Mel-resistant mutants of strain CE1081. Chemically it was considered to be an altered form of matrix protein Ib.

The second *ompC* mutant (CE1151) with an altered protein was isolated as phage Mel resistant. The alteration in this instance reduced by ninety-fold the neutralization of phage Mel by purified protein. The existence of these two mutants is strong evidence for *ompC* being the structural gene for matrix protein Ib, but as is discussed in the introductory discussion of matrix proteins Ia and Ib, these altered proteins could be due to an alteration in a modifying gene rather than in the structural gene for matrix protein Ib. The fact that *ompC II* mutants not only do not produce matrix protein Ib, but also fail to produce protein 2 in suitable lysogens, is in fact strong evidence for *ompC* having a modifying or regulatory role with respect to matrix proteins Ib and 2 (Bassford et al., 1977).

Further support for *ompC* being the structural gene for matrix protein 1b comes from a recent study by M. N. Hall and T. J. Silhavy (personal communication). They constructed *ompC-lac* fusions with β galactosidise expression exhibiting regulatory properties of matrix protein 1b. Thus in these strains an *ompB* mutation reduced β galactosidase levels and growth on various media affected β galactosidase in a manner analogous to the usual effects on matrix protein 1b. It thus seems clear that *ompC* (and presumably *ompF*) are structural genes for matrix proteins.

3.8 The *ompB* Gene (Synonyms *kmt, cry*)

Mutants at this locus at 73.7 min can affect both matrix proteins Ia and Ib, although as we see later, there are mutations affecting either matrix protein Ia *or* Ib, which map at about 74 min, and these mutations are probably also in the same locus. Mutants at this locus were originally given the phenotypic designations of TolIV, TolXIV, and TolXV (Davies and Reeves 1975b) and have also been called *kmt* (Bavoil et al., 1977; von Meyenberg, 1971) and *cry* (Beacham et al., 1973, 1977; Verhoef et al., 1978). Typical *ompB* mutants (Ia^- Ib^-) were mapped at 73.7 min by Sarma and Reeves (1977) and other cotransduction data of other typical and atypical (Ia^+ Ib^- or Ia^- Ib^+) mutations has given compatible results (Chai and Foulds, 1977; Henning et al., 1977 Verhoef et al., 1979).

Usually *ompB* mutants are isolated as colicin or phage resistant. Typical *ompB* mutants are resistant to the same colicins as *ompF* mutants, presumably because of the absence of matrix protein Ia, and are also resistant to colicins E2 and E3 (Davies and Reeves, 1975b; Sarma and Reeves, 1977), which suggests that these colicins require either Ia *or* Ib to be present for their action (see also discussion of protein 2 and new membrane proteins). Mutants selected as colicin tolerant of course include other types in addition to *ompB* mutants, but these can be distinguished by map position and other characteristics. *ompB* mutants, and probably also *ompF* mutants can be isolated as copper or silver resistant (Lutkenhaus, 1977; Pugsley and Schnaitman, 1978a).

Of the typical *ompB* mutants that have been studied, (*ompB*101, *ompB*102, and *ompB*103) 3 were isolated using colicin E3 tolerance (Davies and Reeves, 1975b; Sarma and Reeves, 1977), 20 were isolated using colicin L tolerance (V. Sarma, unpublished data), 1 (cry3-41) was isolated as cryptic for periplasmic enzymes (Beacham et al., 1973; Verhoef et al., 1979) 1 was obtained (*ompB*153) using copper (Pugsley and Schnaitman, 1978a), (*ompB*151 and *ompB*152) were obtained using chloramphenicol (Pugsley and Schnaitman, 1978a), and 1 (CE1107) was obtained with a low level of phage Mel (Verhoef et al., 1979).

The four *ompB* mutants isolated by von Meyenberg (1971) as slow growers on low level substrate were isolated in *E. coli* B, which has only protein Ib. However, these were typical *ompB* mutants in that on transfer to *E. coli* K-12 each mutation led to loss of both matrix proteins Ia and Ib (Bavoil et al., 1977).

Isolation of *ompB* mutants can be achieved using colicin E3, colicin L, or chloramphenicol.

In contrast to the ease with which such mutants are isolated using colicins, they are not readily isolated using bacteriophage. Bassford et al. (1977) failed to obtain any *ompB* mutants with TolIV or TolXIV phenotype using phage PA2 and isolated only *ompC* mutants. Verhoef et al. (1979), using the similar phage Mel, could report only one typical *ompB* and that was after using a low titer of phage. Selection with phage Tu1a has not been reported to yield typical *ompB* mutants, and while mutagenesis followed by Phage PA2 plus colicin E3 selection did yield mutants lacking matrix proteins Ia and Ib (Bassford et al., 1977), at least one of these mutants was a double *ompC–ompF* mutant (Pugsley and Schnaitman, 1978b). It is likely that typical *ompB* mutants (TolIV or TolXIV colicin tolerance phenotype) still produce some matrix protein Ia and Ib, as they retain low sensitivity to bacteriophages.

Selection with bacteriophage does, however, yield mutants, lacking *either* matrix protein Ia or Ib, that map at about 74 min and are currently presumed to be *ompB* mutants. Thus phage Tu1a in one study (Verhoef et al., 1979) yielded only mutants lacking matrix protein Ia but mapping at *ompB*, and in the same study use of low levels of phage Mel yielded a variety of mutants, including atypical *ompB* mutants (lacking only matrix protein Ib). Selection with Tu1a in an *ompC* background (Henning et al., 1977) or with Tu1b in an *ompF* background (Chai and Foulds, 1977) can also yield mutants that when put into a wild-type background, can be shown to affect matrix protein Ia *or* Ib, respectively, and to map in the *ompB* region.

3.9 Variation in *ompB* Mutant Phenotypes

Mutants that are assigned to *ompB* have also been shown to vary in two other ways, although in the absence of complementation studies, we cannot say whether they are all in the same gene.

Davies and Reeves (1975b) classified colicin-tolerant mutants into various phenotypes based on their resistance to colicins, and three of the phenotypes (TolIV, TolXIV, TolXV) were shown to map at *ompB* (Sarma and Reeves, 1977). The variation in colicin tolerance affects essentially the degree of resistance to some of the colicins.

Among the *ompB* mutants lacking only matrix protein Ib, there are some with normal amounts and others with excess matrix protein Ia (Verhoef et al., 1979).

3.10 New Membrane Proteins

Those *ompB* mutant lacking matrix proteins Ia and Ib, and also *ompC–ompF* double mutants, grow poorly on many media, and "revertants" with larger colonies arise quite commonly. Some of these colonies are thought to be intragenic revertants in that they produce the same proteins as the original parent (Bavoil et al., 1977; Pugsley and Schnaitman, 1978b), and the new phenotype maps at *ompB* (Bavoil et al., 1977); others have large amounts of *lamB* protein (Bavoil et al., 1977) and, although no mapping data were reported, they are presumably $malT^c$ mutations.

However, in addition to these, several classes of mutants have been described that produce novel proteins not present in the parent *E. coli* B or K-12 strain. Three loci have been identified as being involved in these secondary mutations.

These mutants have been isolated in several laboratories. They have been obtained during selection for phage Tu1a resistance in an *ompC* mutant (Henning et al., 1977) or for Tu1b or Mel resistance in an *ompF* background (Foulds and Chai, 1978b; Verhoef *et al.*, 1979). In each case proteins Ia and Ib were missing as a result of an *ompB–ompC* or *ompC–ompF* combination, and the presence of the new protein was due to a third mutation. Presumably the growth defect of the selected double mutant was so severe that the third (*nmpA*) mutation was selected during the primary isolation of the double mutant. The new protein has been termed protein Ic (Henning et al., 1977), protein e (van Alphen et al., 1978), protein E (Foulds and Chai, 1978b) and recently new membrane protein A (Pugsley and Schnaitman, 1978b); the last designation is used here.

The mutation maps at a gene *nmpA* at 82.2 min (Foulds and Chai, 1978b; Pugsley and Schnaitman, 1978b). The gene order is thought to be *bglC*, *nmpA*, *ilv*, with *nmpA* closely linked to *bglC*. It is also possible to obtain *nmpA* mutations by SDS selection from an *ompB* mutant (van Alphen *et al.*, 1978; Verhoef et al., 1978). Pugsley and Schnaitman (1978b), after screening revertants of *ompB* or *ompC–ompF* strains, isolated many strains producing new outer membrane proteins and identified two more novel proteins. They mapped the mutations involved, with gene *nmpB* mapping at 8.6 min and the gene order is thought to be *lacY*, *nmpB*, *proC*, *tsx;* the protein is termed new membrane protein B. The *nmpC* mutants, producing new membrane protein C, map at 12 min and the gene order is *purE*, *nmpC*, *fepA*, (*entD*, *entE*, *entF*). It should be noted that there is a discrepancy between this order and that given in the section on *fepA*, which is yet to be resolved.

Strains producing a new membrane protein are given the phenotypic designation Nmp^+ and the mutation is designated an *nmp* (p^+) mutation, *nmp* (p^+) strains produce the new membrane protein in a wild-type background.

Any of these new membrane proteins confers sensitivity to colicins E2 and E3, showing that the requirement for matrix protein Ia or Ib for action of these colicins can be satisfied by any of the three new membrane proteins. $NmpA^+$ and $NmpB^+$ strains are also sensitive to two phages, TC45 and TC23, isolated by Chai and Foulds (1978).

Using selection for resistance to colicin E3 and $AgNO_3$ or to phage TC45, Nmp^- derivatives could be obtained from all three types of Nmp^+ strains. In each case the mutation to Nmp^- could arise at the same locus as the original *nmp* mutation, but in the case of $NmpC^+$ strains, the Nmp^- phenotype did not always map at 12 min and the original mutation *nmpC* (p^+) could be recovered by transduction to another strain. It appears that each of the three genes, *nmpA*, *nmpB*, and *nmpC* can exist in two mutant forms, *nmpA* (p^+) and *nmpA* (p^-) and so on. The *nmpA*$^+$ or wild-type strains have the phenotype $NmpA^-$, with no detectable level of new membrane protein A, a phenotype indistinguishable from that of an *nmpA* (p^-) strain that arises by a second mutation from an *nmpA* (p^+) strain.

The terminology may seem confusing, but it is based on the usual conventions for *E. coli:* Genes have lower case first letters and the wild type is indicated by a superscript plus sign in parentheses, which in this case signifies absence of protein, while phenotypes have an upper case first letter, and in this case the superscript plus sign signifies the presence of a protein. Some of the possible genotypes and phenotypes are given in Table 4. It is of interest that one *nmpC* (p^-) *fepA* mutation is thought to be a deletion (Pugsley and Schnaitman, 1978b).

It was suggested (Pugsley and Schnaitman, 1978b) that the loci *nmpA*, *nmpB*, and *nmpC* are structural genes for the three proteins concerned, the major reason being that Nmp^- mutations also can map at these loci. It was argued (Pugsley and Schnaitman, 1978b) that the *nmp* (p^+) mutations are unlikely to be in regulatory genes, as the effect of mutation was all or none. However, they do not speculate as to the nature of structural gene mutation that would give rise to this all or none effect.

Table 4 Genotype and Phenotype for Selected *nmp* Mutations

	Proteins Present				
Genotype	Ia	Ib	NmpA	NmpB	NmpC
—	+	+	–	–	–
ompB	–	–	–	–	–
ompB, *nmpA* (p+)	–	–	+	–	–
ompB, *nmpB* (p^+)	–	–	–	+	–
ompB, *nmpC* (p^+)	–	–	–	–	+
ompB, *nmpC* (p^+) *nmpC* (p^-)	–	–	–	–	–
ompB$^+$, *nmpA* (p^+)	+	+	+	–	–
ompB, *nmpC* (p^+) *nmp*?[a]	–	–	–	–	–

[a] *nmp*? is the unmapped gene in NmpC (p^-) mutants still carrying *nmpC* (p^+).

To this author it seems most likely that an *nmp* (p^+) mutation is either an operator constitutive mutation adjacent to the structural gene or a repressor minus mutant in a repressor gene. The fact that some of the NmpC$^-$ mutants derived from *nmpC* (p^+) strains retain the *nmpC* (p^+) mutation shows that at least two genes are involved in expression of new membrane protein C, and one at least must be a control gene.

New membrane proteins A, B, and C, are thought to contain different polypeptides and to be coded for by different structural genes (Pugsley and Schnaitman, 1978b). Although this is one possible interpretation of the data, it should be noted that there are reported similarities among these proteins. Only further genetic and chemical work can eliminate the alternative hypothesis that some of the similarities are due to modification of polypeptide chains common to one or more of these proteins. the reported similarities include tryptic peptide patterns for matrix protein Ia and new membrane protein A (Henning et al., 1977), serological cross-reaction between new membrane protein A and matrix protein (Foulds and Chai, 1978b; but see also Pugsley and Schnaitman, 1978), and sensitivity of both NmpA$^+$ and NmpB$^+$ strains to phage TC45 (Pugsley and Schnaitman, 1978b). It should be noted that NmpA and NmpB are very similar (A. P. Pugsley, personal communication). If they are identical, then clearly at least one of the genes *nmpA* and *nmpB* must be a control gene.

3.11 Genetic of Protein 2

Protein 2 is not produced by *E. coli* K-12 but is produced by other strains, including *E. coli* J-5. The ability to produce this protein is determined by a phage, PA2 (Schnaitman et al., 1975), that can lysogenize *E. coli* K-12.

Lysogens of phage PA2 produce protein 2 in addition to the normal K-12 proteins. Protein 2 is a porin and as such can restore sensitivity to silver to an *ompB* mutant. Protein 2 can also substitute for matrix protein Ia or Ib in conferring sensitivity to colicins E2 and E3 (Pugsley and Schnaitman, 1978a). The structural gene for protein 2 is presumably on phage PA2, although this has not been rigorously proven. Selection for silver resistance in a PA2 lysogen of an *ompB* mutant leads to loss of protein 2 (Pugsley and Schnaitman, 1978a). The mutation is in the PA2 genome and the mutant phage does not confer protein 2 production on lysogens. Although not proven, the mutation may well be in the structural gene.

3.12 The Porins of *Salmonella typhimurium*

Salmonella typhimurium has three outer membrane proteins able to function as porins (Nakae and Ishii, 1978, and see also Chapter 11). Recently, bacteriophages have been isolated that use either the 34K or 36K protein as receptor, and this enables isolation of mutants that lack either of these proteins (Nurminen et al., 1976; Johansson et al., 1978). The mutations map at 28 and 46 min and the genes are named *ompD* and *ompC*, respectively; *ompC* is cotransducible with *nalA*, as is the *ompC* gene of *E. coli*, and the gene is thought to be homologous in the two species (M. Nurminen and P. H. Makela, personal communication).

The *ompD* mutation is cotransducible with *trp* and no equivalent mutations have been described for *E. coli;* indeed, *E. coli* K-12 probably has no equivalent to the 34K protein of *Salmonella*.

No *Salmonella* mutants that lack the 35K protein only have been reported, but on the basis of the experiments of Sato and Yura, which are discussed in the section on *ompF*, the 35K protein is thought to be encoded by a *Salmonella ompF* gene homologous with *ompF* of *E. coli*.

It is of interest that *ompB* mutants have been isolated in *Salmonella* using the same phage that selects *ompC* mutants. *Salmonella ompB* mutants lack the 35K and 36K proteins, map at 74 min, and are cotransducible with *nal* mutations (M. Nurminen and P. H. Makela, personal communication).

4 THE *lpp* GENE (SYNONYMS *mlpA*, *lpm*, *lpo*)

Mutants lacking the murein lipoprotein have been isolated in two laboratories, map at 36.5 min, and are thought to be mutated in the one gene now known as *lpp* (B. J. Bachmann, manuscript in preparation), which is the structural gene for the lipoprotein.

Six mutants have been described in which lipoprotein is absent or altered. Four resulted from a complicated selection procedure (see below) (Wu and Lin, 1976; Torti and Park, 1976). One, previously known as *lpm*, was fortuitously also present in a strain of *E. coli* carrying a cell division mutation under study (Suzuki et al., 1976). A second, previously known as *lpo*, was fortuitously present in an F′ plasmid used in the same study (Hirota et al., 1977).

4.1 Isolation of *lpp* Mutants

There are no phages or colicins known that select *lpp* mutants, and the defects present in *lpp* mutants (Hirota et al., 1977; Yem and Wu, 1978) have not been used in devising a selection procedure. The only selection procedure thus far devised involved deprivation of proline and histidine (absent in lipoprotein) to block most protein synthesis, adding labeled arginine during the residual protein synthesis, and then storing the cells at 4°C for "radiation suicide" of wild-type cells (Wu and Lin, 1976). A simlar selection method was used by Torti and Park, 1976).

4.2 Genetic Analysis

Three mutations have been mapped by transduction at about 36.5 min (Suzuki et al., 1976; Hirota et al., 1977; Yem and Wu, 1977).

The amino acid sequence of the lipoprotein is known (see Chapter 1) and one of the *lpp* mutants (*lpp*-1, originally known as *lpm*) has been shown to have cysteine substituted for arginine at position 57 and is clearly a missense mutation (Inouye et al., 1977). Another, originally known as *lpo* (*lpp*-2), completely lacks the protein (no serologically cross-reacting material) and also lacks specific mRNA (Hirota et al., 1977). Since this mutation was fortuitously present in an F′ but can be transduced to the chromosome, it was reasonably suggested (Hirota et al., 1977) that the structural gene was deleted in the F′, as deletions are common in such plasmids.

The third mutation that has been studied in some detail (Yem and Wu, 1978) produces a protein that is deficient in covalently linked diglyceride and that is, to a large extent, not bound to the peptidoglycan (Wu et al., 1977; Yem and Wu, 1978).

Sequencing of the prolipoprotein from the mutant and comparison with the sequence of the wild-type protein revealed a missense mutation converting a glycine to aspartic acid residue at position 14 of the signal sequence (Lin et al., 1978). A merodiploid heterozygous for the *lpp* mutation and the wild-type allele had both forms of the protein (Yem and Wu, 1977), as is expected for a missense mutation. The mutant is sensitive to EDTA and this has been used to select revertants that are due to mutation in the *lpp* gene and are thought to be intragenic suppressors (Yem and Wu, 1978).

4.3 One Atypical Mutant

One mutant that has not been characterized genetically is different from the defined *lpp* mutants. This strain, ST715 (Torti and Park, 1976), was isolated using a selection procedure similar to that used by Wu and Lin (1976) from a parent carrying a temperature-sensitive amber suppressor. The mutant is temperature sensitive for growth and for lipoprotein synthesis and it was suggested that after isolation the strain had an amber suppressible mutation affecting lipoprotein synthesis and that the temperature sensitivity of growth was due to lipoprotein being essential for cell survival.

In support of this hypothesis it was shown that temperature-independent revertants also regained their wild-type rate of lipoprotein synthesis. However, in view of the relatively minor growth defects in defined *lpp* mutants, this hypothesis is now considered unlikely, although it is possible that the genetic background may influence the lethality of an *lpp* mutation. Only further analysis will finally tell whether the strain is mutant in the structural gene or carries a pleiotropic mutation in another gene.

4.4 The *Salmonella typhimurium lkyD* Gene

In *Salmonella*, *lkyD* mutants (Weigand and Rothfield, 1976) resemble in part *E. coli lpp* mutants. Both types of mutants leak periplasmic proteins (Weigand and Rothfield, 1976; Hirota et al., 1977; Yem and Wu, 1978), have an increased proportion of free lipoprotein (not peptidoglycan bound) (Weigand et al., 1976; Wu and Lin, 1976), and form outer membrane blebs (Weigand et al., 1976; Yem and Wu, 1978).

The growth defects in an *lkyD* strain are much greater than those in an *E. coli lpp* mutant completely lacking the lipoprotein, and it is therefore suggested that the growth defects in *lkyD* mutants are not due to the alteration in the lipoprotein binding (Hirota et al., 1977). However, it seems equally possible that the primary defect in *lkyD* mutants is in the structure or peptidoglycan binding of lipoprotein, and the different phenotypes are due to differing genetic backgrounds. It should, however, be noted that *lkyD* mutations map in the *proA–galE* region, well away from the E. coli *lpp* site.

5 THE *tsx* GENE

The *tsx* gene mapping at 9 min was one of the first to be used in bacterial genetics, with mutants easily isolated as resistant to phage T6. However, although phage T6 was thought to utilize a surface protein as receptor, it was only in 1976 that the T6 receptor protein was identified and was shown to be absent in the mutants (Manning and Reeves, 1976a).

The T6 receptor is now known to be involved in nucleoside transport (Hantke, 1976) and mutations at the *tsx* locus were for a while described as *nup* (nucleoside uptake) mutants (McKeown et al., 1976) before it was realized that *nup* and *tsx* were the same.

Isolation of *tsx* mutants has been accomplished using either phage T6, which is highly specific (Hancock and Reeves, 1975) or colicin K, in which case several other classes of colicin-tolerant mutants are also isolated but can be readily distinguished by other characteristics (Davies and Reeves, 1975b).

The isolation of *tsx* mutants using phage T6 resistance yielded very few (1 out of 50) with an altered tsx protein present in detectable levels, but 16% were able to propagate host range T6 phages (Manning and Reeves, 1978). As for the *ompA* protein and phage K3, the conclusion drawn is that alterations to a protein that specifically affect its function as a phage receptor are unlikely, because only a small part of the protein is important in receptor specificity, and also there are many mutants with protein altered in such a way that it is only inserted (or synthesized) at a very low level and is detected by adsorption and propogation of host range mutants of phage T6.

6 THE *lamB* GENE

The outer membrane protein known both as the lambda receptor and as the maltose pore is coded for by the *lamB* gene; *lamB* is clearly the structural gene for the lambda-receptor protein (Thirion and Hofnung, 1972; Randall-Hazelbauer and Schwartz, 1973; Hofnung et al., 1976; Braun-Breton and Hofnung, 1977). It is distal to *malK* in one of two divergent operons mapping in the 90 min region, which together with another operon mapping at 74 min, comprises a regulon of genes concerned with maltose utilization (Bachmann et al., 1976) (Figure 2). Mutants lacking the lamB protein can be selected readily using λ phage and may be mutated within the *lamB* gene, carry a polar mutation in *malK*, or be mutant in *malT*, a positive control

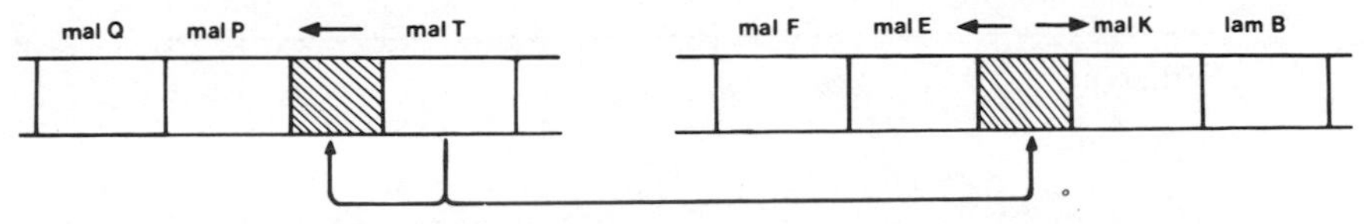

Figure 2 The three operons of the maltose regulon showing the positive control of *malT* on the three *mal* operons. From Hofnung et al. (1976).

gene for the regulon (Hatfield et al., 1969a, b; Hofnung et al., 1971, 1974; Thirion and Hofnung, 1972). Lambda-resistant *malT* or *malK* mutants are unable to utilize maltose, whereas *lamB* mutants grow well on normal levels of maltose; *lamB* mutants are thus easily distinguished on maltose indicator media.

Fine structure gene analysis of the 74 and 90 min operons has been carried out using deletion analysis and it is on this basis that one can say that all resistant mutants that remain Mal^+ on normal experimental levels of maltose map in the one gene known as *lamB*: The *lamB* gene may be divided into 9 regions by the deletions available (Braun-Breton and Hofnung, 1977; M. Schwartz, personal communication) and a total of 30 nonsense mutations and 100 Mu insertion mutations have been mapped in *lamB* (M. Schwartz, personal communication). All *lamB* mutations are shown by complementation analysis to be in the one cistron, but some mutations do show intragenic complementation, suggesting that the protein is oligomeric (Thirion and Hofnung, 1972).

Mutants in *lamB*, isolated as resistant to bacteriophage λ, resemble phage-resistant mutants at the *ompA* or *tsx* loci in that about 50% are able to plaque host range mutants (Hofnung et al., 1976). The ability of a given *lamB* mutant to plaque a host range phage mutant indicates that the *lamB* mutant still has receptor present, altered either in amount or structure so as to allow plaque formation by the host range mutants but not by wild-type phage. As for phage K3 host range mutants and *ompA* mutants, the alternative explanation that the host range phage can now use an entirely different receptor is unsatisfactory, as it does not explain why the host range mutants plaque on only some of the *lamB* mutants.

The host range mutants are of two types, λh and λhh*, and these enable *lamB* mutants to be divided into three classes (Hofnung et al., 1976) (Table 5).

The mutants with the wider host range arise only as two step mutants and

Table 5 Sensitivity of *lamB* Mutants to Host Range Phage (From Hofnung et al., 1976)

	Phage classes		
Bacterial *lamB* Class	λ	λh	λhh*
Wild type	+	+	+
I	−	+	+
II	−	−	+
III	−	−	−

it is thought that at least two mutations are needed to produce a λhh* phage from a wild-type phage.

Class I and class II *lamB* mutations are not suppressible by amber or ochre suppressors (0/23 using amber suppressor su3), whereas one-third (13/39) of the class III mutants were suppressible. Four of those class III mutations not suppressible by su3 were tested with opal and ochre suppressors and all four were suppressible by one or another. It appears then that most of the class III *lamB* mutants carry nonsense mutations and as they probably lack any functional proteins, this accounts for the inability of λh or λhh* phages to plaque on them. The class I and class II mutants, which one may deduce carry missense mutations and produce an altered protein, do in fact produce about normal levels of lamB protein (M. Schwartz, personal communication).

The genetic analysis of the *lamB* gene is considerably further advanced than that of any other gene affecting outer membrane proteins, and likewise its control, as part of the maltose system, is also rather well understood. This knowledge has already been used to produce genetic evidence supporting the existence of a region of the protein involved in its maintenance in the outer membrane (Braun-Breton and Hofnung, 1977) and although the particular hypothesis advanced will require confirmation by biochemical studies of the distribution of the lamB protein, it is clear that for this gene the genetic tools are available for use in dissecting the function of an outer membrane protein.

The *lamB* gene has also been used in an exciting experiment involving gene fusion (Silhavy et al., 1977). Genes were fused such that an NH_2-terminal sequence of the lamB protein was fused to a COOH-terminal sequence of β-galactosidase. Several fusions have been isolated and will enable analysis of those parts of the *lamB* gene and lamB protein involved in locating the protein in the outer membrane; β-galactosidase activity is used as an indicator for the fused protein (see Chapter 7).

7 THE *btuB* GENE (Synonyms *bfe, cer*)

7.1 General Characterization

Strains mutant in *btuB* were first isolated as colicin F resistant (Jenkin and Rowley, 1955) and map at 88 min (Buxton, 1971; Jasper et al., 1972).

Mutants at this locus lack an outer membrane 60K protein involved as vitamin B_{12} receptor and as receptor for E group colicin and phage BF23 (see Chapter 10). It should be noted that colicin A also uses the same receptor (van Vught and Delamarre, 1977) and can thus be thought of as an E group colicin.

btuB mutants are readily isolated as resistant to colicin E or phage BF23 and are readily identified by their phage and colicin resistance pattern (Davies and Reeves, 1975b; Hancock and Reeves, 1975). Either method is satisfactory, but phage BF23 has the advantage that no other class of mutant is selected, whereas the colicins select a wide range of tolerant mutants in addition to the *btuB* receptor mutants. Isolation of *btuB* mutants has also been accomplished following selection for decreased B_{12} utilization (DiGirolamo et al., 1971), but this method is much less productive of mutants. On the basis of differences in the block in B_{12} uptake, *btu* mutants mapping at 88 min were at one time divided into *btuA* and *btuB* (Kadner and Liggins; 1973), but subsequent complementation analysis has shown that mutations in the one gene, *btuB*, can exhibit various phenotypes, and four are recognized (Bassford and Kadner, 1977). Two of these classes are sensitive to colicin E3 and phage BF23, which are clearly not suitable for isolation of these phenotypes. The *btuC* mutants map at 38 min (Bassford and Kadner, 1977) and are not known to be defective in their outer membrane. The btuB protein is present in very low amounts, making it difficult to quantitate or even detect. However, uptake rates double in merozygotes, and binding studies using radioactive colicins suggest that the number of receptors doubles from 220 to 540 per cell (Bassford and Kadner, 1977), indicating that there is gene dosage effect.

7.2 The *res* or *btuB* Gene of *Citrobacter*

The *res* gene (van Vught et al., 1975) of *Citrobacter freundii* has been shown to be homologous to the *btuB E. coli* gene by Van Vught and Delemarre (1977). *Citrobacter* is sensitive to colicin A but not to colicins E1, E2, or E3, whereas *E. coii* K-12, which is very sensitive to E1, E2, and E3, is relatively insensitive to colicin A and is sensitive to phage BF23.

The *E. coli* sensitivities can be added to the *Citrobacter* sensitivities by transfer of a functional *btuB* gene on an F′ plasmid. *Cibrobacter res* mutants resistant to colicin A, map in the region homologous to *btuB*, and vitamin B^{12} protects *Citrobacter* against colicin A, suggesting that as for *E. coli* the two agents share a common receptor (van Vught and Delemarre, 1977). It appears that the homologous gene product can be identified in the two species but is not identical in colicin or phage-receptor function.

7.3 The *bfe* Gene in *S. typhimurium*

Salmonella typhimurium is sensitive to phage BF23 and resistant mutants map at 89 min (Guterman et al., 1975, Sanderson and Hartman, 1978); the gene is presumably homologous to the *E. coli* gene now to be called *btuB* (B. J. Bachmann, personal communication), but previously called *bfe*. The *Salmonella* gene presumably, by analogy with *E. coli*, encodes an outer membrane vitamin B_{12} receptor.

8 GENETICS OF IRON TRANSPORT PROTEINS

Several genes, *tonA*, *fepA*, *cir*, and perhaps *fec*, determine the production of outer membrane proteins known or surmised to be involved in iron transport. Together with secretion of enterochelin, the amounts of these outer membrane proteins are coordinately controlled by availability of iron (Braun et al., 1976; Pugsley and Reeves 1976a, c, 1977).

However, nothing is known of the genetics of this control beyond the fact that mutations at three loci (*tonB*, *exbB* and *exbC*) can affect it (Guterman, 1971; Davies and Reeves 1975a; Pugsley and Reeves 1976a, b; Braun et al., 1976). For *tonB* at least, the effect is probably indirect through an effect on iron uptake and hence on intracellular iron levels (Hancock et al., 1976b; Pugsley and Reeves, 1976a).

8.1 The *tonA* Gene

Among the earliest mutants studied in *E. coli* were *tonA* mutants mapping at 3 min and they have been known since the 1940s to be resistant to phages T1 and T5. They were later shown to be resistant to phage ϕ80 (Wayne and Nielands, 1975), colicin M (Fredericq, 1951; Fredericq and Smarda, 1970), and albomycin, a ferrichrome analogue (Hankte and Braun, 1975a). The tonA protein, missing in most *tonA* mutants, is thought to act as a receptor for phages T1, T5, and ϕ80, for colicin M, ferrichrome, and albomycin; its role in these regards is discussed elsewhere in this book.

8.2 Isolation of *tonA* Mutants

Strains mutant in *tonA* are readily isolated as resistant to phage T1, T5, or ϕ80vir. Phage T5 is preferred, as *tonA* mutants are the only class selected with this phage, whereas T1 and ϕ80 also select *tonB* mutants. Colicin M also selects both *tonA* and *tonB* mutants and, in addition, several classes of colicin-tolerant mutants (Davies and Reeves 1975a). Isolation of *tonA* mutants has been accomplished using albomycin (Hankte and Braun 1975a; Braun et al., 1976), but several other classes of mutants are albomycin resistant and only 20% of albomycin-resistant mutants mapped at *tonA*. There have been no complementation studies reported for *ton* mutants mapping at 3 min, but it is generally assumed that there is but one gene involved.

Most *tonA* mutants lack the tonA protein, regardless of whether phage T1, phage T5, or albomycin selection was used (Braun et al., 1976; Hankte and Braun, 1978). However, in each case some mutants with residual protein can be obtained and some of those isolated as resistant to phage T1 were studied further (Hankte and Braun, 1978). These mutants are partially sensitive to phage T5 although resistant to phage T1. Mutants of the $T1^R$ $T5^S$ phenotype usually map at *tonB* (27 min), but in this study four mapped at 3 min and are presumably *tonA*. Three of those mapping at *tonA* (*tonA*502, *tonA*507, *tonA*509) and five unmapped alleles all had the same phenotype, being fully resistant to phage T1, albomycin, and colicin M and partially sensitive to phage T5. The remaining mutant, B7/11 (*tonA*501), was derived from a different parent, mapped at *tonA* and was partially sensitive to colicin M and albomycin, and of course to phage T5, while resistant to phage T1. The allele numbers are from V. Braun (personal communication).

The *tonA* protein present in these strains is presumably an altered protein resulting from missense mutation. It is not possible to tell from the data if B9/9 (*tonA*502) and the eight other similar derivatives are of independent origin, but it is clear that at least two *tonA* mutations give altered protein. Tlh, a phage mutant able to plaque on *tonB* mutants, was also able to plaque on strain B7/11 (*tonA*501) but not on the other mutants, and T5h, isolated on a T5-resistant *tonA* double mutant with altered protein plaqued efficiently on all these altered protein *tonA* mutants.

Revertants of *tonA* mutants, selected by their ability to use ferrichrome as an iron source, were in general revertant with regard to all aspects of the TonA phenotype (Braun et al., 1976), although some (5/46) exhibited intermediate properties.

8.3 The *fepA* Gene (Synonyms *cbt, cbr, feuB*)

Mapping at 13 min, *fep* mutants have been isolated as defective in ferrienterochelin transport (Cox et al., 1970; I. G. Young, reported by Wookey, 1978). However, most *fep* mutants have been isolated as colicin B or colicin D resistant. The first of these (Davies and Reeves, 1975a) was tolerant to colicins B and D and was originally named *cbt* (colicin B tolerant), but most (27/30) subsequent isolates from the same laboratory lacked receptor and were named *cbr* (colicin B resistant) (Pugsley and Reeves, 1976c), and as both types mapped at about 13 min (Davies and Reeves, 1975a; Pugsley and Reeves, 1977; Pugsley, 1977) they were all considered to carry mutations in the one gene then named *cbr*. Similar colicin-resistant mutants isolated by Hantke and Braun (1975b) were originally called *feuB* (Hancock and Braun, 1976), but are now thought to map at 13 min and be *fepA* mutants.

Typical *fepA* mutants, regardless of how they were isolated, all appear to have the same phenotype. They are defective in ferrienterochelin uptake (Cox et al., 1970; Pugsley and Reeves, 1976a, 1977; Wookey and Rosenberg, 1978) and isolated outer membranes are defective in ferrienterochelin binding (Pugsley and Reeves, 1977) and lack an 81K outer membrane protein (Pugsley and Reeves, 1977; Hancock et al., 1976b).

They are resistant to colicins B and D (Davies and Reeves, 1975a; Hancock et al., 1976a; Pugsley and Reeves, 1976b). Wookey and Rosenberg have measured the level of colicin resistance in their *fepA* mutants and find a hundredfold difference in colicin titer with respect to the parent strain. This level of resistance was reported as sensitivity (Wookey and Rosenberg, 1978). However, their *fepA* mutants and a "*cbr*" (*cbr*-6) and a "*cbt*" (*cbr*-1) mutant from Reeves exhibit the same hundredfold decrease in sensitivity (Wookey, 1978), which is clearly the degree of resistance typical for *fepA* mutants; *fepA* mutants lack colicin receptor activity (Hancock et al., 1976b; Pugsley and Reeves, 1976b) and outer membrane preparations from mutants differ from preparations of *fepA*$^+$ strains in that they do not neutralize colicins B and D (Pugsley and Reeves, 1977).

The 81K outer membrane protein absent in *fepA* mutants is thought, on the basis of the data outlined above, to be a receptor for ferrienterochelin and colicins B and D.

8.4 Isolation of *fepA* Mutants

Since *fepA* mutants do not confer resistance to any known phage, the only simple way to isolate them is by use of colicin B or D. However, it should be noted that these colicins also select several classes of colicin-tolerant mutants, readily distinguished by several properties (Davies and Reeves, 1975a). One *fepA* mutant is thought to be a deletion extending into *nmpC* (Pugsley and Schnaitman, 1978b) and is discussed with the *nmpC* gene.

8.5 Mapping of *fep* Mutants and the Status of *fepB*

Mapping at about 13 min between *lip* and *purE*, *fep* mutants were isolated as defective in ferrienterochelin uptake (Cox et al., 1970; Young et al., 1971; Luke and Gibson, 1971; Langman et al., 1972) and have recently been ordered using Mu insertions in *fep* and *ent* (Woodrow et al., 1975; Laird and Woodrow, 1978). The map order is *fes*, *entD*, *entF*, *fep*, *entC*, *entE* (*entA*, *entB*, *entG*) . . . *lip*.

No complementation studies have been reported for *fep* mutants, but two genes, *fepA* and *fepB* are distinguished both by their cotransduction frequency with *entC* (I. G. Young, reported in Wookey, 1978) and by the nature of the block in ferrienterochelin uptake, *fepA* mutants being defective in transport across the outer membrane and *fepB* being defective in transport across the cytoplasmic membrane. Four *fepB* mutants studied varied in their reaction to colicin, one being fully resistant, two fully sensitive, and one resistant to the same intermediate level as *fepA* mutants. Although it does seem reasonable to recognize two genes, *fepA* and *fepB*, the status of *fepB* cannot be considered to be properly resolved until complementation studies have been carried out: the recent need to merge the comparable *btuB* and *btuA* is one gene *btuB* (Bassford and Kadner, 1977) indicates the need for caution at this stage. The protein composition of *fepB* mutants has not been reported on as yet.

If there are indeed two *fep* genes, then the status of the original "*cbt*" mutation in particular will have to be reconsidered. This mutation (still *cbr*-1 in the Reeves' collection pending resolution of *fep* allele numbers) confers tolerance to colicins B and D (Davies and Reeves, 1975a) and is blocked in ferrienterochelin uptake and binding to the outer membrane (Pugsley and Reeves, 1977). It is thus probably a *fepA* missense mutation giving rise to an altered protein but could perhaps be a *fepB* mutation—its colicin resistance in Dr. Wookey's hands is identical to that of typical *fepA* mutants and some *fepB* mutants.

8.6 The *cir* Gene (Synonym *feuA*)

Colicin-resistant mutants, lacking receptor for colicin I and tolerant to colicin V, have been termed *cir* mutants (Cardelli and Konisky, 1974; Davies and Reeves, 1975a; Pugsley and Reeves, 1977) and map at 44 min.

These mutants lack an outer membrane protein of 74K (Soucek and Konisky, 1977; Pugsley and Reeves, 1977) and this protein is presumed to be the receptor for colicins 1a and 1b, as receptor activity is lacking in cells, outer membrane, and extracts of *cir* mutants (Konisky et al., 1973; Konisky and Liu, 1974; Davies and Reeves, 1975a; Pugsley and Reeves 1977). The colicin I-resistant mutants isolated by Hankte and Braun (1975b) and named *feuA* (Hancock and Braun, 1976) are also *cir* mutants and map at 44 min.

8.7 The *fec* Gene

Mapping at 6 min, *fec* is a gene affecting uptake of iron complexed with citrate (Woodrow et al., 1978) and, by analogy with *tonA* and *fepA*, may well determine the presence of an outer membrane protein receptor for iron citrate, although there are as yet no reports on the protein composition of *fec* mutants. However, it is known that *E. coli* K-12 does produce an outer membrane 81K protein (not the 81K fepA protein) in the presence of citrate (Hancock et al., 1976b) and it is possible that this protein is involved in the uptake of iron citrate and is part of the citrate-inducible uptake system described by Frost and Rosenberg (1973). One might expect *fec* mutants to lack this citrate-inducible outer membrane protein.

8.8 Iron Transport Genes of *Salmonella*

Salmonella typhimurium mutants isolated as resistant to albomycin are defective in ferrichrome uptake and are called *sid* mutants (Luckey et al., 1972). Several genes have been inferred from the varied phenotypes, of which three map at 4 min (Sanderson and Hartmen, 1978). No complementation experiments have been reported to confirm the number of genes present; it is possible that the various phenotypes are all due to mutation in a single gene homologous to *tonA* of *E. coli*, or, alternatively, *E. coli ton A* may encompass several genes. It would be surprising if there were three genes in *Salmonella* corresponding to one in *E. coli*.

Graham and Stocker (1977) have studied the phage and colicin receptor functions of *sid* mutants and compared *E. coli*, *Salmonella typhimurium*, *S. paratyphi B*, and other *Salmonella* species; they concluded that the functionally similar proteins of *E. coli* and *Salmonella typhimurium* differ in their affinity for colicin M and phages T5, T1, and ES18.

9 CONTROL OF OUTER MEMBRANE PROTEINS

Several authors have noted that the amounts of specific outer membrane proteins can be affected by the culture conditions used. The most straightforward examples are control of *lamB*, which as part of the maltose regulon, is inducible by maltose and subject to catabolite repression, and *tsx*, which is also subject to catabolite repression (Manning and Reeves, 1978).

The amounts of the iron transport proteins are also under control and this was discussed above. Although the mechanism is not understood, it is not surprising that the presence of these proteins is repressed by freely available iron.

It is the control of the major outer membrane proteins that is most complex and least understood. The amounts of all the major proteins can be affected by one or more components of the culture conditions. Furthermore, mutants lacking one or more outer membrane proteins may have compensatory amounts of other proteins.

The ratio of matrix protein Ia to matrix protein Ib has been found to vary greatly depending on conditions (Lugtenberg et al., 1976; Bassford et al.,

1977; Manning and Reeves, 1977; van Alphen and Lugtenberg, 1977). Furthermore, when protein 2, NmpA, NmpB, or NmpC is present in addition to matrix proteins Ia and Ib, there is less of matrix proteins Ia and Ib in the outer membrane (Pugsley and Schnaitman, 1978b); the ratio of Ia and Ib is still affected by changes in culture conditions, but these have relatively little effect on proteins 2, NmpA, NmpB, and NmpC. A similar reduction in the amount of matrix proteins Ia and Ib is observed when cells are induced to produce large amounts of lamB protein. The result of these interactions is such that in each case about 30–40% of the total outer membrane protein is peptidoglycan associated (Pugsley and Schnaitman, 1978b). It seems unlikely that the presence of these new proteins affects matrix proteins Ia and Ib through a genetic control system, and the effect may be due to competition for a limited number of sites.

However, the variation in the relative amounts of matrix proteins Ia and Ib is not so readily explained. There does not appear to be competition for sites between these two proteins, as *ompC* mutants do not have increased amounts of matrix protein Ia (Bassford et al., 1977). The change in the relative amounts then is not likely to be due to a direct effect of the environment on the relative ease with which these proteins can insert into some specific site in the membrane.

The control may be at the level of gene function. It may well be that *ompB* is a control gene, but too little is known to indicate if or how it might be involved in regulating the response of proteins Ia and Ib to changes in growth conditions. However, the existence of *ompB* mutations that affect matrix protein Ia and Ib alone, and, in particular, the existence of *ompB* mutants with increased amounts of matrix protein Ia has led Verhoef et al. (1979) to propose that the *ompB* product is a positive control protein acting on *ompC* and *ompF* through different sites on the ompB protein. This model would explain the various *ompB* mutations and could be developed to explain the response of the 1a/1b ratio to environmental changes

The amounts of ompA protein and protein 3b are also subject to considerable variation, and that of protein 3b in particular is affected by growth temperature, the protein being a major protein at 42°C but barely detectable in cells grown at 30°C (Lugtenberg et al., 1976; Manning and Reeves, 1977). Variation in the amount of ompA protein (Lugtenberg et al., 1976; Bassford et al., 1977; Manning et al., 1977) is, like that of protein 3b, not readily explained as yet.

10 CONCLUSION

This chapter is a survey of those genes that are thought to directly affect the outer membrane proteins of *E. coli* K-12 and *Salmonella typhimurium*.

There is no discussion of genes that have an indirect effect. Heptose-less mutants in particular are known to affect the protein composition of the outer membrane (Ames et al., 1974; Koplow and Goldfine, 1974; van Alphen et al., 1974; Hancock and Reeves, 1976), but this effect is outside the scope of this chapter. There is also an increasing interest in plasmid coded outer membrane proteins, which are not discussed in this chapter. Suffice it to say that many of the F sex factor *tra* gene products are outer membrane proteins (Kennedy et al., 1977). The advanced state of genetic analysis of the *tra* operon provides an excellent opportunity to study envelope proteins using genetic techniques.

If we include *ompC* and *ompF*, despite the doubt that still exists, then 10 structural genes namely, *ompA*, *ompC*, *ompF*, *lpp*, *tsx*, *lamB*, *btuB*, *fepA*, *tonA*, and *cir*, have now been identified for outer membrane proteins of *E. coli* K-12. In addition, *nmpA*, *nmpB*, and *nmpC* have been proposed as structural genes (Pugsley and Schnaitman, 1978b) and *fec* may well be the structural gene for an outer membrane protein. Because membrane functions are not readily studied with purified proteins, mutant strains lacking specific proteins have been particularly useful in elucidating the functions of outer membrane proteins, and the relevant studies are discussed elsewhere in this volume.

With the exception of *lamB*, which is part of the well-documented maltose regulon, the control of the expression of the structural genes has not been studied at the molecular level and control genes are not well documented. However, as can be seen in this chapter, the amount of many of the outer membrane proteins does vary considerably depending on growth conditions and it is certain that much will be learned of any genetic basis for these control systems in the near future.

REFERENCES

Achtman, M., S. Schwochow, R. Helmuth, G. Morelli, and P. A. Manning, (1978), *Mol. Gen. Genet.*, **164,** 171–184.

Ames, G. F., Spudich, E. N., and Nikaido, H. (1974), *J. Bacteriol.*, **117,** 406–416.

Bachmann, B. J., Low, K. B., and Taylor, A. L. (1976), *Bacteriol. Rev.*, **40,** 116–167.

Bassford, P. J., and Kadner, R. J., (1977), *J. Bacteriol.*, **132,** 796–805.

Bassford, P. J., Diedrich, D. L., Schnaitman, C. L., and P. Reeves (1977), *J. Bacteriol.*, **131,** 608–622.

Bavoil, P., Nikaido, H., and von Meyenberg, K. (1977), *Mol. Gen Genet.*, **158,** 23–33.

Beacham, I. R., Kamana, R., Levy, L., and Yagic, E. (1973), *J. Bacteriol.*, **116,** 957–964.

Beacham, I. R., Maas, O. and Yagic, E. (1977), *J. Bacteriol.*, **129,** 1034–1044.

Braun, V., Hancock, R. E. W., Hankte, K., and Hartmann, A. (1976), *J. Supramol. Struct*, **5,** 37–58.

Braun-Breton, C., and Hofnung, M. (1977), *FEMS Microbiol. Lett.*, **1**, 371–374.
Buxon, R. S. (1971), *Mol. Gen. Genet.*, **113**, 154–156.
Cardelli, J., and Konisky, J. (1974), *J. Bacteriol.*, **119**, 379–385.
Chai, T., and Foulds, J. (1974), *FEBS Lett.*, **72**, 215–224.
Chai, T-J., and Foulds, J. (1977), *J. Bacteriol.*, **130**, 781–786.
Chai, T-J., and Foulds, J. (1978), *J. Bacteriol.*, **135**, 164–170.
Cox, G. B., Gibson, F., Luke, R. K. J., Newton, N. A., O'Brien, I. G., and Rosenberg, H. (1970), *J. Bacteriol.*, **104**, 219–226.
Datta, D. B., Kramer, C., and Henning, U. (1976), *J. Bacteriol.*, **128**, 834–841.
Datta, D. B., Arden, B., and Henning, U. (1977), *J. Bacteriol.*, **131**, 821–829.
Davies, J. K. (1974), Ph.D. Thesis, Adelaide University.
Davies, J. K., and Reeves, P. (1975a), *J. Bacteriol.*, **123**, 96–101.
Davies, J. K., and Reeves, P. (1975b), *J. Bacteriol.*, **123**, 102–117.
Davies, J. K., and Reeves, P. (1975c), *J. Bacteriol.*, **123**, 372–373.
Demerec, M., Adelberg, E. A., Clark, A. J., and Hartman, P. E. (1966), *Genetics*, **54**, 61–76.
Digirolamo, P. M., Kadner, R. J., and Bradbeer, C. (1971), *J. Bacteriol.*, **106**, 751–757.
Foulds, J. (1974), *J. Bacteriol.*, **117**, 1354–1355.
Foulds, J. (1976), *J. Bacteriol.*, **128**, 604–608.
Foulds, J., and Barrett, C. (1973), *J. Bacteriol.*, **116**, 885–892.
Foulds, J., and Chai, T. (1978a), *J. Bacteriol.*, **133**, 158–164.
Foulds, J., and Chai, T.-J. (1978b), *J. Bacteriol.*, **133**, 1478–1483.
Fredericq, P. (1951), *C. R. Soc. Biol.*, **145**, 930–933.
Fredericq, P., and Smarda, J. (1970), *Ann. Inst. Pasteur.*, **118**, 767–774.
Frost, G. E., and Rosenberg, H. (1973), *Biochim. Biophys. Acta*, **330**, 90–101.
Graham, A. C., and Stocker, B. A. D. (1977), *J. Bacteriol.*, **130**, 1214–1223.
Guterman, S. K. (1971), *Biochem. Biophys. Res. Commun.*, **44**, 1149–1155.
Guterman, S. K., Wright, A. and Boyd, D. H. (1975), *J. Bacteriol.*, **124**, 1351–1358.
Hancock, R. E. W., and Braun, V. (1976), *FEBS Lett.*, **65**, 208–210.
Hancock, R. E. W., and Reeves, P. (1975), *J. Bacteriol.*, **121**, 983–993.
Hancock, R. E. W., and Reeves, P. (1976), *J. Bacteriol.*, **127**, 98–108.
Hancock, R. E. W., Davies, J. K., and Reeves, P. (1976a), *J. Bacteriol.*, **126**, 1347–1350.
Hancock, R. E. W., Hankte, K., and Braun, V. (1976b), *J. Bacteriol.*, **127**, 1370–1375.
Hankte, K. (1976), *FEBS Lett.*, **70**, 109–112.
Hankte, K. (1978), *Mol. Gen. Genet.*, **163**, 131–136.
Hankte, K., and Braun, V. (1975a), *FEBS Lett.*, **49**, 301–305.
Hankte, K., and Braun, V. (1975b), *FEBS Lett.*, **59**, 277–281.
Hankte, K. and Braun, V. (1978), *J. Bacteriol.*, **135**, 190–197.
Hatfield, D., Hofnung, M., and Schwarz, M. (1969a), *J. Bacteriol.*, **98**, 559–567.
Hatfield, D., Hofnung, M., and Schwarz, M. (1969b), *J. Bacteriol.*, **100**, 1311–1315.
Havekes, A. M. (1978), Ph.D. thesis, State University, Utrecht.
Havekes, L. M., and Hoekstra, W. P. (1976), *J. Bacteriol.*, **126**, 593–600.
Henning, U., and Haller, I. (1975), *FEBS Lett.*, **55**, 161–164.

Henning, U., Hindennach, I., and Haller, I. (1976), *FEBS lett.*, **61,** 46–48.

Henning, U., Schmidmayr, W., and Hindennach, I. (1977), *Mol. Gen. Genet.*, **154,** 293–298.

Hirota, Y., Suzuki, H., Nishimura, Y., and Yasuda, S. (1977), *Proc. Natl. Acad. Sci. U.S.*, **74,** 1417–1420.

Hofnung, M., Hatfield, D., and Schwartz, M. (1971), *J. Mol. Biol.*, **61,** 681–694.

Hofnung, M., Hatfield, D., and Schwartz, M. (1974), *J. Bacteriol.*, **117,** 40–47.

Hofnung, M., Jezierka, A., and Braun-Breton, C. (1976), *Mol. Gen. Genet.*, **145,** 207–213.

Ichihara, S., and Mizushima, S. (1978), *J. Biochem.* (*Jap.*), **83,** 1095–1100.

Inouye, S., Lee, N., Inouye, M., Wu, H. C., Suzuki, H., Nishimura, Y., Iketani, H., and Hirota, Y. (1977), *J. Bacteriol.*, **132,** 308–313.

Jasper, P., Whitney, E., and Silver, S. (1972), *Genet. Res.*, **19,** 305–312.

Jenkin, C. R., and Rowley, D. (1955), *Nature* (*Lond.*), **175,** 779.

Johansson, V., Aarti, A., Nurminen, M., and Makela, P. H. (1978), *J. Gen. Microbiol.*, **107,** 183–187.

Kadner, R. J., and Liggins, G. L. (1973), *J. Bacteriol.*, **115,** 514–521.

Kennedy, N., Beutin, L., Achtman, M., and Skurray, R. (1977), *Nature* (*Lond.*), **270,** 580–585.

Konisky, J., and Liu, C-T., (1974), *J. Biol. Chem.*, **249,** 835–840.

Konisky, J., Cowell, B. S., and Gilchrist, M. J. (1973), *J. Supramol. Struct.*, **1,** 208–219.

Koplow, J., and Goldfine, H. (1974), *J. Bacteriol.*, **117,** 527–543.

Krzywy, T., Kucharewycz-Krukowsica, A., and Slopek, S. (1972), *Arch. Immunol. Ther. Exp.*, **19,** 15–45.

Laird, A. J., and Woodrow, G. C. (1978), *Proc. Aust. Biochem. Soc.*, **11,** 69.

Langman, L., Young, I. G., Frost, G. E., Rosenberg, H., and Gibson, F. (1972), *J. Bacteriol.*, **112,** 1142–1149.

Lavoie, M., and Mathieu, L. G. (1975), *Can. J. Microbiol.*, **21,** 1595–1601.

Lin, J. J-C., Kanazawa, H., Ozols, J. and Wu, H. C. (1978), *Proc. Natl. Acad. Sci. U.S.*, **75,** 4891–4895.

Luckey, M., Pollack, J. R., Wayne, R., Ames, B. N., and Nielands, J. B. (1972), *J. Bacteriol.*, **111,** 731–738.

Lugtenberg, B., Peters, R., Bernheimer, H., and Berendsen, W. (1976), *Mol. Gen. Genet.*, **147,** 251–262.

Luke, R. K., J., and Gibson, F. (1971), *J. Bacteriol.*, **107,** 557–562.

Lutkenhaus, J. F. (1977), *J. Bacteriol.*, **131,** 631–637.

McKeown, M., Kahn, M., and Hanawalt, P. (1976), *J. Bacteriol.*, **126,** 814–822.

Manning, P. A., and Reeves, P. (1976a), *Biochem. Biophys. Res. Commun.*, **71,** 466–471.

Manning, P. A., and Reeves, P. (1976b), *J. Bacteriol.*, **127,** 1070–1079.

Manning, P. A., and Reeves, P. (1977), *FEMS Microbiol. Lett.*, **1,** 275–278.

Manning, P. A., and Reeves, P. (1978), *Mol. Gen. Genet.*, **158,** 279–286.

Manning, P. A., Puspurs, A., and Reeves, P. (1976), *J. Bacteriol.*, **127,** 1080–1084.

Manning, P. A., Pugsley, A. P., and Reeves, P. (1977), *J. Mol. Biol.*, **116,** 285–300.

Meyenberg, K. von (1971), *J. Bacteriol.*, **131,** 631–637.

Nakae, F., and Ishii, J. (1978), *J. Bacteriol.*, **133,** 1412–1418.

Nomura, M., and Witten, C. (1967), *J. Bacteriol.*, **94,** 1093–1111.

Nurminen, M., Lounatmaa, K., Sarvas, M., Makela, P. H., and Nakae, T. (1976), *J. Bacteriol.*, **127,** 941–955.

Pugsley, A. P. (1977), *FEMS Microbiol. Lett.*, **2,** 275–277.

Pugsley, A. P., and Reeves, P. (1976a), *J. Bacteriol.*, **126,** 1052–1062.

Pugsley, A. P., and Reeves, P. (1976b), *J. Bacteriol.*, **127,** 218–228.

Pugsley, A. P., and Reeves, P. (1976c), *Biochem. Biophys. Res. Commun.*, **70,** 846–853.

Pugsley, A. P., and Reeves, P. (1977), *Biochem. Biophys. Res. Commun.*, **74,** 903–911.

Pugsley, A. P., and Schnaitman, C. A. (1978a), *J. Bacteriol.*, **133,** 1181–1189.

Pugsley, A. P., and Schnaitman, C. (1978b), *J. Bacteriol.*, **135,** 1118–1129.

Radke, K. L., and Siegel, E. C. (1971), *J. Bacteriol.*, **106,** 432–437.

Randall-Hazelbauer, L., and Schwartz, M. (1973), *J. Bacteriol.*, **116,** 1436–1446.

Reeve, E. C. R., and Doherty, P. (1968), *J. Bacteriol.*, **96,** 1450–1451.

Sanderson, K. E., and Hartman, P. E. (1978), *Bacteriol. Rev.*, **42,** 471–519.

Sarma, V., and Reeves, P. (1977), *J. Bacteriol.*, **132,** 23–27.

Sato, T., Ito, K., and Yura, T. (1977), *Annu. Rep. Inst. Virus Res., Kyoto Univ.*, **20,** 67–68.

Schmitges, C. J., and Henning, U. (1976), *Eur. J. Biochem.*, **63,** 47–52.

Schnaitman, C. A., Smith, D., and Forn De Salsas, M. (1975), *J. Virol.*, **15,** 1121–1130.

Siitonen, A., Johansson, V., Nurminen, M., and Makela, P. H. (1977), *FEMS Microbiol. Lett.*, **1,** 141–144.

Silhavy, T. J., Shuman, H. A., Beckwith, J., and Schwartz, M. (1977), *Proc. Natl. Acad. Sci. U. S.*, **74,** 5411–5415.

Skurray, R. A., Hancock, R. E. W., and Reeves, P. (1974), *J. Bacteriol.*, **119,** 726–735.

Soucek, S., and Konisky, J. (1977), *J. Bacteriol.*, **130,** 1399–1401.

Stocker, B. A. D., and Makela, P. H. (1978), *Proc. R. Soc. Lond. B.*, **202,** 5–30.

Suzuki, J., Nishimura, Y., Iketani, H., Campisi, J., Hirashima, A., Inouye, M., and Hirota, Y. (1976), *J. Bacteriol.*, **127,** 1494–1501.

Thirion, J. P., and Hofnung, M. (1972), *Genetics*, **71,** 207–216.

Torti, S. V., and Park, J. T. (1976), *Nature* (*Lond.*), **263,** 323–326.

Van Alphen, W., and Lugtenberg, B. (1977), *J. Bacteriol.*, **131,** 623–630.

Van Alphen, W., Lugtenberg, B., and Berendsen W. (1976), *Mol. Gen. Genet.*, **147,** 263–269.

Van Alphen, W., van Selm, N., and Lugtenberg, B. (1978), *Mol. Gen. Genet.*, **159,** 75–83.

Van Alphen, L., Lugtenberg, B. von Boxtel, R., Hack, A., Verhoef, C., and Havekas, L. (1979), *Mol. Gen. Genet.*, **169,** 147–155.

Van Vught, A. M. J. J., de Graaff, J., and Stouthamer, A. H. (1975), *Antonie van Leeuenhoek*, **41,** 309–318.

Van Vught, A. M. J. J., and Delemarre, E. C. M. (1977), *Antonie van Leeuenhoek, J. Microbiol. Serol.*, **43,** 7–18.

Verhoef, C., de Graaff, P. J., and Lugtenberg, B. (1977), *Mol. Gen. Genet.*, **150,** 103–105.

Verhoef, C., Lugtenberg, B., van Boxtel, R., de Graaff, P., and Verheij, H. (1979), *Mol. Gen. Genet.*, **169,** 137–146.

Wayne, R., and Nielands, J. B. (1975), *J. Bacteriol.*, **121,** 497–503.

Weigand, R. A., and Rothfield, L. I. (1976), *J. Bacteriol.*, **125,** 340–345.

Weigand, R. A., Vinci, K. D., and Rothfield, L. I. (1976), *Proc. Natl. Acad. Sci. U.S.*, **73,** 1882–1886.

Woodrow, G. C., Young, I. G., and Gibson, F. (1975), *J. Bacteriol.*, **124,** 1–6.

Woodrow, G. C., Langman, L., Young, I. G., and Gibson, F. (1978), *J. Bacteriol.*, **133,** 1524–1526.

Wookey, P. (1978), Ph.D. thesis, Australian National University, Canberra, Australia.

Wookey, P., and Rosenberg, H. (1978), *J. Bacteriol.*, **133,** 661–666.

Wu, H. C., and Lin, J. J-C. (1976), *J. Bacteriol.*, **126,** 147–156.

Wu, H. C., Hou, C., Lin, J. J-C., and Yem, D. W. (1977), *Proc. Natl. Acad. Sci. U.S.*, **74,** 1388–1392.

Yem, D. W., and Wu, H. C. (1977), *J. Bacteriol.*, **131,** 759–764.

Yem, D. W., and Wu, H. C. (1978), *J. Bacteriol.*, **133,** 1419–1426.

Young, I. G., Langman, L., Luke, R. K. J., and Gibson, F. (1971), *J. Bacteriol.*, **106,** 51–57.

Chapter 9

The Cell Cycle-Dependent Synthesis of Envelope Proteins in *Escherichia coli*

MISAO OHKI

Biology Division, National Cancer Center Research Institute, Tokyo, Japan

1 INTRODUCTION

The most striking change that takes place during the bacterial cell cycle, in terms of cell morphology, is the formation of septum. Many interesting observations have been made on this process with the light microscope and the electron microscope. However, studies at the molecular level are extremely limited, and the biochemical basis of the formation of septum is almost unknown. The present state of our knowledge does not allow us to answer even the most fundamental question, that is, whether the septum is formed simply by the extension of preexisting membranes or by the *de novo* assembly of new subunits. One of the widely accepted models is that septum formation is promoted by morphological changes of peptidoglycan layer, which underlies the outer membrane. But another model suggesting that the membrane(s) serves as the primary promoter of the cell division process is still possible. So far, few studies have taken into consideration the possibility that the membrane also should undergo simultaneous changes in its components and structure during the process of septum formation and cell division.

It is known that a number of envelope proteins are synthesized at a constant rate that doubles at a discrete time in the cycle (for example, see James and Gudas, 1976). On the other hand, another class of envelope proteins seems to be synthesized only during a limited period in the cell cycle, shortly before cell division. Therefore, the possibility exists that some of these latter proteins are involved in, or at least in some manner are connected to, the assembly of septum components of the envelope.

The principal purpose of this chapter is to describe how the synthesis of individual membrane proteins and the assembly of membranes are programmed in the cell cycle for the orderly multiplication of bacteria, and how these cell cycle controls are intercalated with each other and with the process of cell division.

2 OUTER MEMBRANE PROTEINS

Recent rapid advances in the studies of bacterial outer membrane, especially the identification of membrane proteins by SDS–acrylamide gel electrophoresis, have made possible the study of the mechanism of biosynthesis of its components in connection with the cell cycle. The major proteins of *E. coli* outer membrane, such as the lipoprotein, heat modifiable proteins (ompA protein and proteins 3b or 3a), and peptidoglycan-associated proteins, that is, porins (except the λ receptor protein as is discussed later), are synthesized and assembled into the envelope throughout a cell cycle (Churchward and Holland, 1976; M. Ohki and H. Nikaido, unpublished work; M. Ohki, I. Yamato, and M. Futai, unpublished work). These major proteins are stable once integrated into the membrane. One known exception is the 76K protein, 70% of which has been shown to become lost from the envelope during the succeeding generation (Churchward and Holland, 1976).

In contrast with the common pattern of continuous increase of total cellular protein, a few proteins have been reported to be synthesized at a particular phase of cell division. The λ receptor protein is an essential component for the adsorption of λ phage and also participates in the permeation of maltose through the outer membrane (Szmelcman and Hofnung, 1975). The *lamB* gene, the structural gene for the λ-receptor protein, is a part of a cluster of the maltose genes and is under the same genetic control as other maltose genes (Schwartz, 1967a, b; Hofnung et al., 1971). The synthesis of λ-receptor protein is induced by maltose, and in fully induced cells, the protein corresponds to 8–10% of total envelope protein (Braun and Krieger-Brauer, 1977). Based on the fact that the λ-receptor protein is the sole element for the specific adsorption of phage λ to the cell, Ryter et al. (1975) approached the problem of the assembly of the λ receptor protein in a unique way. Appearance at the cell surface of the receptor molecules synthesized during brief periods of induction was followed by adsorbtion of λ phage particles to the bacteria and by examination of the infected bacteria under an electron microscope. This specific labeling of the protein with the probe has revealed nonrandom distribution of the labeled cells throughout the population, and two classes of bacteria, that is those beginning to divide and those that had just divided, were especially heavily labeled. Furthermore, phages were concentrated above the central, septum region in the former class and at one of the cell poles in the latter class (Figure 1). These observations thus suggested that insertion into the envelope of the newly synthesized receptor only occurred during a short period of the cell cycle, just before cell division, and that the latter class of cells was generated by the division of cells containing receptors over the septum area. On the basis of this assumption, the observed fraction of 0.17 for the septum labeled cells was equated with the fraction $2^{t/T} - 1$, a fraction whose age is between $T - t$ and T, deduced from the known age distribution function (Powell, 1956), and the value $t/T = 0.23$ was obtained. This indicates that the integration of the receptor into the outer membrane occurs between the cell age of 0.77 and 1.0.

Since the λ receptor protein is a major envelope protein in induced cells, and since its electrophoretic mobility is different from those of other major proteins, the protein can be easily identified by SDS–polyacrylamide gel electrophoresis. Analysis of the λ-receptor protein synthesis by this method has been successfully carried out in the cells labeled at different phases during synchronous growth (M. Ohki and H. Nikaido, unpublished results). As the synchronous culture system, strain N167, a temperature-sensitive *dnaA*-type mutant, was used; this system is highly reproducible and large quantities of synchronous culture can be easily obtained (Ohki, 1972). Exponentially growing cells of N167 were incubated at nonpermissive temperature for about 1 hr and were then returned to a permissive temperature. With a synchronous culture thus obtained, the kinetics of the induction of the λ receptor protein synthesis has been followed by pulse labeling (M. Ohki and H. Nikaido, unpublished work). Figure 2 shows an autoradiogram

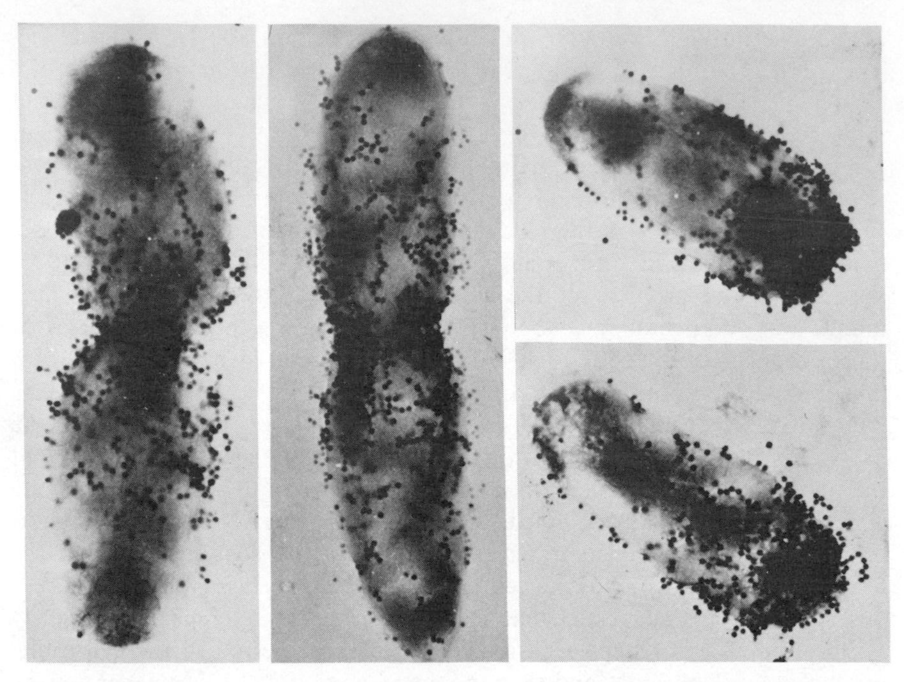

Figure 1 Two types of cells showing the sites of integration of the newly synthesized λ-receptor protein. Exponentially growing cells of *E. coli* K-12 strain 3000 were induced with 10 m*M* maltose for 9 min. The cells were suspended in cold 10 m*M* $MgSO_4$ solution and mixed with λ phages at a multiplicity of infection of 200. After removal of the unadsorbed phage, the cells were fixed and observed under an electron microscope. From Ryter et al. (1975).

of SDS–acrylamide gel and the rates of synthesis of envelope proteins at different stages of the cycle. The receptor protein (48K) was not induced, in spite of the presence of maltose from the beginning of synchronous growth, that is, during the initial 50 min. The first burst of the rate of synthesis was observed between 50 and 70 min. The timing of this burst approximately corresponded to 15–20 min before cell division. The age of the cells at this particular stage of the cycle can be calculated from the pattern of cell division. The age (0.8) was in good agreement with that reported from the integration of the receptor protein, obtained by a completely different approach (Ryter et al., 1975).

The experiment of Ryter et al. (1975) could not provide any definite answer as to whether the cell cycle-dependent appearance of the receptor was due to the cycle-dependent integration of the preexisting molecules into the matrix or reflected cell cycle-dependent net synthesis of the protein. However, experiments to distinguish these possibilities can easily be carried out with synchronous cultures. Figure 3 shows the results of synchronous culture experiments in which cells pulse labeled at different cell phases were chased and the existence of a cellular pool of the λ-receptor protein was examined. As seen by a comparison of the solid lines with the dotted lines, the incorporation of the radioactive receptor protein into the envelope was not observed during chase periods. These results unambiguously indicate that cells have no λ-receptor pool of detectable size and that the protein is indeed synthesized in a cell cycle-dependent manner.

The autoradiogram shown in Figure 2 also indicates that the behavior of a 43K protein is identical to that of the λ receptor protein. Interestingly, this protein appears to be the product of the *malK* gene (P. Bavoil, unpublished observations) and is presumed to be the carrier protein in the active transport of maltose (Hofnung, 1974). The cell cycle-dependent synthesis of this protein is discussed in Section III.

Another outer membrane protein that is known to show cell-cycle dependence in its synthesis is the 80K protein (Gudas et al., 1976). This protein has been reported to be synthesized at the end of the cell cycle near the end of one round of DNA synthesis in acetate-grown cells and at about halfway through the cycle, shortly before the initiation of DNA synthesis when the cells were grown in glucose. A similar tight coupling of the synthesis of this particular protein with DNA synthesis was also observed in the experiments where initiation of DNA synthesis was blocked either by antibiotics or by mutation such as *dnaA* and *dnaB*. Based on these observations,

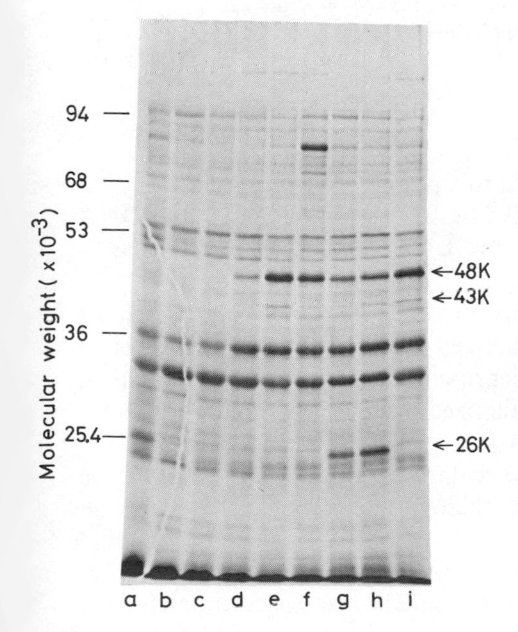

Figure 2 Autoradiogram of a slab gel of envelope proteins from a synchronous culture of *E. coli* strain N167A^{-}3. The λ-receptor protein was induced by adding 10 m*M* maltose at the start of synchronous growth (0 min). Samples taken at indicated times were pulse labeled with [^{3}H]arginine for 12 min, and envelope preparations were applied to an SDS–polyacrylamide slab gel. Thus the intensity of each band of autoradiogram corresponds to the rate of incorporation of the protein relative to that of total envelope proteins. Synchronous cell division took place at 80 and 125 min. Samples were pulse labeled at (*a*) 10, (*b*) 30, (*c*) 40, (*d*) 50, (*e*) 60, (*f*) 70, (*g*) 80, (*h*) 90, and (*i*) 100 min. From M. Ohki and H. Nikaido, (unpublished work.)

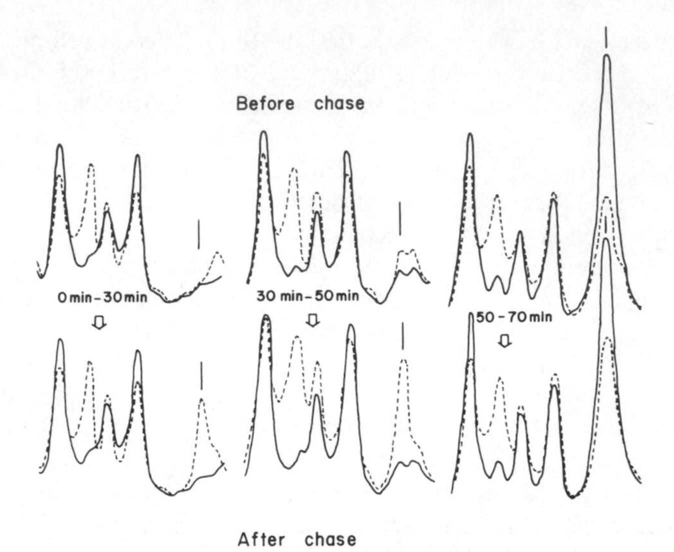

Figure 3 Densitometric tracing of the λ-receptor protein bands from pulse-labeled cells and pulse-chased cells. *E. coli* strain N167A$^-$3 was induced by 10 m*M* maltose at the time of start of synchronous growth. Samples taken at different stages during cell cycle were pulse labeled with [^{14}C]arginine for indicated periods. At the end of the pulse, a large excess of nonradioactive arginine was added, and each culture was divided into two parts. One of them was further incubated in the isotope-free medium for 60 min. Envelope fractions were prepared from the cells before and after chase and were subjected to SDS-polyacrylamide slab gel electrophoresis. Densitometric tracing taken from the stained gel (-----) prior to drying was superimposed on the tracing from the autoradiogram (——). Figures show the position of the tracing surrounding the λ receptor protein, which is indicated by the bars. From M. Ohki and H. Nikaido (unpublished work).

Gudas et al. (1976) have suggested that this protein, called protein D, might act as a metabolic linker of murein synthesis, protein synthesis, and DNA initiation and as an attachment site for DNA to the cell envelope at a specific stage in the cell cycle.

At almost the same time, an envelope protein with a molecular weight of 76,000 has been reported to be synthesized near the time of cell division (Churchward and Holland, 1976). This protein was physiologically unstable, and a large fraction of the newly synthesized protein was rapidly lost from the envelope during subsequent growth.

Because of the similarity in their molecular weights and in their cell-cycle dependence, one would suspect that these proteins might be identical. Recent studies (Boyd and Holland, 1977) have proved this assumption and also that, in contrast to the initial interpretation, protein D (or the 76,000 dalton protein) is the product of the *feuB* gene, a protein essential for the transport of iron–enterochelin complex across the outer membrane (Hantke and Braun, 1975). Furthermore, they have shown that the synthesis of this protein is induced merely by filtering a bacterial culture and resuspending the cells in an identical fresh medium. It appears, therefore, that many of the data used to support the key function of protein D in linking cell division to DNA initiation were filtration artifacts. However, even the work of Boyd and Holland (1977) showed that this protein was synthesized during a limited period in the cell cycle, and a more thorough reinvestigation on this protein seems desirable.

3 INNER MEMBRANE PROTEINS

The inner membrane serves as the site of important and divergent biological functions, such as respiration, ATP production, and the synthesis of lipids and murein. In contrast to the situation with outer membrane proteins, it is almost impossible to identify individual inner membrane proteins by SDS–acrylamide gel electrophoresis of the whole cell envelope, apparently because inner membrane proteins are more numerous and have a rather uniform distribution in terms of both molecular weights and degree of abundance of each species. Therefore, most of the studies in relation to cell cycle have been carried out by assaying the functions of proteins as enzymes.

Cytochrome b_1 is the major cytochrome in *E. coli* and is the first example of a membrane protein shown to increase in a cell cycle-specific manner (Ohki, 1972), as shown in Figure 4. Since the cytochrome was measured in a dual wavelength spectrophotometer measuring the difference in absorbance at 560 nm between the oxidized and reduced states, the obtained values presumably represent the absolute amount of the cytochrome. However, the cytochrome is a complex protein composed of apoprotein and protoheme, and this leads to the problem of whether the synthesis of the cytochrome is regulated in the process of apoprotein synthesis, heme synthesis, or the assembly of the components into the functional molecule. Unfortunately, the answer is unknown.

If it is assumed that an increase in functional activity represents an increase in the amount of the protein, then measurements of activity provides a simple and sensitive way of following the synthesis of specific proteins. The membrane enzymes whose activity have been shown to increase in a cell cycle-dependent manner are D-lactate dehydrogenase (M. Ohki, I. Yamato and M. Futai unpublished observations), succinate dehydrogenase (M. Ohki, I. Yamato and M. Futai unpublished observations), L-α-glycerophosphate dehydrogenase (Ohki and Mitsui, 1974), lactose permease (Ohki and Mitsui, 1974), and the L-α-glycerophosphate permease (Ohki, 1972). As most of the listed members are composed of multiple com-

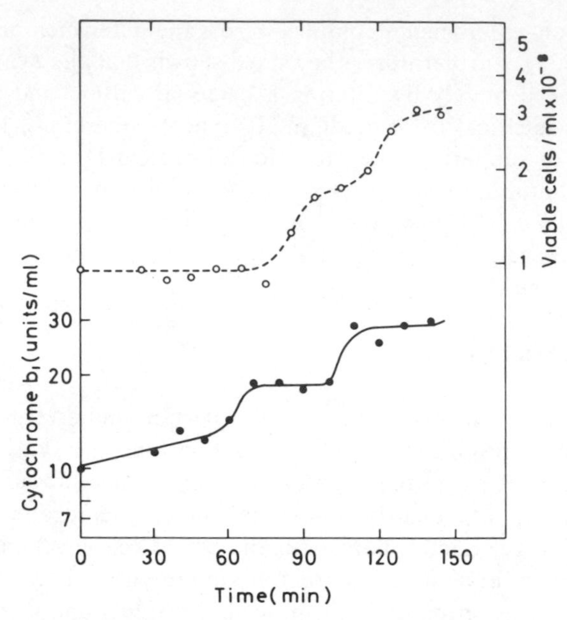

Figure 4 Formation of cytochrome b_1 during synchronous growth of strain N167A$^-$3. From Ohki (1972).

ponents, these results suffer from a problem identical to that mentioned in the case of cytochrome b_1. The situation is the same even for lactose permease, since the accumulation of β-galactosides against a concentration gradient requires a continuing supply of metabolic energy and is obviously a complex process dependent on enzymatic machinery other than the product of the y gene. The present level of our knowledge does not allow us to distinguish between the possibilities mentioned above. In spite of this element of ambiguity, one can see a unique feature among these data, that is, all the enzymes listed above duplicate at 10–20 min before cell division, and the increase in their activity occurs only during particular portions of the cell cycle. Furthermore, accurate measurement of this timing performed by simultaneous assays of the cell cycle-dependent enzymes has revealed that they are duplicated exactly at the same time. This apparent identity in timing leads to the suspicion that their duplication might be regulated by a common mechanism that operates only at a particular point in the cell cycle. It should be noted here that enzymes duplicating at other phases of a cell cycle have not been found.

A limited number of investigations have been carried out to determine whether the cell cycle-dependent duplication is attributable to net synthesis or insertion of preformed molecules of the enzyme protein. Synchronous cultures of *E. coli* were labeled with radioactive amino acids, and the incorporation of radioactivity into D-lactate dehydrogenase molecules was examined by precipitating the enzyme in Triton X-100 with an antiserum prepared against purified D-lactate dehydrogenase. The incorporation of the isotope, and thus the synthesis of the protein, was synchronous with the duplication of the enzymatic activity (M. Ohki, I. Yamato, and M. Futai, unpublished work). Another system studied in this respect was the synthesis of the proteins of the maltose operon (M. Ohki and H. Nikaido, unpublished work). The 43K protein, the product of the *malK* gene, was incorporated at the same time with the λ-receptor protein. This incorporation into the membrane of the 43K protein was shown to be coupled with the net synthesis of the protein molecules, as in the case of the λ receptor protein. The synthesis of the λ-receptor protein and the *malK* gene product occurs at exactly the same time as that of D-lactate dehydrogenase (Figure 5), suggesting the possibility that both are connected to a common regulatory mechanism.

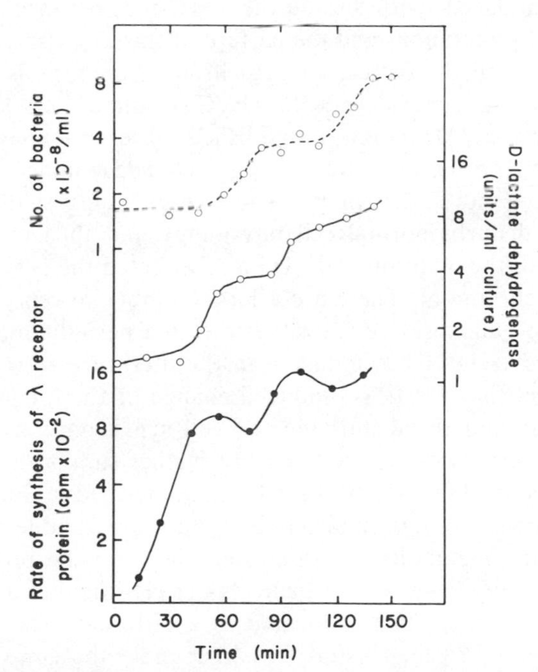

Figure 5 Timing of the synthesis of the λ-receptor protein and D-lactate dehydrogenase during synchronous growth of strain N167A^{-}3. The cells were induced by maltose at the beginning of the synchronous growth and pulse labeled with [^{14}C]arginine at different stages of cell cycle. The rate of synthesis of the λ-receptor protein was measured by counting radioactivity in the band of the protein separated by SDS–polyacrylamide gel electrophoresis. D-Lactate dehydrogenase activity was measured according to the method of Singer and Cremora (1966). From M. Ohki and H. Nikaido (unpublished work).

4 CYTOPLASMIC PROTEINS

All cell cycle-dependent membrane proteins described so far are inducible proteins, except D-lactate dehydrogenase (Futai and Kimura, 1977). The genetic control of some of these proteins is well known. Lactose permease is the product of the y gene of the lactose operon (Fox et al., 1967). The λ receptor and the *malK* gene protein are the products coded for by one operon of the so-called maltose superoperon (Hofnung, 1974), and L-α-glycerophosphate dehydrogenase and L-α-glycerophosphate permease are included in the *glp* "superoperon" system (Cozzarelli et al., 1968). Therefore, one can expect to get important information from the analysis of other enzymes in the same operon as to whether the cell-cycle dependence is due to the posttranscriptional events or not. Some of the soluble enzymes belonging to these superoperons have thus been measured in relation with the cell cycle.

It was found in experiments using synchronous culture of the strain N167 that the basal uninduced synthesis and the TMG-induced synthesis of β-galactosidase were discontinuous and the pattern of the stepwise synthesis coincided completely with the pattern of cytochrome b_1 synthesis. In contrast, the induction of β-galactosidase with IPTG resulted in a more or less continuous synthesis (M. Ohki, unpublished observations). The lactose permease increased in the same way as β-galactosidase under all conditions (M. Ohki, unpublished observations). These results suggest that the induction with IPTG disturbs normal cell physiology and then in turn the cell synchrony; indeed, the addition of IPTG also affected the periodicity of the synthesis of cytochrome b_1. There are a large number of reports examining whether or not; β-galactosidase is synthesized in a periodic manner but the results are controversial. The confusion in the literature seems to come in part from the fact that the IPTG-induced change of the mode of synthesis of β-galactosidase and the disturbance of cell synchrony have been overlooked. The conclusions drawn by Ohki are further supported by the study of the cell cycle-control mutant tsC42, which is described in Section 6.

Maltodextrin phosphorylase and amylomaltase are soluble enzymes that are involved in the metabolism of maltose. They are the products of the genes *malP* and *malQ* in the *malA* gene cluster (Hatfield et al., 1969), one of the maltose operons, which is located at a different locus of the chromosome from the *malB* gene cluster but is still under the same positive control of the *malT* gene product. Figure 6 shows the patterns of their induction during synchronous growth. Both enzymes are induced simultaneously in a periodic manner and the timing of their synthesis coincides with that of the λ-receptor protein and of the *malK* gene product (M. Ohki and H. Nikaido, unpublished work).

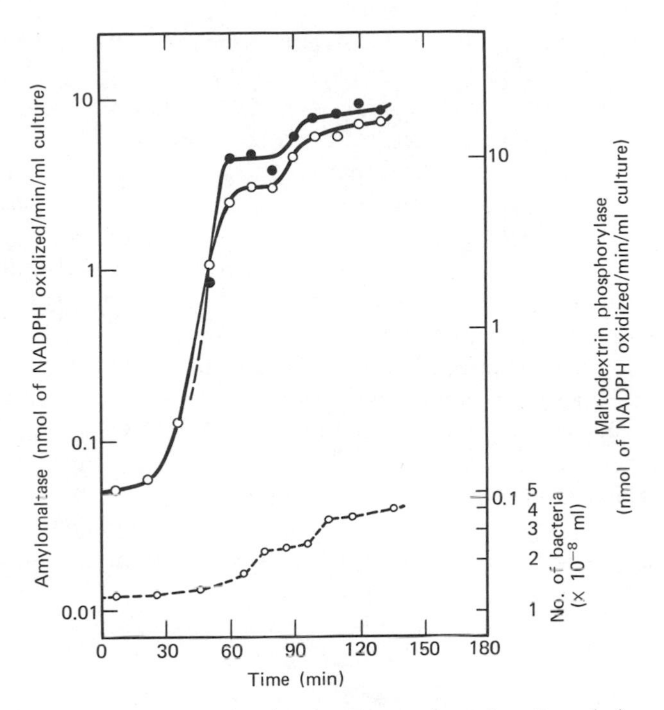

Figure 6 Induction of maltodextrin phosphorylase and amylomaltase during synchronous growth of strain N167A⁻3. Maltose (10 m*M*) was added at the start of synchronous growth (0 min). The samples taken at 5, 20, and 35 min did not show any detectable amylomaltase activity under our assay conditions. (●) Amylomaltase, (○) maltodextrin phosphorylase. From M. Ohki and H. Nikaido (unpublished work).

These two lines of evidence suggest that the periodic synthesis of these enzymes is controlled in the step of gene expression, since the transcription and the translation are coupled in protein synthesis of the inducible and the derepressible operons (Imamoto, 1973; Imamoto and Schlessinger, 1974).

5 PERIPLASMIC PROTEINS

The periplasm, which is the compartment between the cytoplasmic membrane and the outer membrane, contains several percent of the total cellular protein, depending on growth conditions and strains (Oxender and Quay, 1976). There is major interest in the periplasm because it contains a class of proteins called binding proteins. These proteins were shown to play important roles for the transport of amino acids and sugars (Wilson and Smith,

1978; Oxender and Quay, 1976) and also, in some cases, for the chemotactic behavior as sensory receptors (Hazelbauer, 1975).

The correlation between the synthesis of the periplasmic protein and the cell cycle was first demonstrated by Shen and Boos (1973). As shown in Figure 7, the uptake activity for galactose stays constant while the cells are not dividing during synchronous growth, but it abruptly increases at the time of cell division. The synthesis of the periplasmic galactose-binding protein, measured by the incorporation of radioactive amino acids into specific immunoprecipitates, closely followed the pattern of the transport activity. This correlation of the synthesis of the binding protein and cell division has also been observed in mutant BUG-6, which exhibits temperature-sensitive division: The binding protein was normally synthesized at a permissive temperature, but the synthesis immediately stopped when the cells were grown into long filaments at a restrictive temperature. Furthermore, the analysis of the periplasmic proteins of the mutant showed that the cell cycle-dependent synthesis was not unique for the galactose-binding protein, and several other periplasmic proteins were also synthesized in this manner.

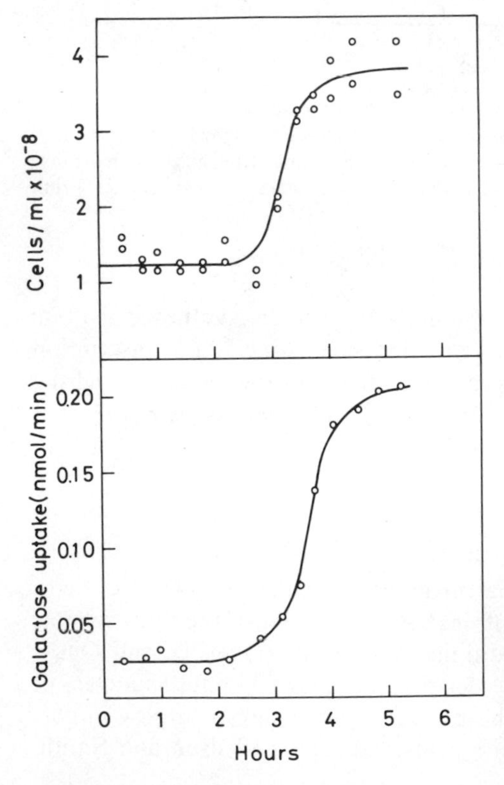

Figure 7 Increase in the activity of galactose transport during synchronous growth. Cells of an *E. coli* strain W3092cy$^-$ were synchronized by a stationary phase culture method of Cutler and Evans (1966). Transport activity was measured using [^{14}]galactose and is expressed as initial rate of galactose uptake. From Shen and Boos (1973).

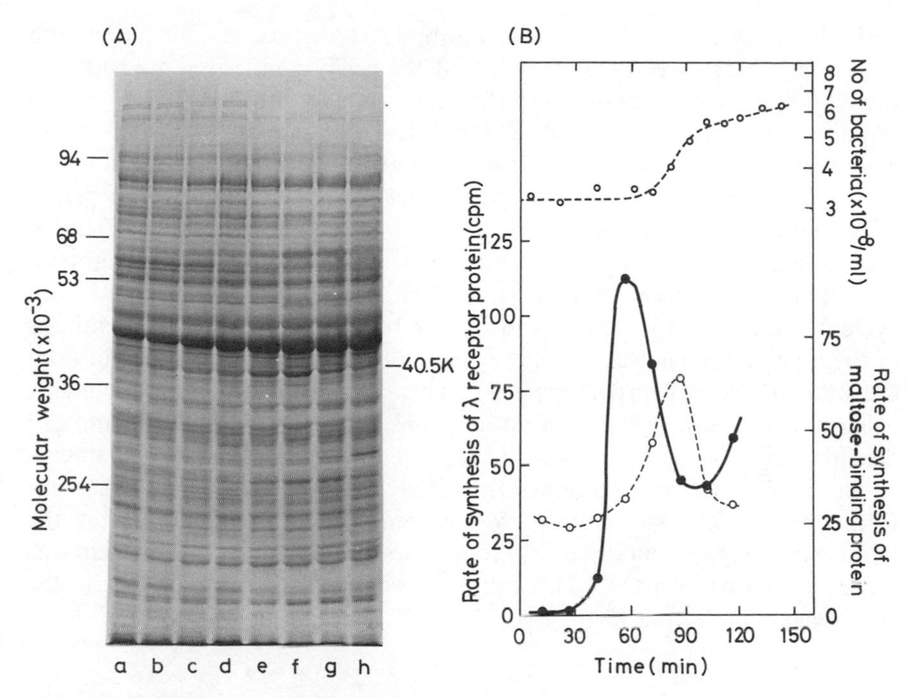

Figure 8 Synthesis of the maltose-binding protein and the λ-receptor protein during synchronous growth of N167A^{-}3. Cells were induced by adding maltose at the start of synchronous growth and were pulse labeled with [^{14}C]arginine for 15 min at different stages of cell cycle. (*A*) Periplasmic fractions were prepared according to the method of Neu and Heppel (1965) and were subjected to SDS–polyacrylamide slab gel electrophoresis. Since an equal amount of radioactivity was applied to each slot, the intensity of each band corresponds to the relative rate of synthesis and its transport, into the perimplasmic space, of the protein. The slots contain samples from the cells pusle labeled at (*a*) 5, (*b*) 20, (*c*) 35, (*d*) 50, (*e*) 65, (*f*) 80, (*g*) 95, and (*h*) 110 min. The bar on the right indicates the position of the maltose-binding protein (40.5K). (*B*) Envelopes were prepared from the same cells after the osmotic shock treatment and the rate of the λ-receptor protein synthesis was measured according to the same method as in Figure 5 (●). The rate of synthesis of the maltose-binding protein (○) was determined from areas under the peak after scanning of the autoradiogram of *A* with a densitometer. From M. Ohki and H. Nikaido (unpublished work).

Recently, an almost identical conclusion was reached by two independent groups (Dietzl et al., 1978; M. Ohki and H. Nikaido, unpublished work) for the maltose-binding protein with respect to the tight coupling between the synthesis and cell division. Dietzel et al. have also demonstrated by electron-microscopic analysis of cells grown on different carbon sources that the induction of cells by maltose results in the formation of large pole caps. These pole caps were considered to arise from enlargement of the periplasmic space and to be filled with the maltose-binding protein. There is no definitive evidence for this interpretation. Nevertheless, in connection with the evidence, discussed in Section 1, that the newly synthesized molecules of the λ receptor are integrated into the septum region, this might become an important problem in the future. Ohki and Nikaido (unpublished work) have shown that the four proteins coded for by the *malP–malQ* and the *malK–lamB* operons of maltose are synthesized only during a brief period of cell cycle, 10–20 min before cell division. In further work, they extended the comparison of the timing of synthesis to the maltose-binding protein. They found, in agreement with Dietzel et al., that the synthesis was cell cycle-dependent, but in addition they found that the timing of the synthesis of the binding protein was clearly separated from that of the λ receptor (and the other three proteins) (Figure 8). This was confirmed by experiments using nalidixic acid, which is a potent inhibitor of DNA synthesis and cell division, and also by experiments using a mutant that forms long filaments under nonpermissive conditions. In both cases, the inhibition of septum formation resulted in a more severely reduced rate of synthesis of the maltose-binding protein, whereas the synthesis of the λ-receptor protein was inhibited to a lesser extent. This cell-cycle control of the second type is discussed again in Section 7.

6 THE CELL CYCLE-CONTROL MUTANT tsC42

On the basis of the earlier observation that the formation of cytochrome b_1 and the L-α-glycerophosphate transport system occurs at the same time, Ohki and Sato (1975) attempted to isolate mutants that had a defect in the synthesis or insertion of membrane proteins at this particular stage in the cell cycle. It was assumed that growth would be temperature sensitive in such mutants and that the formation or insertion of the β-galactoside transport system would be under the control of the corresponding ts mutation. The latter assumption was based on the observation of the cell cycle-dependent formation of the β-galactoside permease described in Section 3.

The mutant tsC42 thus obtained grew at a normal rate at a permissive temperature of 30°C. But the rate of protein synthesis decreased gradually after the shift to the nonpermissive temperature of 42°C, and finally the growth ceased after a 1.7- to 1.8-fold increase in the amount of total protein. The mutant cells at this stage were homogeneous in terms of size, which was, as expected, approximately equivalent to the larger cells in a randomly growing population, that is, to the cells approaching the stage of cell division.

Figure 9 shows the kinetics of formation of cytochrome b_1 in the mutant tsC42 and its parental strain AT713. In the mutant, the rate of formation

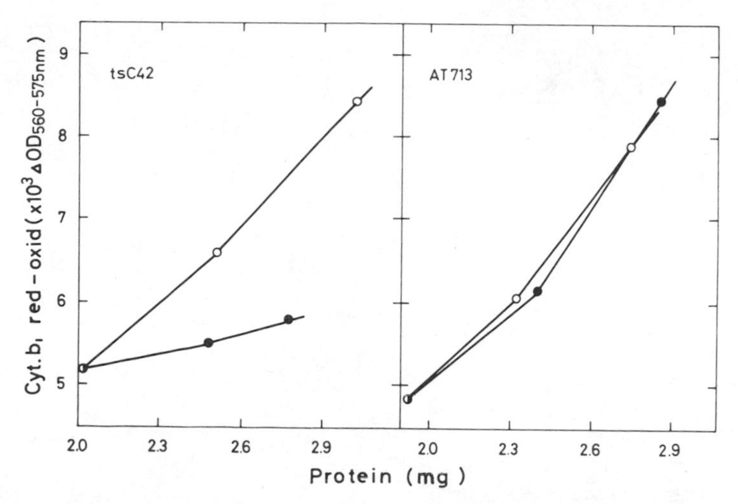

Figure 9 Cytochrome b_1 synthesis in a cell cycle-control mutant tsC42 and its parental strain AT713. Exponentially growing cells of each strain were divided into two parts. One part was further grown at the same temperature (30°C) and the other was shifted to 42°C. Samples were removed after (●) 20 and 40 min at 42°C and (○) after 30 and 60 min at 30°C. The increase in cytochrome b_1 was plotted versus the increase in the total protein. From Ohki and Mitsui (1974).

was greatly reduced at 42°C from the beginning of the incubation, in comparison with the rate at 30°C, whereas the differential rates of synthesis were identical at 42 and 30°C in the parent (Ohki and Mitsui, 1974). Membrane proteins whose synthesis was similarly affected at 42°C in the mutant tsC42 include L-α-glycerophosphate permease, L-α-glycerophosphate dehydrogenase (Ohki and Mitsui, 1974), D-lactate dehydrogenase (I. Yamato and M. Ohki, unpublished), and β-galactoside permease (Ohki and Sato, 1975). Thus the enzymes that have been shown to exhibit the cell-cycle dependence in their synthesis are affected. In another experiment, tsC42 cells were exposed to a nonpermissive temperature for 2 hr to arrest the cells at the specific stage of the cycle. When these cells were returned to a permissive temperature, succinate dehydrogenase was made immediately, and this was followed by cell division in a synchronous manner (Figure 10). The age of cells accumulated during incubation at 42°C, deduced from the pattern of cell division, corresponded to 0.65, a result that is compatible with the cell age at which the formation of the cell cycle-dependent membrane enzyme was observed in the strain N167. It thus appears likely that, because of a specific defect in the mechanism involving the cell cycle-dependent synthesis or insertion of membrane enzymes, the whole population of cells become aligned at a stage of 10–20 min before division during

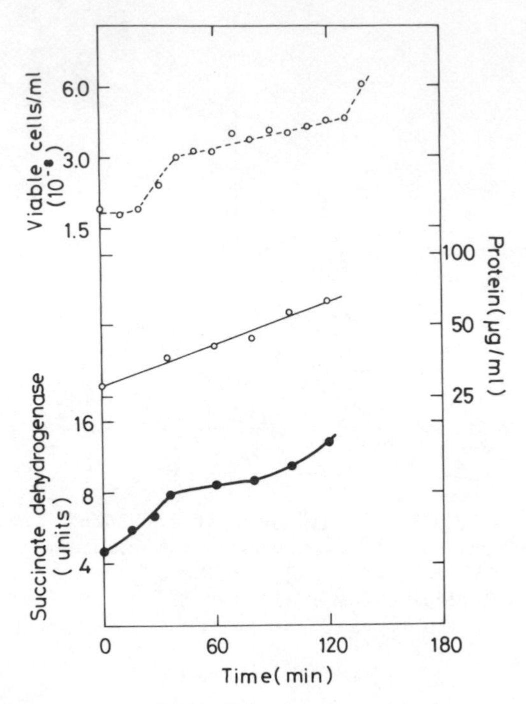

Figure Figure 10 Cell division and formation of succinate dehydrogenase after incubation at a nonpermissive temperature in the mutant tsC42. Exponentially growing cells were incubated at 42°C for 35 min. Then the culture was returned to a permissive temperature (30°C) and the increases of (O----O) viable cells, (●——●) succinate dehydrogenase, and (O——O) total protein were measured. From Ohki and Mitsui (1974).

prolonged incubation at nonpermissive temperatures. The immediate burst of the formation of cell cycle-dependent enzymes after temperature shift from the nonpermissive to the permissive temperatures also has been verified for the L-α-glycerophosphate transport system (M. Ohki, unpublished observations).

Genetic analysis of the mutant tsC42 (Sato et al., 1979) has shown that the mutation responsible for the temperature sensitivity of growth is located at about 21 min on the chromosomal map and is cotransduced with *serC* and *pyrD* at frequencies of 2.6 and 48.7%. The proposed gene order is *serC*-*pyrD*,-tsC42-*pyrC*. Biochemical analysis of temperature-independent revertants, as well as transductants, has indicated that all the mutant phenotypes arose from a single mutational defect. Furthermore, merodiploids obtained by introducting F′ factor carrying the ts^+ allele into ts^- recipient cells behaved like wild-type cells, indicating that the tsC42 mutation is recessive to the wild-type allele.

It has been shown by SDS–acrylamide gel electrophoretic analysis of membrane fractions of the mutant tsC42 that at a nonpermissive temperature, the rate of incorporation of a radioactive amino acid into cytoplasmic membrane proteins was reduced by more than 70% and that into outer membrane proteins was reduced by 50%, whereas the incorporation into soluble proteins was not altered to such a great extent (Yamato et al., 1979). Pulse-chase experiments suggest that the defect lies in the process of protein synthesis, since prelabeled proteins were not inserted into the envelope to a significant extent during the chase period (Yamato et al., 1979).

The reason that the mutant stops growing at about 20 min before cell division is not clear. Possibly the cell cycle control also regulates the synthesis of division proteins. Furthermore, the pattern of DNA synthesis after temperature shift to nonpermissive temperature suggests that this control might also be involved in the synthesis of proteins controlling DNA synthesis (Ohki, unpublished work). The identification of the proteins under the control of the tsC42 gene, therefore, might provide us with important pieces of information for these yet undeveloped fields of study. Apart from these speculations, such a wide-ranging pleiotropy as observed in the synthesis of the membrane proteins has never been described before. The molecular nature of the tsC42 gene product is still unknown, but it is safe to predict that further studies on this and related mutants will shed new light on the field of the regulatory mechanism of protein biosynthesis.

A substantial amount of evidence supports the idea that the mutational defect of tsC42 specifically lies in the cell cycle-dependent control mechanism that operates at 10–20 min before cell division. Since the synthesis (and insertion) of the membrane proteins is inhibited by 50–70% in this mutant at 42°C, one might expect that in a wild-type culture, a burst of synthesis of the majority of membrane proteins occurs 10–20 min before cell division. However, the analysis of membrane proteins in synchronously growing N167 has revealed that such a striking change in the synthesis of membrane protein did not occur (M. Ohki and I. Yamato, unpublished work). The reason for this discrepancy is not clear. It is possible, however, that in tsC42 at a nonpermissive temperature, the blocked synthesis of "cell cycle-dependent" proteins causes secondary inhibitions of the synthesis or insertion of other membrane proteins.

7 MODELS OF CELL-CYCLE CONTROL

7.1 Cell-Cycle Control A

Before discussing the possible mechanisms that underlie the periodic synthesis of a variety of membrane and soluble proteins during a limited period in the cell cycle, it is best to summarize the results that are presented in the preceding sections, with emphasis on the following questions: (1) Is there a common regulatory mechanism for the synthesis or the duplication of the periodic enzymes? (2) At which level does the regulation work?

The maltose system is best suited for the discussion of these problems, since its genetics has been extensively studied and since we could get the most convincing results for most of the enzymes of the system throughout the cell cycle. It has been shown that the expression of three maltose operons (*malPQ* operon located at 74 min, and *malEF* operon and *malK-lamB* operon both located at 90 min) is regulated by the product of the *malT* gene located adjacent to the *malP* gene (Schwartz, 1967a, b; Hofnung et al., 1971). In addition, the expression of these operons is controlled by cyclic AMP and cyclic AMP receptor protein (CAP) (Pearson, 1972; Yokota and Kasuga, 1972). As is discussed in preceding sections, the products of the *malK* and the *lamB* genes are synthesized simultaneously with amylomaltase and maltodextrin phosphorylase, which are soluble enzymes and are coded for by the *malQ* and the *malP* genes. In light of the current understanding that the syntheses of inducible enzymes are tightly coupled with the transcription of the genes, one can reasonably assume that the periodicity of the synthesis of these soluble enzymes, and most probably, that of the simultaneously synthesized membrane proteins, reflects the cell cycle-dependent expression of the genes. The identity of the timing of the synthesis of these proteins with that of another cell cycle-dependent membrane protein is shown in Figure 5.

All the periodic enzymes synthesized or duplicated at a time 10–20 min before cell division are given in Table 1. An important question that can be asked about the periodic synthesis of these enzymes is whether or not their synthesis occurs all at the same stage of the cell cycle. The precise comparison of timing has been achieved by the simultaneous measurement of the known periodically synthesized enzyme and a candidate of interest. Such studies showed that the timing of synthesis or the duplication is nearly identical for all the proteins given in Table 1. It should also be noted that these studies revealed no enzyme that doubles during another stage in the cell cycle, except for the second group of proteins under the control of cell-cycle control B, which is discussed later.

It may be argued that the synthesis of these proteins may occur at the time of replication of their structural genes. This possibility can be ruled out, however, since the structural genes of some of these proteins are widely distributed along the chromosome.

The model assuming the existence of a common cell cycle-dependent control for the synthesis of these proteins is also in agreement with the data obtained from the studies of the mutant tsC42. We therefore propose the

Table 1 Proteins Synthesized Under Cell-Cycle-Control A

Enzyme or protein	Cellular Localization	Reference
Cytochrome b_1	Inner membrane	Ohki (1972)
L-α-Glycerophosphate		
Transport	Inner membrane	Ohki (1972)
Dehydrogenase	Inner membrane	Ohki and Mitsui (1974)
D-Lactate dehydrogenase	Inner membrane	Ohki, Yamato, and Futai (unpublished work)
Succinate dehydrogenase	Inner membrane	Ohki and Mitsui (1974); Ohki, Yamato and Futai (unpublished work)
Lactose permease	Inner membrane	Ohki and Sato (1975); Ohki (unpublished work) Ryter et al. (1975)
λ Receptor protein	Outer membrane	Ohki and Nikaido (unpublished work)
43K malK protein	Inner membrane	Ohki and Nikaido (unpublished work)
β-Galactosidase	Cytoplasm	Ohki (unpublished work)
Maltodextrin phosphorylase	Cytoplasm	Ohki and Nikaido (unpublished work)
Amylomaltase	Cytoplasm	Ohki and Nikaido (unpublished work)

existence of this common regulatory mechanism and call it "cell-cycle control A."

Most of the periodic behavior of the enzymes has been demonstrated by the duplication of their activity. In the early stage of study, one could not distinguish whether the duplication of the activity came from the net synthesis of protein or not. There is currently a small, but growing, amount of information showing that the duplication is caused by the net synthesis of the proteins. First, the synthesis of specific, immunoprecipitable material was shown to be coupled with the duplication of the activity of D-lactate dehydrogenase during cell cycle (M. Ohki, I. Yamato, and M. Futai, unpublished work). Second, the cells of tsC42 did not accumulate any pool of the cell cycle-dependent membrane proteins during the incubation at the nonpermissive temperature, and the immediate synthesis that was observed after returning cells to the permissive temperature seemed to come from newly synthesized proteins (Yamato et al., 1979). These observations, together with the results on the maltose system, lead us to the assumption that the cell-cycle control A regulates not posttranslational process, such as the conversion from inactive to active form and the integration into the membrane matrix, but the net synthesis of the proteins. It should be emphasized, however, that the possibility of regulation at the stage of post-translational modification cannot be ruled out completely, as proteins that fail to be modified could become degraded very rapidly.

There are, furthermore, several lines of evidence suggesting that this control operates at the level of the gene expression. The most convincing evidence is that the cell-cycle control mutant tsC42 fails to synthesize the *lac* mRNA at a nonpermissive temperature, while the total RNA and the *trp* mRNA synthesis continue at almost normal rates, in parallel with the increase of the total protein (see Section 4) (Ohki and Sato, 1975). Other support comes from the results on the maltose system discussed above.

During the last few years, there has been rapid progress in the field of interaction between RNA polymerase (and cyclic AMP receptor protein) and DNA. In the case of the *lac* operon, the entire sequence of DNA of the promoter region has been determined (Gilbert, 1976). Superficially it seems unlikely that another effector is involved in the process of the initiation of mRNA synthesis. Nevertheless, studies on the cell-cycle control have suggested the existence of an unknown element for the expression of the *lac* operon. In fact, additional elements for transcriptional control do not seem impossible at all, in view of the complexity of RNA polymerase–promotor interaction (Chamberlin, 1976).

It is known that the cell becomes committed to divide even in the absence of further RNA, protein, and DNA synthesis at approximately 20 min before cell division (Helmstetter and Pierucci, 1968; Clark, 1968). This transition point seems to coincide well with the time of action of the cell-cycle control A and, furthermore, has been shown to coincide with several important cell cycle-specific events, such as the termination of the oldest round of chromosome replication (Cooper and Helmstetter, 1968), segregation of nuclear bodies (Woldringh, 1976), completion of the period of protein synthesis required for division (Pierucci and Helmstetter, 1969), initiation of termination protein synthesis (Jones and Donachie, 1973), onset of sensitivity to penicillin (Hoffman et al., 1972), and attainment of a particular length of cell size (Donachie et al., 1976). In this chapter cell-cycle control A is discussed in relation to the manner in which envelope proteins are synthesized. However, as is briefly discussed in Section 6, the control mechanism A might also affect the synthesis of cell-division protein(s) and of DNA replication protein(s). Therefore, it is possible that many, if not all, of these cell cycle events are directly or indirectly coupled with cell-cycle control A.

7.2 Cell-Cycle Control B

In Section 5, the maltose-binding protein and the galactose-binding protein are described as being synthesized at a time different from the time of synthesis of proteins under the cell-cycle control A. This cell division-associated synthesis has also been suggested for several other periplasmic proteins in a study of a cell division mutant (Shen and Boos, 1973). An identical conclusion has been drawn from the analysis of the periplasmic protein synthesis during synchronous growth by two-dimensional SDS–acrylamide gel electrophoresis (M. Ohki, unpublished work). Therefore, it seems certain that this regulatory mechanism, called "cell-cycle control B" hereafter, regulates the synthesis of at least several periplasmic proteins.

M. Ohki and H. Nikaido (unpublished work) showed that there exists a difference of 11 min between the times of action of cell-cycle controls A and B in synchronous cultures. An almost identical time difference was observed in randomly growing cultures between the lags before the induction of the λ receptor protein under cell-cycle control A and before that of the maltose binding protein under cell-cycle control B. One cannot expect such a time difference *a priori* in an asynchronous culture, irrespective of the periods in the cell cycle where these proteins are synthesized. Thus the most likely explanation is that a signal turned on earlier is a prerequisite for the synthesis of the maltose-binding protein. A possible linker might be one of the proteins coded for by the *malPQ* operon or the *malK–lamB* operon. Other equally possible candidates are the functioning of cell-cycle control A or a particular protein synthesized under cell-cycle control A.

Shen and Boos (1973) have reported that, when a culture of the cell division mutant BUG-6 that was exposed to a nonpermissive temperature was again shifted to a permissive temperature, cell division was dissociated from the synthesis of the binding protein; cells started immediately to divide, while the synthesis of the glactose-binding protein was delayed for a considerable period. This uncoupling led them to propose that the temperature-sensitive factor in the cell division mutant BUG-6 is not the immediate regulator for the synthesis of galactose-binding protein, but it must regulate cell cycle-specific events which in turn then regulate the synthesis of the binding protein. It seems possible that the mutant BUG-6 is arrested at a stage of cell cycle between cell-cycle controls A and B under the nonpermissive growth conditions. In the light of the finding that cell-cycle control B is not independent of control A, the uncoupling can be explained by assuming that the signal of the cell-cycle control A triggered earlier becomes exhausted during the long period of incubation under the nonpermissive conditions. However, other possibilities cannot be exluded at present.

ACKNOWLEDGMENTS

Some of my recent studies were performed in Dr. H. Nikaido's laboratory at the University of California, Berkeley, and were supported by a U.S.

Public Health Service grant AI-09644 and a grant from the American Cancer Society (BC-20). I thank Dr. Nikaido for his help in the preparation of this manuscript.

REFERENCES

Boyd, A., and Holland, I. B. (1977), *FEBS Lett.*, **76,** 20–42.

Braun, V., and Krieger-Brauer, H. (1977), *Biochim. Biophys. Acta*, **469,** 89–98.

Chamberlin, M. J. (1976), in *RNA Polymerase* (R. Losick and M. J. Chamberlin, eds.), Cold Spring Harbor Laboratory, New York, pp. 159–191.

Churchward, G. G., and Holland, I. B. (1976), *J. Mol. Biol.*, **105,** 245–261.

Clark, D. J. (1968), *J. Bacteriol.*, **96,** 1214–1224.

Cooper, S., and Helmstetter, C. E. (1968), *J. Mol. Biol.*, **31,** 519–540.

Cozzarelli, N. R., Freedberg, W. B., and Lin, E. C. C. (1968), *J. Mol. Biol.*, **31,** 371–387.

Dietzel, I., Kolb, V., and Boos, W., (1978), *Arch. Microbiol.*, **118,** 207–218.

Donachie, W. D., Begg, K. M., and Vicente, M. (1976), *Nature*, **264,** 328–333.

Fox, C. F., Carter, J. R., and Kennedy, E. P. (1967), *Proc. Natl. Acad. Sci. U.S.*, **7,** 698–705.

Futai, M., and Kimura, H. (1977), *J. Biol. Chem.*, **252,** 5820–5827.

Gilbert, W. (1976), *RNA Polymerase* (R. Losick, and M. J. Chamberlin, eds.) Cold Spring Harbor Laboratory, New York pp. 193–205.

Gudas, L. J., James, R., and Pardee, A. B. (1976), *J. Biol. Chem.*, **251,** 3470–3479.

Hantke, K., and Braun, V. (1975), *FEBS Lett.*, **59,** 277–281.

Hatfield, D., Hofnung, M., and Schwartz, M. (1969), *J. Bacteriol.*, **98,** 559–567.

Hazelbauer, G. L. (1975), *J. Bacteriol.*, **122,** 206–214.

Helmstetter, C. E., and Pierucci, O. (1968), *J. Bacteriol.*, **95,** 1627–1633.

Hoffman, B., Messer, W., and Schwartz, U. (1972), *J. Supramol. Struct.*, **1,** 29–37.

Hofnung, M. (1974), *Genetics*, **76,** 169–184.

Hofnung, M. Schwartz, M., and Hatfield, D. (1971), *J. Mol. Biol.*, **61,** 681–694.

Imamoto, F. (1973), *J. Mol. Biol.*, **74,** 113–136.

Imamoto, F., and Schlessinger, D. (1974), *Molec. Gen. Genetics*, **135,** 29–38.

James, R., and Gudas, L. G. (1976), *J. Bacteriol.*, **125,** 374–375.

Jones, N. C., and Donachie, W. D. (1973), *Nature, New Biol.*, **243,** 100–103.

Neu, H. C., and Heppel, L. A. (1965), *J. Biol. Chem.*, **240,** 3685–3692.

Ohki, M. (1972), *J. Mol. Biol.*, **58,** 249–264.

Ohki, M. and Mitsui, H. (1974), *Nature*, **252,** 64–66.

Ohki, M., and Sato, S. (1975), *Nature*, **253,** 654–656.

Oxender, D. L., and Quay, S. C. (1976), *Method Membrane Biol.*, **6,** 183–242.

Pearson, M. L. (1972), *Virology*, **49,** 605–609.

Pierucci, O., and Helmstetter, C. E. (1969), *Fed. Proc.*, **28,** 1755–1760.

Powell, E. O. (1956), *J. Gen. Microbiol.*, **15,** 492–511.

Ryter, A., Shuman, H., and Schwartz, M. (1975), *J. Bacteriol.*, **122,** 295–301.

Sato, T., Ohki, M., Yura, T., and Ito, K., (1979), *J. Bacteriol.*, **138**, 305–312.

Schwartz, M. (1967a), *Ann. Inst. Pasteur* (*Paris*), **112,** 673–700.

Schwartz, M. (1967b), *Ann. Inst. Pasteur* (*Paris*), **113,** 685–704.

Shen, B. H. P., and Boos, W. (1973), *Proc. Natl. Acad. Sci. U.S.*, **70,** 1481–1485.

Singer, T. P., and Cremona, T. (1966), *Methods Enzymol.*, **9,** 302–314.

Szmelcman, S., and Hofnung, M. (1975), *J. Bacteriol.*, **124,** 112–118.

Wilson, D. B., and Smith, J. B. (1978), in *Bacterial Transport* (B. Rosen, ed.), Dekker, New York, Chapter 10.

Woldringh, C. L. (1976), *J. Bacteriol.*, **125,** 248–257.

Yamato, I., Anraku, Y., and Ohki, M., (1979), *J. Biol. Chem.*, in press.

Yokota, T., and Kasuga, T. (1972), *J. Bacteriol.*, **109,** 1304–1306.

Part 2
FUNCTIONS

Chapter 10 Specific Transport Systems and Receptors for Colicins and Phages

JORDAN KONISKY

Department of Microbiology, University of Illinois, Urbana, Illinois

1 INTRODUCTION

Since virulent phages and colicin-producing organisms are abundant in the natural habitat of enteric organisms, the question arises as to what selective pressures are involved in the maintenance of active colicin and virus receptors on the surface of sensitive organisms. It would seem that there has been ample opportunity for the selection of organisms that are insensitive to these lethal agents. Although in the past one could resolve this paradox by using the glib argument that those outer membrane components that serve as such receptors most certainly serve in some important physiological capacity, it is only within the past few years that experimental evidence supporting such a view has been accumulated. The elucidation of outer membrane-mediated transport systems marked an important development in our understanding of how gram-negative bacteria are able to take up nutrient molecules that are too large to permeate the outer membrane sieve barrier of these organisms. A major portion of this chapter is devoted to a description of these systems. Readers are referred to two recent reviews on this subject (Braun and Hantke, 1977; Kadner and Bassford, 1978).

I have limited the description of phage receptors to those that are protein components of the outer membrane. Furthermore, my treatment includes neither a discussion of the detailed mechanism of phage adsorption, nor a summary of what is known about the steps involved in the release of phage nucleic acid from the viral capsid and subsequent transfer of the viral genome across the cell envelope. These topics are discussed in a recent review by M. Schwartz (in press) (see also Braun and Hantke, 1977). Readers seeking information on the properties of other cell surface components serving as phage receptors should consult the review of Lindberg (1973).

2 OUTER MEMBRANE RECEPTORS AS COMPONENTS OF IRON TRANSPORT SYSTEMS

2.1 General Aspects of Iron Assimilation

Because of the low solubility of ferric iron [K_{sp} of $Fe(OH)_3$ at 25°C ≅ $10^{-38.7}$ M], bacteria growing under normal conditions find very little iron in a form that can be easily assimilated. Since bacteria require this trace metal for various metabolic processes, it has been necessary for them to evolve the means both for retrieving iron from insoluble polynuclear complexes and for the delivery of iron ions to the cell, where it can be processed for use in cellular metabolism. In general, the strategy employs the use of low molecular weight iron chelators, termed siderophores (or siderochromes), which are taken up by cells in complex with Fe(III) ion.

Escherichia coli and *Salmonella typhimurium* produce the identical phenolate-type siderophore called enterochelin (O'Brien and Gibson, 1970) and enterobactin (Pollack and Neilands, 1970), respectively. Enterochelin is the trivial term used throughout this chapter. Although enterochelin-mediated iron uptake is the primary system used for iron accumulation by these organisms growing in media containing low levels of iron, both species have the remarkable capability for obtaining iron in the form of a variety of exogeneously supplied hydroxamate-type siderophores produced by other bacteria and fungi. *Aerobacter aerogenes* produces both phenolate-type (enterochelin) and hydroxamate-type (aerobactin) siderophores. Some strains of *E. coli* and *A. aerogenes*, but not *S. typhimurium*, have an iron transport system that utilizes citrate as in iron carrier. Finally, all three species exhibit a low affinity, chelator-independent iron utilization system.

Since my discussion is presented only within the context of the role of the outer membrane in iron uptake, readers seeking a more detailed treatment of siderophore structure and properties may want to consult several excellent reviews (Neilands, 1973; Rodgers and Neilands, 1973). In addition, several reviews deal with the topic of iron transport in bacteria (Lankford, 1973; Rosenberg and Young, 1974; Byers and Arceneaux, 1977).

2.2 The Enterochelin System

Pollack and Neilands (1970) found that *S. typhimurium* growing in a low-iron medium excretes a low molecular weight iron sequestering agent that was given the trivial name enterobactin. A similar compound, given the name enterochelin, was isolated from the culture supernatant of *E. coli* grown under conditions of iron deficiency (O'Brien and Gibson, 1970). Both groups identified this iron ligand as the cyclic triester of 2,3-dihydroxy-*N*-benzoyl-L-serine. The complex contains one atom of ferric iron per enterochelin molecule. The Fe(III) ion is held quite strongly by oxygen atoms that serve as the electronegative donor groups. Because of the high affinity of enterochelin for Fe(III) [the ferric enterochelin formation constant = 10^{52} (J. V. McCandle, A. Avdeef, H. S. Cooper, and K. N. Raymond, cited by Hollifield and Neilands, 1978)] enterochelin functions as an efficient iron scavenger at physiological pH.

Subsequent to the excretion of enterochelin into the medium and chelation of media iron, the ferric–enterochelin complex is taken into cells where the enterochelin ligand is rapidly hydrolyzed to 2,3-dihydroxy-*N*-benzoyl-L-serine prior to the release of iron. Support for this scheme derives from the isolation of mutants that are unable to utilize ferric–enterochelin as a source of iron even though they have normal ability to synthesize enterochelin. Cox et al. were the first to describe a mutant (termed *fep*) that is defective in ferric–enterochelin uptake (Cox et al., 1970; Langman et al., 1972). As is seen later the *fepA* gene plays a critical role in the translocation of the ferric–enterochelin complex across the outer membrane. Langman et al. (1972) have studied mutants lacking ferric–enterochelin esterase activity. Such *fesB* cells are able to take up and, in fact, accumulate ferric–enterochelin but are unable to hydrolyze the ligand; *fep* and *fes* mutants are able to obtain iron for growth by any of the alternative iron uptake systems described in this section. Several other mutants harboring defects in ferric–enterochelin uptake have been partially characterized but as yet remain poorly understood (see Pugsley and Reeves, 1976a, b).

Guterman was the first to demonstrate a relationship between enterochelin and colicin B. Prompted by an earlier discovery that certain colicin B-resistant mutants of *E. coli* excrete an inhibitor of colicin B that was identified as enterochelin (Guterman and Luria, 1969; Guterman, 1971), she was able to determine that enterochelin protects cells from the colicin by acting as an inhibitor of colicin adsorption (Guterman, 1973). During the last several years, a plethora of literature bearing on the relationship of this siderophore to colicin B has accumulated. It is now virtually established that the outer membrane receptor of colicin B is utilized for ferric–enterochelin uptake.

In their studies on siderophore-mediated protection of *E. coli* against several B-type colicins, Wayne et al. (1976) demonstrated that, whereas enterochelin afforded protection against colicins B, V, and Ia when a *fes*$^+$ indicator strain was used, a *fes* mutant strain was protected only against colicin B. These results supported their previous suggestion (Wayne and Neilands, 1975) that colicin B and enterochelin compete for an outer membrane site that functions as a component in the ferric–enterochelin uptake system. They were further able to show that the hydroxamate-type siderophore, ferrichrome, protected cells from all three colicins only in strains able to utilize siderophore iron. Based on these findings, these workers suggested that enterochelin-mediated protection against colicins Ia and V and ferrichrome-mediated protection against colicins Ia, V, and B were, in contrast to enterochelin protection against colicin B, nonspecific and somehow related to iron utilization. Using methods that allowed a determination of both soluble colicin B receptor and ferric–enterochelin binding activity, Hollifield and Neilands (1978) have provided evidence that solubilized outer membrane extracts prepared from colicin B-sensitive cells contain a component for which colicin B and ferric–enterochelin compete for binding. Extracts prepared from a colicin B-resistant strain that manifests neither *in vivo* colicin B binding nor ferric–enterochelin uptake lack this component.

The identification of the ferric–enterochelin receptor has resulted from several lines of investigation. Shifting either *E. coli* or *S. typhimurium* from growth in a high-iron medium to growth in a low-iron medium, a condition that results in a derepression of the enterochelin uptake system (Rosenberg and Young, 1974), leads to a selective increase in the abundance of several outer membrane proteins (see Section 2.7). Such a result constitutes circumstantial evidence that these polypeptides are involved in iron utilization. More direct evidence derives from results correlating the loss of a specific polypeptide with alterations in specific iron transport systems. The requisite genetic manipulation required for such studies is made possible by the fact that siderophores and some colicins share outer membrane receptors, allowing direct selection of mutants defective in the relevant outer membrane polypeptide. Using colicin B as a selecting agent, Hantke and Braun (1975b) isolated a mutant (later given the mnenomic, *feuB*) that is specifically defective in ferric–enterochelin uptake. The mutant does not adsorb colicin B and lacks one of the outer membrane polypeptides (apparent molecular weight 81,000 whose abundance depends on the concentration of iron in the growth medium (Hancock et al., 1976; Ichihara and Mitzushima, 1978). Pugsley and Reeves (1976a) reported the isolation of a colicin B-resistant mutant (mnenomic, *cbr*) with similar properties (Pugsley and Reeves, 1977b).

Whereas the *feuB* mutation was reported to map at approximately 72.5 min on the *E. coli* genetic map (Hancock et al., 1976), the *cbr* locus maps at approximately 13 min (Pugsley and Reeves, 1977b; Pugsley, 1977). New mapping data suggest that *feuB* is identical to *cbr* (Hancock et al., 1977). These genes are probably also identical to the *fepA* gene described by Wookey and Rosenberg (1978). As is mentioned earlier, Hollifield and Neilands (1978) were unable to detect ferric–enterochelin binding activity in membrane extracts prepared from a strain lacking colicin B receptor activity. Examination of the outer membrane polypeptide profile of this mutant growing in a low-iron medium showed it to be lacking a single polypeptide. This polypeptide represents one of the two outer membrane proteins in the parental strain whose abundance was shown to be subject to control by media iron.

In summary, the above results leave little doubt that the *fepA* (*feuB*, *cbr*) gene controls the production of an outer membrane protein that serves as the receptor for both colicin B and ferric–enterochelin.

2.3 The Ferrichrome System

Based on their observation that mutants of *S. typhimurium* blocked in enterobactin (enterochelin) biosynthesis can obtain iron from the hydroxamate-type siderophore ferrichrome, Pollack et al. (1970) suggested that this organism has a transport system for utilization of this iron ligand. Subsequently, direct uptake of [^{3}H]ferrichrome could be demonstrated in strains able to utilize ferrichrome as an iron source (*sid*$^+$ strains), but not in two mutants (*sidA*$^-$, *sidC*$^-$), whose growth rate did not respond to ferrichrome addition (Luckey et al., 1972). Such mutants were generated by selection for resistance to the antibiotic albomycin δ_2, which is a structural analogue of ferrichrome.

Based on the resistance patterns of mutants selected for resistance to phages T1 or T5 or colicin M, it was proposed that these agents share a common receptor, which is encoded by the *tonA* gene (locus 3 min) (Fredericq, 1951; Fredericq and Smarda, 1970). It was later shown that phage ϕ80 utilizes this same receptor (Davies and Reeves, 1975; Hancock and Braun, 1976a). Prompted by the observation that *E. coli tonA* mutants (phages T1, T5, ϕ80, colicin M resistant) are albomycin δ_2 resistant and that the *sid* gene maps at a location (close to *panC*) on the *S. typhimurium* genetic map that corresponds to the position of the *tonA* gene on the *E. coli* map, Wayne and Neilands (1974, 1975) considered the possibility that these agents might utilize a receptor that physiologically functions in *E. coli* ferrichrome uptake. This notion was supported by experiments showing that ferrichrome protects cells from ϕ80 by preventing phage adsorption. Protection was not afforded by several other iron ligands tested, including EDTA, citrate, ferrichrome A, enterochelin, ferroxamine B, and schizokinen. In a subsequent paper, Luckey et al. (1975) demonstrated direct competition between ferrichome and phage T5 for binding to partially purified T5 outer membrane receptor. Ferrichrome also protects cells from colicin M action (Hantke and Braun, 1975a; Wayne et al., 1976). Protection against the colicin was observed in a mutant able to bind ferrichrome but defective in its utilization, providing evidence that the colicin and ferrichrome bind to a common membrane site (Wayne et al., 1976). Using the NaOH extraction method of Weidel et al. (1954), Braun et al. (1973) solubilized the T5 receptor from whole cells of *E. coli* and subsequently purified it to apparent homogeneity. The isolated protein inactivates both phage T5 and colicin M, and evidence was presented that both agents compete for a common site of interaction, confirming that these two agents utilize the same receptor. It was subsequently shown that this same protein could reversibly interact with bacteriophages T1 and ϕ80 (Hancock and Braun, 1976a). When Braun and Wolff (1973) examined the ability of inner and outer membrane preparations to neutralize bacteriophage T5, receptor activity was found to be greatly enriched in the outer membrane fraction. Furthermore, the 78,000 dalton polypeptide identified above as the T5/colicin M receptor is often absent from the outer membrane SDS gel profiles of *tonA* mutants (Braun and Hantke, 1977). Such results establish that the phage/colicin M receptor determined by the *tonA* gene is a component of the outer membrane.

Further evidence that the *tonA* gene of *E. coli* specifies a product involved in ferrichrome uptake derives from the experiments of Hantke and Braun (1975a), who showed that *tonA* mutants are defective in the uptake of ferrichrome iron. They showed, in addition, that the outer membrane of TonA mutants lacks a single polypeptide (apparent molecular weight 78,000) that is present in full revertants of these *tonA* mutants, as well as in *tonA*$^+$ parental strains (Braun et al., 1976). In sum, it is clear that the *tonA* gene of *E. coli* specifies a protein that functions physiologically in ferrichrome uptake and that also serves as the receptor for phages ϕ80, T1, T5, and colicin M. The role of this protein in ferrichrome uptake is discussed in a later section.

With regard to *S. typhimurium*, Luckey and Neilands (1976) have shown that ferrichrome prevents adsorption of phages ES18 and ES18.h1. Furthermore, those albomycin-generated *sid* mutants that can be cotransduced with the *panC* gene are defective in both phage adsorption and ferrichrome uptake. These results led Luckey and Neilands to suggest that the *sid* gene specifies an envelope site that normally serves in ferrichrome binding and in addition serves as the receptor for these phages. Braun et al. (1977) have shown that certain albomycin-and phage ES18-resistant mutants of *S. typhimurium* are missing an outer membrane protein of the same size as that specified by the E. coli *tonA* gene. Mapping data led these workers to conclude that the missing polypeptide was specified by a gene (given the general mnemonic *sid*) located on the *S. typhimurium* genetic map in a region analogous to the *tonA* region of *E. coli*. Unfortunately, the relationship of the above-mentioned *sid* gene to the known *sid* genes described by Luckey et al. (1972) is not known.

Although the *sid* and *tonA* genes determine proteins that mediate ferrichrome and albomycin uptake and, in addition, serve as phage receptors, they are not functionally equivalent. This is evident from the finding that *E. coli* does not adsorb phage ES18, nor does *S. typhimurium* adsorb phages T1 or T5 or colicin M (Graham and Stocker, 1977). That the lack of such cross-reactivity reflects an actual difference in the receptors, and not some species difference that indirectly might affect phage adsorption, is supported by the finding that transfer of the *E. coli tonA* gene into *S. typhimurium* renders the latter sensitive to colicin M and phage T5 (Graham and Stocker, 1977).

2.4 The Citrate System

Based on the finding that citrate stimulates both growth and iron uptake in a *Fep* mutant of *E. coli*, Cox et al., (1970) concluded that this organism possesses a citrate-mediated iron transport system (see also Frost and

Rosenberg, 1973). Since citrate-dependent iron uptake requires prior growth of the strain in the presence of citrate, this system is inducible.

The citrate system is physiologically and genetically distinct from the other high affinity iron transport systems that operate in *E. coli*. (For an important exception see section on *tonB* gene function.) For example, mutants defective in the enterochelin system and in the ferrichrome system are able to obtain iron by way of the citrate system (Cox et al., 1970; Langman et al., 1972; Young et al., 1967; Hantke and Braun, 1975a). Furthermore, mutants (termed *fec*$^-$) carrying mutations affecting the citrate-dependent iron uptake system show normal capacity for enterochelin- and ferrichrome-mediated iron uptake (Woodrow et al., 1978). Finally, of six isolated *fec* mutants, all show a genetic lesion mapping at approximately 6 min on the genetic map, well separated from *tonA* (3 min) and the genes that determine the enterochelin system (13 min).

Hancock et al. (1976) have shown that the outer membrane of strains grown in the presence of 1 m*M* citrate contain a polypeptide (apparent mol. wt, 81,000) that is present to a much lesser extent in cells grown without citrate supplementation. Although it is likely that this protein functions in citrate-mediated iron uptake, its role remains to be elucidated.

2.5 Role of Outer Membrane Receptors

As is discussed in detail in Chapter 11, the diffusion of small molecules across the outer membrane is mediated by a specific class of outer membrane proteins, termed porins. Studies by Nakae and Nikaido (1975) established that whereas the outer membrane is freely permeable to sucrose [mol. wt. 342; Stoke's radius (R_s), 4.6 Å] and raffinose (mol. wt. 504; R_s, 5.7 Å), the membrane is only partially permeable to stachyose (mol. wt. 667; R_s 6.6 Å) and is impermeable to saccharides of molecular weight greater than 900. It is therefore likely that the outer membrane represents a substantial permeability barrier to ferric–enterochelin (801 daltons) and ferrichrome (740 daltons) and thus any system involved in their uptake and utilization must include a means for overcoming this obstacle. It is almost certain that the specific outer membrane receptors described above play a primary role in the translocation of each of these substances across the outer membrane. Support for this notion derives, in part, from the work of Wookey and Rosenberg (1978). These workers have shown that, in contrast to whole cells, uptake of ferrichrome or ferric–enterochelin by spheroplasts does not depend on functional *tonA* or *fepA* genes, respectively. Presumably, spheroplasting leads to disruption of the outer membrane permeability barrier to ferric–enterochelin and ferrichrome, circumventing the need for functional *fepA* and *tonA* gene-encoded siderophore receptors.

The demonstration by Negrin and Neilands (1978) that inner membrane vesicles prepared from an *E. coli* K-12 *tonA* mutant strain are capable of energy-dependent ferrichrome uptake provides further evidence that in whole cells the tonA protein functions to mediate access to the inner membrane.

The mechanism whereby the *tonA* and *fepA* gene products facilitate transport of these substances is unknown. It seems unlikely that they simply serve as aqueous channels of communication between the extracellular milieu and the periplasm or inner membrane. Such a model is not compatible with the high degree of specificity exhibited by these systems. Furthermore, as is discussed earlier, each of these ligands has been shown to bind to its specific receptor protein. A detailed speculative scheme of outer membrane-mediated substrate uptake is presented in a later section.

2.6 The Colicin I Receptor

Colicins Ia and Ib share a common receptor (Konisky and Cowell, 1972). That this protein is a component of the outer membrane was first shown by the studies of Konisky et al. (1973), who demonstrated adsorption of [^{125}I-]colicin Ia to the outer membrane fraction but not to the cytoplasmic membrane of sensitive cells. [^{125}I]Colicin Ia did not adsorb to the outer membrane fraction of a strain previously shown to be resistant to colicin Ia by loss of ability of whole cells to adsorb colicin. When the outer membrane polypeptides isolated from mutants specifically resistant to colicin Ia are analyzed by SDS–PAGE electrophoresis and the resulting profiles are compared to those found for polypeptides isolated from the outer membrane of sensitive cells, it is observed that the outer membrane of such mutants is deficient in a single polypeptide species with a molecular weight of 74,00 (Hancock and Braun, 1976b; Soucek and Konisky, 1977; Pugsley and Reeves, 1977b). The colicin Ia receptor has been solubilized with Triton X-100 plus EDTA (Konisky and Liu, 1974) and purified to homogeneity (Bowles et al., 1976). The receptor is a heat-modifiable protein whose apparent molecular weight as determined by SDS–PAGE is approximately 50,000 (nonheated) or 70,000 (heated in SDS). The purified protein retains full biological activity (neutralization of Ia) and forms a stable complex with pure colicin Ia. Although not unambiguously established, it is likely that the receptor is coded by the *cir* gene, which is located at approximately 41 min on the *E. coli* genetic linkage map (Cardelli and Konisky, 1974; Hancock et al., 1977).

Although it is not possible to ascribe a role for the colicin I receptors in any known *E. coli* iron uptake system, there is sufficient circumstantial evidence to suggest that its physiological function relates to some aspect of iron metabolism. Wayne et al. (1976) found that inclusion of various siderophores in the growth medium of an *E. coli* K-12 strain renders the strain insensitive to colicin Ia. Protection requires the availability or metabolism of siderophore iron. The basis for this finding was provided by the experiments of Konisky et al. (1976), who demonstrated an inverse relationship between the number of colicin I receptors per cell and the ability of *E. coli* K-12 to take up iron from the growth medium. For example, strains carrying mutations in the enterochelin system (*ent*$^-$, *fep*$^-$, or *fes*$^-$) bind a greater amount of [^{125}I]colicin Ia than the parental strain. Furthermore, these mutant strains were shown to be hypersensitive to colicin Ia. The number of receptors per mutant cell is greatly reduced, with a concomitant decrease in colicin sensitivity, by addition of ferrichrome, citrate, or 100 μM ferrous salts to the growth medium of each mutant or addition of enterochelin to the growth medium of the *ent*$^-$ (unable to synthesize enterochelin) strain. The ability of these agents to lower the level of colicin I receptors in these mutants, or, for that matter, in parental cells, requires that the appropriate iron uptake system be operative. For example, enterochelin does not affect receptor levels in a *fep*$^-$ or *fes*$^-$ strain, nor does growth in citrate reduce levels in a strain that is defective in the citrate-mediated iron uptake. As is discussed in a later section, proper activity of the ferrichrome, enterochelin, and citrate systems depends on a functional *tonB* gene; *tonB*$^-$ strains are thus in a condition of iron stress. Such strains exhibit enhanced [^{125}I]colicin Ia adsorption whether or not the growth medium contains enterochelin, ferrichrome, or citrate.

Although the above studies utilized binding of [^{125}I]colicin Ia for enumeration of receptors, the results are fully corroborated by several studies showing that outer membranes prepared from cells grown in low-iron media contain enhanced amounts of the polypeptide determined by the *cir* gene, that is, the colicin I receptor (Hancock and Braun, 1976b; Pugsley and Reeves, 1977b; Ichihara and Mizushima, 1978; Hollifield and Nielands, 1978).

In several respects the relationship between the level of colicin I receptors and the ability of cells to utilize medium iron is analogous to regulation of the enterochelin system (see Section 2.7). Therefore, it is reasonable to entertain the possibility that the colicin I receptor is involved in some aspect of iron uptake. Indeed, there has been a report that a mutant strain missing the colicin I receptor is defective in ferric–enterochelin uptake (Hancock and Braun, 1976b). However, many independent colicin receptor mutants have now been examined, and there is full agreement that mutants specifically lacking the colicin I receptor are normal in their capacity to transport ferric–enterochelin (Wayne and Neilands, 1976; Pugsley and Reeves, 1977b; Soucek and Konisky, 1977; Hancock et al., 1977; Ichihara and Mizushima,

1978). Furthermore, Wayne and Neilands (1976) have reported that the colicin I receptor is not involved in the utilization of several siderophores, including enterochelin, ferrichrome, rhodotorulic acid, schizokinen, ferrioxamine B, and citrate. At least for the present, the colicin I receptor remains a protein in search of a function. It seems likely that this function will involve a siderophore.

2.7 Regulation of Iron Uptake Systems

At least seven *E. coli* genes (*entA* to *entG*) are involved in the synthesis of enterochelin from chorismate and serine (Luke and Gibson, 1971; Young et al., 1971; Woodrow et al., 1975). These map in a cluster at approximately 13 min on the *E. coli* genetic map. Based on the properties of phage Mμ-induced polarity mutants in the *ent* gene cluster, Woodrow et al. (1975) have concluded that at least three *ent* genes (*entA*, *entB*, and *entG*) are part of the same operon. Their results do not rule out the possibility that the entire *ent* cluster might, in fact, comprise a single transcriptional unit. Although *ent*$^-$ strains grow very poorly in media containing less than 2 μM iron, addition of iron to 50–100 μM leads to normal growth. Clearly, enterochelin is required only in low-iron media. Growth of *ent* mutants in low-iron medium can be restored by addition of low levels of enterochelin or by utilization of the separate ferric citrate or ferrichrome iron uptake systems. Pollack et al. (1970) have described *S. typhimurium* mutants, termed *ent*$^-$, that are defective in enterobactin (enterochelin) biosynthesis. Genetic experiments suggest that the genes involved form a cluster that may comprise an operon.

Since enterochelin is not required for iron utilization in high-iron media, it is not surprising that the various genes determining the enterochelin system are subject to regulation by iron. Comparisons of enzyme levels in *E. coli* K-12 grown in normal- and low-iron media have shown that growth in low iron leads to higher levels of the enterochelin-biosynthetic enzymes (Brot and Goodwin, 1968; Young and Gibson, 1969; Bryce and Brot, 1971). Pollack et al. (1970) have reported results indicating that enterobactin (enterochelin) biosynthesis in *S. typhimurium* is regulated by iron. In the case of *E. coli* K-12, increased synthesis of an *ent* gene-determined enzyme, 2,3-dihydroxy-*N*-benzoyl-L-serine synthetase, does not occur until the intracellular iron concentration falls to about 2.2×10^{-18} mole/cell (Bryce and Brot, 1971). The *fepA* and *fesB* genes are closely linked to the ent gene cluster and evidence has been presented that *fep* and *fes* are in a common operon, distinct from the operon encoding the *ent* gene products (Woodrow et al., 1975). Expression of *fes* is regulated by the intracellular concentration of iron (Rosenberg and Young, 1974).

Analysis of the protein profiles of outer membranes prepared from *E. coli* strains fully competent in iron utilization shows that the abundance of several proteins depends on the concentration of media iron. Although some studies have shown that growth in low-iron media leads to the enhanced production of two proteins in *E. coli* B (McIntosh and Earhart, 1976) and *E. coli* K-12 (Uemura and Mizushima, 1975; Pugsley and Reeves, 1976c; Hollifield and Neilands, 1978), improved systems for protein resolution have allowed the detection of three new polypeptides in the outer membranes of *E. coli* K-12 grown in low-iron media (Braun et al., 1976; Pugsley and Reeves, 1977b; Ichihara and Mizushima, 1977; McIntosh and Earhart, 1977). By use of appropriate mutants, two of these could be identified as the polypeptides determined by the *cir* (colicin I receptor, 74,000 daltons) and *feuB/cbr/fepA* (colicin B receptor/ferric enterochelin receptor, 81,000 daltons) genes. The function of the third polypeptide (83,000 daltons) has not been clarified. Mutants that are defective in various aspects of enterochelin-mediated iron uptake produce high amounts of these same outer membrane polypeptides (Davies and Reeves, 1975; Pugsley and Reeves, 1976b; Braun et al., 1976; Konisky et al., 1976; McIntosh and Earhart, 1977).

The results shown in Figure 1 are from a study by Ichihara and Mizushima (1977) and serve to illustrate the regulatory system under discussion. In this experiment outer membranes were prepared from *E. coli* cells growing in media containing various amounts of added ferric chloride. The capacity of such membranes for enterochelin-mediated iron binding was determined, as well as the relative amounts of outer membrane proteins 0-2a (probably corresponding to the 83,000 dalton protein described above), 0-2b (corresponding to the *cbr/fepA/feuA* determined protein), and 0-3 (corresponding to the *cir* determined protein). The results clearly demonstrate a positive correlation between the levels of these proteins and

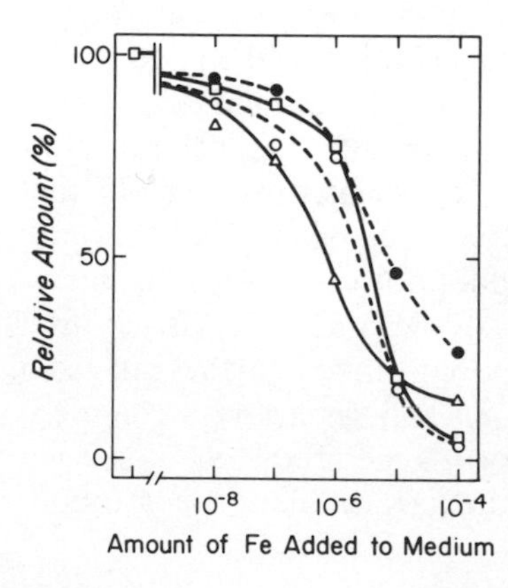

Figure 1 Effect of iron in the growth medium of *E. coli* YA 21 (□) on enterochelin-mediated iron binding to isolated outer membranes; (●) on the amount of outer membrane proteins 0-2a (83K) plus 0-2b (*cbr/fepA/feuA*) (○) on the amount of outer membrane protein 0-3 (cir) and (△) on the amount of 2,3-dihydroxybenzoic acid-containing compounds in the culture medium. From Ichihara and Mizushima (1977).

enterochelin-mediated iron binding activity. In a later paper, these same workers were able to show that protein 0-2b is responsible for such binding of iron and, furthermore, that the number of iron atoms bound was roughly equivalent to the number of 0-2b polypeptides (Ichihara and Mizushima, 1978). These results are in keeping with the previously discussed conclusion that this protein is involved in the binding of ferric–enterochelin to the outer membrane. McIntosh and Earhart (1977) have provided evidence that the synthesis of enterochelin and these outer membrane polypeptides are coordinately regulated by the concentration of intracellular iron. The regulatory level of cell-associated iron was shown to be 2.55×10^{-18} mole/cell, which is in good agreement with the value obtained by Bryce and Brot (1971) in their study on the regulation of enterochelin-biosynthetic enzymes.

As is mentioned in a previous section, growth in citrate leads to the production of an additional outer membrane polypeptide. Since this protein is not synthesized in response to iron stress per se, it must be regulated independently of the three outer membrane proteins discussed above. The elucidation of the role of this protein in citrate-mediated iron uptake, as well as of the regulatory mechanism of this inducible system, must await more extensive genetic analysis.

Reports bearing on the inducibility of the ferrichrome-mediated iron uptake system in *E. coli* are conflicting. Whereas Luckey et al. (1975) and Braun and Hantke (1977) have indicated that synthesis of the *tonA* gene product is not greatly influenced by the level of media iron, Pugsley and Reeves (1976c) demonstrated that growth of *E. coli* K-12 under iron stress results in increased production of the colicin M receptor, which is specified by the *tonA* gene. The basis for this discrepancy remains to be clarified.

In *S. typhimurium* the synthesis of three outer membrane polypeptides was shown to be regulated by the concentration of iron in the medium by Bennet and Rothfield (1976). These proteins, designated OM-1, OM-2, and OM-3, have apparent molecular weights of 82,000, 79,000, and 77,000, respectively. The isolation of a mutant in which all three polypeptides are synthesized constitutively suggests that the production of these polypeptides is coordinately regulated. Although the function of these proteins has not been established, it seems likely that they are involved in some aspect of iron utilization. In a subsequent paper, it was shown that such mutants manifest constitutive expression of (1) enterochelin synthesis, (2) enterochelin-mediated iron uptake, (3) intracellular degradation of ferric-enterochelin, and (4) ferrichrome uptake (Ernst et al., 1978). These results suggest the existence of a gene whose activity mediates the expression of both the ferrichrome- and enterochelin-mediated iron uptake systems in *S. typhimurium*. Ernst et al. considered the possibility that this gene might function in the intracellular processing of iron to a form used in the repression of the high affinity iron uptake systems. It was also considered possible that the gene is responsible for regulating the structural genes of the enterochelin and ferrichrome systems. Whatever the true explanation, an important conclusion is that the regulation of these two systems is coordinate and utilizes at least one common element. These results are also important in that they demonstrate that the uptake system for ferrichrome in *S. typhimurium* is regulated by the iron content of the medium.

The molecular mechanism whereby a condition of iron stress leads to enhanced production of the components of the enterochelin system has not yet been delineated. Several groups have considered the possibility of negative control models in which intracellular iron acts as a corepressor (Brot and Goodwin, 1968; Pollack et al., 1970; Bryce and Brot, 1971; Rosenberg and Young, 1974). However, a positive control model in which intracellular iron is able to influence the activity of an endogenous gene activator molecule is equally tenable. Whatever the mechanism, it is clear that a cytoplasmic regulatory component(s) must interact with structural genes that occupy widely scattered sites on the bacterial chromosome.

3 THE VITAMIN B_{12}/COLICIN E/PHAGE BF23 RECEPTOR

Although vitamin B_{12} is neither synthesized nor required in wild-type enteric bacteria, the auxotrophic requirement of certain mutants defective in methionine biosynthesis can be satisfied by addition of the vitamin to a methionine-free medium. Since the hydrodynamic radius of B_{12} (mol. wt. 1327) presumably precludes its passive diffusion through the aqueous pores of the outer membrane, it is not surprising that a specific transport system utilizing an outer membrane receptor has evolved to facilitate its uptake into cells.

Uptake of B_{12} by cells can occur against a concentration gradient (DiGirolamo and Bradbeer, 1971; Taylor et al., 1972), the energy for accumulation being derived from the proton motive force (Bradbeer and Woodrow, 1976). Kinetic studies have revealed that uptake occurs in two phases—the first being rapid and energy independent, and the second being slower and energy dependent (DiGirolamo and Bradbeer, 1971). Whereas the second phase is thought to reflect active transport of B_{12} across the cytoplasmic membrane, it is clear that the initial rapid phase of uptake reflects interaction of B_{12} with a specific outer membrane receptor protein.

In studies utilizing separated inner and outer membrane fractions, White et al. (1973) demonstrated that B_{12} binding capacity was confined to the outer membrane. The amount of B_{12} that bound to this fraction was close to that expected if all the binding observed in the initial phase of uptake in whole cells represents adsorption to the outer membrane. Since B_{12} does not bind to the outer membrane of mutant cells that lack the initial phase of B_{12} uptake, the interaction between B_{12} and isolated outer membrane represents binding to physiological receptors.

The observation that colicins E1 and K, but not E3, inhibit the energy-dependent phase of B_{12} uptake, whereas colicins E1 and E3, but not K, inhibit the initial energy-independent binding phase (DiMasi et al., 1973), led to experiments that convincingly demonstrated that the outer membrane receptor utilized for B_{12} uptake is identical to the colicin E/phage BF23 receptor. For example, when DiMasi et al. examined the sensitivity of B_{12} transport mutants to colicins E1 and E3, they found that whereas strains possessing functional outer membrane B_{12} receptors are colicin sensitive, strains lacking such receptors are insensitive. The cell envelope fraction isolated from these colicin-insensitive strains fails to neutralize colicins E1 or E3, indicating that they lack colicin receptors. [Sabet and Schnaitman (1971) had previously shown that the colicin E receptor is a component of the outer membrane.] It was also shown that B_{12} is able to protect cells from colicins E1 and E3. Kadner and Liggins (1973) have shown that mutants selected for colicin E1 resistance are defective in B_{12} binding and transport. Furthermore, the colicins and phage BF23 competitively inhibit B_{12} binding to whole cells, outer membranes, and solubilized receptor (DiMasi et al., 1973; Bradbeer et al., 1976). Genetic studies by Kadner and his coworkers have established that the *bfe* gene determining the colicin E/phage BF23 receptor (Buxton, 1971; Jasper et al., 1972) is identical to the *btuB* locus determining the vitamin B_{12} receptor (Kadner and Liggins, 1973; Bassford and Kadner, 1977). Thus there is overwhelming evidence that colicins E2 and E3, phage BF23, and vitamin B_{12} utilize the same outer membrane polypeptide. This protein is present in approximately 200–300 copies/cell and is under control (if not, indeed, encoded) by the *bfuB* gene located at 88 min on the *E. coli* genetic map. There is evidence that the E2/E3 receptor may be but one component of the E1 receptor (Hill and Holland, 1967; Sabet and Schnaitman, 1973; DiMasi et al., 1973).

Using colicin E3 neutralizing activity as an assay, Sabet and Schnaitman (1973) purified the colicin receptor to apparent homogeneity. Since the activity of this 60,000 dalton polypeptide was shown to be sensitive to periodate treatment, it is possible that the receptor is a glycoprotein. The purified protein is fully active in neutralizing colicin E3 but has negligible activity towards colicin E1, vitamin B_{12} and bacteriophage BF23. Substantial activity towards vitamin B_{12} can be restored by inclusion of lipopolysaccharide in the reaction mixture (C. Bradbeer, personal communication).

It seems likely that the vitamin B_{12} receptor plays a critical role in facilitating the translocation of the vitamin across the outer membrane barrier. This view is supported by the finding that mutants thought to be defective in the permeability barrier function of this membrane do not require the *btuB* gene product for vitamin B_{12} uptake (Bassford and Kadner, 1977). Other aspects of receptor-dependent B_{12} uptake and colicin and action are discussed in Section 5.

4 THE COLICIN K/PHAGE T6 RECEPTOR

Early genetic evidence led Fredericq (1949a, b) to conclude that colicin K and phage T6 utilize a common bacterial receptor. He showed that whereas strains sensitive to one of these agents are sensitive to the other, strains selected for resistance to the colicin are cross-resistant to T6, and vice versa. Weltzien and Jesaitis (1971) localized the colicin K receptor to a cell wall fraction prepared from cell spheroplasts. The fact that treatment of this fraction with trypsin destroys its ability to neutralize colicin K led these workers to suggest that the receptor contains protein. Sabet and Schnaitman (1971) showed that the colicin K receptor is a component of the outer membrane and, furthermore, were able to solubilize receptor activity from the cell envelope using Triton X-100 plus EDTA. Thc receptor protein, which is the product of the *E. coli tsx* gene (Manning and Reeves, 1976a), has recently been purified and partially characterized (Manning and Reeves, 1978). Although the receptor (apparent mole wt. 26,000) could be purified in a functional form to apparent protein homogeneity, the final preparation did contain LPS. Preparation of LPS-free protein by means of gel filtration chromatography in SDS led to loss of receptor activity. The LPS-containing receptor protein preparation neutralizes both colicin K- and T6-like bacteriophages.

Based on his finding that colicin K/phage T6-resistant mutants lacking the tsx protein are defective in the uptake of nucleosides, Hantke (1976) has suggested that the phage T6/colicin K receptor may serve as a specific gate for nucleoside uptake. However, further characterization of these mutants is necessary in order to establish the molecular basis for the observed decreased rate of thymidine, uridine, adenosine, and deoxyadenosine uptake.

The production of the T6/colicin K receptor is subject to catabolite repression and requircs the function of both *cya* (adenyl cyclase) and *crp* (cyclic AMP receptor protein) genes (Kumar, 1976; Manning and Reeves, 1978).

5 ROLE OF OUTER MEMBRANES RECEPTORS IN COLICIN ACTION

5.1 Colicin Mode of Action

The mechanisms of action whereby the various colicins kill sensitive *E. coli* involve a sequence of molecular interactions initiated by the adsorption of colicin molecules to specific outer membrane receptor proteins and terminated by the direct interaction of the colicin with its particular intracellular target. Although the intervening steps in the mode of action sequence have been the subject of much discussion and speculation, the molecular details are largely unknown. Although it is likely that the mode of action of each of the many described colicins involves interaction with outer membrane surface components, my treatment here is restricted to those colicins for which the existence of outer membrane receptors has been firmly established. Readers interested in a more comprehensive review of colicin action are referred to two recent reviews (Holland, 1975; Konisky, 1978).

Treatment of either whole cells or isolated ribosomes with colicin E3 leads to ribosome inactivation by means of an identical cleavage of ribosomal 16 S RNA (for a review of E3 action see, Nomura et al., 1974; Holland, 1976). The *in vivo* action of this colicin thus involves direct interaction of the E3 molecule with cellular ribosomes, and the mode of action of this colicin must involve the penetration of all or part of the molecule through the bacterial envelope. Similarly, colicin E2, which has been shown to be a DNA endonuclease (Saxe, 1975; Schaller and Nomura, 1976), must penetrate the cell envelope at least to the extent that the molecule is able to interact with the bacterial chromosome.

Colicins E1, K, and Ia disrupt energy metabolism in treated cells. This is accomplished by colicin-induced membrane depolarization (Gould and Cramer, 1977; Tokuda and Konisky, 1978a; Weiss and Luria, 1978). The molecular basis for colicin E1, K, and Ia action has recently been elucidated. Tokuda and Konisky (1979) have been able to demonstrate that liposome prepared from purified soybean phospholipids become leaky to ions after treatment with colicin Ia. Furthermore, Schein, Kagan, and Finkelstein (1978) have provided evidence that these colicins form ion "channels" in planar phospholipid bilayer membranes. These findings, together with experiments demonstrating that these colicins are active against receptor-free membrane vesicles prepared from bacterial cells (see below), lead to the conclusion that the mode of action of these colicins involves direct interaction of colicin molecules with the cytoplasmic membrane.

Colicin B may have a mode of action similar to that described for colicins

E1, K, and Ia. Guterman (1973) showed that the colicin inhibits DNA, RNA, and protein synthesis in treated cells, while M. Kissil and J. Konisky (unpublished data) have observed colicin B-induced efflux of accumulated thiomethyl-β-D-galactoside. Colicin D, which is structurally related to colicin B (Pugsley and Reeves, 1977a), and, indeed, adsorbs to the same outer membrane receptor protein, has been reported to specifically inhibit protein synthesis (Timmis and Hedges, 1972). It has been suggested that colicin M may interfere with cell wall biosynthesis (Braun et al., 1974); its primary cellular target is unknown.

5.2 Role of Outer Membrane Receptors

That colicin action requires the adsorption of colicin molecules to specific bacterial receptors was first proposed by Fredericq (1946). In studies utilizing radioactive colicins E2, E3, Ia, and Ib, it was directly shown that colicin resistance is often accompanied by loss of ability of whole cells to adsorb colicin (Maeda and Nomura, 1966; Konisky and Cowell, 1972). Furthermore, Konisky et al. (1976) have demonstrated a correlation between cell sensitivity and the number of colicin Ia cell surface receptors.

Although Maeda and Nomura (1966) provided evidence that the colicin E2/E3 receptor is a component of the cell surface layers, the lack of appropriate methods available at that time precluded attempts at more definitive localization. However, with the development of methods for separation of inner and outer membrane (Schnaitman, 1970; Osborn et al., 1972), it has become possible to demonstrate that outer membrane receptor proteins are involved in the action of colicins E1, E2, E3, K, Ia, Ib, B, D, and M.

To gain access to its ultimate cell target, a colicin molecule must overcome the physical barrier presented by the outer membrane. It is almost certain that colicin receptors are utilized for this purpose. The basis for this conclusion rests with observations that conditions leading to the direct access of a colicin to its target alleviate any need for outer membrane receptors.

As is mentioned above, isolated ribosomes and DNA are attacked *in vitro* by colicins E3 and E2, respectively. With regard to those colicins that induce membrane depolarization, there have been several reports of receptor-independent colicin action. Takagaki et al. (1973) found that colicin K inhibits the energy-dependent uptake of [^{32}P]orthophosphate into membrane vesicles prepared from both sensitive and resistant (unable to adsorb the colicin) cells. In this same study these workers reported the isolation of a mutant strain that exhibited receptor-independent sensitivity to colicin K. Starting from a parental strain lacking the colicin K receptor, T4-resistant derivatives were isolated that displayed altered lipopolysaccharide and were hypersensitive to novobiocin. These mutants were colicin K-sensitive, yet did not adsorb the colicin. It seems likely that selection for T4 resistance in this case led to the isolation of mutants in which the outer membrane no longer serves as a permeability barrier to the colicin. Bhattacharyya et al. (1970) showed that colicin E1 inhibits proline accumulation in membrane vesicles prepared from either sensitive or resistant cells. Although the capacity of the resistant strain to adsorb colicin E1 was not determined, it is likely that the strain used lacked the outer membrane receptor for these agents. Tokuda and Konisky (1978b) have been able to demonstrate that interaction of colicin Ia with membrane vesicles known to be lacking the colicin Ia outer membrane receptor leads to membrane depolarization. From such results, it seems appropriate that outer membrane receptors should be viewed as vehicles used by the colicins to overcome the outer membrane permeability barrier.

Colicin receptors also play an essential role in presenting the colicin molecule in the correct orientation to its final target or to the next component in the mode of action sequence. It is well established that adsorption of a particular colicin to its outer membrane receptor is not, in itself, sufficient to bring about the events leading the cell death. For example, selection of colicin-insensitive mutants leads to the isolation of two mutant classes—resistant and tolerant. Whereas resistant mutants are defective in their capacity to adsorb the particular colicin used for selection and lack the outer membrane receptor protein, tolerant mutants both retain the outer membrane receptor and adsorb the colicin but are blocked in some further step of colicin action. Many of these mutants show an alteration in cell envelope structure or function. Studies on the colicin sensitivity of spheroplasts and plasmolyzed cells indicate that colicin action requires a critical structural relationship between inner and outer membrane. For example, although spheroplasts are insensitive to colicins E2 and E3, they adsorb the colicins to the same extent as whose cells (Nomura and Maeda, 1965). Likewise, plasmolysis protects cells from colicins K, E2, and E3 without affecting the capacity of such cells to adsorb colicin (Beppu and Arima, 1967; Holland, 1975).

5.3 Action of Colicins E2 and E3

Maeda and Nomura (1966) showed that whereas the killing action of colicin E2 is single hit, a single killing unit may correspond to 40–300 molecules, depending on experimental conditions. As a possible explanation, they suggested the possibility that colicin E2 receptors are functionally heterogenous such that only a small proportion of adsorptions lead to cell death. Support for this notion comes from the work of Shannon and Hedges (1973), who demonstrated that although a high proportion of E2 adsorptions to the cell surface are reversible, desorption of active colicin molecules does not rescue a cell once it has received a lethal dose. Although other interpretations are possible, these authors suggested a functional heterogeneity in the adsorption of colicin E2 to its specific outer membrane receptor. According to their view, the bulk of colicin binding is reversible and leads to the formation of ineffective colicin–receptor complexes. Lethal colicin action represents irreversible adsorption to a minority of the total receptors to form effective colicin–receptor complexes.

Kadner and Bassford (1978) have recently reviewed experiments bearing on the functional stability of the colicin E2/E3 receptor. These experiments have led to a model in which the functional heterogeneity of this protein is explained in terms of its proposed adhesion site-mediated insertion into the outer membrane. In experiments in which a mutant *bfe* allele was introduced into a bfe^+ recipient by means of conjugation, it was found that colicin E3-resistant recombinants begin to appear with little time lag after the entry of the mutant *bfe* gene (Bassford et al., 1977a). This result is surprising, since one would have expected that many generations of growth would be required for the approximately 200 resident receptors to be diluted out through cell division. It was further shown that the time of onset of colicin resistance preceded the time at which cells were no longer able to adsorb colicin. That is, at early times after the mating, resistant cells adsorb colicin to specific colicin E2/E3 phage Bf23 receptors. These authors further showed that addition of spectinomycin at various times after mating immediately halts further appearance of resistant recombinants despite the fact that the total number of recombinants still increases. Thus protein synthesis is required for the conversion of receptor function from a state that adsorbs the colicin and elicits cell death to a state that adsorbs the colicin but does not lead to killing. Using a strain that contained both an amber mutation in the *bfe* gene and a temperature-sensitive amber suppresor mutation, Bassford et al. (1977b) were able to control the production of the functional *bfe* gene product by simple manipulation of temperature. In an experiment in which cells growing at the permissive temperature were shifted to the restrictive temperature, they observed the rapid appearance of colicin E3 resistance. In spite of this loss of receptor function in terms of colicin function, the temperature shift had no effect on the uptake of vitamin B_{12}, a process that is mediated by the colicin E3 receptor.

The above studies on the functional stability of the *bfe* gene product have led these authors to propose the following scheme (Kadner and Bassford, 1978). According to their notion, insertion of the *bfe* gene product into the outer membrane takes place at inner membrane–outer membrane adhesion sites (see Chapter 6). Furthermore, it is envisaged that the receptor must be at an adhesion site for functional complexes to be formed between the bfe protein and colicins E2 or E3 or bacteriophage Bf23. This alignment is critical for the subsequent interaction of the colicins with their intracellular targets and for phage infection. The scheme further considers that the location of a particular receptor molecule at an adhesion site is transitory since such newly exported proteins would diffuse away from the adhesion site as outer membrane constituents are continuously inserted. Once removed from the region of an adhesion site, any *bfe* gene product could still adsorb colicin and phage, but such interactions would be nonlethal. This model easily accomodates the experimental results showing that only newly synthesized receptors are effective for the lethal adsorption of colicins E2 and E3 and phage BF23. Furthermore, it provides an explanation for the finding that continued protein synthesis is required for the development of rapid colicin resistance in recipient cells in the $bfe^- \times bfe^+$ mating experiment described above, since inhibition of protein synthesis would bring to a halt continued insertion of new outer membrane proteins with subsequent movement of already inserted protein away from adhesion sites. Thus it is to be expected that inhibition of protein synthesis would essentially remove the driving force for movement of newly inserted *bfe* gene product away from adhesion sites and would thus fix the receptor in a state in which colicin or phage adsorption would be functional. Kadner and Bassford (1978) have suggested that the *tolA* gene product may play some role in adhesion site-mediated colicin action or phage infection.

The utilization of the *bfe/bfuB* gene product for vitamin B_{12} uptake is discussed in Section 6.

Holland (1976) has recently reviewed the mechanism of colicin E3 action and has provided a model for colicin E3 penetration of the bacterial envelope. My discussion here relies heavily on his treatment.

Inactivation of ribosomes by colicin E3 requires that colicin molecules (or a derived fragment) must penetrate both the outer and inner membrane. The results of Lau and Richards (1976), who examined the action of colicins E2 and E3 covalently attached to Sephadex G-25 beads, are consistent with this notion. They found that although such colicin retains capacity to adsorb to the cell surface receptors of whole cells, such adsorption does not lead to cell death. This same immobilized colicin E3 was active in *in vitro* inhibition of protein synthesis. These results were interpreted to mean that immobilized colicins E2 and E3 are physically restricted from penetrating the *E. coli* double membrane. Colicin E3 is an elongated molecule (mol. wt. 60,000) whose axial ratio may approach 6:1 (Konisky, 1973). The molecule is therefore of sufficient dimensions to interact with the cytoplasmic membrane while remaining complexed to its outer membrane receptor. In fact, as pointed out by Holland (1976), the receptor may play a crucial role in properly presenting the colicin to the inner membrane. This may explain why treatments that alter inner-outer membrane structural relationships can affect colicin sensitivity. It may further explain certain tolerant phenotypes. Although it may be possible that receptor-complexed colicin molecules are able to penetrate the inner membrane to the extent that each may be able to inactivate a limited number of cellular ribosomes, the single-hit nature of colicin E3 action dictates that a single E3 molecule inactivate enough ribosomes to bring about loss of colony-forming ability. It seems most likely that this is accomplished by complete penetration of the colicin into the cell interior, resulting in interaction with the majority of cellular ribosomes. However, models cannot be ruled out that envisage lateral movement of the colicin along the inner membrane with concomitant ribosome inactivation.

The mechanism whereby colicin E3 might transgress the boundary set by the cell envelope is a fertile topic for speculation. Holland (1976) has raised the possibility that the outer membrane receptor serves to orient the colicin molecule in such a way as to lead to its insertion into the center of a cluster of inner membrane proteins. Such an interaction might then lead to the lateral displacement of the individual subunits of the cluster, generating a hydrophilic pore through which the colicin molecule might pass. He considers it possible that proper gating depends on the presence of an energized membrane, which would serve to explain why colicin E3 action requires energized cells (Jetten and Jetten, 1975). This model easily accommodates mechanisms in which all or part of the colicin molecule enters into the cytoplasm or in which the molecule only partially penetrates the cytoplasmic membrane. Furthermore, the model equally applies to mode of action of colicin E2.

If a bacterium is to use colicin production for the purpose of biological warfare, it must have a mechanism to ensure its own protection against the bactericidal agent that it produces. Such protection is termed "immunity." It is only in the case of E3-colicinogenic cells that the mechanism of immunity is fully understood. The ColE3 plasmid carries the structural gene for a protein (mol. wt. 10,000) that neutralizes colicin E3 activity (Boon, 1971; Bowman et al. 1971; Jakes et al., 1974; Sidikaro and Nomura, 1974). All the evidence suggests that colicin E3 neutralization involves the 1:1 stoichiometric interaction between this "immunity protein" and the colicin. It is likely that immunity towards colicin E2 involves a similar mechanism (see Saxe, 1975; Schaller and Nomura, 1976). Jakes and Zinder (1974) have shown that purified colicin E3 contains about one molar equipment of E3 immunity protein. Removal of the immunity protein leads to increased colicin E3 activity *in vitro*. Indeed, it is likely that the earlier demonstrated *in vitro* activity of colicin E3 is derived from immunity protein-free E3 molecules in the preparation.

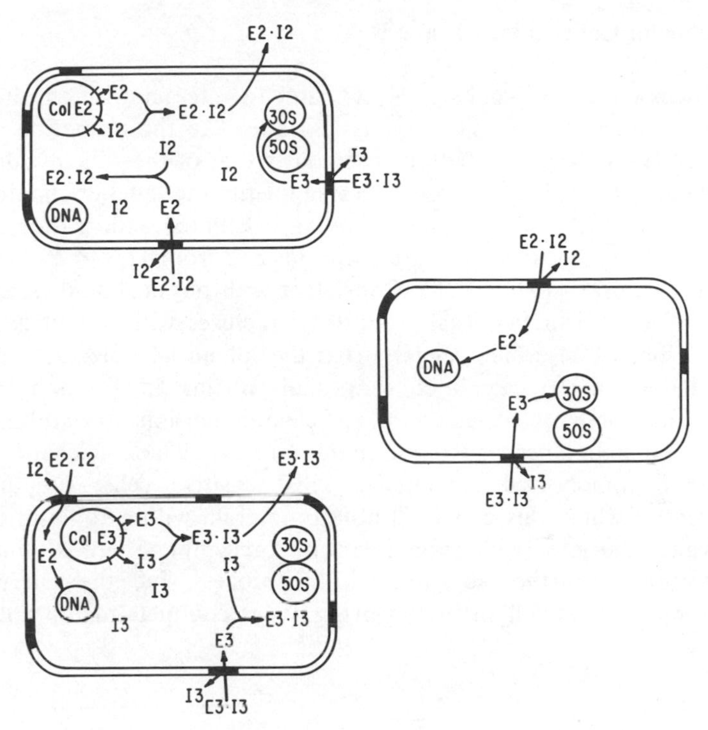

Figure 2 Proposed scheme for colicin E2 and E3 immunity system. I2 and I3 represent colicin E2 and E3 immunity proteins, respectively.

Based on such information it is possible to present the following scheme for the colicin E3 immunity mechanism (Figure 2). It is proposed that upon synthesis, colicin E3 interacts with intracellular immunity protein to form an inactive complex that is subsequently released from the cells. This serves to protect the producing organism from the action of endogenous colicin. Colicin E3-induced ribosome inactivation *in vivo* would then require removal of the immunity protein at some step in the mode of action sequence. Although the model presented in Figure 2 depicts this dissociation occurring at the cell surface, mediated perhaps by the outer membrane receptor, it is equally possible that such activation could occur at some later step in the sequence. Immunity from exogenous colicin E3 would derive from neutralization of such activated colicin molecules by free intracellular immunity protein. The importance of the model within the context of this chapter is that it raises the possibility that the colicin E3 receptor plays a critical and active role in freeing the colicins of neutralizing immunity protein. This scheme also serves in the case of colicin E2.

5.4 Action of Colicins Ia, E1, and K

As is discussed above, colicins E1, K, and Ia interact with the bacterial cytoplasmic membrane in such a way as to lead to the formation of "ion channels." It has been pointed out previously that of the colicins for which physical data are available, those affecting membrane function, namely, E1, K and Ia, are the most elongated, having calculated axial ratios ranging from 9.6 to 15 for idealized prolate ellipsoids and from 11.8 to 20 for oblate ellipsoids (Konisky, 1973, 1978). Consistent with physical studies, electron microscopy of colicin Ia reveals rounded structures with a diameter of 20 nm (Isaacson, 1975), which suggests that the colicin Ia molecule is roughly oblate. Based on the calculated shapes of colicins E1, K, and Ia, it is apparent that these molecules are of sufficient dimensions to simultaneously span the apposed inner and outer membranes. Lau and Richards (1976) found that immobilized colicin E1 kills sensitive cells, supporting a mechanism in which this colicin is able to interact with cytoplasmic membrane while associated with outer membrane receptor. There is no *a priori* reason to propose further steps in the killing process, since the generation of a single ion channel will suffice to bring about complete membrane depolarization.

6 OPERATION OF tonB-DEPENDENT SYSTEMS

6.1 The Requirement for tonB

Gratia (1964, 1966) described the isolation of an *E. coli* mutant (mnemonic *tonB*) that is insensitive to phages T1 and ϕ80 and colicins B, I, and V. Subsequent studies have shown that the *tonB* gene product is also required for the operation of outer membrane-mediated transport systems. Wang and Newton (1971) were the first to demonstrate that the state of the *tonB* gene influences the accumulation of iron. They proposed that the *tonB* gene product itself, or some product regulated by this gene, is an outer membrane component that functions to facilitate the transfer of iron (possibly in a chelate complex) from the external medium to a hypothetical carrier within the cell envelope. It has subsequently been shown that a functional *tonB* gene is required for enterochelin-, citrate-, ferrichrome-, and rhodoturulic acid-mediated iron uptake (Frost and Rosenberg, 1975; Hantke and Braun, 1975a; Pugsley and Reeves, 1976a). With regard to vitamin B_{12}, Bassford et al. (1976) have established that accumulation does not occur in *tonB* mutants. The dependence on tonB is specific to those systems that utilize outer membrane binding proteins. Thus the transport of several substrates (serine, proline, arginine, phosphate, and maltose) that in all likelihood freely diffuse across the outer membrane by way of aqueous pores is not affected in *tonB* mutants (Frost and Rosenberg, 1975; Bassford et al., 1976).

The inability of *tonB* mutants to carry out outer membrane-mediated substrate transport does not derive from a loss of these proteins from the outer membrane. In fact, *tonB* mutants exhibit enhanced levels of all three outer membrane polypeptides whose production is regulated by iron (Davies and Reeves, 1975; Hancock and Braun, 1976b; Braun et al., 1976; Konisky et al., 1976). Such mutants retain the capacity for reversible adsorption of phages T1 and ϕ80 (Hancock and Braun, 1976a), and outer membranes prepared from such strains bind ferric-enterochelin (Pugsley and Reeves, 1976a). Bassford et al. (1976) have shown that *tonB* mutants have near normal levels of B_{12} receptors on their cell surface.

Frost and Rosenberg (1975) reported that whereas exogenously added enterochelin supports neither growth nor iron uptake in a *tonB* mutant strain unable to synthesize enterochelin because of a mutation in the *aroB* gene, addition of precursors of enterochelin did support growth. These precursors did not stimulate growth of a *tonB*$^-$ *aroB*$^-$ *entF*$^-$ strain, demonstrating that rescue by precursor addition requires synthesis of enterochelin. Based on such findings Frost and Rosenberg suggested that whereas the outer membrane of *tonB* mutant strains is impermeable to ferric-enterochelin, such strains are capable of TonB-independent utilization of periplasmic enterochelin generated by its endogenous synthesis from added precursors. They envisaged a model in which the *tonB* gene somehow controls the permeability of the outer membrane to ferric-enterochelin. Hancock et al. (1977) have observed 2,3-dihydroxybenzoate-promoted iron uptake in *E. coli* K-12 *aroB*$^-$ mutants unable to synthesize enterochelin. Such uptake requires neither functional *tonB* nor *feuB* (*fepA*, *cbr*) genes but is dependent on the activity of *entF*, *fesB*, and *fepB* genes. The interpretation was made that the system operating when iron uptake is mediated by a mechanism that circumvents the need for the outer membrane ferric-enterochelin binding protein does not require tonB function. In agreement with other workers (see below) the *fesB* and *fepB* gene functions were assigned to the cytoplasmic membrane.

Although the above results lead to the reasonable proposal that the TonB function is specifically involved in the delivery of the appropriate substrate to the inner membrane, it has recently become clear that its role may not be so simple. Wookey and Rosenberg (1978) have shown that although uptake of ferric-enterochelin by spheroplasts does not depend on the outer membrane *fepA* gene product, it does depend on functional *fepB* and *tonB* genes.

These results show that the latter genes are required for transport across the cytoplasmic membrane, implying that their respective gene products are components of the inner membrane. Further support for the *tonB* gene-determined product having an inner membrane location derives from the finding that uptake of ferric-citrate, ferrichrome, and vitamin B_{12}, but not proline, by spheroplasts is tonB dependent (H. Rosenberg, personal communication).

Although energy is clearly required for outer membrane receptor-mediated uptake of iron chelates and vitamin B_{12} and the irreversible adsorption of phage T1 and ϕ80, the exact nature of the energy-dependent processes in each particular uptake system has not been delineated. Since TonB mutants are defective in the energy-dependent stage of B_{12} uptake and do not exhibit irreversible phage ϕ80 and T1 binding, it has been suggested that tonB may function in energy coupling between the inner membrane and the outer membrane receptors (Hancock and Braun, 1976a; Hancock et al., 1977; Bassford et al., 1977a). These suggestions leave open the question of whether such tonB-dependent coupling is direct or indirect.

Any consideration of possible models for tonB function must take into account the experiments of Kadner and his colleagues, which demonstrate the functional lability of the *tonB* gene product. Using a $tonB_{am}supD_{ts}$mutant (see Section 5) in which the amber mutation is suppressed at 30°C but not at 42°C, it was shown that shift to the nonpermissive temperature leads to a rapid loss in the capacity of such cells to carry out the energy-dependent phase of B_{12} uptake (Bassford et al., 1977b). Shift to nonpermissive temperature also resulted in the rapid loss in sensitivity to colicin D, whose action is known to be tonB dependent (Davies and Reeves, 1975). In a subsequent paper, Kadner and McElhaney (1978) demonstrated that the operation of all known tonB-dependent systems, including sensitivity to colicins B and Ia, irreversible adsorption of phage ϕ80, and siderophore-mediated iron uptake, depends on the continued production of the *tonB* gene product. Based on their finding that the addition of ferrichrome or an enterochelin precursor accelerated the loss of B_{12} uptake in parental cells inhibited in protein synthesis but not in strains lacking the siderophore uptake system for the added siderophore, they suggested that the *tonB* gene product is consumed during its normal functioning in tonB-dependent processes.

Clearly, failure to identify the *tonB* gene product has been a major obstacle in determining its function. However, the recent cloning of the *tonB* gene and the subsequent determination that the gene codes for a 36,000 dalton polypeptide (W. Reznikoff, personal communication) represents a major advance and should provide a direct means of localizing the protein.

6.2 Operation of tonB-Dependent Systems—A Proposal

Although a full understanding of the molecular interactions that occur in those uptake systems that depend on the presence of outer membrane receptor proteins must await further study, it seems of heuristic value to present a speculative model at this time (see Figure 3). An essential feature of the model is that is proposes that efficient siderophore-mediated iron uptake, B_{12} uptake, irreversible adsorption, and subsequent infection by phage T1 and ϕ80, as well as killing by the B-type colicins, are all mediated at sites of apposition between the inner and outer membrane. These sites are distinct from the sites of membrane adhesion (discussed in Chapter 6) that serve as the site of infection for some phages (Bayer, 1968). It is proposed that apposition sites are thermodynamically unstable and that their formation requires energy input in the form of an energized inner membrane. It is postulated that the formation of apposition sites requires the tonB function. Bassford et al. (1976) have previously suggested that tonB may play a role in maintaining the proper orientation of outer membrane receptors with components of the cytoplasmic membrane.

The need for close contact between inner and outer membranes for operation of these systems derives from several considerations. Since the strong binding of the relevant substrates to outer membrane receptors precludes their easy release into the periplasmic space where each might search out the appropriate inner membrane component of its transport system, it is likely that efficient transfer of the substrate from outer to inner membrane requires that the membranes be brought into close proximity. Furthermore, since several cellular nucleases have been localized to the periplasm, it is probable that the process whereby phage nucleic acid gains entry to the cytoplasm avoids contact with periplasmic nucleases. In the case of the various colicins, the need for close contact between outer and inner membranes is discussed earlier (see Section 5).

Although there exists no strong evidence that the *tonB* gene product might be involved in the formation or stabilization of apposition sites, proposal of such a role represents a comfortable extension of many considerations presented in this chapter. Precisely how the *tonB* gene product might function in this capacity is open to speculation. One possibility is that the *tonB* gene product is involved in some hypothetical coupling between the energized membrane state and osmotically driven formation of apposition sites. An alternative role is that the tonB-determined protein might serve in some structural capacity to form or maintain close contact between inner and outer membrane. Thus it may be improper to inquire as to whether the tonB protein is a component of inner or outer membrane. Indeed, it may be transiently and concomitantly associated with both.

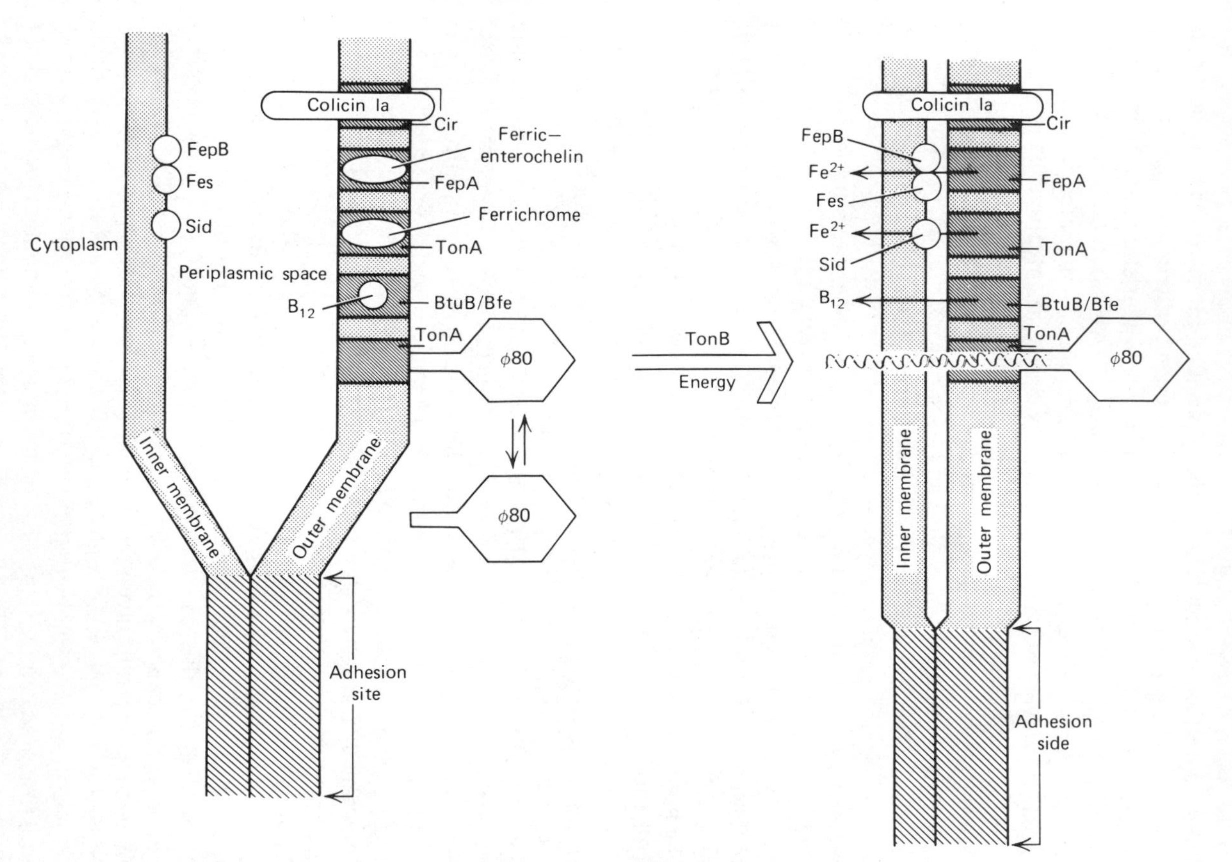

Figure 3 Proposed scheme for operation of tonB-dependent systems. See text for details.

It is possible that there are direct interactions between the TonB protein and the cytoplasmic membrane components of the various receptor-mediated uptake systems described here. This would accommodate findings that the uptake of ferric-enterochelin, ferric-citrate, ferrichrome, and vitamin B_{12} by spheroplasts is tonB dependent. For example, one could envisage that the TonB protein (or part of it) interacts with the fepB or/and fes protein in the cytoplasmic membrane in such a way as to maintain these proteins in a functional conformation. Furthermore, the proposed consumption of the *tonB* gene product during its action is more easily explained if the tonB protein interacts directly with specific components of the systems that require its function. It is altogether possible that the *tonB* gene product serves as a physical link between inner and outer membrane components of each of these systems. As such it may act as an energy-dependent gate whose function is to provide a means whereby outer membrane bound substrates are able to gain access to the inner membrane.

According to the model, at least a part of the energy requirement for colicin Ia action and uptake of siderophores and vitamin B_{12} has a structural basis in the energy-dependent formation of apposition sites. In the case of ferrichrome uptake, energy is further utilized for the active transport of this siderophore across the cytoplasmic membrane (Negrin and Neilands, 1978). It is possible that the translocation of other iron chelates and B_{12} across the cytoplasmic membrane is similarly driven. Bradbeer entertains the possibility that the release of vitamin B_{12} from the receptor is dependent on the proton motive force and a functional *tonB* gene product and may involve direct transfer of the vitamin from the outer membrane to the inner membrane without release into the periplasmic space (Bradbeer et al., 1978). This notion is perfectly consistent with the model presented in Figure 3.

The proposal for functionally distinct sites of adhesion and apposition is supported by studies on the functional stability of the *bfe-btuB* protein. As is discussed in Section 5, there is good evidence that the outer membrane receptor shared by the E colicins, phage BF23, and B_{12} can exist in more than one functional state (see Kadner and Bassford, 1978). It was proposed that whereas newly synthesized (and thus adhesion site-localized) *bfe-btuB* gene product is functional in terms of colicin uptake and phage infection, subsequent movement of such receptors away from zones of adhesion renders them nonfunctional in terms of colicin action and phage infection even though they retain binding capacity. In contrast, it was postulated that such migration allows these very same receptors to enter into a state of interaction with the *tonB* gene product or a structural region specified by this product that renders the capacity for B_{12} uptake. Although the experimental basis for their model does rule out the possibility that B_{12} uptake occurs at adhesion sites, the results do offer the possibility that the operation of tonB dependent systems occurs at sites that are structurally distinct from adhesion sites.

Although phages T1, T5, and ϕ80, as well as ferrichrome and colicin M, utilize the tonA protein as receptor, a clear distinction exists in their mode of interaction with host cells. Thus, whereas ferrichrome uptake, colicin M action, and successful infection by T1 and ϕ80 depends on a functional host *tonB* gene, T5 infection does not. Based on the facts that energy is required for the irreversible adsorption of T1 and ϕ80 and that binding to whole *tonB* cells or to a soluble preparation of the tonA protein is reversible, Hancock and Braun (1976a) have concluded that the functional state of the tonA receptor protein might be directed by the energized state of the cytoplasmic membrane and that the tonB function could serve as a coupling device between the two membranes. This notion can be easily considered within the framework of the model shown in Figure 3. It is simply proposed that the tonB- and energy-dependent formation of apposition sites leads to the functional interaction of these phages and ferrichrome with the inner membrane. This leads to a more stable interaction (operationally irreversible) between the phages and the cell envelope. This proposal is somewhat at odds with the results of Bayer (1968), who in his elegant electron microscopic examination of phage adsorption provided evidence that phage T1 adsorbs to adhesion sites. However, as pointed out in that study, only 1–5% of the total adsorptions survived the fixation and embedding process. It is thus possible that the great majority of interactions are, indeed, at apposition sites but that envelope disruption during preparation for electron microscopy led to phage desorption.

In the case of phage T5 it is proposed that the virus utilizes receptors situated at adhesion sites. This circumvents the need for tonB function. Indeed, Bayer's study (1968) places the site of T5 infection at adhesion sites.

The molecular basis for the functional heterogeneity of the TonA protein is unknown. Although the model proposed here provides the structural basis for the difference in tonB dependence exhibited by colicin M, ferrichrome, and phage T1, ϕ80, and T5, the basic question remains as to why phage T5 utilizes the TonA protein at adhesion sites, whereas (as proposed here) ferrichrome and phages T1 and ϕ80 utilize this same protein at sites of membrane apposition. A reasonable possibility is that the ability of these phages to interact with the tonA protein depends on whether the receptor is in an adhesion site. Such differential recognition might be based on alternative intrinsic structural states of the tonA protein depending on whether the receptor is in an adhesion site or it might be based on specific associations between the tonA protein and particular membrane components. The idea is simply that T1, ϕ80, and ferrichrome interact poorly or not at all with the TonA protein when this protein is in a structural state dictated by its presence at an adhesion site. In contrast T5 is able to recognize precisely this structural state, but it is unable to interact well with this same protein once the protein has migrated from an adhesion site. It is assumed that newly synthesized proteins are inserted into the outer membrane at adhesion sites (see Chapter 6). One prediction from such a suggestion is that it should be possible to isolate host range mutants of T1 and ϕ80 that are able to adsorb at adhesion site-receptors, obviating their dependence on the tonB function. In this regard Hantke and Braun (1978) have reported the isolation of a host range mutant of phage T1 that, although it utilizes the tonA protein as receptor, does not require the tonB function for successful infection.

The model does not exclude the possibility that vitamin B_{12} and the siderophores interact with their appropriate receptors at adhesion sites. In fact, such interactions seem necessary to explain findings that vitamin B_{12} and ferrichrome protect cells against the E colicins and phage T5, respectively. It is even possible that adhesion sites mediate very limited uptake of these substrates.

7 MAJOR OUTER MEMBRANE PROTEINS OF *E. coli* AND *S. typhimurium* AS PHAGE RECEPTORS

As is discussed in Chapter 1, the outer membrane of gram-negative bacteria contains several proteins whose concentration is substantially higher than that of the other 30 or so species of polypeptide found in this membrane. We discuss here how these "major" proteins serve as phage receptors.

Henning and his colleagues have shown that *E. coli* outer membrane protein *II** (nomenclature system of Schmitges and Henning, 1976) is the phage TuII* receptor. The structural gene for this polypeptide has been identified and has been given the mnemonic *ompA* (formerly *con*, *tolG*, *tut*) (Datta et al., 1976; Manning et al., 1976). Phage Tu* was isolated from sewage by screening for phages that form plaques on wild-type *E. coli* K12 but not on a *tolG* derivative (Chai and Foulds, 1974) missing ompA protein (Henning and Haller, 1975). Selection for resistance to the isolated virus leads to the isolation of mutants lacking ompA protein. Although neither isolated ompA protein nor lipopolysaccharide alone inactivates the phage, the interaction of these two outer membrane components *in vitro* results in the generation of phage-neutralizing activity (Datta et al., 1977). Evidence has been provided that indicates it is the protein moiety that serves as the receptor (Datta et al., 1976, 1977). Skurray et al. (1974) found that a class of conjugation-defective mutants of *E. coli* K12 (termed Con$^-$) is defective in the adsorption of phage K3. These mutants lack a membrane protein originally denoted 3A (Manning and Reeves, 1976b), which is identical to ompA protein. Since isolated ompA protein together with lipopolysaccharide inhibits F pilus-mediated conjugation, it is likely that this outer membrane polypeptide acts as a receptor in conjugation (Schweizer and Henning, 1977; van Alphen et al., 1977). Mutants that lack ompA protein can be isolated by selection for phage K3 resistance. Furthermore, an ompA protein–lipopolysaccharide complex has been shown to inactivate the phage (van Alphen et al., 1977). These results indicate that protein II* serves as the receptor for phage K3.

Two sewage phages, TuIa and TuIb, use outer membrane proteins Ia and Ib, respectively, as receptors (Datta et al. 1977). The *in vitro* receptor activity of these proteins requires lipopolysaccharide.

Verhoef et al. (1977) isolated and partially characterized a phage, given the name Mel, which is unable to propagate on *E. coli* K12 strains lacking protein c (protein Ib in the Schmitges-Henning nomenclature). Of 13 independent spontaneous Mel-resistant mutants, none adsorbed the phage, 10 were missing protein C, 2 showed decreased levels of this protein, and 1 mutant contained a normal amount of protein C. Genetic studies show that Mel-sensitive and Mel-resistant transductants contain and lack protein C, respectively. The results strongly suggest that protein C (Ib) serves as the phage receptor.

Schnaitman et al (1975) isolated a temperate phage, named PA-2, that uses protein Ib as receptor (Bassford et al., 1977c). In cells lysogenic for PA-2, outer membrane protein I (Ia plus Ib) is replaced by another polypeptide. Such lysogens display reduced ability to adsorb PA-2. Mutants selected for resistance to virulent or clear-plaque derivatives of PA-2 are altered in matrix protein I. Certain colicin-tolerant mutants whose selection for colicin insensitivity led to the isolation of mutants lacking matrix protein I are resistant to PA-2. Finally, a purified preparation of matrix protein I has been shown to inactivate the phage. Since this preparation contained some LPS, it is possible that the receptor is a complex of matrix protein I and LPS. Bassford et al. (1977c) have provided evidence that the Ib component of matrix protein I is the phage receptor.

Foulds and Chai (1978) have isolated a mutant of *E. coli* that contain a "major" outer membrane protein not present in the parental strain. This polypeptide, termed protein E, is required for the growth of two recently isolated bacteriophages, and evidence has been presented that it is utilized as part of their receptor (Chai and Foulds, 1978).

Studies using phages active on *S. typhimurium* have led to the suggestion that two major outer membrane proteins of this organism can serve as phage receptors (Nurminen et al., 1976). Whereas mutants selected for resistance to phage PH51 lack an outer membrane polypeptide having an apparent molecular weight of 34,000, phage PH105- and P221-resistant mutants lack an outer membrane protein with a molecular weight of 26,000. Furthermore, whereas Triton–EDTA extraction of envelopes prepared from sensitive cells led to solubilization of receptor activity (phage neutralization), similar treatment of envelopes from resistant mutants released no such activity. Analysis of the solubilized material by SDS–polyacrylamide gel electrophoresis showed that the presence of phage PH 105/221 and PH51 receptor activity correlates with the presence of the polypeptides having molecular weights of 36,000 and 34,000, respectively.

8 THE PHAGE LAMBDA RECEPTOR AND MALTOSE TRANSPORT

8.1 The *mal* Genes

The ability of *E. coli* K-12 to metabolize maltose is determined by genes located in one of two regions of the genetic map: *malA* (at 74 min) and *malB* (at 90 min) (Schwartz, 1966). The genes of the *malA* region are essential for the energy-dependent active transport of maltose; this region is comprised of two genes involved in the metabolism of internal maltose and a third gene (*malT*) that specifies a positive regulator for the induction by maltose of all the *mal* genes (Schwartz, 1967; Hofnung et al., 1971; Hofnung, 1974; Hofnung et al., 1974). Genetic studies have led to the conclusion that the MalB region is comprised of two operons—one containing *malE* (*malJ*$_1$) and *malF* (*malJ*$_2$) and the other containing *malK* and *lamB* genes. These operons have adjoining, and perhaps overlapping, promoter regions. Although Hofnung (1974) has suggested that the *malF* gene product may play a specific role in the energy coupling for maltose transport and that the *malK* gene product may be responsible for movement of maltose across the cytoplasmic membrane, such functions have not been confirmed experimentally. In contrast, recent studies have led not only to the identification of the *malE* and *lamB* gene products but have elucidated their respective roles in maltose transport.

The chemotactic response of *E. coli* toward maltose is mediated by a periplasmic chemoreceptor protein that binds the sugar (Hazelbauer and Adler, 1971). This protein (mol. wt. 44,000) has been purified and characterized by Kellermann and Szmelcman (1974), who have shown that it is determined by the *malE* cistron. As expected, the synthesis of this protein is induced by maltose, and fully induced strains contain approximately 30,000 molecules/cell.

The great majority (80%) of spontaneously arising phage λ-resistant mutants of *E. coli* are unable to utilize maltose as the sole carbon source

(Lederberg, 1955). Whereas about 20% of these map within the MalB operon, the remaining map within *malA* at *malT*. In contrast, λ-resistant Mal⁺ mutants (designated *lamB*), which are specifically impaired in phage λ adsorption, map within *malB*, the *lamB* gene being closely linked to the *malK* gene (Thirion and Hofnung, 1972). Evidence for the coordinate regulation of the enzymes of maltose metabolism and λ-receptor synthesis has been presented by Schwartz (1967). Based on his findings that point mutants in the *malT* gene lead to both phage λ resistance and inability to ferment maltose, that maltose (and in some cases trehalose) induces synthesis of both λ receptors and the enzymes of maltose utilization, and that λ-receptor synthesis is constitutive in a strain in which expression of *malA* and *malB* operons are constitutive, he proposed that the *malT* gene product acts as a positive regulator of the *malA* and *malB* operons (Schwartz, 1967; Hofnung et al., 1974). Thus the synthesis of λ receptors depends on functional structural (*lamB*) and regulator (*malT*) genes. Receptor production is also subject to catabolite repression (Howes, 1965; Yokota and Kasuga, 1972). *E. coli* K-12 growing in glucose or maltose as the carbon source elaborate 30 and 6000 λ surface receptors per cell, respectively (Schwartz, 1976). In another study, the outer membrane prepared from maltose-induced cells was estimated to contain approximately 100,000 copies of the λ-receptor polypeptide (Braun and Krieger-Brauer, 1977).

Randall-Hazelbauer and Schwartz (1973) solubilized and partially characterized the λ receptor. The receptor was shown to be determined by the *lamB* cistron and to reside in the outer membrane. Based on the sensitivity of partially purified receptor material to proteases, these workers concluded that a protein moiety is essential for activity. They further showed that the receptor is a component of the bacterial outer membrane. The λ receptor has recently been purified to apparent homogeneity and has been shown to be a single polypeptide chain of 47,000–50,000 daltons (Endermann et al., 1978). This polypeptide is absent from the outer membrane prepared from bacterial mutants that are resistant to phage *L* as a result of a deletion in the λ-receptor structural gene (*lamB*) Enderman et al., 1978).

8.2 The Lambda Receptor in Maltose and Maltotriose Transport

As becomes clear from the following discussion, the λ receptor plays an important role in the delivery of extracellular maltose and its higher analogs to the periplasmic maltose binding protein. Hazelbauer (1975a) noted that although *lamB* nonsense mutants exhibit a chemotactic response to maltose, a comparison with LamB⁺ strains showed that an approximately tenfold higher concentration of maltose was required for maximal activity. A subsequent study showed that such shifts in optimal sugar concentrations are strongly correlated with the reduced ability of *lamB* nonsense mutants to take up maltose at low substrate concentrations (Hazelbauer, 1975b; Szmelcman and Hofnung, 1975). It was suggested that the λ receptor might be involved in the transfer of maltose across the outer membrane and into the periplasmic space where it could interact with the maltose chemoreceptor (Hazelbauer, 1975b). Although *lamB* mutants show no defect in growth rate or yield when grown on 1 m*M* maltose as the sole carbon source, they are defective in maltose metabolism at limiting concentrations of the sugar (Szmelcman and Hofnung, 1975). In fact, these workers were able to demonstrate that *lamB*$^+$ strains enjoy a selective advantage over LamB$^-$ mutants during growth in a chemostat at limiting maltose concentration. No selective advantage was observed when the carbon source was lactose or glycerol. They did note, however, that LamB mutants seemed to be slightly impaired in growth on glucose. This same study convincingly established that the λ receptor is involved in maltose transport. Thus, whereas *lamB* nonsense mutants showed normal maltose transport at high substrate concentrations, the transport rate at low maltose concentrations was 5% (or lower) of that of the wild type. Furthermore, incubation of *lamB*$^+$ cells with serum containing antibodies prepared against purified λ receptor led to a reduction in the maltose transport capacity of such cells.

Further insight into the role of the *lamB* gene product in maltose transport derives from the work of Szmelcman et al. (1976), who determined the maltose and maltotriose transport kinetic parameters in wild-type and *lamB* mutants. They found that whereas mutations in the *lamB* gene do not lead to an alteration in the V_{max} for maltose transport, they cause an increase in the K_m by a factor of 100–500. Based on the finding that the K_m for transport (1 μM) in wild-type cells is identical to the K_d for binding to the maltose-binding protein, these workers concluded that in *lamB*$^+$ cells access to the periplasmic binding protein is not limited by the outer membrane. Since mutations in the *lamB* gene lead specifically to an increase in the K_m for maltose transport and since the λ receptor is an outer membrane protein, the conclusion was drawn that the λ receptor facilitates the diffusion of maltose across the outer membrane. In the absence of this protein the outer membrane is no longer freely permeable to maltose and diffusion across it represents the rate-limiting step in maltose transport at low concentrations. In the case of maltotriose, the K_m of transport (2 μM) in wild-type cells is thirteenfold higher than the K_d for binding of this substrate to the binding protein (0.16 μM). This indicates that the outer membrane is not freely permeable to maltotriose even though it contains the λ-receptor protein. Mutants of *lamB* do not transport maltotriose.

Szmelcman et al. (1976) have considered the possibility that the maltose system has evolved to utilize linear oligosaccharides derived from starch or glycogen rather than maltose. This suggestion finds its basis in several observations. For example, the maltose-binding protein exhibits a higher affinity for maltotriose and other maltodextrins than for maltose. Furthermore, studies with strains containing either missense or nonsense mutations in the *lamB* cistron show that the *lamB* gene product is more stringently required for growth on maltodextrins than on maltose (Szmelcman et al., 1976; Braun and Krieger-Brauer, 1977).

The analogy between the role of porins (see Chapter 11) and the role of the λ receptor in facilitating the diffusion of substrates across the outer membrane is obvious and, in fact, Szmelcman et al. (1976) speculate that the λ receptor may serve as a pore for maltose and higher maltodextrins. Such a function would explain failures to detect binding of maltose to purified λ-receptor protein (Szmelcman et al., 1976). If the λ receptor represents a nonspecific aqueous pore, it should facilitate the passage of molecules of a size comparable to the largest utilized maltodextrins. Whereas this function is normally relegated to the porin proteins, it is reasonable to suppose that the λ receptor might serve as a potential back up system. That such a situation can arise has recently been shown by von Meyenburg and Nikaido (1977), who compared the transport properties of porin-deficient $lamB^-$ and $lamB^+$ strains. These results suggest that the λ receptor is capable of facilitating the diffusion of sugars other than maltose.

Since maltose (344 daltons) and maltotrisoe (504 daltons) are within the exclusion limit (approximately 650 daltons) of the transmembrane pores formed by the porin proteins, the usefulness of a specific system for their movement across the outer membrane is not obvious. One possibility is that the λ receptor and maltose binding protein interact *in vivo* in such a way that access to the maltotriose binding site on the binding protein can be gained only through the lamB protein. Thus, although maltotriose is able to gain entry to the periplasmic space by way of porin proteins, such entry would be "dead end" and would not lead to accumulation of substrate within the cells. In contrast, one would have to argue that such entry of maltose would result in sufficient interaction with the maltose-binding protein to allow some transport and cell growth. As discussed by Kadner and Bassford (1978), whereas maltose molecules diffusing through the *lamB* pore would immediately interact with the binding protein, molecules entering the periplasm by way of porin molecules would probably equilibrate over the entire volume of the periplasmic space before interacting with binding protein. Clearly, direct coupling between transmembrane diffusion and interaction with the specific maltose-binding protein is more efficient.

ACKNOWLEDGMENT

The author expresses his gratitude to K. Frick for his help in the preparation of this chapter.

REFERENCES

Bassford, P. J., and Kadner, R. J. (1977), *J. Bacteriol.*, **132,** 796–805.

Bassford, P. J., Bradbeer, C., Kadner, R. J., and Schnaitman, C. A. (1976), *J. Bacteriol.*, **128,** 242–247.

Bassford, P. J., Kadner, R. J., and Schnaitman, C. A. (1977a), *J. Bacteriol.*, **129,** 265–275.

Bassford, P. J., Schnaitman, C. A., and Kadner, R. J. (1977b), *J. Bacteriol.*, **130,** 750–758.

Bassford, P. J., Jr., Diedrich, D. L., Schnaitman, C. A., and Reeves, P. (1977c), **131,** 608–622.

Bayer, M. E. (1968), *J. Virol.*, **2,** 346–356.

Bennet, R. L., and Rothfield, L. I. (1976), *J. Bacteriol.*, **127,** 498–504.

Beppu, T., and Arima, K. (1976), *J. Bacteriol.*, **93,** 80–85.

Bhattacharyya, P., Wendt, L., Whitney, E., and Silver, S. (1970), *Science*, **168,** 998–1000.

Boon, T. (1971), *Proc. Natl. Acad. Sci. U.S.*, **68,** 2421–2425.

Bowles, L. K., Miguel, A. G., and Konisky, J. (1976), *Bacteriol. Proc.*, 210.

Bowman, C. M., Sidikaro, J., and Nomura, M. (1971), Nature New Biol., **234,** 133–137.

Bradbeer, C., and Woodrow, M. L. (1976), *J. Bacteriol.*, **128,** 99–104.

Bradbeer, C., Woodrow, M. L., and Khalifah, L. A. (1976), *J. Bacteriol.*, **125,** 1032–1039.

Bradbeer, C., Kenley, J., DiMasi, D., and Leighton, M. (1978), *J. Biol. Chem.*, **253,** 1347–1352.

Braun, V., and Hantke, K. (1977), in *Microbial Interactions* (J. L. Reissig, ed.), Chapman and Hall, London, pp. 101–137.

Braun, V., and Krieger-Brauer, H. J. (1977), *Biochim. Biophys. Acta*, **409,** 89–98.

Braun, V., and Wolff, H. (1973), *FEBS Lett.*, **34,** 77–80.

Braun, V., Schaller, K., and Wolff, H. (1973), *Biochim, Biophys. Acta*, **323,** 87–97.

Braun, V., Schaller, K., and Wabl, M. R. (1974), *Antimicrob. Agents Chemother.*, **5,** 520–533.

Braun, V., Hancock, R. E. W., Hantke, K., and Hartmann, A. (1976), *J. Supramol. Struct.*, **5,** 37–58.

Braun, V., Hantke, K., and Stauder, W. (1977), *Mol. Gen. Genet.*, **155,** 227–229.

Brot, N., and Goodwin, J. (1968), *J. Biol. Chem.*, **243,** 510–513.

Bryce, G. F., and Brot, N. (1971), *Arch. Biochem. Biophys.*, **142,** 399–406.

Buxton, R. S. (1971), *Mol. Gen. Genet.*, **113,** 154–156.

Byers, B. R., and Arceneaux, J. E. L. (1977), in *Microorganisms and Minerals* (E. D., Weinberg, ed.) Dekker, New York.

Cardelli, J., and Konisky, J. (1974), *J. Bacteriol.*, **119,** 379–385.

Chai, T., and Foulds, J. (1974), *J. Mol. Biol.*, **85,** 465–474.

Chai, T., and Foulds, J. (1978), *J. Bacteriol.*, **135,** 164–170.

Cox, G. B., Gibson, F., Luke, R. K. J., Newton, N. A., O'Brien, I. G., and Rosenberg, H. (1970), *J. Bacteriol.*, **104,** 219–226.

Datta, D., Kramer, C., and Henning, U. (1976), *J. Bacteriol.*, **128,** 834–841.

Datta, D. B., Arden, B., and Henning, W. (1977), *J. Bacteriol.*, **131,** 821–829.

Davies, J. K., and Reeves, P. (1975), *J. Bacteriol.*, **123,** 96–101.

DiGirolamo, P. M., and Bradbeer, C. (1971), *J. Bacteriol.*, **106,** 745–750.

DiMasi, D. R., White, J. C., Schnaitman, C. A., and Bradbeer, C. (1973), *J. Bacteriol.*, **115,** 506–513.

Endermann, R., Hindennach, I., and Henning, U. (1978), *FEBS Lett.*, **88,** 71–74.

Ernst, J. F., Bennet, R. L., and Rothfield, L. I. (1978), *J. Bacteriol.*, **135,** 928–934.

Foulds, J., and Chai, T. (1978), *J. Bacteriol.*, **133,** 1478–1483.

Fredericq, P. (1946), *C. R. Soc. Biol.*, **140,** 1295–1296.

Fredericq, P. (1949a), *C. R. Soc. Biol.*, **143,** 1011–1013.

Fredericq, P. (1949b), *C. R. Soc. Biol.*, **143,** 1014–1017.

Fredericq, P. (1951), *C. R. Soc. Biol.*, **145,** 930–933.

Fredericq, P., and Smarda, J. (1970), *Ann. Inst. Pasteur*, **118,** 767–774.

Frost, G. E., and Rosenberg, H. (1973), *Biochim. Biophys. Acta*, **330,** 90–101.

Frost, G. E., and Rosenberg, H. (1975), *J. Bacteriol.*, **124,** 704–712.

Gould, J. M., and Cramer, W. A. (1977), *J. Biol. Chem.*, **252,** 5491–5497.

Graham, A. C., and Stocker, B. A. D. (1977), *J. Bacteriol.*, **130,** 1214–1223.

Gratia, J. P. (1964), *Ann. Inst. Pasteur*, (Suppl.) **107,** 132–151.

Gratia, J. P. (1966), *Biken J.*, **9,** 77–87.

Guterman, S. (1973), *J. Bacteriol.*, **114,** 1217–1224.

Guterman, S. K. (1971), *Biochem. Biophys. Res. Commun.*, **44,** 1149–1155.

Guterman, S. K., and Luria, S. E. (1969), *Science*, **164,** 1414.

Hancock, R. E. W., and Braun, V. (1976a), *J. Bacteriol.*, **125,** 409–415.

Hancock, R. E. W., and Braun, V. (1976b), *FEBS Lett.*, **65,** 208–210.

Hancock, R. E. W., Hantke, K., and Braun, V. (1976), *J. Bacteriol.*, **127,** 1370–1375.

Hancock, R. E. W., Hantke, K., and Braun, V. (1977), *Arch. Microbiol.*, **114,** 231–239.

Hantke, K. (1976), *FEBS Lett.*, **70,** 109–112.

Hantke, K., and Braun, V. (1975a), *FEBS Lett.*, **49,** 301–305.

Hantke, K., and Braun, V. (1975b), *FEBS Lett.*, **59,** 277–281.

Hantke, K., and Braun, V. (1978), *J. Bacteriol.*, **135,** 190–197.

Hazelbauer, G. L. (1975a), *J. Bacteriol.*, **122,** 206–214.

Hazelbauer, G. L. (1975b), *J. Bacteriol.*, **124,** 119–126.

Hazelbauer, G. L., and Adler, J. (1971), *Nature New Biol.*, **230,** 101–104.

Henning, U., and Haller, I. (1975), *FEBS Lett.*, **55,** 161–164.

Hill, C., and Holland, I. B. (1967), *J. Bacteriol.*, **94,** 677–686.

Hofnung, M. (1974), *Genetics*, **76,** 169–184.

Hofnung, M., Schwartz, M., and Hatfield, D. (1971), *J. Mol. Biol.*, **61,** 681–694.

Hofnung, M., Hatfield, D., and Schwartz, M. (1974), *J. Bacteriol.*, **117,** 40–47.

Holland, I. B. (1975), *Adv. Microbiol. Physiol.*, **12,** 56–139.

Holland, I. B. (1976), in the *Specificity and Action of Animal, Bacterial and Plant Toxins* (Receptors and Recognition, Series B, Vol. 1) (P. Cuatrecasas, ed.), Chapman and Hall, London, pp. 99–127.

Hollifield, W. C., and Neilands, J. B. (1978), *Biochemistry*, **17,** 1922–1928.

Howes, W. V. (1965), *J. Bacteriol.*, **90,** 1188–1193.

Ichihara, S., and Mizushima, S. (1977), *J. Biochem.*, **81,** 749–756.

Ichihara, S., and Mizushima, S. (1978), *J. Biochem.*, **83,** 137–140.

Isaacson, R. E. (1975) Ph.D. thesis, University of Illinois, Urbana.

Jakes, K., and Zinder, N. D. (1974), *Proc. Natl. Acad. Sci. U.S.*, **71,** 3380–3384.

Jakes, K., Zinder, N. D., and Boon, T. (1974), *J. Biol. Chem.*, **249,** 438–444.

Jasper, P. E., Whitney, E., and Silver, S. (1972), *Genet. Res.*, **19,** 305–312.

Jetten, A. M., and Jetten, M. E. R. (1975), *Biochim. Biophys. Acta*, **387,** 12–22.

Kadner, R. J., and Bassford, P. J. (1978), in *Bacterial Transport* (B. P. Rosen, ed.) Microbiology Series, Dekker, New York.

Kadner, R. J., and Liggins, G. L. (1973), *J. Bacteriol.*, **115,** 514–521.

Kadner, R. J., and McElhaney, G. (1978), *J. Bacteriol.*, **134,** 1020–1029.

Kellermann, O., and Szmelcman, S. (1974), *Eur. J. Biochem.*, **47,** 139–149.

Konisky, J. (1973), in *Chemistry and Functions of Colicins* (L. P. Hager, ed.), Academic Press, New York, pp. 41–58.

Konisky, J. (1978), in *The Bacteria* Vol. 6, (L. N. Ornston and J. R. Sokatch, eds.), Academic Press, London and New York, pp. 71–136.

Konisky, J., and Cowell, B. S. (1972), *J. Biol. Chem.*, **247,** 6524–6529.

Konisky, J., and Liu, C-T (1974), *J. Biol. Chem.*, **249,** 835–840.

Konisky, J., Cowell, B. S., and Gilchrist, M. J. R. (1973), *J. Supramol. Struct.*, **1,** 208–219.

Konisky, J., Soucek, S., Frick, K., Davies, J. K., and Hammond, C. (1976), *J. Bacteriol.*, **127,** 249–257.

Kumar, S. (1976), *J. Bacteriol.*, **125,** 545–555.

Langman, L., Young, I. G., Frost, G. E., Rosenberg, H., and Gibson, F. (1972), *J. Bacteriol.*, **112,** 1142–1149.

Lankford, C. E. (1973), *CRC Crit. Rev. Microbiol.*, **2,** 273–331.

Lau, C., and Richards, F. M. (1976), *Biochemistry*, **15,** 666–671.

Lederberg, E. M. (1955), *Genetics*, **40,** 580–581.

Leong, J., and Neilands, J. B. (1976), *J. Bacteriol.*, **126,** 823–830.

Lindberg, A. A. (1973), *Annu. Rev. Microbiol.*, **27,** 205–241.

Luckey, M., and Neilands, J. B. (1976), *J. Bacteriol.*, **127,** 1036–1037.

Luckey, M., Pollack, J. R., Wayne, R., Ames, B. N., and Neilands, J. B. (1972), *J. Bacteriol.*, **111,** 731–738.

Luckey, M., Wayne, R., and Neilands, J. B. (1975), *Biochem. Biophys. Res. Commun.*, **64,** 687–693.

Luke, R. K. J., and Gibson, F. (1971), *J. Bacteriol.*, **107,** 557–562.

McIntosh, M. A., and Earhart, C. F. (1976), *Biochem. Biophys. Res. Commun.*, **70,** 315–322.

McIntosh, M. A., and Earhart, C. F. (1977), *J. Bacteriol.*, **131,** 331–339.

Maeda, A., and Nomura, M. (1966), *J. Bacteriol.*, **91,** 685–694.

Manning, P. A., and Reeves, P. (1976a), *Biochem. Biophys. Res. Commun.*, **71,** 466–471.

Manning, P. A., and Reeves, P. (1976b), *J. Bacteriol.*, **127,** 1070–1079.

Manning, P. A., and Reeves, P. (1978), *Mol. Gen. Genet.*, **158,** 279–286.

Manning, P. A., Puspurs, A., and Reeves, P. (1976), *J. Bacteriol.*, **127,** 1080–1084.

Nakae, T., and Nikaido, H. (1975), *J. Biol. Chem.*, **250,** 7359–7365.

Negrin, R. S., and Neilands, J. B. (1978), *J. Biol. Chem.*, **253,** 2339–2342.

Neilands, J. B. (1973), in *Inorganic Biochemistry* (G. Eichhorn, ed.), Elsevier, Amsterdam, pp. 167–202.

Nomura, M., and Maeda, A. (1965), *Zentrabl. Bakteriol. Parasitenkd., Abt. 1. Orig.*, **196,** 216–239.

Nomura, M., Sidikaro, J., Jakes, K., and Zinder, N. (1974), in *Ribosomes* (M. Nomura, A. Tissieres, and P. Lengyel, eds.), Cold Spring Harbor Laboratory, New York, pp. 805–814.

Nurminen, M., Lounatmaa, K., Sawas, M., Makela, P. H., and Nakae, T. (1976), *J. Bacteriol.*, **127,** 941–955.

O'Brien, I. G., and Gibson, F. (1970), *Biochim. Biophys. Acta*, **215,** 393–402.

O'Brien, I. G., Cox, G. B., and Gibson, F. (1971), *Biochim. Biophys. Acta*, **237,** 537–549.

Osborn, M. J., Gander, J. E., Parisi, E., and Carson, J. (1972), *J. Biol. Chem.*, **247,** 3962–3972.

Pollack, J. R., and Neilands, J. B. (1970), *Biophys. Res. Commun.*, **38,** 989–992.

Pollack, J. R., Ames, B. N., and Neilands, J. B. (1970), *J. Bacteriol.*, **104,** 635–639.

Pugsley, A. P. (1977), *FEMS Lett.*, **2,** 275–277.

Pugsley, A. P., and Reeves, P. (1976a), *J. Bacteriol.*, **126,** 1052–1062.

Pugsley, A. P., and Reeves, P. (1976b), *J. Bacteriol.*, **127,** 218–228.

Pugsley, A. P., and Reeves, P. (1976c), *Biochem. Biophys. Res. Commun.*, **10,** 846–853.

Pugsley, A. P., and Reeves, P. (1977a), *Antimicrob. Agents Chemother.*, **11,** 345–358.

Pugsley, A. P., and Reeves, P. (1977b), *Biochem. Biophys. Res. Commun.*, **74,** 903–911.

Randall-Hazelbauer, L., and Schwartz, M. (1973), *J. Bacteriol.*, **116,** 1436–1446.

Rodgers, G. C., and Neilands, J. B. (1973), in *Handbook of Microbiology* Vol. II, (A. L. Laskin, and H. A. Lecheuslier, eds.), CRC Press, Cleveland, pp. 823–830.

Rosenberg, H., and Young, I. G. (1974), in *Microbial Iron Metabolism* (J. B. Neilands ed.), Academic Press, New York, pp. 67–81.

Sabet, S. F., and Schnaitman, C. A. (1971), *J. Bacteriol.*, **108,** 422–430.

Sabet, S. F., and Schnaitman, C. A. (1973), *J. Biol.. Chem.*, **248,** 1797–1806.

Saxe, L. S. (1975), *Biochemistry*, **14,** 2058–2063.

Schaller, K., and Nomura, M. (1976), *Proc. Natl. Acad. Sci. U.S.*, **73,** 3989–3993.

Schein, S. J., Kagan, B. L., and Finkelstein, A. (1978), *Nature*, **276,** 159–163.

Schmitges, C. J., and Henning, U. (1976), *Eur. J. Biochem.*, **63,** 47–52.

Schnaitman, C. (1970), *J. Bacteriol.*, **104,** 890–901.

Schnaitman, C. A., Smith, D., and Founde Salsas, M. (1975), *J. Virol.*, **15,** 1121–1130.

Schwartz, M. (1966), *J. Bacteriol.*, **92,** 1083–1089.

Schwartz, M. (1967), *Ann. Inst. Pasteur, Paris*, **113,** 685–704.

Schwartz, M. (1976), *J. Mol. Biol.*, **103,** 521–536.

Schweizer, M., and Henning, U. (1977), *J. Bacteriol.*, **129,** 1651–1652.

Shannon, R., and Hedges, A. J. (1973), *J. Bacteriol.*, **116,** 1136–1144.

Sidikaro, J., and Nomura, M. (1974), *J. Biol. Chem.*, **245,** 445–453.

Skurray, R. A., Hancock, R. E. W., and Reeves, P. (1974), *J. Bacteriol.*, **119,** 726–735.

Soucek, S., and Konisky, J. (1977), *J. Bacteriol.*, **130,** 1399–1401.

Szmelcman, S., and Hofnung, M. (1975), *J. Bacteriol.*, **124,** 112–119.

Szmelcman, S., Schwartz, M., Silhavy, T. J., and Boos, W. (1976), *Eur. J. Biochem.*, **65,** 13–19.

Takagaki, Y., Kunugita, K., and Mitsuhashi, M. (1973), *J. Bacteriol.*, **113,** 42–50.

Taylor, R. T., Norrel, S. A., and Hanna, M. L. (1972), *Arch. Biochem. Biophys.*, **148,** 366–381.

Thirion, J. P., and Hofnung, M. (1972), *Genetics*, **71,** 207–216.

Timmis, K., and Hedges, A. J. (1972), *Biochim. Biophys. Acta*, **262,** 200–207.

Tokuda, H., and Konisky, J. (1978a), *Proc. Natl. Acad. Sci. U.S.*, **75,** 2579–2583.

Tokuda, H., and Konisky, J. (1978b), *J. Biol. Chem.*, **253,** 7731–7737.

Tokuda, H., and Konisky, J. (1979), *Zentrabl. Bakteriol. Parasitenkd., Abt I., Orig.*, in press.

Uemura, J., and Mizushima, S. (1975), *Biochim. Biophys. Acta*, **413,** 163–176.

Van Alphen, L., Havekes, L., and Lugtenberg, B. (1977), *FEBS Lett.*, **75,** 285–290.

Verhoef, C., deGraaff, P. J., and Lugtenberg, B. (1977), *Mol. Gen. Genet.*, **150,** 103–105.

von Meyenburg, K. and Nikaido, H. (1977), *Biochem. Biophys. Res. Commun.*, **78,** 1100–1107.

Wang, C. C., and Newton, A. (1971), *J. Biol. Chem.*, **246,** 2147–2151.

Wayne, R., and Neilands, J. B. (1974), Abstracts, Pacific Slope Biochemistry Conference, p. 55.

Wayne, R., and Neilands, J. B. (1975), *J. Bacteriol.*, **121,** 497–503.

Wayne, R., and Neilands, J. B. (1976), *Fed. Proc.*, **35,** 1453.

Wayne, R., Frick, K., and Neilands, J. B. (1976), *J. Bacteriol.*, **126,** 7–12.

Weidel, W., Koch, G., and Bobosch, K. (1954), *Z. Naturforsch.*, **9b,** 573–579.

Weiss, M. J., and Luria, S. E. (1978), *Proc. Natl. Acad. Sci. U.S.*, **75,** 2483–2487.

Weltzien, H. U., and Jesaitis, M. A. (1971), *J. Exp. Med.*, **133,** 534–553.

White, J. C., DiGorolamo, P. M., Fu, M. L., Preston, Y. A., and Bradbeer, C. (1973), *J. Biol. Chem.*, **248,** 3978–3986.

Woodrow, G. C., Young, I. G., and Gibson, F. (1975), *J. Bacteriol.*, **124,** 1–6.

Woodrow, G. C., Langman, L., Young, I. G., and Gibson, F. (1978), *J. Bacteriol.*, **133,** 1524–1526.

Wookey, P., and Rosenberg, H. (1978), *J. Bacteriol.*, **133,** 661–666.

Yokota, T., and Kasuga, T. (1972), *J. Bacteriol.*, **109,** 1304–1306.

Young, I. G., and Gibson, F. (1969), *Biochim. Biophys. Acta*, **177,** 401–411.

Young, I. G., Cox, G. B., and Gibson, F. (1967), *Biochim. Biophys. Acta*, **141,** 319–331.

Young, I. G., Langman, L., Luke, R. K. J., and Gibson, F. (1971), *J. Bacteriol.*, **106,** 51–57.

Chapter 11 Nonspecific Transport Through the Outer Membrane

HIROSHI NIKAIDO

Department of Microbiology and Immunology, University of California, Berkeley, California

1 MEMBRANE PERMEABILITY

It is common knowledge that certain substances diffuse spontaneously across various membranes. Such "passive" diffusion processes are driven by the differences in chemical potential of solutions bathing the two sides of the membrane, and the rate of diffusion is expected to be proportional to the magnitude of the driving force, that is, the difference in chemical potential, and thus approximately to the difference in concentration (ΔC) of the diffusing solute. It is also obvious that the rate of diffusion should be proportional to the total area of the membrane (A). Thus the "flux" of the solute molecules across the membrane, or the total number of molecules crossing the membrane (dn) in time dt should be proportional to ΔC and A. If we introduce P as a proportionality constant,

$$\frac{dn}{dt} = -P \cdot A \cdot \Delta C \tag{1}$$

The magnitude of P indicates how permeable a given membrane is to a given solute; P is therefore called the "permeability coefficient," and questions related to the permeability of membranes usually resolve themselves into the determination of P. It should be remembered that P is dependent on other conditions, such as temperature and the presence of other solutes.

The description above actually is grossly oversimplified. The correct theoretical analysis of the diffusion process based on irreversible thermodynamics (Kedem and Katchalsky, 1958, 1961) predicts that at least three coefficients or parameters are needed to describe the diffusion of a solute completely. However, we deal mostly with conditions under which the net solvent flow is minimal and the solute concentrations rather low, so that the classical equation (equation 1) is a reasonable approximation.

What is known about the relationship between the properties of solute molecules and their rates of diffusion across biological membranes, or more accurately, the magnitude of P? The early studies by Collander and Bärlund (1933) on the permeability of *Chara ceratophylla* indicated that the more hydrophobic solutes, as judged by their higher partition coefficients in an olive oil/water system, showed higher values of P. However, among molecules of similar hydrophobicity, smaller molecules penetrated more rapidly, and thus there was an effect of molecular size. This is expected, since the solute molecules must traverse the thickness of membrane by diffusion, and the rate of diffusion, or the magnitude of the diffusion coefficient, is expected to be inversely proportional to the square root (for molecules of less than 1000 daltons) or to the cube root (for larger molecules) of the molecular weight (Stein, 1967). One can correct for this size effect by multiplying P by the square root of the molecular weight, $\sqrt{M}$, and Figure 1 shows a reasonably good correlation thus obtained between $P \cdot \sqrt{M}$ and the olive oil/water partition coefficients (Collander, 1949) if very small solutes are excluded from consideration.

These results are certainly consistent with the mechanism in which the solutes cross the membrane by first dissolving in the hydrophobic interior of

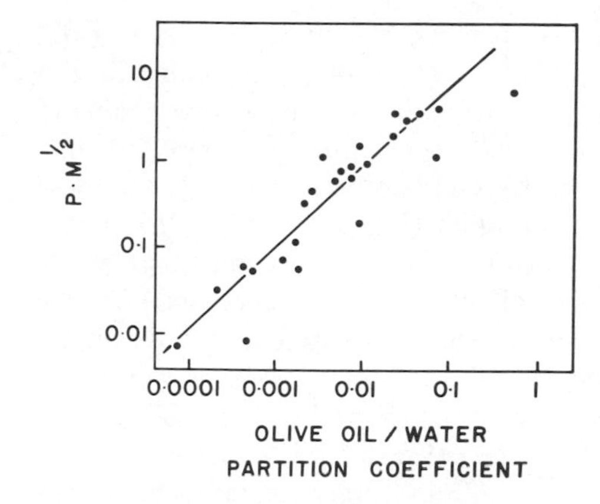

Figure 1 The permeability of *Chara ceratophylla* cytoplasmic membrane to organic nonelectrolytes. Adapted from Collander (1949) by eliminating points corresponding to very small molecules with the molar refraction values of less than 15. P, permeability coefficient; M, molecular weight.

the membrane, then crossing the thickness of this hydrocarbon layer by simple diffusion, and finally redissolving in the aqueous phase bathing the other side of the membrane. On the other hand, very small solute molecules pass through biological membranes much more rapidly than expected from their size and partition coefficients, and their diffusion thus can be explained more easily on the assumption that membranes contain a number of small, water-filled "pores" or "channels" (Solomon, 1968). Subsequently, as a compromise it was proposed that both diffusion through the hydrophobic interior and the passage through pores occurred in a single biological membrane (Wright and Diamond, 1969; Sha'afi et al., 1971). This concept of the coexistence of multiple diffusion pathways has been extremely useful in our analysis of the outer membrane permeability and led us to the unequivocal demonstration of the presence of several diffusion pathways and to the detailed study of their properties (see below). Interestingly, however, to our knowledge the existence of multiple pathways has still not been proven in the more "usual" biological membranes, for example plant cell membranes and red blood cell membranes, for which it was originally proposed. Thus Lieb and Stein (1969) claim that all results are interpretable solely on the basis of diffusion through membrane interior, if a stronger dependence on molecular size, demonstrated to exist in diffusion through synthetic polymer membranes, is assumed, although Sha'afi et al. (1971) claim that their own experimental data do not fit well with the equation of Lieb and Stein.

2 MULTIPLICITY OF DIFFUSION PATHWAYS

The bacterial outer membrane is composed of proteins, phospholipids, and lipopolysaccharides (LPS). LPS, whose structure and biosynthesis have been reviewed extensively (Wright and Kanegasaki, 1971; Lüderitz et al., 1971; Nikaido, 1973; Galanos et al., 1977), is an amphipathic molecule with a hydrophobic end ("lipid A") and a hydrophilic polysaccharide chain. In enteric bacteria, especially in *Salmonella typhimurium*, many mutants defective in almost every step of LPS biosynthesis are known. Since these mutants produce LPS of incomplete structure (Figure 2), it becomes possible to experimentally alter the structure of at least one component of the outer membrane and to study the effect of such alterations on the permeability of the membrane. These studies, initiated by Roantree, Stocker, and their co-workers (Roantree et al., 1969; 1977), as well as by Schlecht, Schmidt, and Westphal (Schlecht and Schmidt, 1969; Schmidt et al., 1969; Schlecht and Westphal, 1970), revealed the following information. (*a*) Wild-type strains of *Salmonella*, *E. coli*, and other related bacteria are naturally sensitive to a number of antibiotics, including neomycin, cycloserine, ampicillin, and cephalothin. The sensitivity to these antibiotics was not affected much by alterations in LPS structure. (*b*) These strains, in contrast, are much more resistant than most gram-positive bacteria to some other antibiotics, including actinomycin D, erythromycin, novobiocin, and rifamycin SV, to such dyes as crystal violet, and to such detergents as bile salts and sodium dodecyl sulfate. Indeed, this difference in dye and detergent sensitivity is the basis for various selective media for gram-nega-

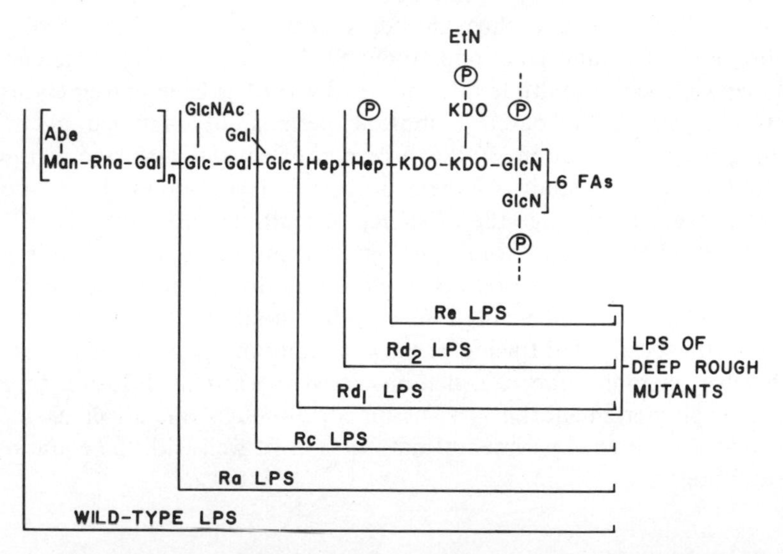

Figure 2 The structure of LPS producted by the wild-type strain, as well as various mutant strains of *S. typhimurium*. Abe, abequose; Man, D-mannose; Rha, L-rhamnose; Gal, D-galactose; GlcNAc, *N*-acetyl-D-glucosamine; Glc, D-glucose; Hep, L-glycero-D-mannoheptose; P, phosphate; KDO, 3-deoxy-D-mannooctulosonic acid (2-keto-3-deoxyoctonic acid); EtN, ethanolamine; and FA, fatty acid.

tive enteric bacteria such as *Salmonella* and *E. coli*, for example, deoxycholate agar and eosinmethylene blue agar. The sensitivity to these agents, however, increased drastically in "deep-rough" mutants of *Salmonella* that produce extremely defective LPS of R_{d_1}, R_{d_2}, and R_e-type (Figure 2) and approached the level seen in gram-positive bacteria. A similar observation was subsequently made with mutants of *E. coli* (Tamaki et al., 1971).

In addition to the genetic approach described above, the structure of the outer membrane can be altered by removing up to 50% of LPS molecules by treating *Salmonella* or *E. coli* cells briefly with EDTA (Leive, 1965c). The removal of LPS by this treatment increases the sensitivity of the organisms toward actinomycin D (Leive, 1965a, b). Subsequent studies, summarized by Leive (1974), showed that the treated cells became sensitive also to novobiocin, rifampicin, long-chain fatty acids, and detergents.

Since the outer membrane is the only subcellular structure that contains LPS in substantial amounts, and since the targets of action of antibiotics and dyes examined are usually located in the cytoplasmic membrane or in the cytoplasm, the results described above suggest that the outer membrane normally acts as a penetration barrier for certain kinds of molecules and that this barrier function can be diminished by the mutational alteration of LPS structure or by the removal of LPS with EDTA. Furthermore, the presence of another group of agents whose efficacy is not influenced by mutations in LPS synthesis suggests the multiplicity of the mechanisms of diffusion across the outer membrane; thus there would be at least one diffusion pathway that is influenced by the structure or the amount of LPS in the outer membrane, and at least another that is not.

How, then, is any compound assigned to one of these pathways? The basis for this assignment had remained unknown for several years, but in 1976, I reviewed the structure of various inhibitors tested and found that most of the compounds whose efficacy was increased in deep-rough mutants and in EDTA-treated cells appeared to be hydrophobic. When the hydrophobicity of these inhibitors was determined quantitatively by measuring their partition coefficients in 1-octanol/phosphate buffer, pH 7.0, it was indeed found that most of these compounds had partition coefficients higher than 0.07 (Table 1). In contrast, the majority of the compounds whose efficacy was unaffected by LPS structure were quite hydrophilic, with partition coefficients lower than 0.02 (Nikaido, 1976) (Table 1). The simplest explanation of these results is as follows. (*a*) There are at least two *general* pathways for the diffusion of small molecules across the outer membrane, one for hydrophilic compounds, and one for hydrophobic compounds. (*b*) The hydrophilic pathway is not affected much by the structure of LPS present. (*c*) The hydrophobic pathway is almost inactive in wild-type strains of enteric bacteria that produce complete LPS; it becomes fully active only

Table 1 Hydrophobicity, Size, and Relative Efficacy Against Deep-Rough Mutants of Various Inhibitory Agents[a]

Agent	Efficacy Ratio (Deep-Rough/ Wild-Type)	Partition Coefficient[b]	Molecular Weight
Actinomycin D		>20	1255
Novobiocin		>20	613
Phenol		>20	94
Crystal violet		14.4	408
Rifamycin SV	≥10[c]	8.8	698
Malachite green		4.2	365
Nafcillin		0.31	414
Oxacillin		0.07	418
Vancomycin		>0.01[d]	≃3300
Penicillin G		0.02	334
Ampicillin		<0.01	349
Cephalothin		<0.01	395
Carbenicillin	0.5–1.0	<0.01	378
Neomycin		<0.01	615
Cycloserine		<0.01	102
Chloramphenicol		12.4[d]	323
Tetracycline		0.07[d]	444

[a] Modified from Nikaido (1976).
[b] Determined in a system containing equal volumes of 1-octanol and 0.05 *M* sodium phosphate buffer, pH 7.0, at 24°C.
[c] Other hydrophobic agents to which deep-rough mutants, EDTA-treated cells, or both become sensitive include fatty acids (Sheu and Freese, 1973), 3,4-benzpyrene (Coratza and Molina, 1978), and a large number of polycyclic carcinogens (Ames et al., 1973).
[d] These compounds do not follow the general tendency for unknown reasons.

in deep-rough mutants or in EDTA-treated cells. The nature of this hydrophobic pathway is described in the next section.

3 BARRIER PROPERTIES AGAINST HYDROPHOBIC MOLECULES

3.1 Measurement of Transmembrane Diffusion Rates

Most biological membranes are known to allow the transmembrane diffusion of small, hydrophobic compounds (Collander and Bärlund, 1933; Collander, 1949; Stein, 1967; Wright and Diamond, 1969); the outer membrane of deep-rough mutants apparently behaves in a similar manner.

In contrast, the results described in the preceding section suggest that the outer membranes of wild-type *S. typhimurium* and *E. coli* constitute an exception in that they are not penetrated by many hydrophobic compounds.

These conclusions, however, were based only on the inhibitory effects measured by using whole cells. A more direct approach was taken by Gustafsson et al. (1973), who showed that the rates of "uptake" of crystal violet by intact cells were dependent on the nature of LPS present. These results, however, were complicated by the massive adsorption of the dye to unidentified cellular constituents, and it was impossible to rigorously rule out the possibility that the differences in uptake were due to different amounts bound to LPS of different structure. We have overcome this difficulty by using a high concentration of a semisynthetic penicillin, nafcillin, as the permeant (Nikaido, 1976). When nafcillin was added to thick suspensions of the cells of deep-rough mutants, its extracellular concentrations at time 0 and at infinite time (actually 5–10 min at 22°C), determined on supernatants after centrifugation, were equal to those predicted on the assumption that nafcillin existed only in the extracellular space at time 0 and was distributed uniformly in extracellular and intracellular spaces at infinite time. Adsorption, degradation, or active transport was therefore negligible under these conditions. Furthermore, the kinetics of diffusion followed a theoretical equation, and first-order rate constants for permeation could thus be calculated. These quantitative diffusion assays led us to the following conclusions (Nikaido, 1976). (*a*) Permeation through the cytoplasmic membrane, measured with EDTA–lysozyme spheroplasts, was at least 10 times as rapid as the penetration rates into intact deep-rough mutant cells. Thus in whole cells the permeation through the outer membrane is the rate-limiting step. (*b*) The permeation rate constants were low (0.005–0.009 min^{-1} at 22°C under our assay conditions) in the wild-type (S) as well as in R_a and R_c mutants (see Figure 2), but were much higher (0.1–1.8 min^{-1}) in deep-rough (i.e., R_{d_1}, R_{d_2}, and R_e) mutants (Figure 3*a*).

It should be emphasized that the permeability of deep-rough mutants to lipophilic compounds serves as a useful tool for microbiologists who want to make *E. coli* or *S. typhimurium* permeable to hydrophobic agents. Thus the Ames carcinogen tester strain of *S. typhimurium* contains a deep-rough mutation so that carcinogens, many of which are hydrophobic, could easily penetrate into the cytoplasm (Ames et al., 1973). In another example, deep-rough mutants were used to supply a hydrophobic intermediate, oleoylglycerol 3-phosphate, to intact cells of *E. coli* (McIntyre and Bell, 1978).

3.2 Mechanism of Diffusion of Hydrophobic Compounds

The mechanism of transmembrane diffusion of hydrophobic molecules has been studied extensively and it is known that the diffusion rates are faster

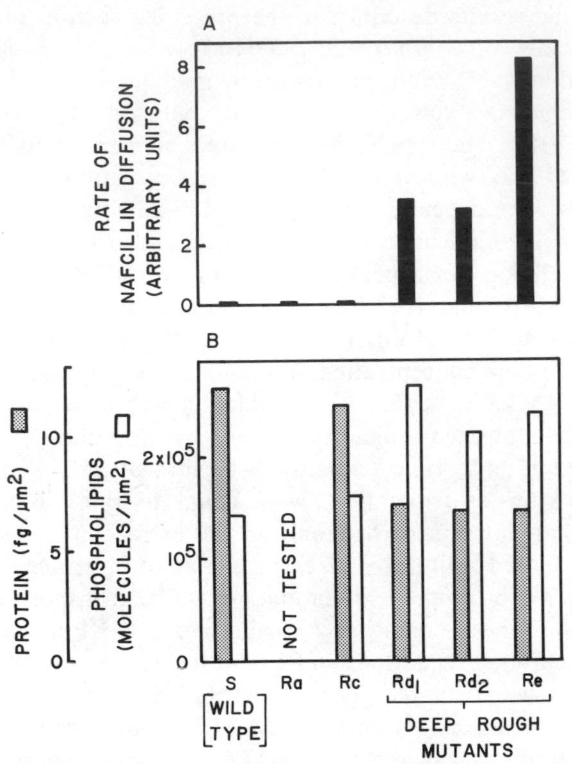

Figure 3 (*A*) Rate of diffusion of a hydrophobic solute, nafcillin, across the outer membranes of *S. typhimurium* strains, and (*B*) the composition of the outer membrane in these strains. Rates of diffusion into intact cells were actually determined, but they reflect the diffusion rates across the outer membrane (for details see Nikaido, 1976). The diffusion rate does not increase significantly even when about 90% of the saccharide residues are lost as in the R_c mutant, but it increases suddenly in correspondence with the increase in phospholipid content (and decrease in protein content) occurring in the deep-rough mutants. The data for part *B* were taken from Smit et al. (1975).

with molecules of smaller size, with those of higher hydrophobicity, and at higher temperature (Collander, 1949; Stein, 1967; Wright and Diamond, 1969; Sha'afi et al., 1971). The critical analysis of these results suggested that the permeant molecules first dissolve in the hydrophobic interior of the membrane, diffuse through the thickness of the hydrocarbon layer, and then cross the membrane by partitioning into the aqueous phase on the other side of the membrane (Stein, 1967). To test whether hydrophobic substances penetrate through the outer membrane of deep-rough mutants by a similar mechanism, we measured the influence of various conditions on diffusion rates and obtained the following results (Nikaido, 1976). (*a*) Among compounds with similar size, the diffusion rate was faster with the more hydrophobic compound. (*b*) The diffusion rates were extremely dependent on temperature, with a Q_{10} value close to 10. (*c*) Compounds of fairly large size apparently penetrated the outer membrane by this mechanism, and there was no indication of a clear-cut size limit. These properties, especially the very high temperature coefficient (Galey et al., 1973), are well known properties of the diffusion process through dissolution into the membrane interior, and we conclude that the hydrophobic compounds penetrate through the deep-rough outer membrane by a similar mechanism.

3.3 Structural Basis for the Lack of Permeability Toward Hydrophobic Compounds

Hydrophobic compounds can diffuse across any biological membrane by the mechanisms just described. In fact they can easily cross artificial phospholipid bilayers, presumably by the same mechanism (Bangham, 1972). The outer membrane of the wild-type *E. coli* or *S. typhimurium* thus seems quite unusual among biological membranes in the sense that they are essentially impermeable to hydrophobic compounds. This property is undoubtedly extremely important for these enteric bacteria, because their normal habitat, the intestinal tract of higher animals, is full of hydrophobic inhibitors, such as bile salts, free fatty acids, and lysophosphatides. Clearly, these organisms would not be able to survive if the outer membrane did not shut out all these noxious compounds.

What unusual structure of the outer membrane makes it impermeable to hydrophobic compounds? Since phospholipid bilayers are permeable as described above, we examined the phospholipid content of the outer membranes of various strains. We were fortunate in having more or less isogenic derivatives of *S. typhimurium*, some of which were impermeable and others of which were permeable to hydrophobic compounds (Figure 3*a*). We first discovered that, in the wild-type *S. typhimurium*, the phospholipid content of the outer membrane *per unit surface area* was quite low and could not completely cover even one side of the membrane (Smit et al., 1975). Furthermore, the phospholipid content remained constant even when most of the polysaccharide chain in LPS was absent as a result of mutation, as in mutants producing R_c LPS (Figure 3*b*). However, only in those "deep-rough" mutants (R_{d_1}, R_{d_2}, and R_e types) showing drastically increased permeability to hydrophobic compounds did we find a significantly increased phospholipid content, which was more than that could be accommodated by one side of the membrane (Figure 3*b*). The outer membrane of the deep-rough mutants contained significantly reduced levels of protein (Figure 3*b*), in confirmation of the earlier results (Ames et al., 1974;

Koplow and Goldfine, 1974), but the number of LPS molecules per unit area of surface remained constant (Smit et al., 1975).

When taken together with the observation that LPS molecules are distributed only on the outer surface of the outer membrane (Mühlradt and Golecki, 1975), these results suggested the following model for the outer membrane of enteric bacteria (Figure 4). (*a*) In the wild-type strains, the outer half of the membrane is almost exclusively occupied by proteins and LPS, and phospholipid molecules are mostly found in the inner half of the membrane. Hydrophobic molecules have difficulty in penetrating this membrane, because they cannot go through the hydrophilic portion of proteins exposed on the surface, and they have difficulty in penetrating the hydrophilic, highly charged "innner core" region of LPS (see Figure 2), as well as through the tightly clustered hydrocarbon chains of LPS. In regard to the last point, it is pertinent that monolayer studies yielded a very small cross-sectional area for LPS hydrocarbons (Romeo et al., 1970), that

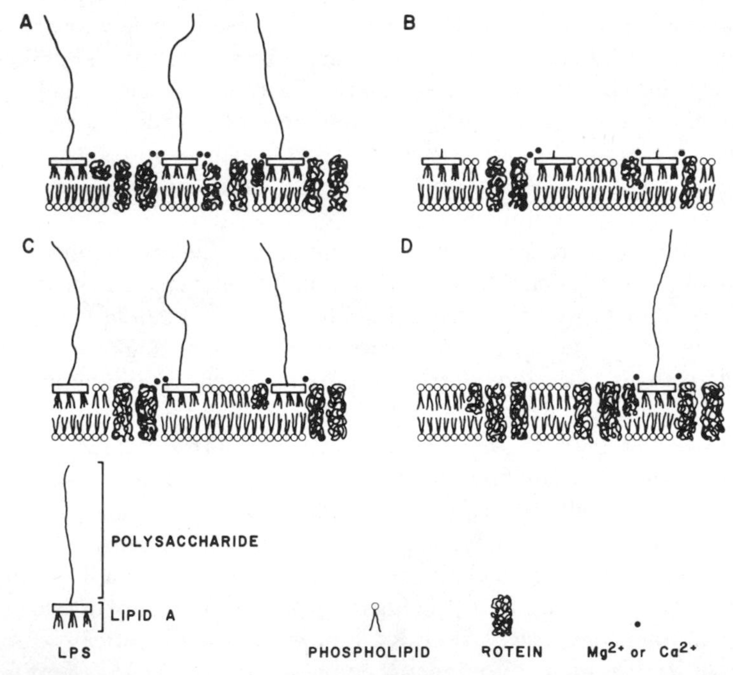

Figure 4 Schematic structure of the outer membrane of *E. coli* and *S. typhimurium*. (*A*) Wild-type strains, (*B*) "deep-rough" mutants producing LPS of R_d- or R_e-type, (*C*) mutants incorporating reduced amounts of proteins into the outer membrane; (*D*) wild-type strains after EDTA treatment. For details see text.

electron spin resonance studies suggested that LPS hydrocarbons are much less mobile than the hydrocarbon chains of phospholipids (Nikaido et al., 1977c), and that a major portion of the resistance in the transmembrane diffusion of hydrophobic compounds arises because of the difficulty in making a hole between the associated hydrocarbon chains (Galey et al., 1973). (*b*) In deep-rough mutants, the incorporation of proteins into the outer membrane is impaired presumably because many of the major proteins interact strongly with LPS (Yu and Mizushima, 1977; Schweizer et al., 1978) and the interaction is rendered difficult because of the extremely defective structure of the mutant LPS. The decrease in protein is compensated by the increased incorporation of phospholipids, which then produce phospholipid bilayer regions in the outer membrane (Figure 4*B*). Hydrophobic molecules can then easily diffuse through these phospholipid bilayer patches. (*c*) The treatment of wild-type cells with EDTA releases LPS or LPS–protein complexes (Leive, 1965c; Rogers et al., 1969; Roberts et al., 1970), presumably because LPS contains many anionic groups and stabilization by divalent cations is essential for maintaining LPS in the membrane. This release is presumably followed by the flip-over of phospholipids from the inner half, or by the lateral diffusion of phospholipids from the cytoplasmic membrane in a manner similar to that described by Jones and Osborn (1977). Thus phospholipid bilayer regions are again generated (Figure 4*D*), and the outer membrane becomes permeable to hydrophobic molecules.

An alternative interpretation of the hydrophobic permeability of the outer membrane of deep-rough mutants is that the saccharide chains of LPS correspond to the major barrier against lipophiles and that the deletion of this carbohydrate layer is the primary factor in making the deep-rough strains permeable. Such a hypothesis, however, predicts a gradual increase in permeability as the saccharide chains become shorter and shorter. A sudden and discontinuous rise in permeability from R_c to R_{d_1} mutants (Figure 3*a*), accompanying a loss of only one glucose residue (see Figure 2), cannot be easily explained by this hypothesis. Furthermore, mutants with a decreased amount of outer membrane proteins but with apparently normal LPS (Figure 4*C*) were shown to be sensitive to hydrophobic inhibitors (Ames et al., 1974), and this result shows conclusively that saccharide layers of LPS are not the primary factor that makes the outer membrane impermeable for lipophilic compounds. (However, they can certainly have quantitative influence on permeability, as is discussed later.)

The model presented in Figure 4 has been tested by three different approaches, and so far the results obtained are completely consistent with the model.

1 Most of phospholipid found in the outer membrane is phosphatidylethanolamine (Osborn et al., 1972). We have devised a covalent-labeling reagent for amino groups, which cannot penetrate through the outer membrane (Kamio and Nikaido, 1976). When the whole cells of wild-type *S. typhimurium* were treated with this reagent, there was practically no labeling of phosphatidylethanolamine, a result suggesting that the head groups of phospholipids are not exposed on the outside surface of the outer membrane. This is not due to the steric hindrance by the polysaccharide chains, because identical results were obtained by using cells of R_c mutants (see Figure 2) that are missing 80–90% of the sugar residues found in the wild-type LPS. Furthermore, with cells of deep-rough mutants, which lack only a few additional sugar residues, a significant fraction of phosphatidylethanolamine reacted with the reagent, just as predicted by the model (Figure 4*B*).

2 Freeze-fracture of the outer membrane of *E. coli* (van Gool and Nanninga, 1971) and *S. typhimurium* (Smit et al., 1975) outer membrane reveals a concave fracture face densely filled with particles, the major constituent of which was shown to be proteins (Smit et al., 1975). In *Pseudomonas aeruginosa* similar particles seen on the concave fracture face appear to be protein–LPS complexes, since EDTA treatment, which releases such complexes from the whole cell (Rogers et al., 1969; Roberts et al., 1970), results in the disappearance of a major fraction of these particles (Gilleland et al., 1973). Thus it seems likely that the outer half of the outer membrane (corresponding to the concave fracture face) of *E. coli* and *Salmonella* is filled with protein–LPS complexes (or "particles"). It should be emphasized that smooth faces corresponding to the phospholipid monolayer were found abundantly on the convex face but not on the concave face. In contrast, the concave fracture face of the outer membrane of deep-rough mutants showed much smooth area, as predicted by the model.

3 Electron spin resonance studies with spin-labeled probes show that the hydrocarbon chains of LPS have rather low fluidity. Thus artificial mixed bilayers, with LPS molecules interdigitated among phospholipid molecules, exhibit significantly lower fluidity than the bilayers made of phospholipids alone (Nikaido et al., 1977c). Yet phospholipid hydrocarbon chains in the outer membrane show a level of fluidity very similar to that in the cytoplasmic membrane (Overath et al., 1975; Nikaido et al., 1977c), a result indicating that phospholipids in the outer membrane occupy a domain completely separated from LPS. This conclusion was also supported by the titration of the size of domains by exchange broadening of ESR line widths (Nikaido et al.,

1977c). In another electron spin resonance study Rottem and Leive (1977) concluded that the amount, as well as the length of the saccharide chain, of LPS affects the fluidity of hydrocarbons in the outer membrane. However, the membranes were prepared by lysozyme digestion in this study, and it is likely that extensive reorganization of membrane components took place (see below). Indeed, the comparison of their spectra with their own earlier results (Rottem et al., 1975) and ours indicates very strongly that in their new preparation the phospholipid hydrocarbons had a much more restricted mobility, presumably owing to the formation of artifactual, LPS–phospholipid mixed bilayer regions.

We therefore believe that our model (Figure 4) is a fairly accurate representation of the overall structure of the outer membrane in enteric bacteria. Not all gram-negative bacteria, however, live in an environment full of hydrophobic inhibitors, and there is no reason to believe that this very asymmetric structure, presumably evolved in enteric bacteria as an adaptation to an unusual environment, should be universally found. In fact, many nonenteric gram-negative bacteria appear to produce outer membranes that are permeable to hydrophobic molecules. for example, many *Pseudomonas* species utilize a number of very hydrophobic molecules as carbon and energy sources (Stanier et al., 1966). *Neisseria gonorrhoeae* is extremely sensitive to such hydrophobic agents as erythromycin, rifampin, acridine orange, ethidium bromide (Maness and Sparling, 1973), and free fatty acids (Miller et al., 1977). Furthermore, a hydrophobic dye, crystal violet, was shown to diffuse as rapidly into *N. gonorrhoeae* as into deep-rough mutants of *E. coli* (Wolf-Watz et al., 1975).

It is possible, and even probable, that this permeability to hydrophobic compounds is mainly due to the presence of porin channels of large diameter (Hancock and Nikaido, 1978; R. E. W. Hancock, L. Zalman, and H. Nikaido, unpublished results) (see also Section 4). On the other hand, there is no reason for these lipophile-permeable outer membranes to maintain the asymmetric structure observed in the enteric bacteria, and indeed freeze-fracture electron microscopy of *Pseudomonas aeruginosa* revealed large, structureless patches on the concave fracture face (Gilleland et al., 1973), suggesting a structure of the outer membrane similar to that shown in Figure 4*C*).

Finally, it should be emphasized that our model is only a first approximation, and a great deal of refinement is needed to understand fully the structure–function relationship in the penetration of the outer membrane by lipophiles. For example, we have always emphasized the significance of phospholipid content per unit area of the membrane, but this does not mean that other factors do not affect the hydrophobic permeability. In fact, our model assumes that LPS and proteins act as barriers agains hydrophobic compounds, and thus it is expected that alterations in these molecules would have some influence on permeability. Coleman and Leive (1978; also personal communication) found that two classes of lipophile-permeable mutants could be isolated from *E. coli*. One is the conventional deep-rough mutant that maps in the *rfa* region and becomes sensitive to most hydrophobic compounds. Another class, which apparently is identical to *acrA*, is impaired in its ability to phosphorylate the proximal part of LPS and interestingly becomes hypersensitive to hydrophobic, cationic inhibitors, a result suggesting that the high concentration of negative charges on the surface of outer membrane acts as a barrier for cationic lipophiles, presumably by binding to them in the aqueous phase and thereby decreasing the lipid/water partition coefficient of these inhibitors. In this connection, it is interesting to recall that *Pseudomonas aeruginosa*, which produces an LPS with a much higher phosphorus content that *E. coli* or *S. typhimurium* LPS (Fensom and Gray, 1969), is in some cases highly resistant to cationic detergents (MacGregor and Elliker, 1958). As another example, *S. typhimurium* strains that fail to synthesize the enterobacterial common antigen and the O-antigen portion of LPS owing to the deltion in *rfb* genes were found to be hyperesensitive to several anionic detergents but not to cationic dyes (Mäkelä et al., 1976). This altered sensitivity, however, is not the direct result of the absence of the polysaccharide chains mentioned above, since the *rff-rfb* double mutants, which are still unable to synthesize these polysaccharide chains, are completely resistant to the detergents. Possibly some other components of the outer membrane are altered, or the inner membrane becomes hypersensitive.

4 NONSPECIFIC PERMEABILITY TOWARD SMALL, HYDROPHILIC MOLECULES

4.1 Equilibrium Studies With Intact Cells

Gram-negative bacteria are sensitive to a number of hydrophilic antibiotics (Table 1) that must penetrate through the outer membrane to reach their targets of action. Similar diffusion is required also for nutrients, including sugars, amino acids, and inorganic salts, as well as for waste products of metabolism. It seems very unlikely that these compounds penetrate by way of the "hydrophobic pathway" described above, since this pathway is essentially absent in wild-type organisms and since, even if the pathway were open, the diffusion of these compounds would be extremely slow owing to their small partition coefficients. These considerations led us to postulate the existence of a separate penetration mechanism for hydrophilic compounds, a mechanism presumed to operate fully in the wild-type strains, as well as in deep-rough mutants. We studied the properties of this postulated pathway by using oligosaccharides as permeants (Decad et al., 1974; Decad and Nikaido, 1976).

To study their penetration through the outer membrane, it was necessary to use oligosaccharides that neither penetrated through nor were actively transported across the cytoplasmic membrane. Oligosaccharides of the sucrose–raffinose series fulfilled these conditions in *S. typhimurium* and were used in most of our experiments. Raffinose (galactosyl-sucrose), stachyose (galactosyl-galactosyl-sucrose), and verbascose (galactosyl-galactosyl-galactosyl-sucrose) have an added advantage that they can be easily labeled with tritium by oxidation of C-6 of the terminal galactose residue with galactose oxidase, followed by its reduction by NaB^3H_4. Since the space between the outer membrane and the cytoplasmic membrane (i.e., the "periplasmic space") is small, the diffusion into this space was difficult to measure. [A recent paper by Stock et al. (1977) concludes that the periplasmic space corresponds to 20–40% of the total cell volume in normal, unplasmolyzed *S. typhimurium* and *E. coli*. In our hands, however, the space never exceeded 5% of the cell volume, except in starved cells or stationary phase cells in which up to 13% of the cell volume became the periplasmic space (Decad, 1976). The cause for this discrepancy is not clear at present.] We therefore expanded the periplasmic space by plasmolyzing cells in 0.3–0.5 *M* NaCl or in 0.5 *M* sucrose (Decad and Nikaido, 1976). Under these conditions, 40–50% of the cell volume was occupied by the periplasmic space, and the extent of penetration of oligosaccharides into this space could easily be measured by centrifuging down plasmolyzed cells after incubation with the radioactive oligosaccharides and by determining the concentration of the radioactive substance both in the supernatant and in the pellet. Corrections for intercellular space in the pellet were made by adding to the incubation mixture, large, uncharged polymers (e.g., [^{3}H]dextran), which were assumed to be impermeable through the outer membrane.

These studies showed the existence of a clear size limit in the penetration process through this pathway (Decad and Nikaido, 1976). In *E. coli* and *S. typhimurium*, sucrose (342 daltons) and raffinose (504 daltons) penetrated fully into the periplasmic space after 5 min of incubation at room temperature, but stachyose (666 daltons) and large oligosaccharides showed only partial penetration. The partial penetration was apparently due to the heterogeneity of the cell population, as prolonged incubation did not increase the degree of penetration. Furthermore, much of the heterogeneity seemed to have been caused by damages produced during the preparation of plasmolyzed cells. Thus in experiments in which the cells were prepared rapidly and the degree of plasmolysis was not extreme, stachyose penetrated little (less than 25% penetration), whereas verbascose (828 daltons), inulins, and dextrans did not penetrate at all. It seems likely that the slight penetration of stachyose represents diffusion into damaged cells, and we conclude that the cell wall of *E. coli* and *S. typhimurium* is permeable to a disaccharide (sucrose) and a trisaccharide (raffinose) but is essentially impermeable to tetrasaccharides and higher oligosaccharides. There thus seems to be a sharp exclusion limit around 550–650 daltons for the raffinose series of oligosaccharides.

It was gratifying to note that these results were consistent with the results of a pioneering study by Payne and Gilvarg (1968), who found that peptides larger than about 600–650 daltons could not enter *E. coli* cells and postulated the presence of a barrier "external to the peptide transport system." However, neither the study of Payne and Gilvarg (1968) nor our results described above told us which of the two layers of the cell wall, the outer membrane and peptidoglycan, acted as the limiting barrier for the penetration of hydrophilic molecules. Our later studies, however, clearly showed that the outer membrane acted as limiting molecular sieve (Nakae and Nikaido, 1975). In one series of experiments the peptidoglycan layer was degraded by lysozyme treatment or by growth in the presence of penicillin. The plasmolyzed cell or spheroplasts produced from these bacteria still showed the same molecular sieving properties seen before, in spite of the absence of the intact peptidoglycan layer. In another series of experiments, the outer membrane was isolated in the form of closed vesicles. These vesicles apparently had a number of "cracks" probably produced as a result of incomplete resealing of membrane fragments. Heating in the presence of Mg^{2+}, however, was found to seal most of these cracks and produced vesicles having permeability properties resembling those of plasmolyzed cells.

How do the hydrophilic molecules pass through the outer membrane? As is stated earlier, diffusion by dissolution into membrane interior seems impossible, as this would require breaking a great many hydrogen bonds between the oligosaccharides and water molecules, thermodynamically a very unfavorable operation. The only possible mechanisms seem to be diffusion through water-filled pores and carrier-mediated diffusion. It is not possible to distinguish absolutely between these two mechanisms with the data described above. Even the theoretical line of demarcation is sometimes unclear, as the latter mechanism can be considered as diffusion through specialized pores (Singer, 1974). We found, however, that diffusion of oligosacchrides through the outer membrane was extremely rapid even at 0°C (Decad and Nikaido, 1976), and this observation, as well as the identical size limit for widely different kinds of compounds, seemed to favor the aqueous pore mechanism. We sought a more definitive answer to this question through reconstitution studies described in the next section.

4.2 Reconstitution Studies

When the outer membrane is dissociated by detergents and the detergents are then slowly removed in the presence of Mg^{2+}, one observes the reformation of structures morphologically resembling the outer membrane. Experiments of this type were performed on *E. coli* outer membrane first by DePamphilis (1971) by using dissociation by Trition X-100 and EDTA, and by Bragg and Hou (1972), as well as by Sekizawa and Fukui (1973), who used SDS as the dissociating agent and more recently by Nakamura and Mizushima (1975), who used partially purified components. These experiments established that the supramolecular structure of the outer membrane was ultimately determined by the structure and properties of the component molecules. However, they did not throw much light on the nature of interaction between the component molecules or on the functions of individual components.

In our laboratory, reconstitution experiments were performed to identify the components that are responsible for the characteristic hydrophilic permeability of the outer membrane. Since both the hydrophilic permeability and the presence of LPS are unique attributes of the outer membrane, we first examined the possibility that LPS might be involved in the formation of pores. We prepared liposomes containing both LPS and phospholipids in the bilayer structure by resuspending a dried film of phospholipids in an aqueous suspension of LPS, as described by Kinsky's group (Kataoka et al., 1971). These liposomes bounded by the mixed phospholipid–LPS bilayer were, however, as impermeable toward hydrophilic compounds as the liposomes made only of phospholipids (Nikaido and Nakae, 1973), that is, both of them allowed a fairly rapid penetration of glycerol and a slow diffusion of erythritol but were essentially impermeable to glucose, sucrose, and higher oligosaccharides. These results led us to the alternative hypothesis that outer membrane proteins play an essential role in the formation of pores. This hypothesis was tested in our laboratory by Nakae (1975), who made liposomes in the presence of outer membrane proteins. For this purpose, the outer membrane proteins were extracted using 0.7 *M* lithium diiodosalicylate. Removal of the chaotropic agent by extensive dialysis resulted in the precipitation of proteins, which were then recovered by centrifugation and resuspended by sonication in an aqueous buffer. Membrane vesicles were reconstituted by adding this protein suspension and an aqueous suspension of LPS to a dried film of phospholipids. When the reconstitution medium contained both [^{3}H]dextran and [^{14}C]sucrose, both were presumably present in intravesicular, as well as extravesicular space. When the vesicles were then separated from the medium by gel filtration through Sepharose 4B, a large portion of intravesicular [^{14}C]sucrose diffused out through the protein-containing membrane during the filtration step, so that the ^{3}H/^{14}C ratio of the recovered liposome preparation was usually several times higher than the ^{3}H/^{14}C ratio of the initial reconstitution mixture. A control mixture lacking proteins produced liposomes essentially impermeable to sucrose, that is, with hardly any increase in the ^{3}H/^{14}C ratio. In another control experiment, proteins from sheep red blood cell membranes were completely inactive in the production of sucrose-permeable pores. We conclude from these results that outer membrane proteins are essential in the formation of water-filled pores.

We then used partially purified preparations of outer membrane proteins for reconstitution to identify the protein(s) involved in pore formation (Nakae, 1976a). As was originally shown by Rosenbusch (1974) in *E. coli* B, treatment of *S. typhimurium* cell envelope with SDS at 37°C leaves behind an insoluble complex that contains peptidoglycan with covalently linked lipoprotein, as well as the "peptidoglycan-associated" major proteins of 34,000–36,000 daltons. We found that the SDS-insoluble complex was much more active than the SDS-solubilized proteins in producing the pores. We then degraded the former with lysozyme and fractionated the products by gel filtration in SDS. This produced a protein aggregate containing only the 34,000, 35,000, and 36,000 dalton (34K, 35K, and 36K) proteins from *S. typhimurium* (Nakae, 1976a), that containing only approximately 36,000 dalton proteins from *E. coli* K-12 (Nakae, unpublished results), and a similar oligomer composed only of the 36,500 dalton protein from E. coli B (Nakae, 1976b), all of which were extremely active in producing pores in the reconstitution experiments (Table 2). Because of this activity, the name, "porin" was proposed for these proteins (Nakae, 1976b).

The pores produced by the addition of porins were of the expected size, in that they allowed a free diffusion of sucrose and raffinose yet essentially excluded stachyose and larger oligosaccharides. The vesicles formed were also permeable to a wide variety of small, hydrophilic molecules, including galactose, glucosamine, glucose 1-phosphate, lysine, tryptopham, uridine, UMP, GDP, and polyethyleneglycol of 600 daltons, but not to polyethyleneglycol of 1540 daltons (Nakae, 1976a). Since complexes containing only one to three protein species can produce permeability to such a wide variety of substances, the mechanism of diffusion cannot be the facilitated diffusion that requires a specific carrier protein for each compound. This lack of specificity is indeed one of the strongest pieces of evidence suggesting the involvement of aqueous pores.

Table 2 Results of a Reconstitution Experiment[a]

	^3H (Dextran) (cpm)	^{14}C (Sucrose) (cpm)
Added to the reaction mixture	240,000	260,000
Recovered in vesicles		
Exp. I (without porins)	2,400	2,100
Exp. II (with 10 μg porin)	3,350	<30

[a] Membrane vesicles were reconstituted from 0.8 mg LPS and 0.8 mg phospholipids, with or without 10 μg of porin oligomer ["Complex I" of Nakae (1976a) from *S. typhimurim*], in the presence of [^3H]dextran and [^{14}C]sucrose. Vesicles were recovered after gel filtration, which removed radioactive saccharides in the extravesicular space, as well as those able to diffuse out through the vesicle membrane. For details see Nakae (1976a).

The channel-forming ability of the porin has recently been confirmed by a different approach. Benz et al. (1978) added an extremely dilute (e.g. 0.5 ng/ml) solution of *E. coli* B porin to the aqueous solution in which a planar lipid bilayer membrane was bathed. Interestingly, the conductance increased in a stepwise manner, and the single incremental step of 1.7 nS (nanosiemens) (in 1 *M* KCl), suggesting the opening of a channel of about 1 nm in diameter, was interpreted to arise from the insertion of one molecule or oligomer of the porin into the membrane. At slightly higher (e.g., 20 ng/ml) concentrations of porin, the conductance eventually increased about 10^5-fold, clearly indicating that the effect of porin is distinct from the nonspecific effect of many hydrophobic proteins that increase conductance only up to a hundredfold (Montal, 1976). In a similar approach, Schindler and Rosenbusch (1978) have incorporated *E. coli* B porins into planar phospholipid bilayers in the absence of detergents or hydrocarbons. This made the membrane permeable to ions and a nonelectrolyte, glucose. The most interesting feature of this work was the suggestion that the opening of the pores was controlled by the voltage across the membrane. Thus voltages in excess of 140 mV must be applied to open up the pores initially, and even after this "initiation," the conductivity through the pore shows a very interesting dependence on voltage in that the current decreases with increasing voltage if the latter exceeds 140 mV. Possibly these results suggest that the permeability through the pore is physiologically regulated, perhaps by means of the potential across the outer membrane. However, the potentials that have been measured (Stock et al., 1977) are far lower than those effective in this reconstituted system.

Reconstitution of hydrophilic pores has been achieved also by using outer membrane proteins of species other than *E. coli* or *S. typhimurium*. Decad (1976) showed that proteins in the "SDS-nonextractable fraction" of *Proteus morganii* cell envelope, as well as the Triton X-100–EDTA-extracted proteins of *Pseudomonas aeruginosa* cell envelope, could be used to reconstitute sucrose-permeable membranes. More recently, Nixdorff et al. (1977) achieved similar results with 39,000 dalton outer membrane protein from *Proteus mirabilis*, extracted with 50% acetic acid.

A partial reconstitution from isolated phospholipids, LPS, and a small amount of outer membrane fragments has been performed by using components from *P. aeruginosa* (Hancock and Nikaido, 1978). The results suggested, interestingly, that the exclusion limit of *P. aeruginosa* pores is several thousand daltons, much higher than in the enteric bacteria. This is in agreement with our earlier observation with plasmolyzed cells (Decad and Nikaido, 1976), although data obtained with plasmolyzed cells were thought to be unreliable because of the possibility of plasmolysis-induced damages. In any case, reconstituted vesicles give much better defined cutoff curves (Nakae, 1976a) and thus seem to contain much fewer nonspecific "cracks"; these data are therefore more reliable. The *P. aeruginosa* porin has recently been purified by ion-exchange chromatography in Triton X-100 and was identified as thc 35,000–37,000 dalton major protein(s) (Hancock et al., 1979).

4.3 Rate Studies With Intact Cells

If all the pores in the outer membrane are surrounded by porin molecules, the mutational loss of the porin is expected to be lethal. However, the loss of these proteins (Davies and Reeves, 1975; Foulds 1976; Nurminen et al., 1976), or of all "major" proteins (Henning and Haller, 1975), reportedly did not produce any marked reduction in growth rates. The solution to this puzzle was provided in 1977 by studies utilizing two different methods of approach. Both of these methods enabled us to estimate the *rates* of diffusion across the outer membrane, rather than the *equilibrium distribution* of solutes described in Sections 4.1 and 4.2, and gave us new insights into the function of the outer membrane pores.

As is mentioned earlier, the diffusion rates across the outer membrane were too rapid to measure with the equilibrium method described earlier. Zimmermann and Rosselet (1977) recently devised an elegant method to measure these rates. They measured the rates of hydrolysis of β-lactam antibiotics by intact cells, as well as by broken cells of *E. coli* harboring an R factor. The β-lactamase coded for by the R factor is located in the periplasmic space. Thus with intact cells, the rate of diffusion across the outer membrane should be balanced, at the steady state, by the rate of hydrolysis in the periplasmic space. Zimmermann and Rosselet (1977) made a very important contribution in showing that the rates of hydrolysis by intact cells indeed followed this prediction and could be calculated by assuming that the diffusion and hydrolysis respectively follow Fick's law and Michaelis-Menten kinetics. It is also necessary to point out that this quantitative treatment is absolutely required for the discussion of outer membrane permeability. For example, determination of "crypticity" (ratio of rates of hydrolysis in intact cells and in broken cells) at an arbitrarily chosen substrate concentration, a procedure unfortunately still followed by some workers, means very little because the mathematical equations show us that in any system the numerical values of "crypticity" can be altered freely, depending on the substrate concentration used (Zimmermann and Rosselet, 1977, see also below).

We have followed the Zimmermann-Rosselet approach and have determined the permeability coefficients of the outer membrane in various strains of *S. typhimurium* (Nikaido et al., 1977b). The results showed clearly that the permeability of the outer membrane toward a hydrophilic β-lactam of 415 daltons, cephaloridine, decreased significantly when one of the three porins was lost through mutation, and that mutants with reduced levels of all three porins usually had permeability coefficients less than 10% of that found in the wild type. Similar results were independently obtained with *E. coli* K-12 by workers who approached the problem from the opposite direction, that is, by examining membrane alterations in "cryptic" mutants. These mutants, studied by Beacham et al. (1977), showed an increased K_m but unaltered V_{max} for the hydrolysis, by intact cells, of nucleotides by periplasmic nucleotidases. When the outer membrane of these mutants was analyzed, it was found to have drastically reduced levels of porins (Beacham et al. 1977).

In the approach described above, the inherent difficulty in the determination of diffusion rates across the outer membrane was solved by the use of enzymes that hydrolyzed substrates rapidly as they came into the periplasmic space. The requirement for the assay of outer membrane permeability is thus the efficient removal of diffusing substrates from the periplasmic space; only under these conditions the overall rates of the process are determined predominantly by the diffusion rates across the outer membrane, and thus the latter rates become measurable. The removal of substrates can also be performed by active transport systems located in the cytoplasmic membrane, and this forms the basis of the second approach to be described, that is, the study of overall transport kinetics and "apparent transport K_m" values. (It should be emphasized again that uptake assays at a single, arbitrary concentration do not give much information.) The diffusion rate (V)

across the outer membrane is determined by Fick's law, and if we denote by C_o and C_p the substrate concentrations in the outside medium and in the periplasmic space, respectively, then

$$V = P \cdot A \cdot (C_o - C_p) \tag{2}$$

where P and A are permeability coefficient and the area of the membrane, respectively. The rate of active transport follows Michaelis-Menten kinetics, so that

$$V = \frac{C_p \cdot V_{\max}}{K_m + C_p} \tag{3}$$

At steady state the two rates are equal, and we can eliminate C_p. Thus if we substitute the value of C_p obtained from equation 2 into equation 3 we get

$$V = \frac{(P \cdot A \cdot C_o - V) \cdot V_{\max}}{P \cdot A \cdot K_m + P \cdot A \cdot C_o - V} \tag{4}$$

We can solve this equation for V, and this leads us to

$$V = \tfrac{1}{2} \{V_{\max} + P \cdot A \cdot K_m + P \cdot A \cdot C_o - [(V_{\max} + P \cdot A \cdot K_m + P \cdot A \cdot C_o)^2 - 4 \cdot C_o \cdot P \cdot A \cdot V_{\max}]^{1/2}\} \tag{5}$$

This equation looks rather unwieldy, but it can be simplified to more familiar forms in extreme cases. First, let us consider the case where the outer membrane permeability (P) is so high in comparison with V that the outer membrane practically does not offer any resistance in the overall transport process. For this situation, it is more convenient to rewrite equation 4 as follows

$$\frac{1}{V} = \frac{P \cdot A \cdot K_m + (P \cdot A \cdot C_o - V)}{(P \cdot A \cdot C_o - V) \cdot V_{\max}} = \frac{1}{V_{\max}} + \frac{K_m}{V_{\max}} \cdot \frac{P \cdot A}{(P \cdot A \cdot C_o - V)} \tag{6}$$

Now, from equation 2, $P \cdot A \cdot C_o - V = P \cdot A \cdot C_p$. Since the diffusion through the outer membrane is not limiting, $C_p \doteqdot C_o$ and $(P \cdot A \cdot C_o - V) \doteqdot P \cdot A \cdot C_o$. This simplifies equation 6 into the familiar Lineweaver-Burk form of the Michaelis-Menten equation,

$$\frac{1}{V} = \frac{1}{V_{\max}} + \frac{K_m}{V_{\max} \cdot C_o} \tag{7}$$

Thus, as expected, the kinetics of the overall transport process are determined almost exclusively by the properties of the active transport system in the inner membrane, and the effect of the outer membrane barrier is hardly noticeable.

Let us next consider another extreme case, in which the permeability of the outer membrane is quite low. In this case, in order to get half-maximal rate of V or even 10% of the maximal rate of V, one has to use a very high concentration of C_o, so in the regions of our interest, generally $C_o \gg K_m$ and $P \cdot A \cdot C_o \gg P \cdot A \cdot K_m$. From equation 2 $V > P \cdot A \cdot C_o$, so that $V_{max} > P \cdot A \cdot C_o \gg P \cdot A \cdot K_m$. Thus in equation 5, terms containing $P \cdot A \cdot K_m$ can be neglected. This leads to

$$V = \tfrac{1}{2} \{V_{max} + P \cdot A \cdot C_o - [V^2_{max} - 2 \cdot P \cdot A \cdot C_o \cdot V_{max} + (P \cdot A \cdot C_o)^2]^{1/2}\} = P \cdot A \cdot C_o \quad (8)$$

Lineweaver-Burk plots in this case yield a straight line, which, when extrapolated, intersects the y-axis not at $1/V_{max}$, but at the origin. The slope of the line is directly related to $P \cdot A$.

Lastly, at very high values of C_o, the C_p becomes quite high regardless of the absolute values of P, according to equation 2. At high values of C_p, V approaches V_{max} according to equation 3. Thus we arrive at an important prediction, that the changes in the permeability of the outer membrane will not affect the maximal rates of overall transport process, although they will alter the slopes of Lineweaver-Burk plots, and thus the value of "apparent K_m," which is defined as the value of C_o giving the half-maximal rate of transport. It should be noted here that the traditional method of extrapolating the Lineweaver-Burk plots and determining the intercepts with the x-axis does not work, because, as is shown earlier, the extrapolated straight line tends to intersect the origin if the outer membrane constitutes a significant barrier for penetration.

The results of calculations, based on the values of K_m, P, and V_{max} assumed for the glucose transport in *E. coli* are shown in Figure 5. It is seen that the decrease in outer membrane permeability (P), expected to be found in porin-deficient mutants, increases the "apparent K_m of the overall transport" (K'_m) without affecting the V_{max}.

These considerations suggest that the effect of porin deficiency on transport rates becomes apparent only at low substrate concentrations (or high values of $1/C_o$). Furthermore, the effect will be difficult to detect even at low values of C_o if the V_{max} of the active transport system in the inner membrane happens to be low.

Actually, mutants with the predicted change, that is, increased "K_m" values for the transport of a number of substrates, have been isolated some time ago by von Meyenburg (1971) without the expectation of any outer membrane defect. These mutants were recently shown to be deficient in porins, in the 36.5K porin in *E. coli* B/r and in the two porins (proteins 1a and 1b) in *E. coli* K-12 (Bavoil et al., 1977). Since the porin-deficient

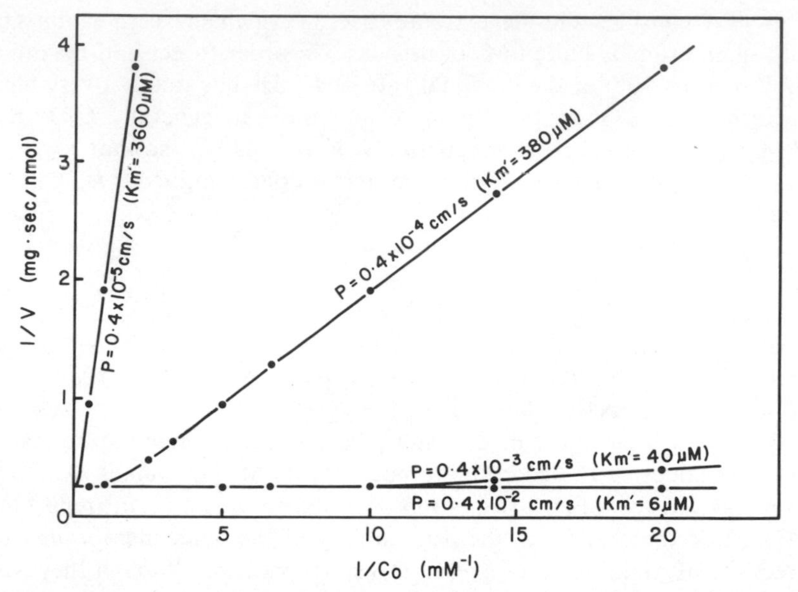

Figure 5 Theoretically predicted behavior of overall transport systems. Shown is the predicted behavior of a high V_{max} (4 nmole/mg dry wt. sec), low K'_m (2 μM) transport system in the wild-type strain, as well as in mutants producing 10, 1, and 0.1% of the normal amount of porin. In the wild-type cell, where an outer membrane permeability of 0.4×10^{-2} cm/sec was assumed, the presence of the outer membrane raises the overall K'_m from 2 to only 6 μM. In mutants, apparent K'_m values are increased, but V_{max} does not change. The plots approximately follow straight lines except at very high values of C_o. Because of this deviation from linearity, the apparent K'_m values were obtained as C_o values giving the half-maximal transport (or growth) rates, rather than by extrapolation of the linear portion. The values of V_{max}, P, and K_m used approximate those obtained experimentally or assumed for the transport of glucose in *E. coli* B/r (Bavoil et al., 1977; H. Nikaido and E. Y. Rosenberg, unpublished results). The value of V_{max} is a minimal estimate, and K_m was obtained by adjusting its values so that the "overall transport K_m" (K'_m) would equal the observed value (6 μM).

mutants have an increased K_m for the transport of various sugars, a sugar alcohol, a sugar phosphate, amino acids, a pyrimidine, and even inorganic anions like phosphate and sulfate (von Meyenburg, 1971), this approach establishes the role of porins in the transmembrane diffusion of a very wide range of hydrophilic substrates. In addition, if we assume that the active transport systems in the cytoplasmic membrane have very high affinities (or low K_ms) for the substrate, we can estimate the minimal values of permeability coefficient toward various substrates (Bavoil et al., 1977).

The transport studies and the β-lactamase approach in our laboratory have so far produced the following information.

1 The permeability coefficient (P) for glucose in the wild-type *E. coli* B/r strain was calculated to be larger than 1.4×10^{-2} cm/sec (Bavoil et al., 1977), but this estimate should be changed to larger than 0.3×10^{-2} cm/sec in view of more recent data (H. Nikaido and E. Y. Rosenberg, unpublished data). The significance of this figure is explained as follows. If the outer membrane did not offer any resistance to the diffusion of glucose, the value of P can be estimated from $P = D/d$, where D and d are the diffusion coefficient of glucose in water and the thickness of the membrane, respectively. Thus at 25°C, $P = 0.673 \times 10^{-5}$ cm^2/sec $\div$ (5×10^{-7}) cm $=$ 13.5 cm/sec. The observed value is lower by a factor of at least 10^3, and we see that the outer membrane does constitute a significant barrier, as predicted.

Since the outer membrane is permeable to raffinose, the hydrated radius of which is 0.60 nm (Schultz and Solomon, 1961), as a first approximation we can assume the presence of pores 0.6 nm in radius. Since there are about 10^5 molecules of porin per cell (Rosenbusch, 1974), the total area of the cross section of pores per cell is $(0.6 \times 10^{-3})^2 \times 3.14 \times 10^5$ μm^2 if one molecule of porin is assumed to produce one pore. Since the surface area of the average *E. coli* cell is on the order of 3 μm^2, the expected permeability coefficient can be calculated by multiplying the "free diffusion" permeability coefficient described above, by the relative proportion of the pore area. Thus P (expected) $=$ 13.5 cm/sec $\times$ 0.11 μm^2 $\div$ 3 μm^2 $=$ 0.5 cm/sec. The observed value is lower than this. But even molecules with radii smaller than the pore radius are significantly hindered in their diffusion because of their collision with the rims of the pore and of the viscious drag along the pore wall. According to the so-called Renkin equation (Renkin, 1954), the area *available for diffusion* is obtained by multiplying the geometrical area by a factor, $(1 - a)^2 (1 - 2.104a + 2.09a^3 - 0.95a^5)$, where a is the ratio of the solute radius to the radius of the pore. If we calculate this correction factor by using the hydrated radius of glucose, 0.42 nm (Schultz and Solomon, 1961), and correct, by this factor, the expected P value obtained above, a value of 0.37×10^{-2} cm/sec is obtained, which is very close to the "observed" value of more than 0.3×10^{-2} cm/sec.

2 The temperature dependence of the diffusion rate was found to be quite low, in contrast to the diffusion of hydrophobic substances described in Section 3. Diffusion of a β-lactam antibiotic, cephaloridine, gave a Q_{10} value of about 2 (H. Nikaido and E. Y. Rosenberg, to be published); this small temperature dependence is strong evidence that the permeation occurs through the water-filled pores. The simplest explanation for the observation that the Q_{10} value was slightly higher than that expected for free diffusion in water, that is, $Q_{10} = 1.2$, is that the water inside the pore is more structured or more strongly associated than the bulk water (see below).

3 The most remarkable observation is the presence, among permeable solutes, of a very large difference in permeability coefficients. Thus in wild-type cells, the permeability coefficient of $\geq 0.3 \times 10^{-2}$ cm/sec for glucose can be contrasted with at least a hundredfold lower coefficient for 6-aminopenicillanic acid (1.4×10^{-5} cm/sec), whose size (216 daltons) is not much different from that of glucose (180 daltons). We do not yet fully understand the factors that influence the rate of penetration through these pores, but a few parameters are already known to be important. The first factor is obviously the size of the solute. In the "apparent transport K_m" experiments, the calculation of outer membrane P values on the basis of data on wild-type cells gives only minimal values with a large margin of uncertainty. In contrast, the overall transport rates in intact cells of porin-deficient (*ompB*) mutants are limited essentially by diffusion rates through the outer membrane and thus allow a fairly accurate calculation of P values (Bavoil et al., 1977). Since the *ompB* gene is most probably a regulatory gene (Ichihara and Mizushima, 1978), the mutants are expected to produce smaller numbers of unaltered pores. Thus if we assume that hydrophilic solutes diffuse through the mutant outer membrane by way of the remaining porin channels, the data shown in Figure 6 indicate clearly the effect of solute size, and furthermore there is a remarkably good fit with the theoretical values calculated on the basis of the Renkin equation, assuming the presence of pores of 0.6 nm in radius.

The second factor is the hydrophobicity. Hydrophobic antibiotics could not rapidly penetrate through the outer membrane of wild-type *E. coli* or *S. typhimurium* even when their size was clearly below the exclusion limit of the pores (Nikaido, 1976). This inability to rapidly penetrate the outer membrane suggested that the pores tend to exclude hydrophobic solutes, and this idea was tested directly by Zimmermann and Rosselet (1977). These workers compared the outer membrane permeability of *E. coli* to that of a number of semisynthetic cephalosporin derivatives, and their data, shown in graphic form in Figure 7, confirm the inverse correlation between hydrophobicity and transmembrane diffusion rates (through pores), at least for a series of monoanionic cephalosporins. Thus, in this series, an increase of hydrophobicity producing a tenfold increase in partition coefficient is seen to produce a sixfold decrease in permeability coefficient. At present it is not known why pores exlude hydrophobic molecules. An obvious possibility is that these hydrophobic antibiotics cannot penetrate into the pores because they form micelles. However, although these substances do indeed form micelles at very high concentrations, they seem to exist mostly as monomers at the concentrations used for permeability study (H. Nikaido and P. Bavoil, unpublished results). Perhaps it is significant that the diameter of the pore is not much different from the diameter of solute

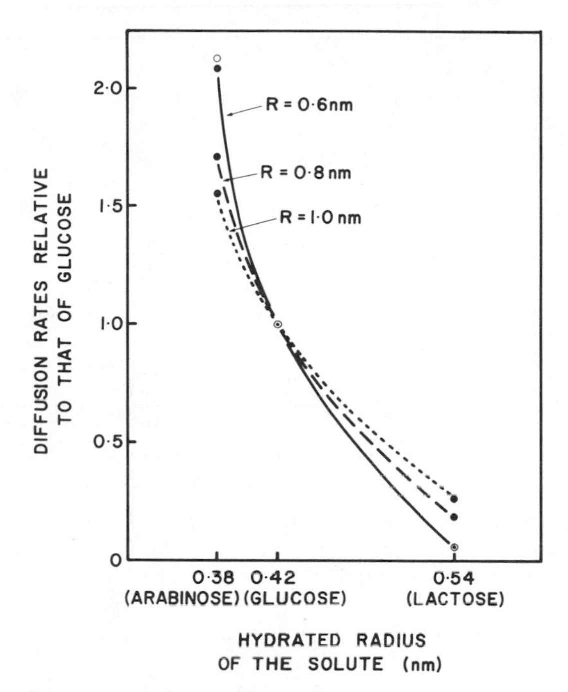

Figure 6 Theoretical and observed permeability toward solutes of various sizes. The theoretical values (●) were calculated by multiplying the free diffusion coefficients for arabinose, glucose, and lactose by appropriate Renkin factors calculated for various pore sizes. The value of hydrated radius used were determined according to Schultz and Solomon (1961). The observed permeability (○) was calculated from the "growth K_m" of porin-deficient (*ompB*) mutant CM7 on these carbon sources (H. Nikaido and E. Y. Rosenberg, to be published) and from the growth rate and growth yields (for details, see Bavoil et al., 1977). *R*, pore radius.

molecules. Thus the walls of the pore are probably lined with hydrophilic or hydrogen-bonding structures, so that in the "empty" pore, these structures are rather strongly hydrogen bonded to water molecules in the pore. This makes the water in the channel more structured or more strongly hydrogen bonded than the bulk water. Penetration of hydrophobic molecules into the pore breaks this structure and thus may be energetically unfavorable. In contrast, hydrophilic molecules may be able to get into the pore, because the hydrogen bonds that exist between the wall and water molecules can be replaced by the hydrogen bonds between the wall and solutes.

In this connection, the apparently larger pores found in *P. aeruginosa* may play a significant role in allowing the easier diffusion of hydrophobic molecules (Hancock and Nikaido, 1978). It is known that *P. aeruginosa* can grow on a number of organic acids and alcohols (Stanier et al., 1966), and

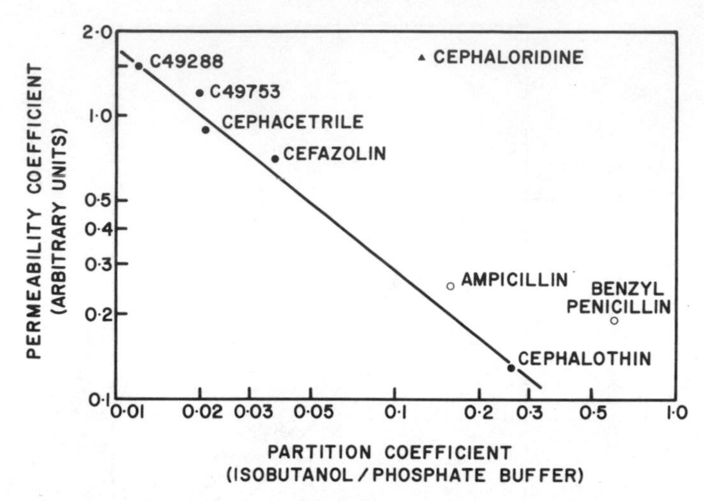

Figure 7 Hydrophobicity and the rate of diffusion across the outer membrane of β-lactam antibiotics. Based on the results of Zimmerman and Rosselet (1977). The partition coefficients were determined in isobutanol/0.02 *M* phosphate buffer, pH 7.4, containing 0.9% NaCl, at 37°C.

even on paraffinic hydrocarbons and camphor in the presence of a proper "degradative" plasmid (Chakrabarty et al., 1973). Possibly a larger pore may permit the passage of hydrophobic molecules enveloped in a "cage" of structured water molecules, or even of micelles. "Crypticity" measurements in *Haemophilus influenzae* containing R plasmids suggest that the outer membrane of this species permits a much more efficient diffusion of β-lactams (Medeiros and O'Brien, 1975); it would be most interesting to determine the exclusion limit of pores in this organism.

The third factor is the electrical charge. In the experiments of Zimmermann and Rosselet (1977), the zwitterionic cephaloridine penetrated through the outer membrane much faster than the monoanionic cephalosporins of comparable hydrophobicity (Figure 7). It seems possible that the pore walls contain a number of anionic groups; in this connection we recall the observation that porins are the most acidic proteins found in the outer membrane, both in *E. coli* (Rosenbusch, 1974; Schmitges and Henning, 1976) and in *S. typhimurium* (G. F. L. Ames, personal communication). However, a significant part of the resistance against anionic solutes may come from the fact that the periplasmic space contains many negatively charged macromolecules, and thus there exists a Donnan equilibrium favoring the exit of diffusible anions from the periplasmic space (Stock et al., 1977).

4 These considerations help us understand the functions of porins in a physiological context. A very large number (about 10^5) of porins exist in a single cell. If we consider cells growing in a commonly used medium containing a typical concentration, say 0.2%, of glucose, the porins are of course present in a large excess (perhaps 10^3-fold) over what is needed for the inflow of glucose at a rate needed for exponential growth at the doubling time of about 45 min (for calculation, see Bavoil et al., 1977). But the production of these large numbers of porins is not wasteful, because they are needed under two conditions. One is for the diffusion of less permeable compounds. The Renkin equation predicts that the diffusion rate of disaccharides will be about 9% of that of glucose. Hydrophobicity and electric charges, for example, in some peptides that are a major source of nutrition for enteric bacteria in their natural habitat, further reduce the diffusion rates by several orders of magnitude. It thus seems that for the diffusion of certain classes of compounds the presence of 10^5 copies of porins is indeed necessary. Another important factor is the effect of the concentration of nutrients. We are so used to active transport systems that it is often hard for us to understand the simple fact that the passive diffusion rate is *proportional* to the concentration difference across the membrane. Thus, although 10^5 copies of porins might seem an unnecessary excess in a medium containing 0.2% (roughly 10^{-2} *M*) glucose, all these porins become necessary if *E. coli* tries to keep growing at its maximal rate in the presence of 10^{-5} *M* glucose.

These calculations also suggest why specific transport systems have to be present in the outer membrane for some solutes that are able to penetrate through nonspecific pores. Since disaccharides diffuse through porin pores at about one-tenth of the rate of glucose, the diffusion of maltose is expected to become limiting at an outside concentration of 10^{-4} *M*, a prediction corroborated by the measurement of overall transport rates in *lamB* mutants lacking the specific outer membrane transport system (Szmelcman et al., 1976). Thus *E. coli* cells need the *lamB* specific transport system (see Chapter 10) in order to grow at maximal rates with dilute ($<10^{-4}$ *M*) concentrations of maltose and to take advantage of its active transport system, which has a K_m of about 10^{-6} *M* (Szmelcman et al., 1976). In another example, active transport of nucleosides is the highest volume transport system in *E. coli*, because nucleosides are split and only the ribose portion is used as carbon and energy sources; most of the bases are excreted (Peterson et al., 1967; Koch, 1971). Such a high throughput system becomes quickly limited by the diffusion step through the outer membrane, and thus for low outside concentrations of nucleosides, a specific protein, T6 receptor, is needed to allow their rapid diffusion (Hantke, 1976).

It is obvious that special mechanisms are needed for compounds that are excluded by the pores. These specific systems are discussed in Chapter 10.

4.4 Porins as Proteins

The "major proteins" of *E. coli*, containing both porins and the ompA protein (also called 3a, II,* tolG, etc.), were fractionated in the laboratories of Rothfield (Moldow et al., 1972), Schnaitman, (1973b, 1974a), Henning (Garten and Henning, 1974; Hindennach and Henning, 1975; Garten et al., 1975), and Mizushima (Uemura and Mizushima, 1975). The only porin (matrix protein) in *E. coli* B/r was isolated by Rosenbusch (1974) by a remarkably simple procedure that took advantage of the fact that only porins (and lamB protein that is absent in this strain) remain noncovalently attached to the peptidoglycan layer when the cell envelope was extracted with SDS solutions at 60°C. Porins of *S. typhimurium* LT2 were isolated by Nakae (1976a) and Nakae and Ishii (1978). The SDS–polyacrylamide gel electrophoretic patterns of the major proteins, including porins, are shown schematically in Figure 8.

During these fractionation studies some common properties of porins came to be recognized. (*a*) They are strongly associated with the underlying

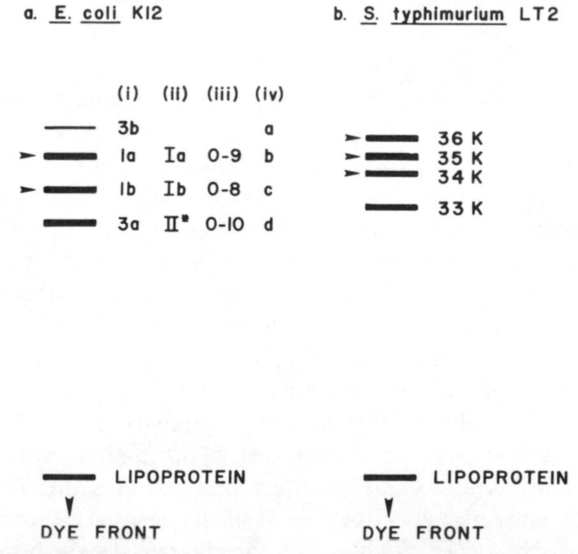

Figure 8 SDS–acrylamide gel patterns of "major proteins" of *E. coli* and *S. typhimurium*. In *E. coli*, the gel system of Lugtenberg et al. (1975) was used; in *S. typhimurium* the system was that of Ames (1973). Arrows indicate the porins. Nomenclature of *E. coli* proteins: (*i*) Schnaitman, (*ii*) Henning, (*iii*), Mizushima, and (*iv*) Lugtenberg system.

peptidoglycan (Rosenbusch, 1974). Electrostatic forces apparently play a major role in this association, as porins can be released from peptidoglycan in the presence of 0.5 *M* NaCl (and SDS) (Hasegawa et al., 1976). The ompA protein, in contrast, does not show a strong affinity toward peptidoglycan. (*b*) Porins tend to remain oligomeric even in SDS (Palva and Randall, 1978; Nakae, 1976a, b), so that they tend to migrate at a much slower rate in SDS–acrylamide electrophoresis than expected from their molecular weights (Schnaitman, 1973a). One has to heat them in SDS to cause them to dissociate into monomers. In contrast, the ompA protein migrates faster in SDS–acrylamide electrophoresis than expected from its size, and heating in SDS decreases its mobility (Schnaitman, 1973a). (*c*) Both porins and ompA protein contain lysine residues that have been oxidized to allysine (α-amino-adipic acid semialdehyde) (Diedrich and Schnaitman, 1978). It seems possible that these modified residues play a role in the stabilization of the structure of the outer membrane and cell wall, through the formation of covalent cross-links.

The amino acid composition of the porins is not noticeably hydrophobic (Rosenbusch, 1974; Garten and Henning, 1974; Ichihara and Mizushima, 1978; Tokunaga et al., 1978a). This is not surprising, since the protein must have large areas that are in contact with water. Nevertheless, the isolated protein is insoluble in water and is reported to be soluble in benzene or toluene (Schindler and Rosenbusch, 1978). A very interesting feature is the abundance of amino acid residues with hydroxyl groups. Thus tyrosine content is very high (6–9% of all residues), and threonine content is also high (6–8%). The intact outer membrane is very rich in proteins containing β-sheet structure (Nakamura et al., 1974), and apparently the oligomeric porin isolated in SDS still retains this "native" conformation, as it shows an almost "pure" β-sheet circular dichroism spectra (Rosenbusch, 1974; Nakamura and Mizushima, 1976) and binds very little SDS (Rosenbusch, 1974).

4.5 Presence of Multiple Species of Porins and the Regulation of Their Synthesis

Both *S. typhimurium* LT2 and *E. coli* K-12 synthesize more than one species of "peptidoglycan-associated" major proteins (Figure 8 and Table 3). In the case of *S. typhimurium* the three species (36K, 35K, and 34K) have been purified from mutants described by Nurminen et al. (1976) and Smit and Nikaido (1978), and have been shown to differ extensively in their primary structures (Tokunaga, et al., 1979a). Yet each of these porins produces channels allowing the diffusion of β-lactam antibiotics (Nikaido et al, 1977b), as well as sugars (Nakae and Ishii, 1978), and with an indistin-

Table 3 Comparison of *E. coli* and *S. typhimurium Porins*

E. coli K12	*S. typhimurium* LT2	Location of Gene[a]	Physiological Control of Synthesis
1a		21	Repressed by salts (Nakamura and Mizushima, 1976; van Alphen and Lugtenberg, 1977)
	35K	21 (?)[b] (T. Sato, K. Ito, and T. Yura, personal communication)	Repressed by salts (Smit and Nikaido, 1978)
1b		48	Production enhanced in Trypticase-Soy broth; decreased in Difco Nutrient Broth (Bassford et al., 1977)
	36K	47 (P. H. Mäkelä, personal communication)	Decreased in Difco Nutrient Broth (P. Bavoil and H. Nikaido, unpublished work)
2		Unknown (coded for by a prophage)	cAMP-Dependent (Schnaitman, 1974b; Schnaitman et al., 1975)
	34K	34 (P. H. Mäkelä, personal communication)	cAMP-Dependent (B. Rotman and G. F.-L. Ames, personal communication)

[a] Map location of *E. coli* genes is based on the 100 min map of Bachmann et al. (1976). To facilitate comparison between the two species, map locations of *S. typhimurium* genes were converted to a 100 "unit" map.
[b] Although a *Salmonella* porin gene is known to be located in this region, the identity of the porin is not known. We suspect that it is the 35K porin gene, because genes for other porins are located elsewhere.

guishable exclusion limit (Nakae and Ishii, 1978). In *E. coli* K-12, the *N*-terminal amino acid sequences are similar but not identical in porins 1a and 1b (Ichihara and Mizushima, 1978). Yet the presence of either protein 1a or 1b alone gives the mutants significant permeability to amino acids (Lutkenhaus, 1977; Pugsley and Schnaitman, 1978a). Lysogenization of K-12 strains with a temperate phage PA-2 (Schnaitman et al., 1975) causes the appearance of a new peptidoglycan-associated protein, protein 2, which was recently shown to confer permeability to amino acids, carbohydrates, and sulfate (Pugsley and Schnaitman, 1978a), and thus should be classified as a porin. Again, the amino acid sequence of this protein is different from that of other porins (Schnaitman, 1974a).

In addition to these porins, the synthesis of "hidden" porins can become "derepressed" if strong selective pressure is applied. Such a condition exists in mutants deficient in normal porins: they frequently produce "new" porins that enable them to grow faster in the presence of dwindling supply of nutrients (Bavoil et al., 1977; Henning et al., 1977; Foulds and Chai, 1978; van Alphen et al., 1978). In a detailed study, Pugsley and Schnaitman (1978b) showed that each of the three species of "new" proteins, all with molecular weights around 35,000, can more or less function as porins.

Why are all these multiple species of porins produced? One possible answer was suggested by Pugsley and Schnaitman (1978b). They argue that porins are essential for survival of gram-negative bacteria, yet are used as targets by phages and bacteriocins because of their exposure on the cell surface. Thus it would be advantageous to carry several genes for different porins, some of which might not be expressed until a genetic change occurs. The finding, by Chai and Foulds (1978), that there are phages utilizing some of the "new" porin proteins of K-12 as receptors suggests that the synthesis of these proteins is "derepressed" in a significant fraction of *E. coli* population in nature.

Another possible answer to this question is related to the slightly different permeability properties of channels produced by different species of porin. This is suggested by two kinds of observations. The first kind shows that pores made of different species of porins may exert some discriminating influences on the diffusion of certain classes of solutes. Thus when mutants less permeable to Cu^{2+} or chloramphenicol are selected, they usually lack porin 1a but not 1b (Lutkenhaus, 1977; Reeve and Doherty, 1968; Foulds, 1976; Chai and Foulds, 1977), a result suggesting that porin 1a produces a more efficient pore for these solutes. Furthermore, van Alphen et al. (1978) showed that adenosine 5′-monophosphate penetrates much more readily into *E. coli* K-12 mutants containing porin 1a than into those producing only 1b. Similarly, one of the three "new" porins expressed in pseudorevertants of porin-deficient mutants of *E. coli* K-12 seemed to be less efficient in allowing the passage of adenosine 5′-monophosphate and chloramphenicol (Pugsley and Schnaitman, 1978b). Unfortunately, some authors overinterpreted these results and suggested that some porin channels are *specific* (van Alphen et al., 1978). In our view, this is a very misleading concept. The basic difficulty with these approaches is that all transport *rate* assays have to be done with intact cells of mutants, which often show compensating increases in remaining species of porins whenever one or more porins are absent. For example, in the study of van Alphen et al. (1978), the alleged absence of function for 1b is based on the high permeability of 1a$^+$ 1b$^-$ mutants and on the low permeability of 1a$^-$ 1b$^+$ mutants. However, in the former case, the effect of the mutational loss of 1b might have been obscured by increases in 1a, and the effect of the presence of 1b in a 1a-deficient background could have escaped detection because of the use of an unfavorable substrate concentration. In fact, more recent results from the same group of workers suggest that porin 1b does act as a low efficiency pore for adenosine 5′-monophosphate (B. Lugtenberg, personal communication).

The currently available data thus suggest that the porin species produce channels of grossly similar properties, but there are probably subtle differences between them; these subtle differences may become significant in the natural habitat, where the diffusion through the outer membrane could act as a considerable barrier because of the low external concentration of substrates available. What is needed now is a careful study on how the diffusion through each porin channel is affected by various parameters (size, charge, hydrophobicity, etc.) of the solute, rather than hasty conclusions on the "specificity" of each porin channel.

The second line of evidence suggesting somewhat different permeability properties or "functions" or each class of porin comes from the observation that the *production* of the various porin species is controlled by a different set of physiological conditions. Similar parameters seem to control the synthesis of each of the porins of *E. coli* and *Salmonella*, and these regulatory properties, as well as the genetics, suggest rather strongly the correspondence between each porin species in these two organisms (Table 3). Furthermore, some hints on the possible physiological functions of each porin species may be obtained from these results. Thus the "salt-repressible" porin (35K or 1a) appears to be most effective in the diffusion of a number of solutes (van Alphen et al., 1978; Lutkenhaus, 1977; Nikaido et al., 1977b) and could form the basic, all-purpose channels. This idea fits also with the observation that *E. coli* B/r, a strain that appears to have lost many of the functions that are not vital for survival under laboratory conditions, produces only one species of porin (Rosenbusch, 1974; Nakae, 1976b), which seems to be the equivalent of 1a (Schmitges and Henning, 1976). How do we interpret the repression of synthesis of this porin by high ionic strength? When enteric bacteria are thrown out of their protective environment, that is, the intestinal tract of warm-blooded animals, the salt concentration of the environment would increase as a result of the drying out of feces. This could act as a signal to tell *E. coli* cells of a change in the environment, so that they can shut down the most effective pores and survive in a state of minimal metabolic activity or with minimal exchange of material with the outside world. Under certain conditions, in contrast, the rise in cellular cyclic AMP concentrations tells *E. coli* that it is becoming starved and should expand its catabolic abilities. The increased production of 34K porin and protein 2 in the presence of the cyclic AMP thus suggested that these are porins designed to take in additional nutrients. The increased production of 36K and 1b porin in Trypticase-Soy medium probably reflects also certain physiological functions of this species, although they cannot yet be predicted as we do not know what components or parameters in this medium is responsible for this change. It must, of course, be borne in mind that these considerations are highly speculative.

An interesting possibility is that the cells may become somewhat resistant to antibiotics through the loss or decrease in porins. At least in the laboratory, selection of *E. coli* mutants with low-level resistance to chloramphenicol enriches for mutants lacking 1a ("*cmlB*") (Reeve and Doherty, 1968; Chai and Foulds, 1977). Similarly, "adaptation" of *P. aeruginosa* to media containing an increasing concentration of polymyxin results in a population with drastically reduced levels of a 35–37K major outer membrane protein (H. E. Gilleland, Jr., personal communication), which recently was identified as the porin (Hancock et al., 1979). It is also possible that some of the R factors increase the host resistance by repressing the synthesis of host porins. Thus Rajul Iyer and her coworkers have made the exciting discovery that some R factors belonging to the N compatibility group drastically reduce the level of the porin when introduced into *E. coli* B/r (Iyer, 1977; Iyer et al., 1978). Although these workers favor the interpretation that preexisting porin-deficient *mutants* acted as better recipients of these conjugative plasmids, we are most attracted to the possibility of the "repression" of porin synthesis by the plasmid genome. Interestingly, some plasmid-bearing strains produce a porin with a slightly different gel electrophoretic mobility; perhaps these are secondary mutants derepressed for the synthesis of an alternative porin because the complete repression of the synthesis of the only remaining porin of *E. coli* B/r presents too strong a selective disadvantage.

4.6 Outer Membrane Proteins Not Yet Known to Produce Pores

By no means do all outer membrane proteins produce pores. Although it was proposed, on the basis of the amino acid sequence, that Braun lipoprotein may form oligomers with a central channel, which would then serve as transmembrane pores (Inouye, 1974), so far no experimental evidence favoring this hypothesis has appeared. On the contrary, all efforts to reconstitute the channel-containing membrane by the use of Braun lipoprotein have been unsuccessful (Nakae, 1976a, b). Recently, by using a mutant lacking the lipoprotein (Hirota et al., 1977), we have shown that the absence of this protein did not produce any reduction in the hydrophilic permeability of the outer membrane, measured by the diffusion of 6-aminopenicillanic acid (Nikaido et al., 1977a).

Another case in point is the "heat-modifiable protein" 3a (ompA protein), which has been claimed to produce pores (Manning et al., 1977). This conclusion, however, is based on reduced transport rates of only two amino acids, tested at a single concentration, in a mutant lacking this protein. As we emphasize earlier, the decrease in outer membrane permeability produces an increase in "apparent K_m" values of the overall transport process; thus in order to detect any difference between the wild type and mutants, substrate concentrations in the range of K_m or lower must be used (Figure 5). Although amino acid transport in *E. coli* usually has K_m values around 1 μM or lower (Kaback and Hong, 1973; Oxender and Quay, 1975), Manning et al. (1977) used a very high substrate concentration of 500 μM and still found a large difference, a result not expected for mutants with reduced outer membrane permeability. Since a large fraction of the mutant population was not viable, it seems much more plausible that the low transport rates simply reflect the low proportion of viable cells. We thus conclude that so far there is no reliable evidence that ompA protein forms pores. In *S. typhimurium*, the mutational loss of 33K protein, corresponding in its gel electrophoretic behavior and the location of its structural gene (B. A. D. Stocker, personal communication) to ompA protein, did not produce any change in the transmembrane diffusion rates of cephaloridine (Nikaido et al., 1977b).

5 PENETRATION OF MACROMOLECULES

Under certain conditions, macromolecules can apparently cross the outer membrane barrier. Thus increased leakage of periplasmic enzymes out into the medium has been reported in a large number of mutants, including the "deep-rough" mutants with very incomplete LPS (Lindsay et al., 1973), mutants lacking lipoprotein (Hirota et al., 1977; Nikaido et al., 1977a), and mutants with defects of unknown nature (Lopes et al., 1972; Weigand and Rothfield, 1976). Many of these mutants are also sensitive to lysozyme (Tamaki and Matsuhashi, 1973); their outer membrane thus also allows the inward penetration of macromolecules from the external medium.

The molecular mechanism of this "leakage" process is not presently understood. Although misleading pictures have been published (e.g., Witholt et al., 1976), it seems very unlikely that these proteins pass through "pores," in view of the size of the pores in the *E. coli* outer membrane and of the diverse nature of mutants with increased macromolecular permeability. At present, the most likely possibility seems to be the mechanism involving transient rupture and resealing of the membrane. If this is so, the penetration by macromolecules would be enhanced by conditions favoring the destabilization of membrane organization. This seems to be the case, as deep-rough mutants produce membranes where normally strong interactions between LPS and "major proteins" (see Section 3) are weak, and the outer membrane of lipoprotein-deficient mutants does not get stabilized by the anchoring to the peptidoglycan layer. Furthermore, destabilization of the outer membrane by the removal of divalent cations enhances the "leakiness" even in wild-type cells, as evidenced by the well-known "osmotic shock" procedure (Nossal and Heppel, 1966) used to release the periplasmic enzymes into the suspending medium. Finally, a major cause of the instability of outer membrane structure, the electrostatic repulsion between the negatively charged groups, can be minimized by adding excess Mg^{2+} to the medium. This treatment is known to decrease the leakage in deep-rough mutants (Lindsay et al., 1973), lipoprotein-deficient mutants (Nikaido et al., 1977a), and porin-deficient mutants (Nikaido et al., 1977b).

6 STRUCTURE–FUNCTION RELATIONSHIP IN THE OUTER MEMBRANE

The striking lack of permeability of the *E. coli* or *S. typhimurium* outer membrane to hydrophobic compounds can be explained very well by the asymmetric construction of this membrane (Section 3). Thus the low fluidity of the LPS hydrocarbon chains, comprising the outer half of the membrane, presumably hinders the dissolution and diffusion of hydrophobic molecules into the membrane interior. In addition, the presence of a high concentration of charged groups in the proximal region of the LPS saccharide chain, and possibly the presence of the polysaccharide chain "layer," will further interfere with the diffusion process.

In many nonenteric bacteria, however, the outer membrane appears to allow the penetration of various hydrophobic molecules. In these strains, the structure of the membrane may be less asymmetric, and one electron microscopic study suggests the presence of phospholipids both in the outer and inner leaflets of *P. aeruginosa* outer membrane (Section 3). Thus the absence of phospholipid bilayer regions in the outer membrane of enteric bacteria is probably an exception rather than the rule among gram-negative bacteria and could be the consequence of the adaptation, by *E. coli* and its relatives, to an unusual environment filled with such hydrophobic inhibitors as bile salts and free fatty acids.

Another remarkable property of the outer membrane is the nonspecific permeability toward small, hydrophilic substances. As is seen in Section 4, for most substances this permeability can be explained almost completely by the presence of a special class of proteins, namely, porins. The function of these proteins suggests that they span the thickness of the membrane. The available evidence is indeed compatible with this assumption. Thus porins are exposed on the outside surface of the membrane as shown by labeling with a nonpenetrating labeling reagent (Kamio and Nikaido, 1977) and ferritin-labeled antibodies (Smit and Nikaido, 1978) and by their activity as phage receptors (Chai and Foulds, 1977; Datta et al., 1977). Furthermore, the affinity of porins toward peptidoglycan (Rosenbusch, 1974; Hasegawa et al., 1976) suggests strongly that porins are in contact with the underlying peptidoglycan layer in intact cells.

How are the porins organized in the membrane? Porins solubilized by SDS from *E. coli* and *S. typhimurium* outer membranes exist mostly as trimers (Palva and Randall, 1978; Nakae et al., 1979; Tokunaga et al., 1979b). They also seem to exist as trimers in the outer membrane. Thus cross-linking studies by Reithmeier and Bragg (1977b) with bifunctional reagents with cell envelopes or outer membrane produced dimers and trimers of porin. What was thought to be a "tetramer" in this study was later found to be another form of the trimer by Palva and Randall (1978), who also found that among higher oligomers, hexamers and nonamers are preponderant. [Chopra et al. (1977) found that larger amounts of stable porin oligomers were produced in the presence of lysozyme. Our interpretation of this observation is that the basic protein (lysozyme) stabilizes the oligomer of acidic proteins, porins, by binding to it. Their results do not prove their conclusion that all porin oligomers are artifacts, a conclusion that is certainly untenable in view of the various studies mentioned above.]

Does each monomer of porin contain a channel or does a channel get formed in the central space after the association of three subunits? All attempts to reconstitute sucrose-permeable membranes by using monomeric porins have failed so far (T. Nakae, personal communication). On the other hand, some published results are at least consistent with the one-porin one-channel hypothesis. For example, Benz et al. (1978) calculate that the 1.7 nS increment in conductivity in 1*M* KCl indicates the presence of a single pore 0.93 nm in diameter. On the surface this seems to favor the hypothesis of one central channel per porin trimer. However, these workers did not take into account the fact that hydrated ions have diameters that are not negligible in comparison with the diameter of the pore. If we use the hydrated ionic radius of 0.2 nm, calculated from the hydration number (2) given by Bockris (1949) for both K^+ and Cl^-, and make corrections according to the Renkin equation, assuming a pore radius of 0.6 nm, we find that *three* pores of this radius will give a conductivity of 1.4 nS in 1 *M* KCl, which is in reasonable agreement with the observed value of 1.7 nS. Furthermore, although the most common incremental step corresponded to 1.7 nS, many increments of smaller magnitude were seen (Benz et al., 1978). Possibly the latter were caused by the insertion of porin monomers or dimers. In a similar experiment by Schindler and Rosenbusch (1978), porins incorporated into planar lipid bilayers in the absence of detergents and organic solvents were shown to produce a "unit" increment in conductance that actually consisted of three equal steps. This results strongly suggets that each subunit in the trimer has a capacity of producing a channel, but the "opening" of channel within a trimer is a highly cooperative phenomenon.

The existence of the "triplet of pores" is also suggested by structural studies. When the cell envelope of *E. coli* B is extracted with SDS at room temperature, peptidoglycan sheets with hexagonally arranged oligomers of porin are left behind (Rosenbusch, 1974). Because of this regular arrangement, Steven et al. (1977) were able to use image-enhancement techniques in electron microscopy, and they obtained the negatively stained picture schematically shown in Figure 9. The proteins (porins) clearly exist as trimers, and a triplet of holes is produced by each monomeric unit contributing a hole or indentation of approximately 2 nm in diameter. Although

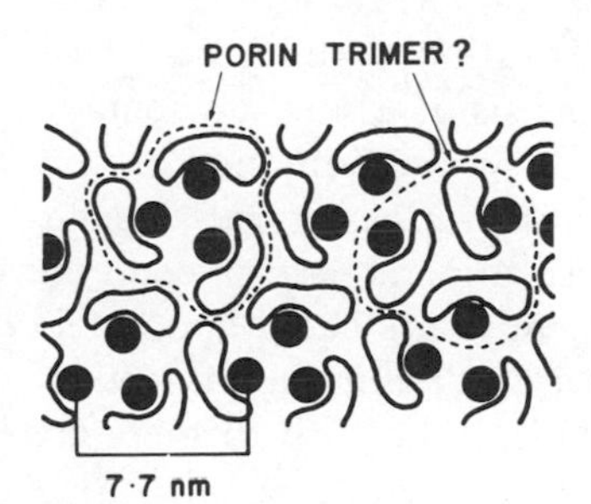

Figure 9 Negatively stained image of the porin–lipoprotein–peptidoglycan complex, after image enhancement. After Steven et al. (1977).

Steven et al. argued that these holes were probably occupied by lipoproteins, this appears unlikely because the wall of the holes would then have to be hydrophobic and would repel the hydrophilic particles of the negative stains. Ueki et al (1979) examined the small-angle X-ray scattering by oriented layers of intact outer membranes of *S. typhimurium*. Such a method is advantageous in that extraction artifacts can be avoided. The study revealed a series of in-plane reflections that appeared to correspond to protein aggregates of about 11 nm in diameter. The most impressive feature is the regularity and rigidity of the structure within each particle, which most probably corresponds to a trimer of porins.

The porin trimers in the outer membrane are probably associated with LPS, in view of the known affinity of porin to LPS (Yu and Mizushima, 1977). However, there is so far no decisive evidence that the trimer is strongly associated with any other protein, or that "stoichiometric aggregates" containing all major proteins, proposed some time ago (Henning et al., 1973b), actually exist. Cross-linking studies (Palva and Randall, 1976; 1978; Reithmeier and Bragg, 1977b) consistently resulted in failure to cross-link porins to any other protein. DeMartini and Inouye (1978) observed that the porin molecules become dissociated more easily from the complex composed of porins, lipoproteins, and peptidoglycan sheets if the lipoprotein-to-peptidoglycan association is broken with trypsin, which does not cleave any bonds in the porin. This observation, which is actually a confirmation of earlier work [see Figure 1 of Nakae (1976b)] lead the authors to the conclusion that the porins are associated with Braun lipoprotein molecules *in situ*. This is quite plausible, although it is still possible that such associations are artifacts caused by the coexistence of two hydrophobic substances in an aqueous environment.

How are the porin trimers or trimer–LPS complexes arranged in the plane of membrane? Rosenbusch (1974) found that "particles," which most probably correspond to porin trimers (Steven et al., 1977), are arranged in a regular hexagonal array on peptidoglycan sheets if most of phospholipids, LPS, and other proteins are extracted away from the cell envelope of *E. coli* B/r with 2% SDS at 60°C. The inherent tendency of porins to form hexagonal arrays has since been confirmed by reconstitution experiments. Thus Yamada and Mizushima (1978) showed that dialysis of a solution containing *E. coli* K-12 porin and LPS in 1% SDS results in the formation of vesicles, which are bounded by "membranes" composed of hexagonal arrangement of porin–LPS complexes. When the reconstitution is performed in the presence of peptidoglycan sheets with covalently attached lipoprotein molecules, the porin–LPS assemblies are formed on the surface of peptidoglycan sheets. These results suggest the possibility that porin trimers are arranged in this regular manner in the intact outer membrane.

One technique for observing the distribution and organization of proteins in the plane of the membrane is freeze-fracture electron microscopy. The structure shown in Figure 4*A* is expected to be quite resistant to freeze-fracture cleavage, because so many protein molecules traverse the membrane. Indeed the outer membranes of wild-type *E. coli* or *S. typhimurium* are quite difficult to cleave, but those of deep-rough LPS mutants are easily cleaved (Smit et al., 1975; Irvin et al., 1975; Bayer et al., 1975), possibly because much fewer transmembrane proteins are present (see Section 3). In any case, the freeze-fracture analysis showed a number of densely packed particles on the concave fracture face, and very few particles on the convex side (van Gool and Nanninga, 1971; Smit et al., 1975). Smit et al. (1975) showed, by using deep-rough mutants, that there was a proportionality between the reduction in the total protein content of the outer membrane and the reduction in particle density. This observation suggests very strongly that at least the major component of the "particles" is protein. Since a 1 μm^2 area of *S. typhimurium* outer membrane was found to contain about 9000 particles (Smit et al., 1975), whereas it is expected to contain about 10,000 porin trimers, on the basis of Rosenbusch's data of 10^5 porin monomers per cell (Rosenbusch, 1974), it seems likely that at least a major fraction of these particles should contain porins. However, in contrast to what is expected from the studies discussed above, these particles were not arranged in a regular, hexagonal array (Smit et al., 1975).

One possible explanation of this discrepancy is that porins are arranged in a regular array, but during freeze-fracture they associate randomly with neighboring molecules of LPS and other proteins, so that the arrangement of these aggregates or "particles" becomes less regular. If this explanation is correct, then X-ray diffraction patterns of intact outer membrane should reflect the regular arrangement of porin oligomers. However, such studies suggest that, if porin trimers are arranged in a hexagonal lattice in the intact membrane, the arrangement has enough irregularity so that typical lattice patterns are not easily recognizable (T. Ueki, personal communication). In conclusion, the porin trimers undoubtedly have an inherent tendency to form a hexagonal lattice, but it seems likely that in the actual outer membrane the lattice is perturbed frequently, perhaps by the presence of other proteins and other components of the outer membrane.

7 CONCLUSIONS

The following, highly speculative, hypothesis is suggested by the survey of outer membrane permeability presented in this chapter. The primary function of the outer membrane appears to be the protection of the underlying peptidoglycan layer against peptidoglycan-degrading enzymes and perhaps antibody and antibodylike agents. This concept is very strongly supported by the observation that gram-positive bacteria, which do not produce outer membranes, have undergone extensive diversification of peptidoglycan structure during evolution, presumably in order to avoid the attack by the prevailing agents, whereas all gram-negative bacteria produce peptidoglycan of a simple, identical structure (Schleifer and Kandler, 1972).

The outer membrane must have channels for the transport of nutrients. Larger channels are advantageous since they allow the uptake of larger peptides and oligosaccharides, as well as hydrophobic compounds, and since they allow faster diffusion rates. We therefore believe that the large pores found in *P. aeruginosa* (Hancock and Nikaido, 1978) are close to the prototype of pores in gram-negative bacteria. In fact, a survey of several nonenteric genera of gram-negative bacteria suggested the presence of similar, larger pores in all cases (R. E. W. Hancock, L. Zalman, and H. Nikaido, unpublished results).

When, however, the enteric bacteria adapted themselves to their highly specialized haibtat, these large pores had to be eliminated because bile salts would diffuse through them and solubilize the cytoplasmic membrane, and hydrophobic inhibitors, such as free fatty acids, would also reach their sites of action. Thus we can speculate that the narrow pores are special features of the enteric bacteria, primarily designed to keep out the hydrophobic compounds. The alteration in pore diameter, in turn, necessitated the production of specific transport systems in the outer membrane, because the narrow pores were incapable of transporting large molecules at sufficient rates. Furthermore, since hydrophobic compounds can penetrate through phospholipid bilayer regions of any membrane, it was necessary to eliminate these areas by making the structure of the membrane extremely asymmetric (Section 3). Thus we predict that the virtual absence of phospholipid molecules from the outer leaflet is also a special adaptive feature confined to enteric bacteria.

The investigation of the permeabilty of the bacterial outer membrane sheds some light on the permeability of other biological membranes. Although the coexistence of "diffusion through pores" and "penetration by way of the hydrocarbon interior of phospholipid bilayers" was proposed for other biological membranes (Wright and Diamond, 1969; Sha'afi et al., 1971), these proposals were based on kinetic data that were open to alternative interpretations (Lieb and Stein, 1969). Thus the isolation of a specific protein with the capability of forming water-filled pores, to our knowledge achieved for the first time in bacterial outer membrane, is conclusive proof that the transmembrane pores do exist at least in some biological membranes. It is hoped that reconstitution studies will be carried out with many other biological membranes, so that the permeability properties, as well as the structure–function relationship of these membranes will become more sharply defined. We have already obtained evidence indicating that protein(s) in the outer membranes of animal and plant mitochondria produce large, water-filled pores, roughly of the size found in *P. aeruginosa* outer membrane (H. Nikaido, L. Zalman and Y. Kagawa, unpublished data).

ACKNOWLEDGMENTS

The author thanks all his past and present associates for their ideas, discussion, and hard work. Research in the author's laboratory has been supported by research grants from the U.S. Public Health Service (AI-09644) and from the American Cancer Society (BC-20).

REFERENCES

Ames, G. F.-L. (1974), *J. Biol. Chem.*, **249,** 634–644.

Ames, B. N., Lee, F. D., and Durston, W. E. (1973) *Proc. Natl. Acad. Sci. U.S.*, **70,** 782–786.

Ames, G. F.-L., Spudich, E. N., and Nikaido, H. (1974), *J. Bacteriol.*, **117,** 406–416.

Bachmann, B. J., Low, K. B., and Taylor, A. L. (1976), *Bacteriol. Rev.*, **40,** 116–167.

Bangham, A. D. (1972), *Chem. Phys. Lipids*, **8,** 386–392.

Bassford, P. J., Jr., Diedrich, D. L., Schnaitman, C. A., and Reeves, P. (1977), *J. Bacteriol.*, **131,** 608–622.

Bavoil, P., Nikaido, H., and von Meyenburg, K. (1977), *Mol. Gen. Genet.*, **158,** 23–33.

Bayer, M. E., Koplow, J., and Goldfine, H. (1975), *Proc. Natl. Acad. Sci. U.S.*, **72,** 5145–5149.

Beacham, I. R., Haas, D. and Yagil, E. (1977), *J. Bacteriol.*, **129,** 1034–1044.

Benz, R., Janko, K., Boos, W., and Läger, P. (1978), *Biochim. Biophys. Acta*, **511,** 309–315.

Bokris, J. O. (1949), *Q. Rev.*, (*Lond.*) **3,** 173–180.

Bragg, P. D., and Hou, C. (1972), *Biochim. Biophys. Acta*, **247,** 478–488.

Chai, T.-J, and Foulds, J. (1977), *J. Bacteriol.*, **130,** 781–786.

Chai, T.-J, and Foulds, J. (1978), *J. Bacteriol.*, **135,** 164–171.

Chakrabarty, A. M., Chou, G., and Gunsalus, I. C. (1973), *Proc. Natl. Acad. Sci. U.S.*, **70,** 1137–1140.

Chopra, I., Howe, T. G. B., and Ball, P. R. (1977), *J. Bacteriol.*, **132,** 411–418.

Coleman, W. G., Jr., and Leive, L. (1978), *Fed. Proc.*, **37,** 1393.

Collander, R. M. (1949), *Physiol. Plant.*, **2,** 300–311.

Collander, R., and Bärlund, H. (1933), *Acta Bot. Fenn.*, **11,** 1–114.

Coratza, G., and Molina, A. M. (1978), *J. Bacteriol.*, **133,** 411–412.

Datta, D. B., Arden, B., and Henning, U. (1977), *J. Bacteriol.*, **131,** 821–829.

Davies, J. K., and Reeves, P. (1975), *J. Bacteriol.*, **123,** 96–101.

Decad, G. (1976), Ph.D. thesis, University of California, Berkeley.

Decad, G., and Nikaido, H. (1976), *J. Bacteriol.*, **128,** 325–336.

Decad, G., Nakae, T., and Nikaido, H. (1974), *Fed. Proc.*, **33,** 1240.

DeMartini, M., and Inouye, M. (1978), *J. Bacteriol.*, **133,** 329–335.

DePamphilis, M. L. (1971), *J. Bacteriol.*, **105,** 1184–1199.

Diedrich, D. L., and Schnaitman, C. A. (1978), *Proc. Natl. Acad. Sci. U.S.*, **75,** 3708–3712.

Fensom, A. H., and Gray, G. W. (1969), *Biochem. J.*, **114,** 185–196.

Foulds, J. (1976), *J. Bacteriol.*, **128,** 604–608.

Galanos, C., Lüderitz, O., Rietschel, E. T., and Westphal, O. (1977), in *International Review of Biochemistry*, Vol. 14, Biochemistry of Lipids, II (T. W. Goodwin, ed.), University Park Press, Baltimore, pp. 239–335.

Galey, W. R., Owen, J. D., and Solomon, A. K. (1973), *J. Gen. Physiol.*, **61,** 727–746.

Garten, W., and Henning, U. (1974), *Eur. J. Biochem.*, **47,** 343–352.

Garten, W., Hindennach, I., and Henning, U. (1975), *Eur. J. Biochem.*, **59,** 215–221.

Gilleland, H. E., Jr., Stinnett, J. D., Roth, I. L., and Eagon, R. G. (1973), *J. Bacteriol.*, **113,** 417–432.

Gustafsson, P., Nordstrom, K., and Normark, S. (1973), *J. Bacteriol.*, **116,** 893–900.

Hancock, R. E. W., and Nikaido, H. (1978), *J. Bacteriol.*, **136,** 381–390.

Hancock, R. E. W., Decad, G., and Nikaido, H. (1979), *Biochim. Biophys. Acta*, in press.

Hantke, K. (1976), *FEBS Lett.*, **70,** 109–112.

Hasegawa, Y., Yamada, H., and Mizushima, S. (1976), *J. Biochem. (Tokyo)*, **80,** 1401–1409.

Henning, U., and Haller, I. (1975), *FEBS Lett.*, **55,** 161–164.

Henning, U., Höhn, B., and Sonntag, I. (1973b), *Eur. J. Biochem.*, **39,** 27.

Henning, U., Schmidmayr, W., and Hindennach, I. (1977), *Mol. Gen. Genet.*, **154,** 293–298.

Hindennach, I., and Henning, U. (1975), *Eur. J. Biochem.*, **59,** 207–213.

Hirota, Y., Suzuki, H., Nishimura, Y., and Yasuda, S. (1977), *Proc. Natl. Acad. Sci. U.S.*, **74,** 1417–1420.

Ichihara, S., and Mizushima, S. (1978), *J. Biochem. (Tokyo)*, **83,** 1095–1100.

Inouye, M. (1974), *Proc. Natl. Acad. Sci. U.S.*, **71,** 2396–2400.

Irvin, R. T., Chatterjee, A. K., Sanderson, K. E., and Costerton, J. W. (1975), *J. Bacteriol.*, **124,** 930.

Iyer, R. (1977), *Biochim. Biophys. Acta*, **470,** 258–272.

Iyer, R., Darby, V., and Holland, I. B. (1978), *FEBS Lett.*, **85,** 127.

Jones, N. C., and Osborn, M. J. (1977), *J. Biol. Chem.*, **252,** 7405–7412.

Kaback, H. R., and Hong, J. (1973), *CRC Crit. Rev. Microbiol.*, **3,** 333–376.

Kamio, Y., and Nikaido, H. (1976), *Biochem.*, **15,** 2561–2570.

Kamio, Y., and Nikaido, H. (1977), *Biochim. Biophys. Acta*, **464,** 589–601.

Kataoka, T., Inoue, K., Lüderitz, O., and Kinsky, S. C. (1971), *Eur. J. Biochem.*, **21,** 80–85.

Kedem, O., and Katchalsky, A. (1958), *Biochim. Biophys. Acta*, **27,** 229–246.

Kedem, O., and Katchalsky, A. (1961), *J. Gen. Physiol.*, **45,** 143–179.

Koch, A. L. (1971), *Adv. Microb. Physiol.*, **6,** 147–217.

Koplow, J., and Goldfine, H. (1974), *J. Bacteriol.*, **117,** 527–543.

Leive, L. (1965a), *Proc. Natl. Acad. Sci. U.S.*, **53,** 745.

Leive, L. (1965b), *Biochem. Biophys. Res. Commun.*, **18,** 13–17.

Leive, L. (1965c), *Biochem. Biophys. Res. Commun.*, **21,** 290–296.

Leive, L. (1974), *Ann. N.Y. Acad. Sci.*, **235,** 109.

Lieb, W. R., and Stein, W. D. (1969), *Nature*, **224,** 240–243.

Lindsay, S., Wheeler, B., Sanderson, K. E., Costerton, J. W., and Chen, K.-J. (1973), *Can. J. Microbiol.*, **19,** 335–343.

Lopes, J., Gottfried, S., and Rothfield, L. (1972), *J. Bacteriol.*, **109,** 520–525.

Lüderitz, O., Westphal, O., Staub, A.-M., and Nikaido, H. (1971), in *Microbial Toxins* Vol. 4, (G. Weinbaum, S. Kadis, and S. J. Ajl, eds.), Academic Press, New York, pp. 145–233.

Lugtenberg, B., Meijers, J., Peters, R., van der Hock, P., and van Alphen, L. (1975), *FEBS Lett.*, **58,** 254–258.

Lutkenhaus, J. F. (1977), *J. Bacteriol.*, **131,** 631–637.

MacGregor, D. R., and Elliker, P. R. (1958), *Can. J. Microbiol.*, **4,** 499–503.

McIntyre, T. M., and Bell, R. M. (1978), *J. Bacteriol.*, **135,** 215–226.

Mäkelä, P. H., Schmidt, G., Mayer, H., Nikaido, H., Whang, H. Y., and Neter, E. (1976), *J. Bacteriol.*, **127,** 1141–1149.

Maness, M. J., and Sparling, P. F. (1973), *J. Infect. Dis.*, **128,** 321–330.

Manning, P. A., Pugsley, A. P., and Reeves, P. (1977), *J. Mol. Biol.*, **116,** 285–300.

Medeiros, A. A., and O'Brien, T. F. (1975), *Lancet*, **1,** 716–718.

Miller, R. D., Brown, K. E., and Morse, S. A. (1977), *Infect. Immunol.*, **17,** 303–312.

Moldow, C., Robertson, J., and Rothfield, L. (1972), *J. Membrane Biol.*, **10,** 137–152.

Mühlradt, P. F., and Golecki, J. R. (1975), *Eur. J. Biochem.*, **51,** 343–352.

Nakae, T. (1975), *Biochem. Biophys. Res. Commun.*, **64,** 1224–1230.

Nakae, T. (1976a), *J. Biol. Chem.*, **251,** 2176–2178.

Nakae, T., and Ishii, J. (1978), *J. Bacteriol.*, **133,** 1412–1418.

Nakae, T., and Nikaido, H. (1975), *J. Biol. Chem.*, **250,** 7359–7365.

Nakae, T., Ishii, J., and Tokunaga, M. (1979), *J. Biol. Chem.*, **254,** 1457–1461.

Nakamura, K., and Mizushima, S. (1975), *Biochim. Biophys. Acta*, **413,** 371–393.

Nakamura, K., and Mizushima, S. (1976), *J. Biochem.* (*Tokyo*), **80,** 1411–1422.

Nakamura, K., Ostrovsky, D. N., Miyazawa, T., and Mizushima, S. (1974), *Biochim. Biophys. Acta*, **332,** 329–335.

Nikaido, H. (1973), in *Bacterial Membranes and Walls* (L. Leive, ed.), Dekker, New York, pp. 131–208.

Nikaido, H. (1976), *Biohim. Biophys. Acta*, **433,** 118–132.

Nikaido, H., and Nakae, T. (1973), *J. Infect. Dis.*, (*Suppl.*), **128,** 30–34.

Nikaido, H., Bavoil, P., and Hirota, Y. (1977a), *J. Bacteriol.*, **132,** 1045–1047.

Nikaido, H., Song, S. A., Shaltiel, L., and Nurminen, M. (1977b), *Biochem. Biophys. Res. Commun.*, **76,** 324–330.

Nikaido, H., Takeuchi, Y., Ohnishi, S.-I., and Nakae, T. (1977c), *Biochem. Biophys. Acta*, **465,** 152–164.

Nixdorff, K., Fitzer, H., Gmeiner, J., and Martin, H. H. (1977), *Eur. J. Biochem.*, **81,** 63–69.

Nossal, N. G., and Heppel, L. A. (1966), *J. Biol. Chem.*, **241,** 3055.

Nurminen, M., Lounatmaa, K., Sarvas, M., Mäkelä, P. H., and Nakae, T. (1976), *J. Bacteriol.*, **127,** 941–955.

Osborn, M. J., Gander, J. E., Parisi, E., and Carson, J. (1972), *J. Biol. Chem.*, **247,** 3962–3972.

Overath, P., Brenner, M., Gulik-Krzywicki, T., Shechter, E., and Letellier, L. (1975), *Biochim. Biophys. Acta*, **389,** 351–369.

Oxender, D. L., and Quay, S. (1975), *Ann. N.Y. Acad. Sci.*, **264,** 358–371.

Palva, E. T., and Randall, L. L. (1976), *J. Bacteriol.*, **127,** 1558–1560.

Palva, E. T., and Randall, L. L. (1978), *J. Bacteriol.*, **133,** 279–286.

Payne, J. W., and Gilvarg, C. (1968), *J. Biol. Chem.*, **243,** 6291–6299.

Peterson, R. N., Boniface, J., and Koch, A. L. (1967). *Biochim. Biophys. Acta* **135,** 771–783.

Pugsley, A. P., and Schnaitman, C. A. (1978a), *J. Bacteriol.*, **133,** 1181–1189.

Pugsley, A. P., and Schnaitman, C. A. (1978b), *J. Bacteriol.*, **135,** 1118–1129.

Reeve, E. C. R., and Doherty, P. (1968), *J. Bacteriol.*, **96,** 1450–1451.

Reithmeier, R. A. F., and Bragg, P. D. (1977b), *Biochim. Biophys. Acta*, **466,** 245–256.

Renkin, E. M. (1954), *J. Gen. Physiol.*, **38,** 225–243.

Roantree, R. J., Kuo, T., MacPhee, D. G., and Stocker, B. A. D. (1969), *Clin. Res.*, **17,** 157.

Roantree, R. J., Kuo, T.-T., and MacPhee, D. G. (1977), *J. Gen. Microbiol.*, **103,** 223–234.

Roberts, N. A., Gray, G. W., and Wilkinson, S. G. (1970), *Microbios*, **2,** 189–208.

Rogers, S. W., Gilleland, H. E., Jr., and Eagon, R. G. (1969), *Can. J. Microbiol.*, **15,** 743–748.

Romeo, D., Girard, A., and Rothfield, L. (1970), *J. Mol. Biol.*, **53,** 475–490.

Rosenbusch, J. P. (1974), *J. Biol. Chem.*, **249,** 8019–8029.

Rottem, S., and Leive, L. (1977), *J. Biol. Chem.*, **252,** 2077–2081.

Rottem, S., Hasin, M., and Razin, S. (1975), *Biochim. Biophys. Acta*, **375,** 395.

Schindler, H., and Rosenbusch, J. P. (1978), *Proc. Natl. Acad. Sci. U.S.*, **75,** 3751–3755.

Schlecht, S., and Schmidt, G. (1969), *Zentralbl. Bakteriol. Parasitenkd. Infektionskr. Hyg. I. Orig.* **212,** 505–511.

Schlecht, S., and Westphal, O. (1970), *Zentralbl. Bakteriol. Parasitenkd. Abt. I. Orig.*, **213,** 356–381.

Schleifer, K. H., and Kandler, O. (1972), *Bacteriol. Rev.*, **36,** 407–477.

Schmidt, G., Schlecht, S., and Westphal, O. (1969), *Zentralbl. Bakteriol. Parasitenkd. Abt. I. Orig.*, **212,** 88–96.

Schmitges, C. J., and Henning, U. (1976), *Eur. J. Biochem.*, **63,** 47–52.

Schnaitman, C. A. (1973a), *Arch. Biochem. Biophys.*, **157,** 541–552.

Schnaitman, C. A. (1973b), *Arch. Biochem. Biophys.*, **157,** 553–560.

Schnaitman, C. A. (1974a), *J. Bacteriol.*, **118,** 442–453.

Schnaitman, C. A. (1974b), *J. Bacteriol.*, **118,** 454–464.

Schnaitman, C. A., Smith, D., and de Salsas, M. F. (1975), *J. Virol.*, **15,** 1121–1130.

Schultz, S. G., and Solomon, A. K. (1961), *J. Genl Physiol.*, **44,** 1189–1199.

Schweizer, M., Hindennach, I., Garten, W., and Henning, U. (1978), *Eur. J. Biochem.*, **82,** 211–217.

Sekizawa, J., and Fukui, S. (1973), *Biochim. Biophys. Acta*, **307,** 104–117.

Sha'afi, R. I., Gary-Bobo, G. M., and Solomon, A. K. (1971), *J. Gen. Physiol.*, **58,** 238–258.

Sheu, C. W., and Freese, E. (1973), *J. Bacteriol.*, **115,** 869–875.

Singer, S. J. (1974), *Annu. Rev. Biochem.*, **43,** 805–833.

Smit, J., and Nikaido, H. (1978), *J. Bacteriol.*, **135,** 687–702.

Smit, J., Kamio, Y., and Nikaido, H. (1975), *J. Bacteriol.*, **124,** 942–958.

Solomon, A. K. (1968), *J. Gen. Physiol.*, **51,** 335s–364s.

Stanier, R. Y., Palleroni, N. J., and Doudoroff, M. (1966), *J. Gen. Microbiol.*, **43,** 159–271.

Stein, W. D. (1967), *The Movement of Molecules Across Cell Membranes*, Academic Press, New York, pp. 65–125.

Steven, A. C., ten Heggeler, B., Muller, R., Kistler, J., and Rosenbusch, J. P. (1977), *J. Cell Biol.*, **72,** 292–301.

Stock, J. B., Rauch, B., and Roseman, S. (1977), *J. Biol. Chem.*, **252,** 7850–7861.

Szmelcman, S., Schwartz, M., Silhavy, T. J., and Boos, W. (1976), *Eur. J. Biochem.*, **65,** 13–19.

Tamaki, S., and Matsuhashi, M. (1973), *J. Bacteriol.*, **114,** 453–454.

Tamaki, S., Sato, T., and Matsuhashi, M. (1971), *J. Bacteriol.*, **105,** 968–975.

Tokunaga, H., Tokunaga, M., and Nakae, T. (1979a), *Eur. J. Biochem.*, in press.

Tokunaga, M., Tokunaga, H., Okazima, Y, and Nakae, T. (1979b), *Eur. J. Biochem.*, in press.

Ueki, T., Mitsui, T., and Nikaido, H. (1979), *J. Biochem. (Tokyo)*, **85,** 173–182.

Uemura, J., and Mizushima, S. (1975), *Biochim. Biophys. Acta*, **413,** 163–176.

Van Alphen, W., and Lugtenberg, B. (1977), *J. Bacteriol.*, **131,** 623–630.

Van Alphen, W., van Selm, N., and Lugtenberg, B. (1978), *Mol. Gen. Genet.*, **159,** 75–84.

Van Gool, A. P., and Nanninga, N. (1971), *J. Bacterol.*, **108,** 474–481.

Von Meyenburg, K. (1971), *J. Bacteriol.*, **107,** 878–888.

Weigand, R. A., and Rothfield, L. I. (1976), *J. Bacteriol.*, **125,** 340–345.

Witholt, B., van Heerikhuizen, H., and de Leij, L. (1976), *Biochim. Biophys. Acta*, **443,** 534–544.

Wolf-Watz, H., Elmros, T., Normark, S., and Bloom, G. D. (1975), *Infect. Immunol.* **11,** 1332–1341.

Wright, A., and Kanegasaki, S. (1971), *Physiol. Rev.*, **51,** 749.

Wright, E. M., and Diamond, J. M. (1969), *Proc. R. Soc. B.*, **172,** 227–271.

Yamada, H., and Mizushima, S. (1978), *J. Bacteriol.*, **135,** 1024–1031.

Yu, F., and Mizushima, S. (1977), *Biochem. Biophys. Res. Commun.*, **74,** 1397–1402.

Zimmermann, W., and Rosselet, A. (1977), *Antimicrob. Agent. Chemother.*, **12,** 368–372.

Chapter **12**

Cell-to-Cell Interactions in Conjugating *Escherichia coli*: the Involvement of the Cell Envelope

PAUL A. MANNING AND MARK ACHTMAN

Max-Planck Institut für molekulare Genetik, Berlin, West Germany

1 INTRODUCTION

Bacterial conjugation was first described when Lederberg and Tatum (1946a, b) demonstrated the formation of prototrophic recombinants from mixed cultures of auxotrophic strains of *Escherichia coli* K-12. Soon after, the requirement for cell-to-cell contact that distinguished conjugation from other forms of bacterial genetic exchange (DNA transformation, bacteriophage-mediated transduction) was also demonstrated (Davis, 1950).

During the 1950s the basic features of conjugation, which have proven so useful to bacterial geneticists, were elucidated (for review see Hayes, 1964). DNA transfer occurs from F^+, F′ or Hfr donor cells unidirectionally to F^- recipient cells. The donor cells carry a sex factor, the F plasmid, which is cytoplasmic in F^+ or F′ cells and integrated into the bacterial chromosome in Hfr cells. Although it is plasmid DNA that is transferred from F^+ or F′ donors versus chromosomal DNA from Hfr donors, the basic mechanism of DNA transfer is the same in all cases, with the differences being due to the covalent linkage of F DNA to chromosomal DNA in the Hfr donors. It is this basic mechanism of conjugation to which we address the remainder of this review.

Conjugation is not unique to *E. coli* or to the F sex factor. Conjugation has been described in numerous gram-negative (and a few gram-positive) bacterial species, and many different sex factors, each of which converts host cells to donors of plasmid DNA, have been described. For technical reasons, plasmid DNA transfer is usually difficult to demonstrate unless the plasmid is covalently linked to DNA that changes the phenotype of the host. Many sex factors from nature are linked to such DNA and confer on their bacterial hosts the ability to produce colicins (Col factors; for review see Hardy, 1975; Rowbury, 1977), antibiotic resistance (R factors; for review see Falkow, 1975), or other diverse properties (summarized by Novick, 1974). Cells carrying these sex factors can transfer the sex factor DNA to other bacteria, occasionally across species barriers. As a consequence of the clinical and industrial usage of antibiotics, the rare cells have been selected that have acquired an R factor such that now numerous bacteria of clinical significance are resistant to the very antibiotics that were formerly used for therapy.

Analysis of these various sex factors has revealed that although some (the F-like sex factors) share extensive DNA homology with the F plasmid (Sharp et al., 1973), most are unrelated. The nature of DNA transfer has been extensively investigated only for F-like and for the unrelated (but apparently similar) I-like sex factors. Genetic analyses have been reported only for F-like and for the unrelated P sex factors. For the others the only information available currently is that the plasmid DNAs are all quite large (larger than 25 megadaltons) and that some encode the ability to synthesize pili, which may be involved in conjugation. Most of the information available on conjugation has been gained with the F sex factor, and we concentrate here on *E. coli* K-12 cells and F-mediated conjugation.

Several reviews have recently appeared that are related to various aspects of conjugation: F pili (Brinton, 1971; Tomoeda et al., 1975); sex factor genetics (Willetts, 1972; Achtman, 1973a; Rowbury, 1977; Achtman and Skurray, 1977); Col factors (Hardy, 1975); and the mechanism of DNA transfer (Curtiss, 1969; Curtiss et al., 1977; Achtman and Skurray, 1977).

However, it is only since 1976 that information on the proteins involved in conjugation has become available. Many of these proteins are associated with the *E. coli* cell envelope, and we use this opportunity to present our current concepts and information on these proteins.

2 OVERVIEW

Cells carrying the F plasmid only transfer F DNA when *tra* (*tra*nsfer) cistrons on the F DNA are transcribed and translated to yield tra proteins. Some of these tra proteins are needed for synthesis (and possibly assembly) of F pili, the organelles that recognize and allow contact with F^- cells. Other tra proteins are needed for forming stable contacts with the F^- recipient cells, and yet others are needed to trigger DNA synthesis and to allow DNA transfer during conjugation. Although very little is known about the detailed biochemical function of any of these tra proteins, most of them have now been identified by SDS–polyacrylamide gel electrophoresis and have been assigned to their encoding *tra* cistrons. In addition, certain chromosomally encoded proteins seem to be important for the efficient functioning of the F-encoded tra proteins in the donor cell.

The recipient cell probably plays a specific role in each of the substages of conjugation. However, the *ompA*-encoded protein, which we occasionally refer to as ompA protein, is the only one that has been assigned to a substage. The *ompA*-encoded protein is required in the outer membrane of the F^- cell to allow stabilization of mating contacts with F plasmid-carrying donor cells. A second requirement for recipient ability is that cells not carry an F factor. Cells carrying an F-like plasmid are *not* good recipients in conjugation with other cells carrying a closely related plasmid because of the plasmid-encoded phenomenon of surface exclusion. The F plasmid encodes two proteins, which are incorporated into the cell envelope and one of which prevents contact stabilization and the other triggering of DNA transfer.

Of the proteins just mentioned, some are functionally equivalent among different F-like plasmid transfer systems, whereas others are plasmid specific and only perform their role in conjugation to allow transfer of very closely related plasmids.

3 THE RECIPIENT IN CONJUGATION

Recent analyses have defined a class of mutants of F^- *E. coli* K-12 cells referred to as $ConF^-$ (*con*jugation defective recipients with *F* carrying donors) or $ConI^-$ (*con*jugation defective recipients with *I* factor-carrying donors). The $ConF^-$ mutants known at present fall into two classes: those affected in the *ompA* cistron (Manning et al., 1976; Henning et al., 1976) and those affected in lipopolysaccharide (LPS) structure. Mutants in the *ompA* gene have been mapped at 21.5 min on the *E. coli* K-12 linkage map (Bachmann et al., 1976). The *ompA* gene was formerly referred to as *tolG* (Foulds and Barrett, 1974; Foulds, 1974), *con* (Skurray et al., 1974; Manning and Reeves, 1976a, b; Manning et al., 1976), or *tut* (Henning et al., 1976), but these designations have all been shown to refer to the *ompA* gene (Manning and Reeves, 1976a). The *ompA* gene is the structural gene (Datta et al., 1976; Manning et al., 1976) for the major heat modifiable outer membrane protein referred to variously as II* (Henning et al., 1973), 3a (Schnaitman, 1974), B* (Reithmeier and Bragg, 1974), G (Chai and Foulds, 1974), and d (Lugtenberg et al., 1975). The outer membranes of *ompA* mutants often lack ompA protein (Figure 1, tracks *1* and *2*).

Various means have been used for isolating *ompA* mutants, including resistance to colicin L (Foulds and Barrett, 1974; Davies and Reeves, 1975a, b), resistance to bacteriophages K3 or TuII* (Skurray et al., 1974; Manning et al., 1976; Henning and Haller, 1975; Henning et al., 1976), and screening for defective recipients after using zygotic induction (Havekes and Hoekstra, 1976). Recipient ability of *ompA* mutants often is reduced by up to 5000-fold (Skurray et al., 1974; Manning and Reeves, 1975; 1976a, b; Manning et al., 1976; Havekes and Hoekstra, 1976). Using a temperature-sensitive *ompA* mutant, Manning and Reeves (1976b) found a marked increase in the ability to produce transconjugants associated with an increased amount of ompA in the outer membrane.

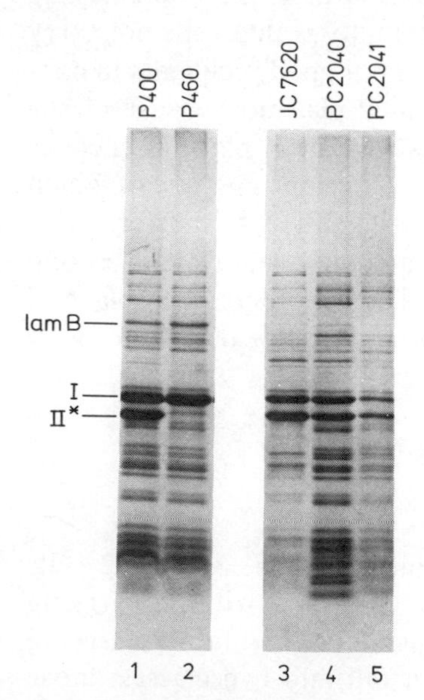

Figure 1 SDS–polyacrylamide gel electrophoresis of outer membranes. The Triton X-100 insoluble components of the cell envelopes from comparable numbers of cells were obtained and analyzed as described by Achtman et al. (1978c). (*1*) P400, (*2*) P460, an *ompA* mutant of P400, (*3*) JC7620, (*4* and *5*) PC2040 and PC2041, respectively, the heptose-less LPS mutants of JC7620.

The correlation among *ompA* mutants of resistance to bacteriophage K3, tolerance to colicins K and L, and defective recipient ability in conjugation is not absolute (Manning et al., 1976). Mutants that are defective in conjugation but still sensitive to bacteriophage K3 have been isolated (Havekes and Hoekstra, 1976). Bacteriophage K3-resistant mutants with minimal effects on conjugation or colicin sensitivity have also been described (Manning et al., 1976). On examining a number of *ompA* mutants or LPS mutants apparently affecting ompA protein function (see below), it was found that the correlation between colicin L-tolerance and defective recipient ability in conjugation was absolute, suggesting that the same active site of ompA is used in killing by colicin L and in conjugation (Achtman et al., 1978c).

Other evidence has implicated lipopolysaccharide in recipient ability. Mutants that were ConF$^-$ but were not mutated at *ompA* appear to be affected in their LPS composition. These include the high level ampicillin-resistant and ϕW-resistant mutant of Monner et al. (1971), the mutants of Reiner (1974), which were isolated as being resistant to the single stranded DNA phage ST-1, and the mutants of Havekes et al. (1976), which were isolated as being resistant to bacteriophages T3, T4, and T7. In a survey of a large number of mutants selected as resistant to at least one of a series of colicins or bacteriophages, only LPS mutants (aside from *ompA* mutants) were defective as recipients in conjugation (Manning and Reeves, 1975, 1977; Manning, 1977). Most of these mutants have been shown to have a heptose-less LPS (Hancock and Reeves, 1976).

It has been shown in *Salmonella typhimurium* (Ames et al., 1974) and in *E. coli* (Koplow and Goldfine, 1974; van Alphen et al., 1976) that heptose-less LPS mutants are also affected in the protein composition of their outer membrane. An example of such effects can be seen in the mutants of Havekes et al. (1976) (Figure 1, tracks *3–5*). PC2040 (track *4*) showed an increase in the amounts of a large number of minor proteins, whereas PC2041 showed some minor protein increases but also many reductions in the amounts of other proteins (Achtman et al., 1978c). Manning (1977) also observed that there was a very good correlation between increase in LPS defect and decreases in recipient ability. Thus it would seem that LPS alterations can have a wide range of effects on the outer membrane protein composition. These LPS mutations probably also alter specific interactions of the different intramembrane components.

It has recently been shown that ompA protein both is linked to the peptidoglycan and interacts with the LPS (Enderman et al., 1978; Schweizer et al., 1978). *In vitro*, LPS protects isolated ompA protein (in solution) from proteolytic cleavage to exactly the same extent as it does in the cell envelope. LPS can also reverse the denaturation caused by boiling ompA protein in SDS solutions. This latter protein has been shown to be the receptor for bacteriophages K3 and TuII* (van Alphen et al., 1977; Schweizer and Henning, 1977; Manning, 1977); for this activity the presence of LPS is required. Thus it is likely that there is a specific interaction between ompA protein and LPS in the outer membrane. These considerations raise the possibility that the role of LPS in conjugal recipient ability is indirect and dependent on function of ompA protein.

Achtman et al. (1978c) gathered a wide range of ConF$^-$ mutants affecting ompA protein or LPS and compared them with each other. They all behaved quite similarly. When analyzed for recipient ability in liquid medium, the mutants were defective, as reported by others. However, if the matings were performed on the surface of membrane filters, this defect was almost totally reversed. In all cases, no mating contact was observed using physical analysis by a Coulter counter. Thus it was concluded that all the mutants were affected in the formation of stable mating contacts. Furthermore, it is likely that all these mutants were ConF$^-$ as a result of an effect on ompA protein; that is, they were either direct effects as in the case of *ompA* mutants, or they were indirect effects as in the case of the LPS mutants, which were probably affecting the intramembrane interactions of ompA protein and LPS.

In a normal mating mixture consisting of equal numbers of donor (F′ or Hfr) and recipient (F$^-$) cells, mating contacts are quickly formed that hold conjugating cells together in mating aggregates despite pipetting, centrifugation, or mild shear forces (Achtman, 1975). Within 15 min at 37°C, 70% of all cells may be held together in such mating aggregates. Approximately half of the mating cells are in pairs (one donor and one recipient), but the other half are in larger aggregates (Figure 2) that may contain up to 50 cells and have variable proportions of donors and recipients (Achtman, 1975; Achtman et al., 1978b). Despite this heterogeneity, the interactions can be analyzed in terms of mating contacts between cells.

Both F pilus-to-wall and wall-to-wall mating contacts are detectable (Ou and Anderson, 1970; Achtman et al., 1978b). The full role of the detectable F-pilus contacts is still uncertain; they are clearly needed for the initial interactions between conjugating cells and are not needed for DNA transfer or to hold cells together once more stable contacts have formed (Achtman et al., 1978b).

It seemed possible that the *ompA* mutants were blocked in binding of donor-borne F pili or subsequent to F-pilus binding but before the formation of stable wall-to-wall contact. Therefore, the prototype *ompA* mutant, P460 (Skurray et al., 1974; Manning and Reeves, 1976a), was examined in more detail (Achtman et al., 1978c). As was the case for the other ConF$^-$ mutants tested, the formation of mating aggregates in liquid matings was not detectable with the Coulter counter technique. However, when liquid

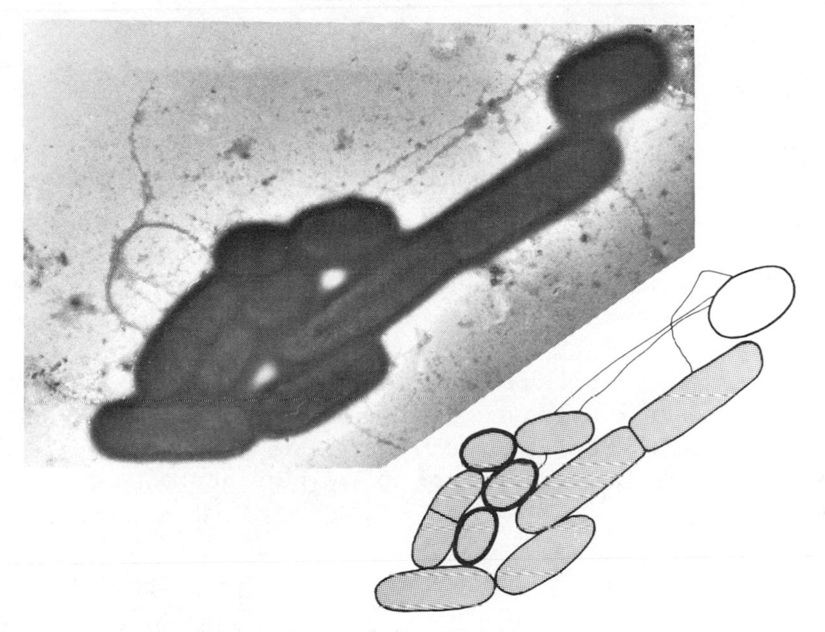

Figure 2 A mating aggregate from an *E. coli* K-12 Hfr × *E. coli* C F$^-$ mating. The same aggregate is demonstrated diagrammatically: the thin walled cells represent the Hfr cells and the small, thick walled cells represent the F$^-$ cells. The thin lines connecting cells are F pili; only F pili-connecting cells are indicated in the diagram. The unshaded cells is probably also F$^-$ cell. From Achtman et al. (1978b), with permission.

matings were analyzed by electron microscopy, under conditions where the shear forces inherent in the Coulter counter method are greatly diminished, normal levels of mating aggregation were detected. Both F pilus-to-wall and wall-to-wall contacts were seen (Achtman et al., 1978c). P460 also resembled the other ConF$^-$ mutants in that it was 5000-fold worse as a recipient than its *ompA*$^+$ parent in liquid matings but only fourfold worse in matings conducted on a membrane filter. Thus P460 is blocked in conjugation beyond the substages of F-pilus binding and unstable wall-to-wall contact formation.

Inhibition of conjugation has also been achieved by adding low concentrations of purified ompA protein (Schweizer and Henning, 1977; van Alphen et al., 1977; Achtman et al., 1978c). As was observed in the bacteriophage receptor studies, LPS is essential for activity. Furthermore, the effect of externally added ompA protein + LPS mimics that of *ompA* mutants, since the formation of stable mating aggregates (measured with a Coulter counter) is inhibited at the concentrations that inhibit conjugation, while the formation of unstable mating aggregates (measured by electron microscopy) is only inhibited at higher concentrations (Figure 3). We conclude that ompA protein is needed in the LPS environment of the outer membrane of the recipient for conversion of unstable to stable mating aggregates (Achtman et al., 1978c).

In contrast to these results with F factor-carrying donors, other sex factors are not necessarily dependent on ompA protein for conjugation. Only F and closely related F-like plasmids transfer poorly to *ompA* mutants, while other F-like plasmids such as R100-1 and I-like plasmids transfer efficiently to *ompA* mutants (Manning and Reeves, 1975, 1976a, 1977). It seems likely that conjugation with these plasmids is dependent on other components of the cell envelope outer membrane for contact formation. This component may be LPS itself for the I-like plasmids, since mutants detected as ConI⁻ were all LPS mutants (Havekes et al., 1977b; Manning and Reeves, 1977). One mutant was shown to be blocked in the formation of stable mating

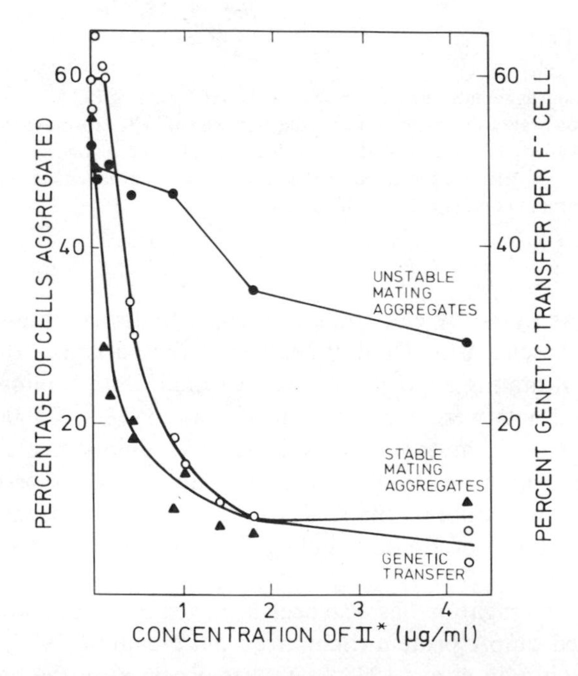

Figure 3 Effects of purified protein II* (ompA protein) on conjugation. Mating mixtures of an F′ *lac* donor strain and an F⁻ recipient were analyzed in the presence of various concentrations of purified protein II* dissolved in LPS. (●) electron microscopy, (○) genetic transfer, (▲) Coulter counter. Modified from Achtman et al., (1978c).

aggregates with an I factor-carrying donor (Achtman et al., 1978c). The ConI$^-$ mutants were normal recipients with F factor-carrying donors. Furthermore, low concentrations of purified LPS, which had no effect on conjugation with F factor-carrying donors, inhibited DNA transfer from I factor-carrying donors (Havekes et al., 1977b). Havekes et al. (1977b) suggested that LPS fulfills a role in I plasmid conjugation equivalent to that of ompA protein in F plasmid conjugation. However, it has not yet been demonstrated that LPS is involved in the same substage of conjugation as is ompA protein.

The difference between F conjugation and I conjugation is not surprising, since these two plasmids are totally unrelated. However, there are also differences among F-like plasmids that are, nevertheless, similar enough that they all encode F-like pilus synthesis and share extensive homology with F within their *tra* DNA. R100-1, one of these F-like plasmids, transfers to *ompA* or LPS mutants at normal levels and seems to use yet another outer membrane component (Manning and Reeves, 1975). Con$^-$ mutants relatively specific for R100-1 have been isolated (Havekes et al., 1977a), but it is still unclear what component is affected. These specificities among F-like plasmids are discussed in more detail below after the discussion on *tra* cistrons.

The observations summarized above have introduced new complexities. Just a few years ago, it seemed possible that the recipient cell would have one surface component to which sex pili would bind and that the recipient's subsequent role in conjugation was restricted to DNA entry. Now, the only cell components so far identified that are used in conjugation prove to be specific for an intermediate stage in contact formation. The receptor(s) for sex-pilus binding remain to be identified, as do the membrane components involved in unstable contact formation and DNA entry. Clearly, much remains to be done in this area.

4 THE DONOR IN CONJUGATION

F factor DNA is a double stranded circle, 94.5 kb (kilobases or thousands of base pairs, equals 62 megadaltons) in size (Sharp et al., 1972). This is sufficient DNA to encode about 100 proteins. Genetic analyses have revealed that approximately one-third of this DNA, the *tra* region, encodes proteins involved in conferring donor ability on the host cells. The *tra* region encodes F-pilus synthesis, contact formation, and poor recipient ability (surface exclusion). These are now discussed in some detail. Only very few other functions (replication, incompatibility, restriction of certain bacteriophages, and chromosomal integration) have been identified, and these genes can be assigned to the remaining two-thirds of F DNA.

4.1 The F Pilus

The classifical property of cells carrying an F or F-like plasmid is that they possess F or F-like sex pili. These can be recognized immunologically (Ørskov and Ørskov, 1960; Ishibashi, 1967; Knolle and Ørskov, 1967) or by electron microscopy (Crawford and Gesteland, 1964). F pili can be readily distinguished from other surface appendages (common pili, flagellae) since they adsorb male-specific RNA bacteriophages along their sides (Crawford and Gestland, 1964; see Figure 4) and DNA bacteriophages at their tips (Caro and Schnös, 1966). *E. coli* cells growing at 37°C in nutrient broth possess an average of one to three F pili, but growth in minimal medium, into stationary phase or at lower temperatures can drastically reduce the level of piliation (Curtiss et al., 1969; Novotny and Fives-Taylor, 1974; Helmuth and Achtman, 1978). Cell associated F pili are variable in length, with the average length being about 2 μ, but with some pili growing up to 20 μ long (Brinton, 1965). F pili seem to grow from random points on the cell surface. However, Bayer (1975) has found that F pili are anchored at adhesion sites between the inner and outer membranes (Figure 5). Although no

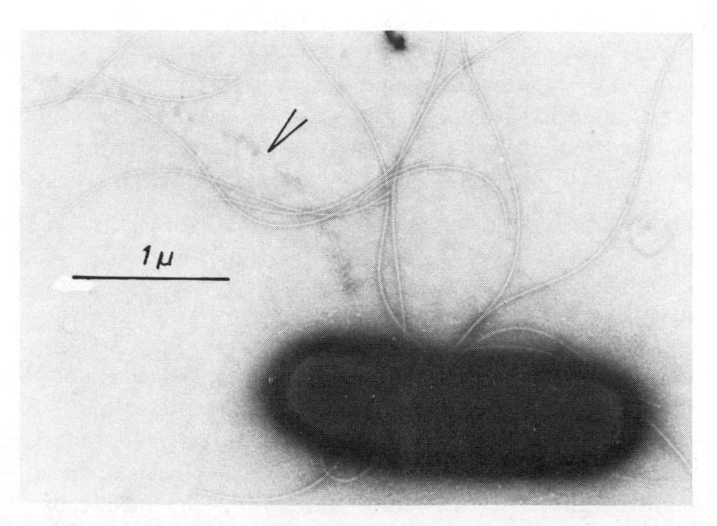

Figure 4 The F pilus. The arrow indicates an F pilus projecting from an F^+ cell and labeled with the pilus-specific RNA phage, M12. The thicker filaments are flagellae. Photograph courtesy of G. Morelli.

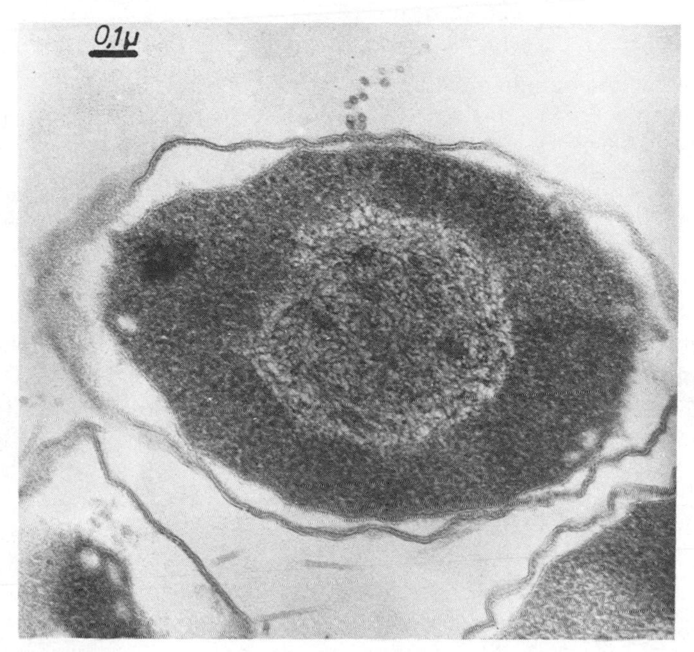

Figure 5 Emergence of an F pilus from a site of adhesion of the inner and outer membrane. Photograph courtesy of M. Bayer.

specific basal structure has yet been detected, analogy with flagellae (DePamphilis and Adler, 1971) raises the possibility that F pili are anchored by an organelle into the cell envelope. Since adhesion sites have been implicated in bacteriophage infection (Bayer, 1968a, b) it is possible that the association between F pili and adhesion sites is relevant to transmembrane transport of nucleic acids in conjugation and in bacteriophage infection.

A number of different sex pili encoded by different sex factors have now been identified [see Achtman and Skurray (1977) for a recent summary]. In no case is the information available yet as complete as that for F pili. In fact, the limit of our knowledge of these other pili is that they serve as adsorption organelles for specific RNA- and/or DNA-containing bacteriophages. No serological cross-reactions or common sensitivity to a bacteriophage have yet been reported for the different sex pili described. In contrast, F-like plasmids all encode sex pili that are at least partially similar to F pili. These F-like pili can be differentiated from F pili since they are not serologically identical and since cells carrying them are not necessarily as efficient in allowing F-specific bacteriophage multiplication as cells carrying F pili (Dennison and Hedges, 1972). The *traA* cistron is responsible for this specificity (see below; Willetts, 1971). F-like pili are similar enough to F pili that cells carrying the F factor plus an F-like plasmid synthesize mixed pili containing subunits of each of the two types (Lawn et al., 1971).

The role of F pili in conjugation has recently become somewhat clearer. F pili are essential on donor cells for the initial step(s) of conjugation. They can be removed from donor cells by high speed blending without loss of viability (Novotny et al., 1969). These cells are then incapable of mating unless new F pili are synthesized. Achtman et al. (1978b) showed that low SDS concentrations dissolve F pili from cell surfaces and prevent mating aggregate formation and DNA transfer (Figure 6). Mating aggregate formation (and DNA transfer) can also be blocked by addition of DNA bacteriophages that absorb to the F-pilus tip (Ou, 1973). Thus mating aggregate formation is dependent on F pili and these probably are involved in recognition of a suitable receptor on the recipient cell surface. It has also

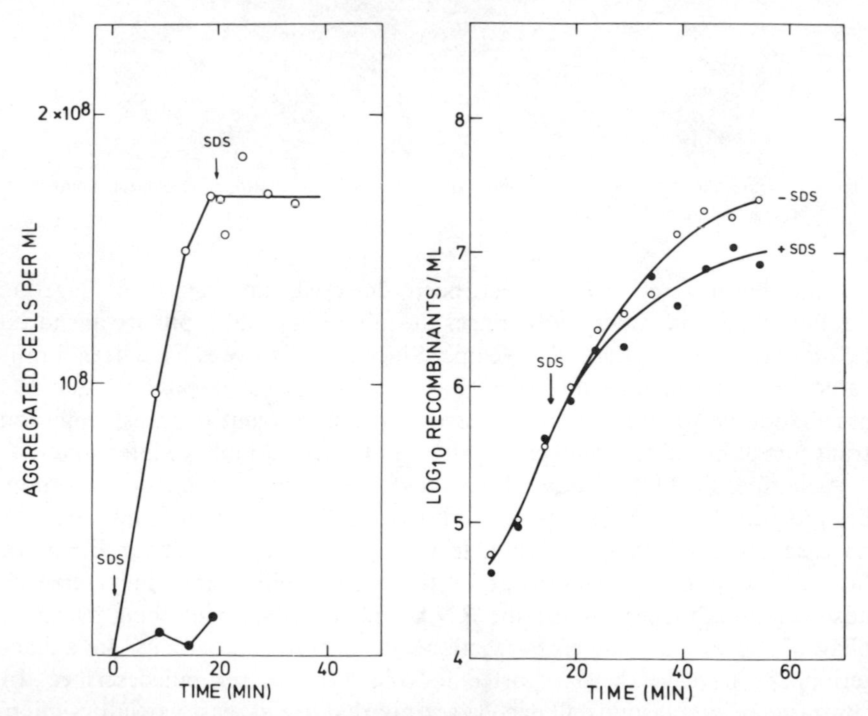

Figure 6 (*a*) Effect of SDS on mating aggregation. Mating mixtures containing equal proportions of Hfr donor and F⁻ recipient cells (total cell concentration 2.5×10^8/ml) were placed at 37°C at 0 min. SDS (final concentration 0.01%) was added at time 0 (●) or 19 min (○). (*b*) Effect of SDS on DNA transfer. Equal volumes of Hfr and F⁻ cells were incubated with gentle aeration and at the time indicated, SDS (final concentration 0.01%) was added. At the times indicated genetic recombinants were scored. Data from Achtman et al. (1978b).

been suggested that DNA transfer occurs through the F pili connecting donor and recipient cells (Brinton, 1965, 1971). In agreement, DNA transfer did occur between cells only loosely connected, without direct wall-to-wall contact (Ou and Anderson, 1970). However, as shown in Figure 6, once mating aggregates were formed, SDS did not dissolve the mating contact and did not efficiently prevent DNA transfer even though all extended F pili connecting conjugating cells had been dissolved (Achtman et al., 1978b). Thus the primary contact within mating aggregates is not an extended F pilus, and DNA transfer is not dependent on extended F pili once mating aggregates have formed. One possibility is that the only role of F pili is to establish initial contacts between conjugating cells and that subsequent events involve direct interactions among the cell envelopes (Achtman and Skurray, 1977). However, it has been proposed that F pili retract during conjugation or bacteriophage infection (Curtiss, 1969; Marvin and Hohn, 1969). Some evidence for retraction has been obtained. Upon infection with a DNA-containing bacteriophage, F pili became shorter and fewer (Jacobson, 1972). NaCN or low temperatures led to fewer F pili on the cell surface (Novotny and Fives-Taylor, 1974). Thus F pilus retraction might bring the cell surfaces into intimate contact and the retracted pili might possibly be used as a bridge to allow DNA transfer. Such retracted pili might well be resistant to SDS and might be short enough to escape detection by electron microscopy. However, the evidence for retraction of F pili occurring during conjugation is not yet conclusive and other possibilities should still be considered.

F pili have been purified in milligram amounts in three laboratories (Brinton, 1971; Date et al., 1977; Helmuth and Achtman, 1978), as have F-like pili encoded by the F-like R factor R1drd19 (Beard et al., 1972). These pili can adsorb RNA- and DNA-containing bacteriophages, although the binding is reversible (Helmuth and Achtman, 1978). Binding of RNA bacteriophages was cooperative since at any one multiplicity of bacteriophages, only some pili adsorbed bacteriophages, but those pili were densely covered. With increasing bacteriophage multiplicity, more and more F pili (up to 50%) were densely covered. The purified F pili also bound to *E. coli* surfaces (Figure 7), but no significant binding was detected to *Bacillus* or *Pseudomonas* cells (Helmuth and Achtman, 1978). Up to 50 pili could bind to one *E. coli* cell, and under standardized conditions, 20% of the F pili present bound to the cell surface. The receptor for the F pilus is unknown, but it is not ompA protein (Achtman et al., 1978c). No details on the kinetics of binding are yet available and this system still remains difficult to analyze. F pili aggregate at low concentrations (1 μg/ml) and stick nonspecifically to most substances. Attempts at radiochemical labeling *in vitro* have inactivated binding activity (R. Helmuth, personal communication).

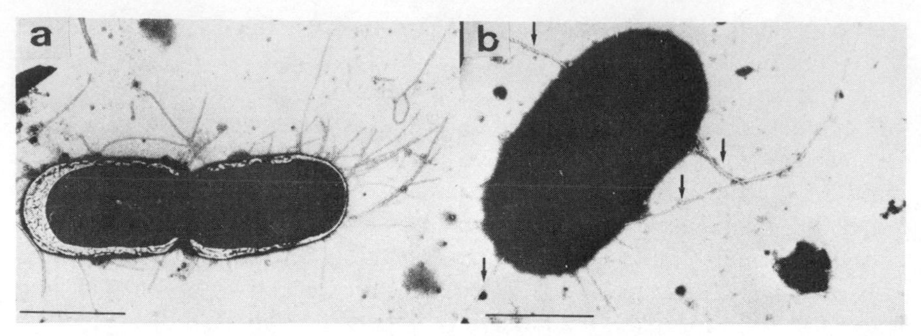

Figure 7 Binding of F pili to F^- cells of *E. coli* K-12. (*a*) Cells to which 5 μg/ml of F pili had been added followed by one washing. (*b*) The same as in *a* except that the cells received two cycles of washing. Scale = 1 μm. Modified from Helmuth and Achtman (1978).

The purified F pili have only one detectable subunit of 10,750–11,800 daltons (Brinton, 1971; Date et al., 1977; Helmuth and Achtman, 1978). The subunits are dissolved by SDS but are resistant to Triton X-100 or 6*M* guanidine HCl (Date et al., 1977; Helmuth and Achtman, 1978). Aggregated pilus crystals can be dissolved with 80% surcrose or 6*M* guanidine·HCl, but the pili reaggregate once the sucrose or guanidine is removed (Helmuth and Achtman, 1978). The F pilin subunit contains a single D-glucose and two phosphate residues (Brinton, 1971). F pili have 69% α-helical content (Date et al., 1977).

The same purified F pili that can bind to *E. coli* cells have been analyzed by optical and X-ray diffraction. As a result, certain basic parameters have been elucidated (Folkhardt et al., 1979). F pili are hollow cylinders 80 Å wide with a central hole of about 20 Å, although these dimensions vary slightly depending on water content. The shell contains a hydrophobic, electron-poor region that may be important for conformation of the pilus. The subunits are arranged in helices, with four helices per pilus and a helix repeat of 128 Å (Figure 8). Folkhardt et al. (1979) have proposed that retraction might be achieved by a conformational transition from ordered helix to sheet structure, which could then collapse and possibly be resorbed into the cell envelope.

4.2 Genetics of the F Factor

As is stated earlier, the F factor is large enough to encode 100 average sized proteins. This indicates the complexity of the processes that the F factor must carry out to survive in a host cell and to be able to transfer itself. Willetts (1972) and Achtman (1973a) reviewed the genetics of the F factor. Since that time a more detailed knowledge of the genetic structure has developed, warranting an expanded discussion here. The current map of the F factor is presented in Figure 9.

A number of properties other than those involved in conjugation have been mapped on the F factor. Morrison and Malamy (1971) defined two *pif* cistrons that are responsible for the restriction or inhibition of growth of ϕII and T7-like bacteriophages. Incompatibility (*inc*) and all the genes associated with the replication of the F-factor DNA (*frp*), as well as the origin of vegetative replication (*oriV*), lie on a single restriction fragment which has been mapped (Guyer et al., 1976;). Several insertion sequences (IS) on F have also been identified and these are involved in the integration of F into the bacterial chromosome that gives rise to Hfr strains (Davidson et al., 1975). As can be seen from the map (Figure 9), there are still a number of regions with no known function and it is not known whether or not these regions represent genetically silent DNA.

Many *tra* cistrons that are involved in DNA transfer have been identified and mapped. Numerous point mutations were isolated after mutagenesis of F'*lac* or F'*gal* (Ohtsubo et al., 1970; Achtman et al., 1971, 1972; Miki et al., 1978). These were analyzed by complementation analyses and assigned to *tra* cistrons by various, somewhat complicated techniques (Ohtsubo et al., 1970; Achtman et al., 1972; Willetts and Achtman, 1972; Willetts, 1973; Achtman et al., 1978d; Miki et al., 1978). These techniques were necessitated by an incompatibility that prevents the stable maintenance of two independent F factors in the same cell. The more recently developed and easy techniques involve using chimeric plasmids carrying F DNA cloned on a small plasmid compatible with F (Skurray et al., 1976; Achtman et al., 1978d) or lambda bacteriophages carrying F DNA (λd*tra*; Miki et al., 1978) as a source of complementing genes.

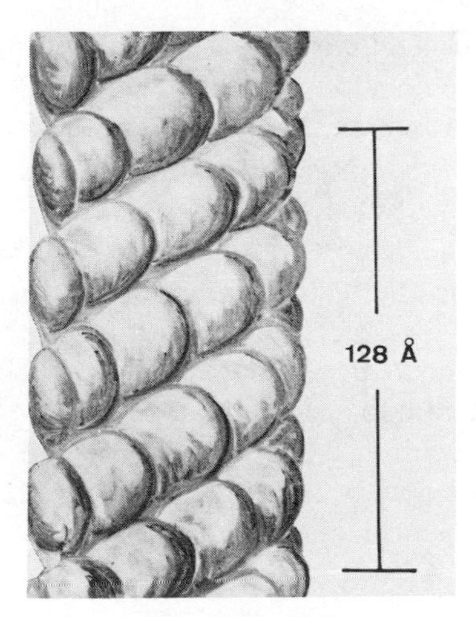

Figure 8 Schematic model of F-pilus structure, based on information from electron microscopy and X-ray diffraction. The 128 Å line is the repeat distance for each of the four helices of F-pilin subunits.

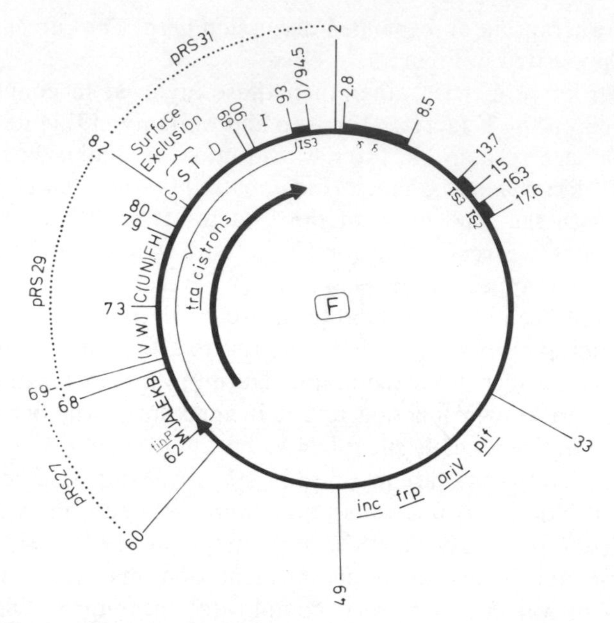

Figure 9 A current map of the F factor. Mnemonics: *inc* for *inc*ompatibility; *frp* for *F rep*lication; *oriV* for *ori*gin of *v*egetative replication; *pif* for *p*hage *inf*ection inhibition; IS2, IS3, and $\gamma - \delta$ are insertion sequences; *tra* for *tra*nsfer operon cistrons; *fin*P is the F-specific component of the FinOP repression system (Willets, 1972). The genes are located according to Skurray et al. (1976), Achtman et al. (1978d), Thompson and Achtman (1978), Manning et al. (1979). Chimeric plasmids pRS27, pRS29, and pRS31 contain pSC101 plus the regions of the factor as shown; *traG* straddles an EcoRI restriction site at 82 kb. The heavy arrow indicates the direction of transcription of the *tra* operon. More precise mapping of the *tra* cistrons is shown in Figure 11.

The physical order of *tra* cistrons has been determined using complementation and marker rescue analyses with deletion mutants (Ohtsubo, 1970; Ippen-Ihler et al., 1972; Thompson and Achtman, 1978; Manning et al, 1979) or λd*tra*'s containing various lengths of F DNA. Insertion mutants have been generated with Mu-1 bacteriophage. Since Mu-1 insertions are absolutely polar, these mutants have allowed the demonstration that most of the cistrons known, from *traA* to *traI*, are in one operon transcribed as shown in Figure 9 (Helmuth and Achtman, 1975). More recent evidence suggests that a weak promoter maps between *traD* and *traI*, although *traI* is still transcriptionally included in the *tra* operon (Achtman et al., 1978d).

Three of the cistrons in Figure 9 are not required for DNA transfer and were elucidated by other techniques. The cistrons *traS* and *traT* are involved in surface exclusion, the poor recipient ability in conjugation of F factor-carrying cells. Cells carrying *traS*$^-$ or *traT*$^-$ mutants are better recipients than cells carrying a wide-type F factor; they remain good donors (Achtman et al., 1977). The *traS* cistron is plasmid specific and only acts to prevent transfer from donor cells carrying closely related sex factors (Willetts and Maule, 1974; Achtman et al., 1977).

The *finP* cistron is involved in transcriptional regulation of the *tra* cistrons and is also plasmid specific. Although the *tra* cistrons of F are derepressed naturally, many F-like sex factors are repressed and can repress the F factor as well (Meynell et al., 1968). Repression mediated by the FinOP system involves the *finO* gene of the F-like sex factor plus the *finP* gene of F whose gene products act together at a site, *fisO*, mapping very near *traJ* to turn off *traJ* transcription (Finnegan and Willetts, 1971, 1972; Willetts 1977a; Willetts and Finnegan, 1972). The *traJ* cistron is needed for transcription of the other *tra* cistrons known (except *traM*) (Achtman et al., 1972; Achtman, 1973b; Willetts, 1977a) and seems to act as a positive control gene. Mutants of *traJ* are deficient in DNA transfer and surface exclusion. Cells carrying F plus a repressing F-like plasmid have the same phenotype as *traJ* mutants. However, *finP* mutants of F are transfer proficient and surface exclusion proficient, even when a repressing F-like sex factor, which supplies *finO*, is present in the same cell as the *finP*⁻ F*lac* mutant (Finnegan and Willetts, 1971, 1972). The F factor is naturally mutant in *finO* and so is normally derepressed for transfer even though it possesses *finP*.

The various other *tra* cistrons are involved in various aspects of DNA transfer. The cistrons *traJ*, *A*, *L*, *E*, *K*, *B*, *V*, *W*, *C*, *U*, *F*, *H* and the promoter proximal part of *traG* are essential for F-pilus synthesis. Only two of these 13 cistrons, *traJ* and *traA*, vary in specificity between F-like plasmids, and mutations in any of the others can be complemented by an F-like plasmid or chimeric plasmids carrying part of the F-like plasmid (Willetts, 1971; Willetts, 1973; Miki et al., 1978; Achtman et al., 1978a). In contrast, *traA* is responsible for the sensitivity to infection by male-specific bacteriophages (Willetts, 1971). Cells carrying both an F-like plasmid and a *traA* mutant of F*lac* are only as sensitive to these bacteriophages as cells carrying the F-like plasmid alone, and *traJ* is not complemented by most F-like plasmids.

The other cistrons, *traM*, *traN*, the promoter distal portion of *traG*, *traD*, and *traI* are not needed for F-pilus synthesis but are needed for mating aggregate formation (*traG* and *traN*) or triggering of DNA transfer (*traI* and *traM*) or for DNA transfer itself (*traD*) (Achtman and Skurray, 1977; Kingsman and Willetts, 1978; our unpublished data). Of these, *traM* and *traI* are plasmid specific (Willetts, 1971; Achtman et al., 1978a, Miki et al., 1978; M. Achtman, R. Thompson and B. Kusecek, unpublished observations).

An interesting result arises when one compares the plasmid specificities of the cistrons just mentioned with that of *ompA* (Table 1). It seems as if to a large extent, plasmid specificity for individual cistrons among F-like plasmids has evolved independently. Thus four surface exclusion (*traS*) specificity groups are involved in the five plasmids listed (Willetts and Maule, 1974), while only three F-pilus (*traA*) specificity groups and only two *traI* specificity groups are involved. It seems as if the various gene products have not evolved coordinately and cannot be mutually involved in a common enzymatic pathway.

4.3 Identification of the tra Proteins

The cloning of restriction endonuclease fragments of the F factor has allowed the identification of various transfer gene products. Skurray et al. (1976) described the cloning of large segments of F factor DNA on the small plasmid vector pSC101 by using partial digests generated with the

Table 1 Specificity of Conjugation Proteins for F-like Sex Factors

	Recipient	Donor				
Sex Factor	*ompA*[a]	F Pili[b]	J,M,6c[c]	I[c]	*finP*[d]	Sfx[e]
F′ or Hfr	A	A	A	A	A	A
ColV2	A	A	A	A	A	B
ColVBtrp	A	A	A	A	B	C
R1drd19	B	B	B	A	C	C
R100-1	C	C	Not A	B	D	D

Note: The letter assignments for specificity groups are arbitrary and were used to permit a cross comparison.

[a] Group A sex factors are transferred at $\leq 10^{-3}$ of normal efficiency to *ompA* mutants; those of group B are transferred at 0.1 and those of group C at 1.0 (Manning and Reeves, 1976a).

[b] Group A encode F pili that allow infection by RNA-containing bacteriophages at an efficiency of 1.0; group B encode at 0.1 and group C at 0.01 (Willetts, 1977b).

[c] Group A sex factors can complement *traI*, *traJ*, and *traM* mutants of F*lac* and can mobilize chimeric plasmids containing *oriT* but not encoding a functional TraMp or 6c. Others cannot (M. Achtman, R. Thompson, and B. Kusecek, unpublished observations.)

[d] Sex factors in each specificity group encode a FinP protein, which together with FinO protein can repress other members of that same group (Finnegan and Willetts, 1972; Willetts, 1977b).

[e] Sfx = surface exclusion. Specificity groups are discussed in the text.

restriction endonuclease EcoRI. By this means, the whole transfer operon has been cloned in a number of pieces to yield pRS27 (EcoRI fragments f6 and f15), pRS29 (f15 and f1), and pRS31 (f17, f19 and f2) (Figure 10). The same EcoRI fragments have also been individually cloned on the ColE1-derived plasmid RSF2124 (Achtman et al., 1978d). Although these chimeric plasmids are not self-transmissible, one can transfer them into different strains by transformation with the purified DNA (Cohen et al., 1972). In this way the *tra* cistrons have been assigned to individual EcoRI restriction fragments (Skurray et al., 1976; Achtman et al., 1978d; Manning et al., 1979).

These chimeric plasmids have also been introduced into minicell-producing strains so that the encoded proteins could be identified. Minicells are small cells derived by abnormal cell division and have no chromosomal DNA. However, the small chimeric plasmids readily segregate into the minicells, so that it is possible to examine the gene products of this DNA in the absence of those encoded by chromosomal DNA (for review on minicells see Frazer and Curtiss, 1975). The chimeric plasmids are smaller than F and thus the DNA contains fewer cistrons and encodes fewer proteins. Furthermore, F DNA is somewhat difficult to purify, whereas the chimeric plasmid DNA is relatively easy to purify. The purified DNA can also be added to an *in vitro* DNA-dependent protein-synthesizing system where proteins encoded by the plasmid DNA can be recognized. Numerous polypeptide bands encoded by the cloned DNA have been identified (Kennedy et al., 1977) *in vitro* and in minicells. To assign these to genetically known *tra* cistrons, the chimeric plasmid DNA was mutagenized *in vitro* with hydroxylamine according to Humphreys et al. (1976), and defined *tra* mutants of the chimeric plasmids were isolated (Kennedy et al., 1977). Other mutations were moved genetically from F*lac* mutants to the chimeric plasmids. Most of these mutations led to the lack of synthesis of one protein each, both *in vitro* and in minicells. As a result, most of the *tra* cistron gene products were identified. Still other proteins were encoded by the cloned DNA for which no cistrons were then known, (Kennedy et al., 1977). Therefore, deletions were generated within the chimeric plasmids and the physical extent of the deletions correlated with expression of *tra* cistrons genetically as well as with protein synthesis in minicells. These results allowed the mapping and identification of most of the proteins for which no *tra* cistrons are yet known (Thompson and Achtman, 1979a; Manning et al., 1979). The current map of the *tra* region is presented in Figure 11 and the individual proteins that have been identified are given in Table 2. The left-hand and right-hand parts of the *tra* region are essentially saturated with cistrons, while the central portion still has large gaps.

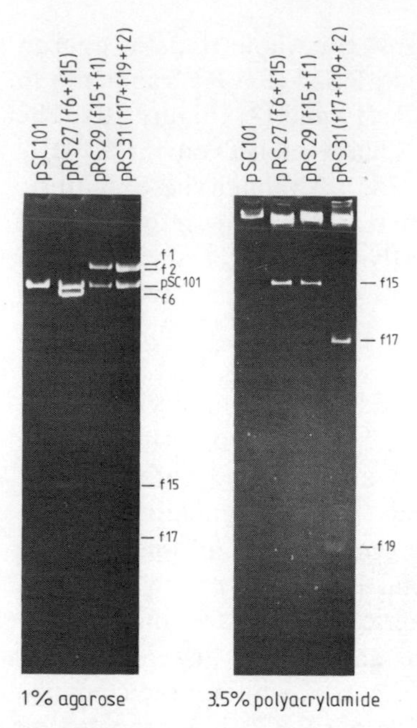

Figure 10 Analysis of the EcoRI restriction fragments of the chimeric plasmids pRS27, pRS29, and pRS31 (vector plasmid pSC101) examined on: (*a*) 1% agarose gel, (*b*) 3.5% polyacrylamide gel. The restriction fragments were visualized under ultraviolet light following staining with ethidium bromide.

4.4 Localization and Precursors of the tra Proteins

Kennedy et al. (1977) separated the cell envelope from the cytoplasm of minicells in which tra proteins had been radioactively labeled during synthesis. Many of the tra proteins were associated with the cell envelope. It therefore became interesting to ask which proteins were located in the inner membrane and which were in the outer membrane. Minicells are somewhat similar to cells in the stationary phase of growth and it is notoriously difficult to achieve good purification of inner and outer membranes by sucrose density gradient centrifugation of minicell membrane vesicles (Goodell et al., 1974; Levy et al., 1974; Gayda and Markovitz, 1978). However, we developed a modification of the techniques of Osborn et al. (1972) that yielded separation adequate enough to allow assignments to be made (Achtman et al., 1979). An example of such a membrane separation is shown in Figure 12, where minicells carrying the chimeric plasmid pRS31 were used. In addition to the evidence that distinctive protein species were enriched in certain fractions (stained gel), analyses of the inner membrane enzymes NADH oxidase and lactate dehydrogenase revealed that these were enriched in the inner membrane fractions. The assignments are summarized in Table 2. Several proteins were found predominantly in the outer membrane (TraJp, TraAp, TraKp, TraLp, TraBp, 6e, and TraTp), while others were in the inner membrane (TraEp, TraMp and TraSp). Interestingly, two proteins (6d and TraDp) were predominantly associated with the unseparated M band and were found in considerable amounts in both inner and outer membranes. (Could these be localized in the adhesion sites?) These results allow us to conclude that F-pilus synthesis involves *tra* proteins localized in both inner and outer membranes (TraJp, TraAp, TraKp, TraLp, TraBp, and TraEp). The same conclusion applies to surface exclusion (TraSp and TraTp).

It was also important to know whether the tra proteins existed in a precursor form. Several workers have convincingly demonstrated that some proteins exported to the periplasm or outer membrane have an extra series of about 20 amino acids at the *N*-terminal end (Inouye et al., 1977; Sekizawa et al., 1977; Inouye and Beckwith, 1977; Chang et al., 1978; Randall et al., 1978). If this were true for the *tra* proteins, then the gene sizes indicated in Figure 11 would be too small. Therefore, the proteins synthesized *in vitro* were compared by SDS–gel electrophoresis with the membrane-bound proteins synthesized in minicells (Figure 13). In all cases, the *in vitro* and *in vivo* products migrated identically, indicating that processing is not necessary to incorporate these proteins into the inner and outer membrane of the minicells. Thus the cistron sizes in Figure 11 represent the actual DNA needed to encode the tra proteins.

The results described above were only possible because the proteins were selectively radioactively labeled in the absence of other cell protein syn-

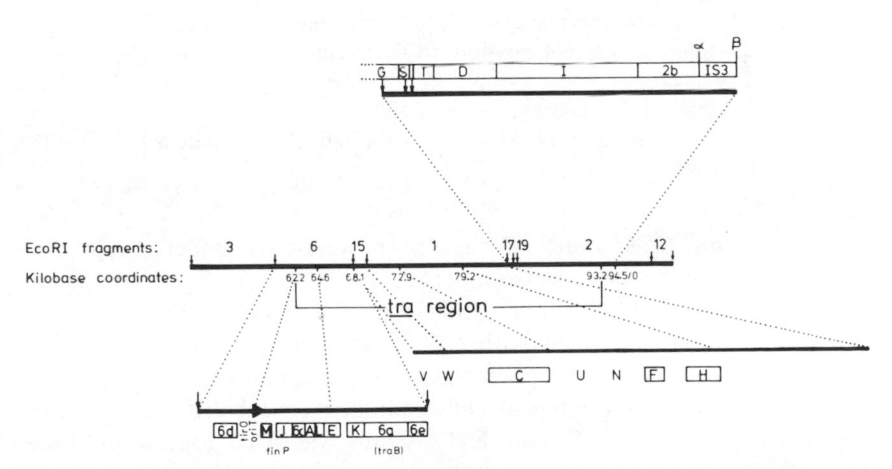

Figure 11 A physical map of the *tra* region. The cistrons encoding the various tra proteins have been assigned to the DNA as shown by a combination of DNA cloning, deletion analysis, and genetic complementation (Kennedy et al., 1977; Achtman et al., 1978a, d; Thompson and Achtman, 1978, 1979a; and Manning et al., 1979). The boxes represent the amount of DNA required to encode the appropriate protein.

Table 2 Tra Operon Proteins

Gene or Protein[a]	Mol. wt. ($\times 10^{-3}$)	Location[b]
6d[c]	29	IM and OM
traM[c]	13	IM
traJ	23.5	OM
6c (*traY*)	16	CE
traA	13.7	OM
traL	11	OM
traE	19	IM
traK	24	OM
6a (*traB*)	55	OM
6e	23.5	OM
traC	78	?
traF	25	CE
traH	40	CE
traS	18	IM
traT	25	OM
traD	77	IM and OM
traI	170	Cyto
2b	79	Cyto

[a] The products of *traV*, *traW*, *traU*, *traN* and *traG* have not yet been identified.

[b] Im, inner membrane, OM, outer membrane; CE, cell envelope (it was not possible to determine whether IM or OM location); Cyto, cytoplasm; ? indicates that no unambiguous result has been obtained.

[c] Proteins 6d and TraMp are to the left of the main *tra* operon.

thesis. In whole cells, most of the tra proteins are undetectable upon SDS–polyacrylamide gel electrophoresis. However, TraTp is detectable (Achtman et al., 1977; Minkley and Ippen-Ihler, 1977). TraJp was just detectable by staining of gels with Coomassie blue when the outer membranes of minicells or of their parental strain were analyzed. TraJp was also detectable by autoradiography of cells radioactively labeled with [^{35}S]methionine when a normal *E. coli* K-12 strain was used (our unpublished results) but not when coomassie blue stain was used. Since the synthetic rates of TraJp and TraTp are approximately equal both in minicells and *in vitro* (Kennedy et al., 1977), it is possible that TraJp is degraded continuously in growing cells. The same conclusion may apply to the other tra proteins.

In this context, a *traT* mutant of the chimeric plasmid pRS31 is interesting. The mutant carries an amber *traT* mutation and synthesizes an amber fragment in minicells with a molecular weight of about 20,000 instead of the wild type, which has a molecular weight of 25,000 (Achtman et al., 1977; Kennedy et al., 1977). This polypeptide is incorporated into the outer membrane of minicells at the same rate as the wild-type protein. It is also detectable upon staining gels of the minicell outer membranes. However, it is totally undetectable in normal cells and may normally be degraded rapidly.

4.5 F-pilus Biosynthesis and the *traJ* Gene

As is mentioned earlier, the gene products of *traJ*, *traA*, *traL*, *traE*, *traK*, *traB*, *traV*, *traW*, *traC*, *traU*, *traF*, *traH* and part of the *traG* gene are needed for F-pilus biosynthesis. The *traJ* gene is also needed for expression

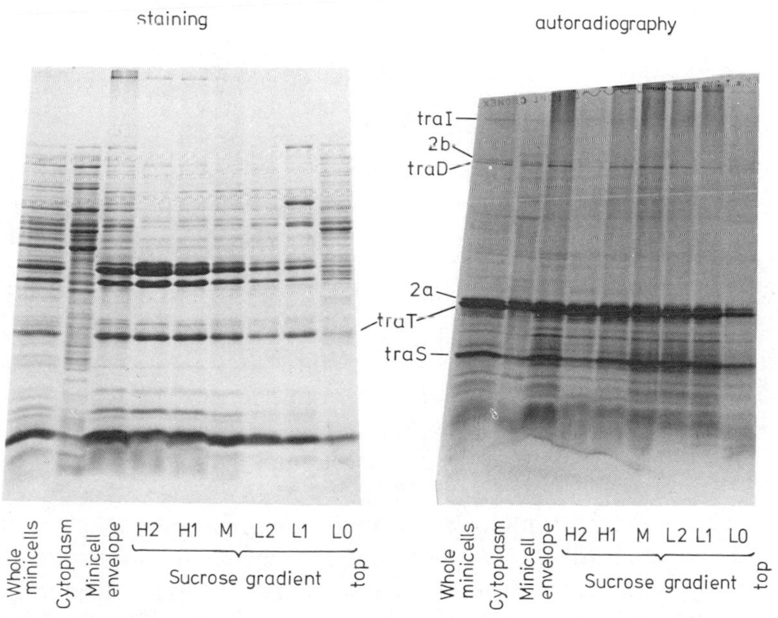

Figure 12 Membrane separation of minicells containing the chimeric plasmid pRS31. Mini cells were purified by sucrose gradient centrifugation and labeled with [^{35}S] methionine, and the membrane vesicles were separated on a sucrose gradient. These membrane fractions were then analyzed on a 15–25% polyacrylamide gradient gel, stained with coomassie blue G-250, and subjected to autoradiography. Two outer membrane fractions (H2 and H1), an intermediate density membrane band (M), and three inner membrane fractions (L2, L1 and LO) were recognized (Achtman et al., 1979).

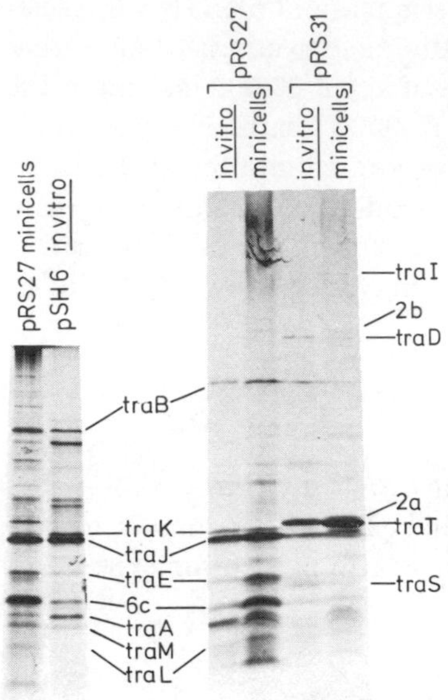

Figure 13 Comparison of the proteins produced in minicells and *in vitro* by the chimeric plasmids pRS27 and pRS31. Samples were analyzed by (*a*) 15–25% or (*b*) 11–20% SDS–polyacrylamide gel electrophoresis of proteins labeled with [^{35}S]methionine (minicells) or [^{14}C]amino acids (*in vitro*) followed by autoradiography. Plasmids pRS27 and pSH6 possess only the f6 EcoRI fragment of F DNA in common (see Figure 14).

of surface exclusion. Since the surface exclusion genes, *traS* and *traT*, are not needed for F-pilus biosynthesis, it seemed likely that the *traJ* gene product was needed for the positive control of expression of the *tra* operon cistrons (Finnegan and Willetts, 1973; Achtman, 1973b). The isolation of *traJ*$^-$, J-independent revertants from a *traJ*$^-$ mutant, supported this idea (Achtman, 1973b). When it was shown that function of *traJ* was needed for transcription of *tra* operon cistrons (Willetts, 1977a), it seemed fairly clear that *traJ* played a purely regulatory role in control of transcription.

Other evidence seemed equally convincing that *traA* was the structural gene for F pilin. The *traA* cistron is responsible for the specificity of F-like pili with respect to infection by male-specific bacteriophages (Willetts, 1971). Isolation of some transfer-proficient *traA* mutants has been accomplished but they are defective in infection by RNA-containing male-specific bacteriophages (C. C. Brinton, Jr., personal communication; N. S. Willetts, personal communication). Most conclusively, suppression of a *traA* amber mutation by amber suppressors seemed to change the amino acid sequence of the F pilin synthesized in those cells (Minkley et al., 1976). The function of the other 11 cistrons is not clear, but they might be involved in phosphorylation or glucosylation of F pilin (Brinton, 1971), in assembly of F pili, or in forming a (speculative) basal structure to anchor F pili in the cell envelope. In agreement with these ideas, Beard and Connolly (1975) reported detecting pilin subunits within the outer membrane of cells carrying the F-like plasmid R1-19.

Recent data from this laboratory throw doubt on many of the conclusions just drawn. Examination by SDS–polyacrylamide gel electrophoresis of cell envelopes prepared from cells carrying F or F-like plasmids revealed the TraTp protein (Minkley and Ippen-Ihler, 1977; Achtman et al., 1977) and occasionally the TraJp protein (our unpublished results). A third plasmid-encoded protein whose function is unknown has also been detected in the outer membrane, but no trace of F pilin or the TraAp protein was seen (Achtman et al., 1977). Thus it is unlikely that large pools of F pilin (of the molecular weight found in F pili) are present in the cell. Furthermore, TraJp was not needed for transcription and translation of *tra* operon proteins in minicells or *in vitro* (Kennedy et al., 1977). Thus TraJp cannot be an essential transcriptional factor. TraAp has a different molecular weight (13,500) from F pilin (about 11,000). If *traA* were the structural gene for F pilin, processing would have to occur to convert TraAp to F pilin. However, TraAp synthesized *in vitro* was not precipitated by anti-F pilus antiserum (Kennedy et al., 1977). Instead, TraJp was preferentially precipitated (R. Helmuth, personal communication) (Figure 14). This result suggested that TraJp might be the F-pilin precursor. Therefore, TraJp and TraAp synthesized *in vitro* were purified and their tryptic fingerprints were compared with those of purified F pilin (R. Helmuth, personal communication). Only the fingerprint of TraJp showed similarity to that of F pilin, and numerous peptides seemed to migrate identically on thin layer chromatography. We tentatively conclude that *traJ* is the structural gene for F pilin. Since TraJp is a 23,000 dalton protein that is found as such in cell envelope outer mem-

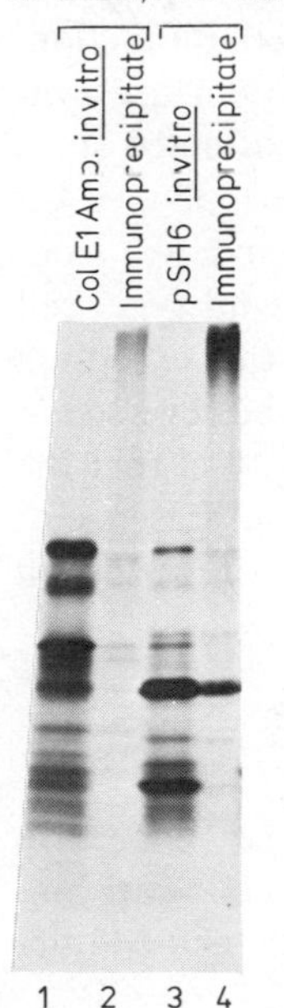

Figure 14 Immunoprecipitation of TraJp by anti-F pilus antiserum. (*1*) Protein synthesized *in vitro* using ColE1Amp. (*2*) The same as *1* but precipitated with anti-F pilus, antiserum. (*3*) Proteins synthesized *in vitro* using pSH6 (EcorRI, fragment f6 of the F factor cloned in ColE1Amp). (*4*) The same as *3* but precipitated with anti-F pilus antiserum. (R. Helmuth, personal communication).

branes, there does seem to be a pool of F pilin-precursor molecules present in the cell membrane, but not of the processed F-pilin size (11,000 daltons). Presumably, processing occurs during or just prior to F-pilus assembly.

The manner in which *traJ* is involved in regulation is not clear. It is still possible that it is a transcriptional factor. However, the outer membrane localization of TraJp, coupled with its nonessentially for transcription in minicells or *in vitro*, suggests that an unrecognized protein monitors the amount of TraJp in the cell envelope and triggers transcription of the *tra* operon when TraJp is present in unprocessed form. This mechanism does not simultaneously explain how the number of F pili per cell is regulated or how processing of TraJp occurs. It is also much more complicated than the simple regulation known for the *lac* or *mal* operons. However, it does make certain predictions that should be biochemically testable.

4.6 Chromosomal Mutants

Up to this point the discussion is concerned with F-encoded proteins necessary for donor ability. It is intuitively clear that many functions of the living cell related to its normal growth are necessary before conjugation can occur. However, recently mutants were isolated whose properties suggest that certain chromosomally encoded cell components play specific roles in donor ability. The *sfr* (Beutin, 1978) and *cpx* mutants (P. Silverman, personal communication) are pleiotropic mutants affecting F-factor expression; *sfrB* and *cpxA* mutants also result in several changes in cell envelope properties. The *sfrB* mutants do not allow F-pilus synthesis, flagellar function, or surface exclusion even though flagellae and the TraT protein are synthesized (Beutin, 1978). The LPS structure also seems to be affected since the sensitivity pattern of the *sfrB* cells to LPS-specific bacteriophages C3 and U21 is altered. However, the protein pattern of the cell envelope from *sfrB* mutants is not dramatically different from that of wild-type cell envelopes on SDS–polyacrylamide gel electrophoresis with the exception that the amount of TraTp is reduced. This effect however can be reversed in a *rho* mutant (our unpublished results). One exciting possibility is that the *sfrA* gene product is the cellular component that recognizes the presence of unprocessed TraJp in the cell envelope and triggers transcription of the *tra* operon.

4.7 Surface Exclusion

Surface exclusion is the phenomenon whereby cells carrying a sex factor are reduced in their ability to act as recipients with other donor cells carrying the same or a related sex factor (Lederberg et al., 1952; Willetts and Maule,

1974). Thus a pure culture of cells carrying an F factor will not transfer DNA from one cell to a second at any reasonable efficiency. Among the F-like sex factors, four surface exclusion specificity systems have been elucidated (Willetts and Maule, 1974; see Table 1). Cells carrying sex factors of one specificity group can transfer DNA efficiently to cells carrying a sex factor from any of the three other groups.

The mechanism of surface exclusion is still not totally clear. However, experiments using cells carrying either an F factor (Achtman et al., 1971; Achtman, 1977) or a chimeric plasmid (Achtman et al., 1977) have shown that surface exclusion is correlated with the inhibition of mating aggregation. Further analyses were greatly aided by cloning the surface exclusion genes from the F factor on the vector plasmid pSC101 (Skurray et al., 1976). The resulting plasmid pRS31 (see Figure 10) carries part of *traG*, as well as *traS*, *traT*, *traD*, *traI*, and the cistron encoding the protein 2b (Manning et al., 1979). Transcription occurs from a promoter on pSC101 and, for reasons that are still not totally clear, results in elevated expression of surface exclusion proteins (Skurray et al., 1976; Achtman et al., 1977). Thus, where cells carrying an F factor are approximately 200-fold worse recipients with Hfr donors than are F^- cells, cells carrying pRS31 are 10,000-fold worse recipients than F^- cells. This increase in surface exclusion was correlated with increased levels of TraT protein in the outer membrane (Achtman et al., 1977). Cells carrying the F factor had 29,000 TraTp molecules per cell; cells carrying pRS31 and 71,000 copies/cell; and cells carrying both F plus pRS31 had 84,000 copies/cell.

Mutants of pRS31 that do not synthesize TraTp and others that do not synthesize TraSp have been isolated (Achtman et al., 1977; Kennedy et al., 1977). These mutants demonstrated that two cistrons, *traS* and *traT*, were each partially and independently responsible for surface exclusion (M. Achtman, P. A. Manning, S. Schwuchow, B. Kusecek and N. Willetts, manuscript in preparation) and that surface exclusion consisted of two partially independent phenomena (Achtman et al., 1977). Presence of the TraT protein in the outer membrane of cells is sufficient to prevent those cells from forming stable mating aggregates with F factor-carrying donors. Presence of TraSp in the cell envelope inner membrane has a minor effect on mating aggregate formation but effectively prevents DNA transfer into the *traS*$^+$ cells even within the stable mating aggregates that are formed. The effects of possessing TraTp plus TraSp are cooperative and result in much poorer recipient ability than the sum of the individual effects (Achtman et al., 1977).

TraTp was readily detected within cell envelopes of F factor-carrying cells upon SDS–polyacrylamide gel electrophoresis followed by coomassie blue staining; TraSp has not been detected by this technique (Achtman et al.,

1977; Figure 15). When cells carrying F-like sex factors of the different surface exclusion specificity groups were examined, a protein comigrating with TraTp was found (Minkley and Ippen-Ihler, 1977), although in lower concentrations than with F-carrying cells (Achtman et al., 1977, 1978a). This observation needs to be correlated with the plasmid specificity of surface exclusion (Willetts and Maule, 1974). Either the TraT proteins encoded by these F-like sex factors have the same molecular weight but different enzymatic specificity, or the apparent plasmid specificity is due to TraSp being plasmid specific plus TraTp being synthesized in low amounts that have little biological effect. These F-like sex factors exerted some low level complementation of *traT*$^-$ mutants of F DNA; none was detected for *traS*$^-$ mutants (Achtman et al., 1977; 1978a). This supports the second possibility but does not exclude the first. Possibly both are true: TraTp is partially plasmid specific but still retains common functional aspects and can cooperatively stimulate surface exclusion with different TraSp proteins. In one attempt to detect the TraSp of an F-like sex factor, chimeric plasmids carrying *traS* and *traT* from R6-5 were introduced into minicells and the radioactive proteins synthesized were examined. Although TraTp was readily detectable, no protein comigrating with the F factor's TraSp was seen (Achtman et al., 1978a). Thus TraSp encoded by the F-like sex factors is probably of a different molecular weight, as well as specificity, as confirmed by recent work which shows that the F-like plasmid R100-1 produces

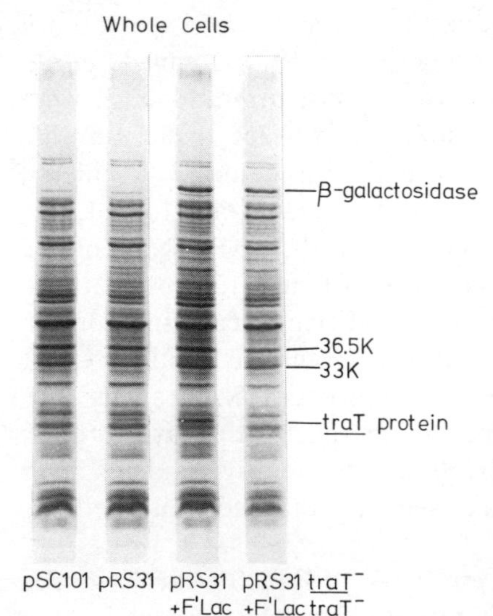

Figure 15 SDS–polyacrylamide gel electrophoresis of whole cells stained with coomassie blue G-250. Chimeric plasmid pRS31 contains EcoRI restriction fragments f17, f19 and f2 of the F-factor cloned into the vector plasmid pSC101. The mutation carried by pRS31 and F*lac* in the right most track is *traT*246, an amber mutation.

a TraS protein of different molecular weight to that of F (~15000 instead of 18000) (unpublished results).

Since TraTp is readily detectable upon SDS–polyacrylamide gel electrophoresis of cell envelopes we were able to examine some of its properties and compare it with major outer membrane proteins. TraTp resembles protein I in that it is associated with the peptidoglycan and is rendered only partially soluble by SDS at low temperatures (P. A. Manning and M. Achtman, manuscript in preparation). Furthermore, the solubilized material is aggregated and can only just enter 10% acrylamide gels; once boiled in SDS, this material is mostly converted to the 25,000 dalton monomeric form. TraTp is also exposed on the cell surface since it can be coupled to high molecular weight, CNBr-activated dextran.

It has been suggested that TraTp exerts its function in the cell envelope by masking an unknown receptor needed in the cell envelope outer membrane for stable mating aggregate formation (Achtman et al., 1977). This cannot be the F-pilus receptor, since purified F pili bind normally to cells possessing TraTp (R. Helmuth, personal communication). It is unlikely to be ompA protein since the plasmid specificity to TraTp differs from ompA protein, and TraTp-containing cells do not efficiently form unstable mating aggregates, whereas cells lacking ompA protein do. Thus we conclude that if TraTp masks a conjugation receptor, that receptor is still undetected. In this regard, it is interesting that Beard and Bishop (1975) found that an intact peptidoglycan in the donor cell was essential for surface exclusion (of the R1-19 type) in the recipient cell to be functional. This observation suggests that the peptidoglycan of the donor must normally be at least partially degraded for conjugation to occur and that surface exclusion in the recipient prevents triggering of the degradation in the donor.

4.8 Triggered Events in Conjugation

Evidence has accumulated that two events, initiation of DNA transfer and of disaggregation, may be triggered during conjugation. These are now discussed, and our current speculation on mechanisms is presented.

The replication of F DNA that occurs during conjugation is normally masked by vegetative replication of both chromosomal and plasmid DNA. However, *dnaB*$_{ts}$ mutants of *E. coli* do not undergo this vegetative replication at nonpermissive temperatures but are still capable of conjugational DNA replication (Marinus and Adelberg, 1970). This observation yielded a method for selectively assaying conjugational DNA replication. Conjugational DNA transfer and conjugational DNA replication are normally correlated. However, either can occur without the other (Sarathy and Siddiqi,

1973; Wilkins and Hollom, 1974; Ou, 1975; Kingsman and Willetts, 1978). Ou (1975) reconciled this discrepancy by proposing that a mating signal is generated during conjugation that triggers both DNA transfer and DNA replication. Once triggered, transfer and replication proceed independently of each other. Thus it was possible to assay triggering by measuring DNA replication even under conditions where DNA transfer did not occur. Neither DNA replication nor DNA transfer occurred when $dnaB_{ts}$ donor cells were mixed with cells carrying an F factor and expressing surface exclusion (Ou, 1975). However, minicells derived from F^+ parental cells apparently inherited only part of the surface exclusion proteins, since mixing these minicells with $dnaB_{ts}$ donor cells did trigger DNA replication but not DNA transfer in the donor cells (Ou, 1975). Thus apparently surface exclusion specifically inhibits triggering and independently inhibits DNA transfer.

What cell envelope components are involved in generating the mating signal? Kingsman and Willetts (1978) examined $dnaB_{ts}$ mutants carrying transfer-deficient mutants of F*lac*. They found that DNA replication (but not DNA transfer) was triggered by F^- cells with $traD^-$, $traG^-$, and $traN^-$ mutants but not with $traI^-$ or $traM^-$ mutants. Stable mating aggregates are formed by $traI^-$ and $traM^-$ mutants (Achtman et al., 1971; M. Achtman, G. Morelli, and P. A. Manning, manuscript in preparation), and the TraI and TraM proteins may be directly involved in triggering DNA replication plus DNA transfer. TraIp is cytoplasmic, whereas TraMp is associated with the cell envelope. Recent data have implicated TraIp and TraMp as antagonists of the DNA transfer inhibitors that are normally present in F^+ cells and that prevent DNA transfer (M. Achtman, R. Thompson, and B. Kusecek, unpublished observations). Thus TraIp and TraMp may be involved in recognizing mating contact and transducing this contact into DNA transfer by inactivating the transfer inhibitors.

At what stage of conjugation does triggering occur? DNA replication was triggered by F-piliated $traD^-$, $traG^-$ and $traN^-$ mutants. Whereas $traD^-$ mutants form stable mating aggregates, $traG^-$ and $traN^-$ mutants do not efficiently form stable mating aggregates either by Coulter counter or electron microscopic analysis (Achtman and Skurray, 1977; M. Achtman, G. Morelli and P. A. Manning, manuscript in preparation). Thus triggering must be an early event in conjugation, occurring among cells that are still only unstably aggregated. Since $traI^-$ and $traM^-$ mutants do form stable mating aggregates but do not trigger DNA replication, it seems that mating aggregation can proceed through the stabilization stage even when triggering does not occur. Therefore, triggering and stabilization are simultaneous events in conjugation that can proceed independently of each other.

The enzymes involved in DNA replication and DNA transfer are unknown. However, the origin of DNA transfer, *oriT*, has been localized to a 200 bp (base pairs) segment of F DNA by analysis of deletion mutants (Thompson and Achtman, 1979a). Most of this region has been sequenced (Thompson and Achtman, 1979b), although it is still not certain which parts of the sequence are relevant to DNA transfer. The deletion mutants have also been used to map where the transfer inhibitors (which TraIp and TraMp counteract) act. A site, *tirO*, that is essential for action of the transfer inhibitors has been mapped 400 bp to the left of *oriT*, as shown in Figure 11.

Speculations on mechanisms involved in DNA transfer have been made elsewhere (Achtman and Skurray, 1977) and we therefore forego such speculation here. In essence, no details are available on the enzymes involved or on the cell components that permit it. The only F factor encoded proteins that may play a role are TraDp, which is not needed for mating aggregate formation or for triggering (Achtman et al., 1971; Kingsman and Willetts, 1978), and possibly the 6c protein, which is plasmid specific (Thompson and Achtman, 1978; M. Achtman, R. Thompson, and B. Kusecek, unpublished observations). During DNA transfer, a single strand of F DNA is transferred 5′ end first (Ohki and Tomizawa, 1968; Rupp and Ihler, 1968), apparently starting from *oriT* (Willetts, 1972; Crisona and Clark, 1977; Thompson and Achtman, 1979b). It is not clear whether transfer of the molecule stops at *oriT* again or continues beyond in a rolling circle fashion (see Achtman and Skurray, 1977 for discussion). However, at least a complete single stranded molecule is transferred, and quite quickly (a few minutes at 37°C for the 94,000 bp long F DNA).

Therefore, successful DNA transfer triggers disaggregation (Achtman et al., 1978b). Disaggregation was observed with conjugation involving all F-like and the one I-like plasmid tested and may be a generalized sequel to successful DNA transfer. Disaggregation occurred rather slowly but was essentially complete within 30 min at 37°C after DNA transfer was complete (Achtman et al., 1978b). The fact that disaggregation is triggered was inferred from the observation that mating aggregates between Hfr and F^- cells did not disaggregate. Disaggregation was also not observed with *traD*, *traI* or *traM* mutants (Achtman et. al., manuscript in preparation). The basis for the discrepancy between cells carrying an autonomous sex factor and cells carrying a sex factor integrated into the bacterial chromosome is not understood. However, the discrepancy did reveal that stable mating aggregates do not decay spontaneously and therefore that disaggregation is a specific event. Without knowing the molecular basis of the mating contact, it is difficult to speculate on the events involved in disaggregation. Disaggregation was not fully inhibited by treatment with CN^- or chloramphenicol and was accelerated by low concentrations of SDS (Achtman et al., 1978b).

5 THE MATING CYCLE

The complexities of conjugation have been summarized by a working model, the mating cycle (Achtman and Skurray, 1977). Here we present a refined version (Figure 16) that includes triggering of DNA replication and reassigns certain tra proteins to stages of conjugation other than those to which they were assigned earlier. Cells carrying the F factor are shown with small internal circles to represent plasmid DNA, while F pili are

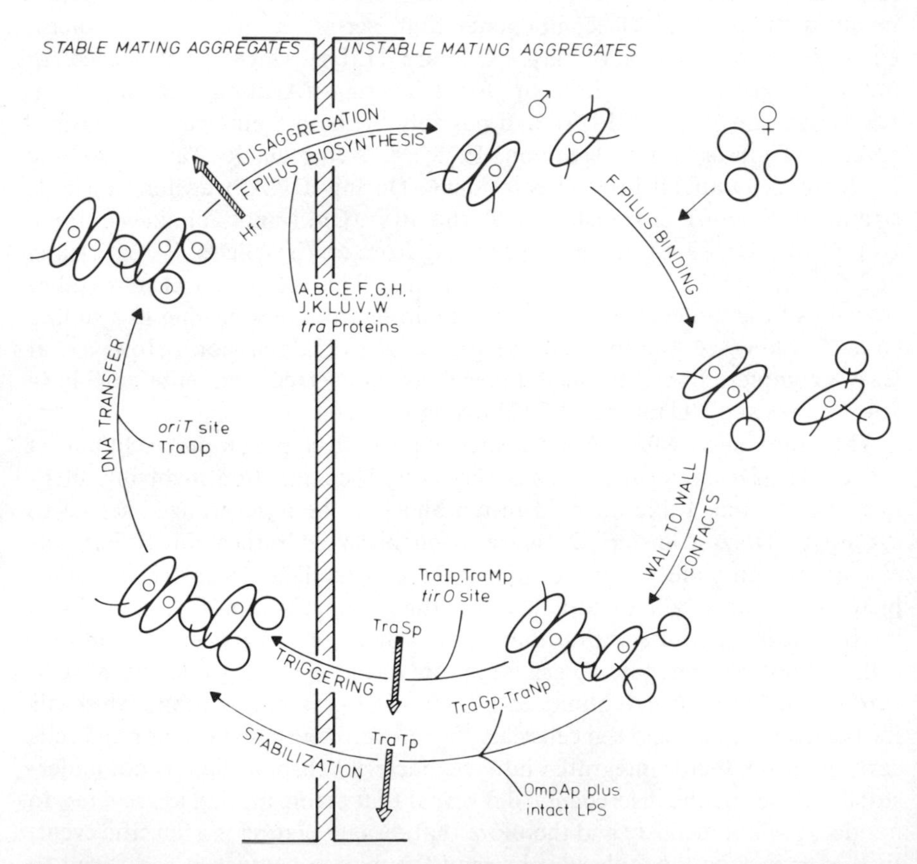

Figure 16 The mating cycle. Proteins needed for conjugation by the donor cell are indicated inside the circle. The ompA protein and an intact LPS are required by the recipient. Hatched arrows indicate blocking. See text for further explanations.

represented by lines projecting from the cell surface. The original donors in a mating mixture are represented as ovals, while the original recipients are drawn as circles. During the mating cycle these recipients acquire an F factor and express *tra* cistrons, thus becoming converted to potential donors that can reinitiate the mating cycle once fresh recipients are encountered.

It may seem confusing that we represent some donor cells in wall-to-wall contact with each other. How can we distinguish between donor-to-donor and donor-to-recipient contacts within our scheme? In fact, this represents a real and unsolved technical problem, since donor–donor interactions cannot be distinguished by electron microscopy from donor–recipient mating contacts (Achtman et al., 1978b). We know of no barrier to F-pilus binding or to wall-to-wall contact between donor cells. Instead, the model assumes that, after these initial interactions, the *traS* and *traT* gene products first act to prevent stabilization and triggering.

The individual stages of the mating cycle are based on direct observation or good evidence. F-pilus binding, wall-to-wall contact formation, stabilization and triggering, DNA transfer, disaggregation, and expression are all discussed in some detail above. However, in no case is the molecular nature of these stages known. The receptor for F-pilus binding is undefined other than that it is present on both F^+ and F^- *E. coli* cells and absent in *Pseudomonas* and *Bacillus* cells. The molcular components responsible for the formation of unstable wall to wall contacts are unknown. We presume that there is a donor receptor that interacts with ompA protein, as well as recipient receptors that interact with TraIp and TraMp, TraGp and TraNp, but these remain to be identified. We believe that they exist by analogy with the specificity of receptors for bacteriophages and colicins. What is unprecedented is the large number of matching receptors for sequential stages that we are postulating to be involved in conjugation.

The assignment of tra proteins to certain stages is somewhat speculative and we try here to justify our choices.

Clearly, ompA protein is involved in stabilization, since dilution-sensitive mating aggregates have been detected with *ompA*$^-$ mutants or after addition of purified ompA protein. No data are available on whether ompA protein is required for triggering as well, and we have arbitrarily restricted its role to stabilization.

Triggering has not previously been assigned a location in the mating cycle. We were forced to place it early, since *traG*$^-$ and *traN*$^-$ mutants trigger DNA replication without efficiently forming stable mating aggregates. Thus triggering probably occurs at a stage before or simultaneous with stabilization. Triggering might occur as a result of F-pilus binding and might be parallel to unstable wall-to-wall contact formation. Our choice of depicting it parallel with stabilization is arbitrary.

We have assigned TraGp, TraNp, and TraTp to stabilization in lieu of a better possibility. The presence of TraTp in the recipient cell does prevent efficient stable mating aggregate formation but does not fully prevent some F-pilus and wall-to-wall contacts between donor cells. Possibly the presence to TraTp in the recipient cell maintains cell contacts in a reversible form, since donor–donor contacts are very quickly dissolved by low SDS concentrations (Achtman et al., 1978b). The analysis of *traG*$^-$ and *traN*$^-$ donors yielded results similar to those of *traT*$^+$ recipients. Few mating aggregates were formed, but these were dilution resistant and involved both F-pilus and wall-to-wall contacts. Currently, the simplest solution is to assign all three proteins to the stabilization stage even though their function is not identical to that of ompA protein.

Our assignment of TraIp and TraMp to triggering is demanded by the observations of Kingsman and Willetts (1978) coupled with our observation that *traI*$^-$ and *traM*$^-$ mutants form stable mating aggregates and therefore cannot be blocked in stabilization. If triggering occurs earlier in the mating cycle than drawn, then TraIp and TraMp would still be involved in triggering but at the earlier stage. The assignment of TraSp to triggering is speculative and is based on the observations of Ou (1975). He found that cells carrying an F factor and expressing both *traS* and *traT* did not trigger DNA replication in donor cells, whereas minicells derived from F$^+$ parental cells did trigger DNA replication when used as recipients. This suggests that either TraSp or TraTp (or both) is not functional in the minicells. TraTp is a stable major outer membrane protein for which no evidence of turnover exists, while TraSp apparently undergoes rapid turnover, since we have never detected the protein by coomassie blue staining even though its synthetic rate is reasonably high. Thus it seems likely that TraSp is nonfunctional in the minicells and we have therefore assigned TraSp to the triggering stage. This interpretation implies that surface exclusion prevents the parallel stages of stabilization and triggering, and the concept of parallel stages may explain why two surface exclusion proteins have evolved. We note that this assignment also implies that triggering is normally needed not only for DNA replication, but also for DNA transfer, since TraSp does prevent DNA transfer within stable mating aggregates (Achtman et al., 1977). Once triggered, apparently replication and transfer can proceed independently.

TraDp is clearly required for DNA transfer itself since triggering of DNA replication does occur with *traD*$^-$ mutants (Kingsman and Willetts, 1978) and stable mating aggregates are formed (Achtman et al., 1971; Achtman and Skurray, 1977). TraDp is also needed for infection (but not for adsorption) of F-piliated cells by the RNA-bacteriophage f2 (Achtman et al., 1971). It is not clear that the origin of transfer, *oriT*, is needed at this stage. Conceivably, triggering occurs at *oriT* and therefore *oriT* should be assigned to triggering rather than DNA transfer itself. However, this uncertainty, as the others listed above, does not seriously detract from the advantage of having a working model that makes specific predictions.

Based on the arguments presented above, the model in Figure 16 predicts: (1) *ompA*$^-$ mutants trigger DNA replication and the addition of purified ompA protein does not prevent triggering; (2) *traT*$^+$ *traS*$^-$ recipients trigger DNA replication, whereas *traT*$^-$ *traS*$^+$ recipients do not; (3) it should be possible to isolate mutants of the recipient cell blocked in F-pilus binding or in wall-to-wall contact formation, those that do not allow DNA triggering or those that do not allow DNA transfer; (4) it should be possible to isolate mutants of the donor cells that do not allow the formation of unstable wall-to-wall contacts; and (5) if TraSp, an inner membrane protein, prevents triggering, then contact between donor cell envelope and the inner membrane of the recipient cell is a relatively early event in conjugation.

6 SUMMARY

Bacterial conjugation between gram-negative bacteria involves complex membrane interactions with involvement of specific molecules in both parental cell envelopes. The complexity of these interactions is only starting to become apparent with the discovery of numerous genes and the subdivision of conjugation into various stages. However, many of the genes are conveniently clustered within the F factor's *tra* operon, which has now been mapped to near saturation. Most of the *tra* gene products have now been identified and some of them have been assigned to intracellular locations. Two of the more interesting components, F pili and ompA protein, have been purified and exert biological effects *in vitro*. Analysis of conjugation allows the interaction of genetical, physical, and biochemical techniques and may eventually shed light on certain functions of the cell envelope. However, very few biochemical details are yet available on which envelope components interact with one another and on the changes in the cell envelope that are needed for transport of DNA. These latter two areas represent the potentially most exciting ones for future research.

REFERENCES

Achtman, M. (1973a), *Curr. Top. Microbiol. Immunol.*, **60,** 79–123.

Achtman, M. (1973b), *Genet. Res., Camb.*, **21,** 67–77.

Achtman, M. (1975), *J. Bacteriol.*, **123,** 505–515.

Achtman, M. (1977), in *Plasmids: Medical and Theoretical Aspects*, 3rd International Symposium on Antibiotic Resistance S. Mitsuhashi, L. Rosival, and V. Krcmery, eds.), Avicenum, Prague, pp. 117–125.

Achtman, M., and Skurray, R. (1977), in *Microbiol. Interactions* (Receptors and Recognition, Series B., Vol. 3) (J. L. Reissig, ed.), Chapman and Hall, London, pp. 233–279.

Achtman, M., Willetts, N., and Clark, A. J. (1971), *J. Bacteriol.*, **106,** 529–538.

Achtman, M., Willetts, N., and Clark, A. J. (1972), *J. Bacteriol.*, **110,** 831–851.

Achtman, M., Kennedy, N., and Skurray, R. A. (1977), *Proc. Natl. Acad. Sci. U.S.*, **74,** 5104–5108.

Achtman, M., Kusecek, B., and Timmis, K. (1978a), *Mol. Gen. Genet.*, **163,** 169–179.

Achtman, M., Morelli, G., and Schwuchow, S. (1978b), *J. Bacteriol.*, **135,** 1053–1061.

Achtman, M., Schwuchow, S., Helmuth, R., Morelli, G., and Manning, P. A. (1978c), *Mol. Gen. Genet.*, **164,** 171–183.

Achtman, M., Skurray, R. A., Thompson, R., Helmuth, R., Hall, S., Beutin, L., and Clark, A. J. (1978d), *J. Bacteriol.*, **133,** 1383–1392.

Achtman, M., Manning, P. A., Edelbluth, C., and Herrlich, P. (1979), *Proc. Natl. Acad. Sci. U.S.*, in press.

Ames, G. F.-L., Spudich, E. N., and Nikaido, H. (1974), *J. Bacteriol.*, **117,** 406–416.

Bachmann, B. J., Low, K. B., and Taylor, A. L. (1976), *Bacteriol. Rev.*, **40,** 116–167.

Bayer, M. E. (1968a), *J. Virol.*, **2,** 346–356.

Bayer, M. E. (1968b), *J. Gen. Microbiol.*, **53,** 395–404.

Bayer, M. E. (1975), in *Membrane Biogenesis*, (A. Tzagoloff, ed.), Plenuum Press, New York, pp. 393–427.

Beard, J. P., and Bishop, S. F. (1975), *J. Bacteriol.*, **123,** 916–923.

Beard, J. P., and Connolly, J. C. (1975), *J. Bacteriol.*, **122,** 59–65.

Beard, J. P., Howe, T. G. B., and Richmond, M. H. (1972), *J. Bacteriol.*, **111,** 814–820.

Beutin, L. (1978), Ph.D. thesis, Freie Universität Berlin.

Brinton, C. C., Jr. (1965), *Trans. N.Y. Acad. Sci.*, **27,** 1003–1054.

Brinton, C. C., Jr. (1971), *Crit. Rev. Microbiol.*, **1,** 105–160.

Caro, L. G., and Schnös, M. (1966), *Proc. Natl. Acad. Sci. U.S.*, **56,** 126–132.

Chai, T., and Foulds, J. (1974), *J. Mol. Biol.*, **85,** 465–474.

Chang, N. C., Blobel, G., and Model, P. (1978), *Proc. Natl. Acad. Sci. U.S.*, **75,** 361–365.

Cohen, S. N., Chang, A. C. Y., and Hsu, L. (1972), *Proc. Natl. Acad. Sci. U.S.*, **69,** 2110–2114.

Crawford, E. M., and Gesteland, R. F. (1974), *Virology*, **22,** 165–167.

Crisona, N. J., and Clark, A. G. (1977), *Science*, **196,** 186–187.

Crosa, J. H., Luttrop, L. K., Heffron, F., and Falkow, S. (1975), *Mol. Gen. Genet.*, **140,** 39–50.

Curtiss, R. III (1969), *Annu. Rev. Microbiol.*, **23,** 69–136.

Curtiss, R. III, Caro, L. G., Allison, D. P., and Stallions, D. R. (1969), *J. Bacteriol.*, **100,** 1091–1104.

Curtiss, R. III, Fenwick, R. G., Jr., Goldschmidt, R., and Falkinham, J. O. III (1977), in *R-factor: Drug Resistance Plasmid* (S. Mitsuhashi, ed.), University of Tokyo Press, Tokyo, pp. 109–134.

Date, T., Inuzuka, M., and Tomoeda, M. (1977), *Biochemistry*, **16**, 5579–5585.

Datta, D. B., Krämer, C., and Henning, U. (1976), *J. Bacteriol.*, **128**, 834–841.

Davidson, N., Deonier, R. C., Hu, S., and Ohtsubo, E. (1975), in *Microbiology 1974* (D. Schlessinger, ed.), ASM Press, Washington, D.C., pp. 56–65.

Davies, J. K., and Reeves, P. (1975a), *J. Bacteriol.*, **123**, 102–117.

Davies, J. K., and Reeves, P. (1975b), *J. Bacteriol.*, **123**, 372–373.

Davis, B. D. (1950), *J. Bacteriol.*, **60**, 507–580.

Dennison, S., and Hedges, R. W. (1972), *Genet. Res., Camb.*, **70**, 55–61.

DePamphilis, M. L., and Adler, J. (1971), *J. Bacteriol.*, **105**, 396–407.

Enderman, R., Krämer, C., and Henning, U. (1978), *FEBS Lett.*, **86**, 21–24.

Falkow, S. (1975), *Infectious Multiple Drug Resistance*, Dion, London.

Finnegan, D. J., and Willetts, N. S. (1971), *Mol. Gen. Genet.*, **111**, 256–264.

Finnegan, D. J., and Willetts, N. S. (1972), *Mol. Gen. Genet.*, **119**, 57–66.

Finnegan, D. J., and Willetts, N. S. (1973), *Mol. Gen. Genet.*, **127**, 307–316.

Folkhardt, W., Lconard, K. R., Malscy, S., Marvin, D. A., Dubochct, G., Engcl, A., Achtman, M., and Helmuth, R. (1979), *J. Mol. Biol.*, in press.

Foulds, J. (1974), *J. Bacteriol.*, **117**, 1354–1355.

Foulds, J., and Barrett, C. (1974), *J. Bacteriol.*, **116**, 885–892.

Frazer, A. C., and Curtiss, R. III (1975), in *Curr. Top. Microbiol. Immunol.*, **69**, 1–84.

Gayda, R. C., and Markovitz, A. (1978), *J. Bacteriol.*, **136**, 369–380.

Goodell, E. W., Schwarz, U., and Teather, R. M. (1974), *Eur. J. Biochem.*, **47**, 567–572.

Guyer, M. S., Figurski, D., and Davidson, N. (1976), *J. Bacteriol.*, **127**, 988–997.

Hancock, R. E. W., and Reeves, P. (1976), *J. Bacteriol.*, **127**, 98–108.

Hardy, K. (1975), *Bacteriol. Rev.*, **39**, 464–515.

Havekes, L., and Hoekstra, W. (1976), *J. Bacteriol.*, **126**, 593–600.

Havekes, L. M., Lugtenberg, B. J. J., and Hoekstra, W. P. M. (1976), *Mol. Gen. Genet.*, **146**, 43–50.

Havekes, L., Hoekstra, W., and Kempen, H. (1977a), *Mol. Gen. Genet.*, **155**, 185–189.

Havekes, L. Tommassen, J., Hoekstra, W., and Lugtenberg, B. (1977b), *J. Bacteriol.*, **129**, 1–8.

Hayes, W. (1964), *The Genetics of Bacteria and Their Viruses*, Wiley, New York.

Helmuth, R., and Achtman, M. (1975), *Nature*, **257**, 652–656.

Helmuth, R., and Achtman, M. (1978), *Proc. Natl. Acad. Sci. U.S.*, **75**, 1237–1241.

Henning, U., and Haller, I. (1975), *FEBS Lett.*, **55**, 161–164.

Henning, U., Hoehn, B., and Sonntag, I. (1973), *Eur. J. Biochem.*, **39**, 27–36.

Henning, U., Hindennach, I., and Haller, I. (1976), *FEBS Lett.*, **61**, 46–48.

Humphreys, G. O., Willshaw, G. A., Smith, R., and Anderson, E. S. (1976), *Mol. Gen. Genet.*, **145**, 101–108.

Inouye, H. I., and Beckwith, J. (1977), *Proc. Natl. Acad. Sci. U.S.*, **74**, 1440–1444.

Inouye, S., Wang, S. S., Sekizawa, J., Halegoua, S., and Inouye, M. (1977), *Proc. Natl. Acad. Sci. U.S.*, **74**, 1004–1008.

Ippen-Ihler, K., Achtman, M., and Willetts, N. (1972), *J. Bacteriol.*, **110**, 857–863.

Ishibashi, M. (1967), *J. Bacteriol.*, **93**, 379–389.

Jacobson, A. (1972), *J. Virol.*, **10,** 835–843.

Kennedy, N., Beutin, L., Achtman, M., Skurray, R., Rahmsdorf, U., and Herrlich, P. (1977), *Nature* (*Lond.*), **270,** 580–585.

Kingsman, A., and Willetts, N. (1978), *J. Mol. Biol.*, **122,** 287–300.

Knolle, P. and Ørskov, I. (1967), *Mol. Gen. Genet.*, **99,** 109–114.

Koplow, J., and Goldfine, H. (1974), *J. Bacteriol.*, **117,** 527–543.

Lawn, A. M., Meynell, E., and Cooke, M. (1971), *Ann. Inst. Pasteur*, **120,** 3–8.

Lederberg, J., and Tatum, E. L. (1946a), *Cold Spring Harbor Symp. Quant. Biol.*, **II,** 113–114.

Lederberg, J., and Tatum, E. L. (1946b), *Nature* (*Lond.*), **158,** 558.

Lederberg, J., Cavalli, L. L., and Lederberg, E. M. (1952), *Genetics*, **37,** 720–730.

Levy, S. B., McMurry, L., and Palmer, E. (1974), *J. Bacteriol.*, **120,** 1464–1471.

Lugtenberg, B., Meijers, J., Peters, R., van der Hoek, P., and van Alphen, L. (1975), *FEBS Lett.*, **58,** 254–258.

Manning, P. A. (1977), Ph.D. thesis, University of Adelaide, South Australia.

Manning, P. A., and Reeves, P. (1975), *J. Bacteriol.*, **124,** 576–577.

Manning, P. A., and Reeves, P. (1976a), *J. Bacteriol.*, **127,** 1070–1079.

Manning, P. A., and Reeves, P. (1976b), *Biochem. Biophys. Res. Commun.*, **72,** 694–700.

Manning, P. A., and Reeves, P. (1977), *J. Bacteriol.*, **130,** 540–541.

Manning, P. A., Puspurs, A., and Reeves, P. (1976), *J. Bacteriol.*, **127,** 1080–1084.

Manning, P. A., Thompson, R., and Achtman, M. (1979), in *Proceedings of the 4th European Meeting on Bacterial Transformation and Transfection* (S. W. Glover and L. O. Butler, eds.) in press.

Marinus, M. G., and Adelberg, E. A. (1970), *J. Bacteriol.*, **104,** 1266–1272.

Marvin, D. A., and Hohn, B. (1969), *Bacteriol. Rev.*, **33,** 172–209.

Meynell, E., Meynell, G. G., and Datla, N. (1968) *Bacteriol. Rev.*, **32,** 55–83.

Miki, T., Horiuchi, T., and Willetts, N. S. (1978), *Plasmid*, **1,** 316–323.

Minkley, E. G., Jr., and Ippen-Ihler, K. (1977), *J. Bacteriol.*, **129,** 1613–1622.

Minkley, E. G., Jr., Polen, S., Brinton, C. C., Jr., and Ippen-Ihler, K. (1976), *J. Mol. Biol.*, **108,** 111–121.

Monner, D. A., Jonsson, S., and Boman, H. G. (1971), *J. Bacteriol.*, **107,** 420–432.

Morrison, T. G., and Malamy, M. H. (1971), *Nature* (*Lond.*), **231,** 37–41.

Novick, R. P. (1974), in Handbook of Microbiology, Vol. 4 (A. I. Laskin and H. A. Lechevalier, eds.) CRC Press, Cleveland, pp. 537–586.

Novotny, C. P., and Fives-Taylor, P. (1974), *J. Bacteriol.*, **117,** 1306–1311.

Novotny, C., Carnahan, J., and Brinton, C. C., Jr. (1969), *J. Bacteriol.*, **98,** 1294–1306.

Ohki, M., and Tomizawa, J. (1968), *Cold Spring Harbor Symp. Quant. Biol.*, **33,** 651–658.

Ohtsubo, E. (1970), *Genetics*, **64,** 189–197.

Ohtsubo, E., Nishimura, Y., and Hirota, Y. (1970), *Genetics*, **64,** 648–654.

Ørskov, I., and Ørskov, F. (1960), *Acta Pathol. Microbiol. Scand.*, **48,** 37–46.

Osborn, M. J., Gander, J. E., Parisi, E., and Carson, J. (1972), *J. Biol. Chem.*, **247,** 3962–3972.

Ou, J. T. (1973), *J. Bacteriol.*, **114,** 1108–1115.

Ou, J. T. (1975), *Proc. Natl. Acad. Sci. U.S.*, **72,** 3721–3725.

Ou, J. T., and Anderson, T. F. (1970), *J. Bacteriol.*, **102,** 648–654.

Randall, L. L., Hardy, S. J. S., and Josefsson, L.-G. (1978), *Proc. Natl. Acad. Sci. U.S.*, **75,** 1209–1212.

Reiner, A. M. (1974), *J. Bacteriol.*, **119,** 183–191.

Reithmeier, R. A. F., and Bragg, P. D. (1974), *FEBS Lett.*, **41,** 195–198.

Rowbury, R. J. (1977), *Prog. Biophys. Mol. Biol.*, **31,** 271–317.

Rupp, W. D., and Ihler, G. (1968), *Cold Spring Harbor Symp. Quant. Biol.*, **33,** 647–650.

Sarathy, P. V., and Siddiqui, O. (1973), *J. Mol. Biol.*, **78,** 443–451.

Schnaitman, C. A. (1974), *J. Bacteriol.*, **118,** 442–453.

Schweizer, M., and Henning, U. (1977), *J. Bacteriol.*, **129,** 1651–1652.

Schweizer, M., Hindennach, I., Garten, W., and Henning, U. (1978), *Eur. J. Biochem.*, **82,** 211–217.

Sekizawa, J., Inouye, S., Halegoua, S., and Inouye, M. (1977), *Biochem. Biophys. Res. Commun.*, **77,** 1126–1133.

Sharp, P. A., Hsu, M. T., Ohtsubo, E., and Davidson, N. (1972), *J. Mol. Biol.*, **71,** 471–497.

Sharp, P. A., Cohen, S. N., and Davidson, N. (1973), *J. Mol. Biol.*, **75,** 235–255.

Skurray, R. A., Hancock, R. E. W., and Reeves, P. (1974), *J. Bacteriol.*, **119,** 726–735.

Skurray, R. A. Nagaishi, H., and Clark, A. J. (1976), *Proc. Natl. Acad. Sci. U.S.*, **73,** 64–68.

Thompson, R., and Achtman, M. (1978), *Mol. Gen. Genet.*, **165,** 295–304.

Thompson, R., and Achtman, M. (1979a), *Mol. Gen. Genet.*, **169,** 49–57.

Thompson, R., and Achtman, M. (1979b), in *Proceedings of the 4th European Meeting on Bacterial Transformation and Transfection* (S. W. Glover and L. O. Butler, eds.) in press.

Tomoeda, M., Inuzuka, M., and Date, T. (1975), *Prog. Biophys. Mol. Biol.*, **30,** 23–56.

Van Alphen, L., Havekes, L., and Lugtenberg, B. (1977), *FEBS Lett.*, **75,** 285–590.

Van Alphen, W., Lugtenberg, B., and Berendsen, W. (1976), *Mol. Gen. Genet.*, **147,** 263–269.

Wilkins, B. M., and Hollom, S. E. (1974), *Mol. Gen. Genet.*, **134,** 143–156.

Willetts, N. S. (1971), *Nature, New Biol.*, **230,** 183–185.

Willetts, N. (1972), *Annu. Rev. Genet.*, **6,** 257–268.

Willetts, N. S. (1973), *Genet. Res., Camb.*, **21,** 205–213.

Willetts, N. (1977a), *J. Mol. Biol.*, **112,** 141–148.

Willetts, N. (1977b), in *R. Factors* (S. Mitsuhashi, ed.) University of Tokyo Press, Tokyo, pp. 89–107.

Willetts, N., and Achtman, M. (1972), *J. Bacteriol.*, **110,** 843–851.

Willetts, N. S., and Finnegan, D. J. (1972), in *Bacterial Plasmids and Antibiotic Resistance* (V. Kromery, L. Rosival, and T. Watanabe, eds.), Avicenum, Prague pp. 173–177.

Willetts, N., and Maule, J. (1974), *Genet. Res., Camb.*, **24,** 81–89.

Chapter **13**

The Outer Membrane and Chemotaxis: Indirect Influences and Secondary Involvements

GERALD L. HAZELBAUER

Membrane Group, Wallenberg Laboratory, University of Uppsala, Uppsala, Sweden

1 INTRODUCTION

When Masayori Inouye asked if I would write a chapter entitled "Mechanism of Chemotaxis: Role of the Outer Membrane," my initial reply was that such a chapter would be very short. It could be a single statement: "There is none." Upon consideration the response seemed a bit too abrupt and I agreed to produce a chapter. An outer membrane cannot be an absolute requirement for motility or chemotaxis, since gram-positive species perform both functions with no apparent difficulty. That fact was the basis for my suggestion of a one-sentence chapter. Yet the structural integrity and various functions of the outer membrane surely do have an influence on motility and taxis of gram-negative bacteria. Whether that influence should be included under the label "mechanism" is another question, but in any case outer membrane enthusiasts might well be interested in incorporating the phenomena of motility and chemotaxis into their studies of the roles of the outer membrane.

A secondary motivation for writing the chapter is to take the opportunity to correct a misimpression that sometimes seems to be made by the title of a paper I once published: "Role of the Receptor for Bacteriophage Lambda in the Functioning of the Maltose Chemoreceptor of *Escherichia coli*." My conclusion as summarized in the final sentence of the abstract was: "Thus the lambda receptor protein would participate in maltose chemoreception only indirectly, through its role in maltose transport." With some regularity references to that paper seem to emphasize the implication contained in the title, that is, participation of the lambda receptor in the process of chemotaxis toward maltose, rather than the indirect effect of the lambda-receptor function on maltose response. In one of the following sections the indirect nature of the participation is emphasized as a counterweight to the title's misleading implication.

In this chapter I present a short review of bacterial motility and chemotaxis in which the existence of outer membrane is hardly acknowledged and then a discussion of the influence of outer membrane on those phenomena.

2 BACTERIAL MOTILITY AND CHEMOTAXIS

Motile bacteria are sensory cells, responding to changes in concentration of active compounds by altering the frequency of tumbles (Macnab and Koshland, 1972; Berg and Brown, 1972; Brown and Berg, 1974), a pattern of behavior that can be considered analogous to a sensory nerve responding to chemical excitation by altering the frequency of action potentials (Goy and Springer, 1978). Certainly one of the stronger motivations for studying bacterial chemotaxis is grounded in a belief that the unity of biology extends beyond the "central dogma" and macromolecular synthesis to include the principles of receptor function. The feeling is that there are untapped areas of the *coli*/elephant parallelism. This all sounds slightly like the "wider significance" section of a grant proposal. Yet what is just a little amazing is that recent progress in the field has so strongly emphasized the validity of such beliefs. In this brief summary I hope to provide the broad outlines of bacterial chemotaxis, colored to some extent by my own interest, and to indicate the significance of recent advances. A number of reviews are already available that cover the beginnings of modern interest in the area (Adler, 1975; Berg, 1975a; Koshland, 1975a) or consider certain aspects in more detail (Berg, 1975b; Parkinson, 1977; Silverman and Simon, 1977c; Koshland, 1977b; Hazelbauer and Parkinson, 1977; Goy and Springer, 1978).

2.1 Motility

Bacteria move by means of bacterial flagella (Figure 1). A flagellum is primarily a filament, a left-handed helical polymer of flagellin that generates cell motion by rotating like a propeller (Berg and Anderson, 1973; Larsen et al., 1974b; Silverman and Simon, 1974). Rotation originates in the basal end, a central shaft surrounded by a series of rings that are embedded in the layers of the cell envelope (Figure 2). The shaft and filament are connected by a flexible hook that probably functions to allow the flagella to bend and form a bundle behind the cell body (Macnab, 1977). Rotation is powered by proton-motive force (Larsen et al., 1974a; De Jong et al., 1976; Manson et al., 1977) and is presumed to originate in relative movement of the two filament-distal rings (Berg, 1974; Läuger, 1977). Counterclockwise rotation of the filament generates a pushing force on the cell and in the case of a peritrichously flagellated bacterium, the left-handed filaments form an untangled bundle that exerts a concerted force on the cell, resulting in the forward motion of smooth swimming (Macnab, 1977). The flagellar motor can also rotate clockwise, in which case complex deformations of the filaments occur and the bundle dissociates (Macnab and Orston, 1977). Transitory clockwise rotation of individual flagella does not exert a coordinated force and the cell tumbles, making no forward progress. Reversal to counterclockwise rotation reestablishes smooth swimming, pointed in a new, randomly chosen direction (Berg and Brown, 1972). Tumbles occur every few seconds, creating a swimming path that resembles a random walk (Berg and Brown, 1972). The chemotactic sensory system controls the balance between counterclockwise and clockwise rotation and thus the frequency of tumbles. Favorable changes enhance counterclockwise rotation, which in turn enhances smooth swimming; unfavorable changes induce clockwise rotation and thus tumbles (Larsen et al., 1974b; Macnab and Koshland,

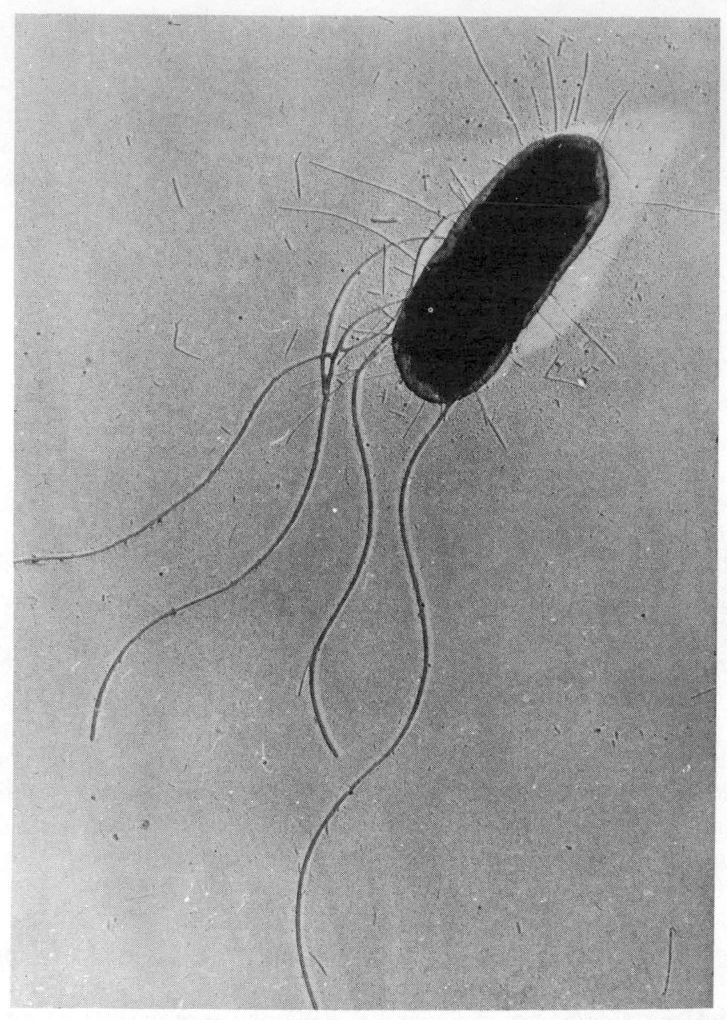

Figure 1 *Escherichia coli* shadowed with platinum. The large filaments are flagella and the smaller are type I pili. (Electron micrograph courtesy of H. Berg and G. Fonte, University of Colorado).

1972; Tsang et al., 1973). Chemotaxis, movement in response to a chemical gradient, occurs as a biased random walk.

These aspects of motility constitute the most striking differences between bacterial chemotaxis and sensory–motor phenomena in eukaryotes. The flagellar basal end is presently the only example of a rotary motor in biology (however, see Tamm, 1978), and taxis by modulation of turning frequency, a strategy properly termed klinokinesis (Fraenkel and Gunn, 1940), may not occur in eukaryotes (Carlile, 1975).

2.2 Temporal Sensitivity and Adaptation

Bacteria respond to changes in concentrations of active compounds over time as well as over distance (Macnab and Koshland, 1972; Brown and Berg, 1974). In fact, behavior in spatial gradients is probably a reflection of temporal changes in the surroundings of a swimming cell. Bacterial response to temporal stimulation emphasizes features of bacterial behavior that are common to most sensory systems. Cells alter the pattern of flagellar rotation and reversal in response to *changes* in the chemical environment; behavior is essentially not influenced by constant levels of active

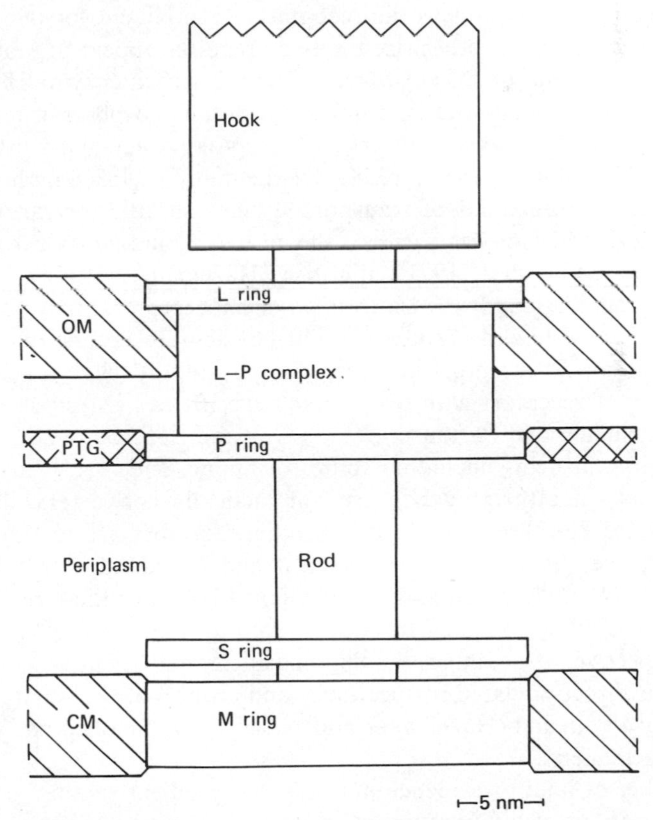

Figure 2 A model of the basal end of the *E. coli* flagellum showing interactions with the layers of the cell envelope. OM, outer membrane; CM cytoplasmic membrane; PTG peptidoglycan. Adapted from DePhamphilis and Adler (1971b, c).

compounds, even at high absolute concentration. Furthermore, the response is *transient*, that is, the pattern of flagellar rotation returns to the unstimulated state after an interval of some seconds to several minutes, depending on the type and magnitude of the temporal change. Thus cells adapt to the new concentration of the active chemical, again emphasizing that the stimulus is not the presence of a compound per se but a change in its concentration. Temporal sensitivity and adaptation are closely linked and interrelated properties of the bacterial sensory system.

2.3 Receptors

Bacterial attractants and repellents are recognized by specific receptors (Adler, 1969). Specific mutants have been identified for almost all the dozen sugar receptors of *E. coli* (Hazelbauer and Parkinson, 1977). In each case the receptor protein serves as the recognition component for two functions, taxis and transport, but otherwise the two processes appear to function independently. In *E. coli* or *S. typhimurium* probably each enzyme II, the membrane-associated, sugar-specific component of the phosphotransferase transport systems (PTS), serves as a chemoreceptor (Adler and Epstein, 1974; Lengler, 1975; Melton, et al., 1978). Of the many periplasmic binding proteins that are components of transport systems specific for various sugars, amino acids, and ions, only three, the proteins binding galactose-glucose (Hazelbauer and Adler, 1971), maltose (Hazelbauer, 1975a), and ribose (Aksamit and Koshland, 1974), are also chemoreceptors. It is possible that some amino acid-binding proteins might be chemoreceptors, but identification of amino acid receptors has been complicated by what appears to be a multiplicity of receptors with overlapping specificities (Mesibov and Adler, 1972; Hazelbauer and Parkinson, 1977; Goy and Springer, 1978). However, some binding proteins, including some for amino acids are clearly not chemoreceptors since their ligands are not tactically active (Hazelbauer and Adler, 1971). The three periplasmic protein receptors are readily available as pure species (Anraku, 1968; Aksamit and Koshland, 1972; Willis and Furlong, 1974; Kellermann and Szmelcman, 1974). For these receptors correlations of binding of ligand to pure receptor *in vitro* with *in vivo* response to the respective attractants show that specificity and sensitivity of the tactic response is directly related to specificity and affinity of the isolated receptor protein for the ligand (Hazelbauer and Adler, 1971; Spudich and Koshland, 1975; Hazelbauer, 1977).

A number of lines of evidence indicate that gradient sensing, comparison of concentrations at different times, is a central function, not one performed at the level of receptor protein (Hazelbauer and Parkinson, 1977). Receptors can be considered as reporter molecules sending information about concentration of their ligand, as reflected in percent occupancy of binding sites, to the tumble regulator (see below). It is reasonable to expect that ligand binding generates an informational signal by inducing a conformational change in the receptor protein (Zukin et al., 1977). Since association of periplasmic binding proteins to other components of the transport or taxis systems must be very weak or very transitory (no one has detected specific association of a binding protein to a membrane component), a simple model suggests that only ligand-occupied receptor has affinity for the taxis component, (Strange and Koshland, 1976) (or probably for the transport component) and thus the informational signal is simply the creation of a functional binding site for the next component in the taxis pathway, presumably a cytoplasmic membrane protein. Thus it is interaction of receptor and transducer that generates a signal no longer directly linked to the stimulus molecule itself. This type of pattern is reminiscent of certain other receptor systems, particularly ones in which hormone receptors interact with adenylate cyclase (Bennet et al., 1976).

2.4 Signal Transducers

Information about signal transducers, the components that interact with occupied receptors, is still rather sparse, but it is possible to sketch a relatively coherent picture by making a few assumptions to fill in hazy areas. I outline such a scheme in this section.

An unusual class of taxis mutants called *trg* was mentioned by Ordal and Adler (1974) and Strange and Koshland (1976). These mutants appeared to be specifically defective in response to ribose and galactose, two attractants recognized by independent receptors, and thus could be transducer mutants. Isolation of null mutations in *trg*, generated by insertion of transposons into the gene (Harayama et al., 1979), make it possible to demonstrate that the *trg* product is required for response to ribose or galactose, but its absence does not affect other responses, the production of ribose and galactose receptors, or the functioning of those proteins in their respective transport systems (Hazelbauer and Harayama, 1979). Cross-inhibition studies have shown that saturation of one of these receptors can drastically reduce the sensitivity of the other (Adler et al., 1973; Strange and Koshland, 1976). A simple interpretation of these data is that the *trg* product is the transducer for both the ribose and galactose receptors, and both receptors, when occupied by ligand, interact directly with the trg protein at functionally overlapping sites (Ordal and Adler, 1974; Adler, 1975; Strange and Koshland, 1976; Hazelbauer and Harayama, 1979).

The existence of a transducer for the ribose and galactose receptors implies that there are transducers for groupings of other receptors, of which there are approximately 20 in *E. coli.* I would suggest that there are probably only a few more transducers, perhaps only three in addition to the *trg* product, and that two have already been identified in a somewhat different context.

Mutations in *tsr* eliminate sensitivity to a group of amino acids (Hazelbauer et al., 1969; Mesibov and Adler, 1972) that are probably recognized by a number of different receptors (Springer et al., 1977b), as well as sensitivity to five different receptor classes of repellents (Tso and Adler, 1974), without significantly reducing responses to sugars. Mutations in *tar* eliminate sensitivity to a complementary group of amino acids (Mesibov and Adler, 1972), to maltose, and to a different group of repellents (Springer et al., 1977b), again without significantly reducing responses to sugars. Response to some amino acids is only somewhat reduced in either mutant but is missing in a *tsr-tar* double mutant, probably because the compounds are recognized by receptors belonging to both the *tsr* and *tar* groupings. Determination of the tactic phenotype of *tsr-tar* double mutants is difficult because their tumble frequency is significantly reduced (Springer et al., 1977b; Silverman and Simon, 1977b). Such strains are inefficient in making progress up a spatial gradient, so it is not surprising that they do not show taxis toward sugars in assays of response to spatial gradients (Springer et al., 1977b; Goy and Springer, 1978). Yet double mutants possessing some tumbling activity respond to temporal gradients of ribose or galactose (Goy and Springer, 1978).

The large number of receptors linked to the *tsr* and *tar* products, as well as other observations discussed in later sections, suggest that the two proteins constitute the final step in passage of tactic signals from the relevant receptors to the central tumble regulator (Springer et al., 1977b; Silverman and Simon, 1977b). It might also be that the *tsr* and *tar* products interact directly with receptor proteins (Parkinson, 1977), which would mean that those two proteins would also be the initial transducers for their grouping of receptors. Then the tsr and tar proteins would each constitute a single step, direct linkage between many receptors proteins and the tumble regulator, as the trg protein is the link for its receptors. Analogous roles for products of *tsr*, *tar*, and *trg* would imply that the biochemical properties of the first two proteins (discussed in the following section) would be shared by the *trg* product.

These considerations do not account for one group of receptors, the enzyme II proteins that recognize the remainder of sugar attractants; *trg* mutations do not affect responses to those attractants (Hazelbauer and

Harayama, 1979). The effects of *tsr* or *tar* mutations on such responses have not been investigated in detail, but neither mutation reduces response to glucose significantly (Mesibov and Adler, 1972) [unfortunately a complicated case since three receptors, galactose, glucose, and mannose, all contribute to glucose taxis (Adler and Epstein, 1974)]. Thus there may be at least one more transducer, serving the enzyme II receptors. A single pathway for signals from all those receptors is suggested by the many instances of cross-inhibition among attractants recognized by the various enzyme II receptors (Adler et al., 1973).

2.5 The Tumble Regulator

It appears that signals reporting occupancy levels of all receptors must focus to a central machinery that integrates information from all receptors, detects changes, that is, gradients, regulates rotation of the flagellar motor, and adapts (Adler, 1975). The components involved in these interrelated functions can be collectively termed the tumble regulator.

Components of the tumble regulator are defined genetically by the phenotype of *che* mutants, defective response to all stimuli but still motile (Armstrong et al., 1967). Six *che* genes are known in *E. coli* (Silverman and Simon, 1977a; Parkinson, 1978) and seven are known in *S. typhimurium* (Warrick et al., 1977). Each *che* product appears to be intimately involved in the control of the balance between the two directions of flagellar rotation since all *che* mutants are drastically perturbed in that balance, exhibiting either almost exclusively smooth swimming or almost exclusively tumbly behavior (Parkinson, 1976).

Flagellar function and the control of tumble frequency by the tactic machinery can be separated conceptually and genetically, but there are numerous indications that the two are aspects of a single, integrated pheonomenon of directed motility. Motility in the absence of a functional sensory system appears to be of no advantage for *E. coli* in its natural milieu of the gut, while motility directed by the taxis system clearly is (Allweiss et al., 1977). There are approximately 20 *fla* genes required for production and assembly of flagellar components (Iino, 1977, Silverman and Simon, 1977c). These genes, as well as *mot* genes, whose products are required for flagellar rotation (Armstrong and Adler, 1969), *che* genes, and *tsr-tar* are for the most part clustered on the genetic map and are subjected to the same complex gentic controls (Parkinson, 1977; Silverman and Simon, 1977c; Iino, 1977). Che phenotypes can be generated by certain specific mutations in some *fla* genes (Silverman and Simon, 1973; Parkinson, 1976; Collins and Stocker, 1976; Warrick et al., 1977). Analysis of patterns of genetic complementation and supression among various *che* mutations indicates interaction between some gene products including a fla-che pair (Parkinson, 1976,1977, 1978).

All six *che* genes in *E. coli* make products that appear in the soluble fraction of broken cells (Silverman and Simon, 1977a; Matsumura et al., 1977), indicating that those proteins are essentially hydrophilic. In contrast, the *tsr* and *tar* products, as well as flagellar basal ends, are all firmly embedded in the cytoplasmic membrane (Kort et al., 1975; Ridgeway et al., 1977; DePhamphilis and Adler, 1971c). Thus the che proteins, the core of the tumble regulator, can be viewed as a link of soluble components between two complexes of cytoplasmic membrane proteins. As would be expected, some che proteins appear to have affinity for membrane, presumably for proteins of those two complexes (Ridgeway et al., 1977), and could thus be considered peripherial membrane proteins.

All functions of the tumble regulator are intimately associated with a specific kind of protein methylation reaction that is clearly central to the mechanism of chemotaxis (Kort et al., 1975). Glutamic acid residues of the *tsr* and *tar* products are carboxymethylated by donation of a methyl group from *S*-adenosylmethionine to the ϵ-carboxyl group (Kleene et al., 1977; van der Werf and Koshland, 1977). Carboxymethylation is catalyzed by a specific enzyme, the product of *cheR* in *S. typhimurium* and probably *cheX* in *E. coli* (Springer and Koshland, 1977; Stock and Koshland, 1978). Demethylation is mediated by a different enzyme, the *cheX* product in *Salmonella* and the *cheB* product in *E. coli* (Stock and Koshland, 1978). There appears to be a basal level of methylation of the tsr and tar proteins (called methylated chemotaxis proteins, MCP I and MCP II in this context) (Kort et al., 1975; Springer et al., 1977b). Tactic stimulation results in changes in the level of methylation, favorable gradients (attractant increases, repellent decreases) yielding an increase, and unfavorable gradients yielding a decrease (Kort et al., 1975; Springer et al., 1977b; Silverman and Simon, 1977b).

The time courses of these changes correspond to adaptation times, and a number of lines of evidence strongly link adaptation to the methylation reactions (Kort et al., 1975; Springer et al., 1975; 1977a; Goy et al., 1977). It appears as if methylated MCP is also critical for the clockwise, tumble-mode rotation, since methinonine-starved cells cease tumbling (Springer et al., 1975) and *tsr-tar* double mutants exhibit minimal tumbling activity (Springer et al., 1977b; Silverman and Simon, 1977b). In general, *tsr*-mediated stimuli result in changes in MCP I methylation and *tar*-mediated stimuli result in changes in MCP II methylation (Springer et al., 1977b; Silverman and Simon, 1977b). However, methylation may be even more complicated since each MCP runs as multiple bands on SDS–polyacrylamide gels and methylation of some of those bands may increase at the same time that methylation of others decreases. Both MCP mutants (Muskavitch et al., 1978) and certain *che* alleles of a *fla* gene (Rubik and Koshland, 1978) exhibit an unusual behavioral pattern of reversed polarity, responding to attractants as repellents and vice versa. If the *trg* product performs a role analogous to the tsr and tar proteins (see above), then it might also be methylated. Thus there could be a minimum of four MCPs, tsr, tar, trg, and a transducer, for enzyme II receptors.

Protein carboxymethylation is a widespread biochemical reaction, detectable in the many mammalian tissues examined and particularly evident in endocrine organs and some neural tissues (Kim et al., 1975; Diliberto and Axelrod, 1976). The function of these methylations is still not clear, but recent studies indicate that carboxymethylation of protein on the membranes of secretory vesicles is part of the mechanism of hormone release from cells of the adrenal medulla (Diliberto et al., 1976), while other work implicates carboxymethylation in chemotaxis by leucocytes (O'Dea et al., 1978; Pike et al., 1978). Like protein phosphorylation, it may well be that protein carboxymethylation is often part of the mechanism that links receptors to a cellular response in organisms from *E. coli* to mammals.

2.6 Ions and Chemotaxis

Significant activity has gone into the study of possible roles in chemotaxis for certain ions, ion fluxes, or changes in membrane potential. It is clear that proton-motive force, the gradient of hydrogen ions across the cytoplasmic membrane (Mitchell, 1973), is the source of energy for flagellar rotation (Larsen et al., 1974a; De Jong et al., 1976; Manson et al., 1977), and thus in the absence of proton-motive force, which includes a component of membrane potential, cells do not swim. The possibility of transient changes in potential or related fluxes of certain ions being involved in the taxis mechanism is unclear since studies of the gram-negative *E. coli* suggest such participation (Szmelcman and Adler, 1976), while work with the gram-positive *Bacillus subtilis* argues against it (Miller and Koshland, 1977). These differences are considered in a later section.

Studies by Ordal (1976, 1977) suggest a role for Mg^{2+} and Ca^{2+} in determining the direction of flagellar rotation. As in the case of *Paramecium* (Eckert, 1972; Naitoh and Eckert, 1969), internal calcium concentrations above a certain level appear to induce one mode of flagellar function, while at concentrations below that level the other mode occurs. Ordal and Fields (1977) suggest that favorable stimuli would cause binding of free calcium, while unfavorable stimuli would result in release of bound ion. Adaptation would occur upon reestablishment of the initial level of free calcium ion.

The widespread involvement of calcium ion in receptor-mediated phenomena makes participation of that ion in the bacterial sensory mechanism a particularily interesting possibility.

3 THE OUTER MEMBRANE IN MOTILITY AND TAXIS

This brief summary provides a basis for considering the influence of outer membrane structure and function on motility and taxis. Three aspects are considered, the flagellum and the outer membrane, the outer membrane as a barrier, and the outer membrane and the periplasm.

3.1 The Flagellum and Outer Membrane

3.1.1 Structure The gram-positives include motile species, for example, *Bacillus subtilis*. Thus an outer membrane cannot be required for flagellar function per se. However, in gram-negatives the flagellar basal end includes special structures that mediate insertion into the outer membrane and perhaps passage through the peptidoglycan layer. There is an intimate structural interaction between basal end and outer membrane, and perturbation of outer membrane integrity disturbs flagellar function. Observations pertaining to basal end–outer membrane interactions are summarized in the following paragraphs, and the possibilities of functional interrelationship are considered.

DePhamphilis and Adler (1971a, 1971b) found that the structure of intact flagella, that is, the complete filament–hook–basal end complex, is very similar in a representative gram-positive (*B. subtilis*) and a representative gram-negative (*E. coli*) bacterium. The difference is two extra rings found proximal to the hook on gram-negative basal ends (Figure 2). The outermost (*L*) ring is embedded in the outer membrane, while the inner (*P*) ring may align with the peptidoglycan monolayer (DePhamphilis and Adler, 1971c). In considering models for flagellar rotation, Berg (1974) suggested that these two outer rings could be considered simply as bushing required for passage of the rod through the envelope. The absence of bushing structures in gram-positive basal ends may indicate that the peptidoglycan and teichoic acid complexes of the gram-positive cell wall do not abut the flagellar rod so closely that there is interference with its rotation. Berg (personal communication) has also suggested that the outer rings might be necessary to maintain the penetration barrier established by the outer membrane.

The pair of outer rings is connected to form a single unit that is resistant to the Triton X-100–EDTA treatment, which dissolves outer membrane, as well as to the low pH treatment, which dissociates filament, hook and rod (DePhamphilis and Adler, 1971b). The outermost *L* ring can be considered a large outer membrane protein complex. The perimeter of the ring appears to be rather hydrophobic; basal ends aggregate by circumference-to-circumference association of L rings (DePhamphilis, 1971; DePhamphilis and Adler, 1971b). The ring has a specific affinity for LPS. Vesicles reformed from detergent-dissolved, purified LPS or from total outer membrane incorporate basal ends by inclusion of *L* rings in the reconstituted membrane (DePhamphilis, 1971). Basal ends can serve as catalysts for reconstitution of outer membrane from dissolved components. In such experiments, each reconstituted vesicle contains a basal end inserted by its *L* ring (DePhamphilis, 1971).

The properties of the complex of the two outer rings, *L* and *P* (DePhamphilis, 1971; DePhamphilis and Adler 1971b, c) parallel many recent observations about the peptidoglycan-associated proteins of the outer membrane (see Chapter 1). The L–P ring complex is resistant to mild detergents [as far as I know no one has examined its stability in SDS at 60°C where matrix protein trimers persist (Rosenbusch, 1974; Palva and Randall, 1978)] and exhibits an affinity for LPS, as do many other outer membrane proteins (Nakae, 1976; Datta et al., 1977). The complex spans the outer membrane and provides a hole across its width for the passage of the flagellar rod. DePhamphilis and Adler (1971c) suggested that the *P* ring might associate with peptidoglycan, although direct evidence for that interaction has not been reported. If we consider the L–P complex to resemble an enlarged porin-like complex [a 22.5 nm diameter complex with a 10 nm pore for the basal end structure versus on 8 nm complex (Rosenbusch, 1974) with a 1 nm pore (Schindler and Rosenbusch, 1978; Benz et al., 1978) for porin], then we would expect the *P* ring to associate with peptidoglycan, and basal ends might well be found on isolated sacculi.

3.1.2 Function In light of the intimate structural relation between basal ends and the outer membrane, it is not very surprising that disruption of outer membrane eliminates normal flagellar function. About half the LPS of the cell is lost upon exposure to Tris–EDTA (Levy and Leive, 1968), resulting in loss of the integrity of the penetration barrier usually provided by the outer membrane. Motile cells suspended in 20% sucrose–Tris–EDTA [Stage I of the Neu and Heppel (1965) osmotic shock procedure] become essentially nonmotile, presumably because of injury to the outer membrane. Motility is restored to cells removed from that solution by incubation at 35°C in tryptone broth or in buffer containing a tenfold concentrated supernant after cells are removed from the stage I solution (Hazelbauer and Adler, 1971). In either case, inhibition of protein synthesis by chloramphenicol does not prohibit restoration. A similar pattern of loss and restoration of motility is observed upon treatment with only Tris–EDTA (Szmelcman and Adler, 1976). Probably restoration of motility is the result of replacement of LPS lost from the outer membrane, either using the energy and material supplied by tryptone broth to synthesize LPS endogenously or incorporating LPS micelles present in the stage I supernant. This implies that the outer membrane must be intact for flagella to propel the cell. At present we cannot distinguish between the possibilities that outer membrane disruption halts flagellar rotation or that disruption only uncouples rotation from exertion of a propelling force.

3.1.3 Biosynthesis Mutants of *S. typhimurium* and *E. coli* defective in the core polysaccharide of lipopolysaccharide have outer membranes perturbed both by reduced amounts of several major proteins (Ames et al., 1974; Koplow and Goldfine, 1974) and by rearrangement of its hydrophobic bilayer (Nikaido, 1976). Lack of the outer LPS saccharides in *gal*E or rough mutants does not result in such pleiotrophic defects (Komeda et al., 1977; Ames et al., 1974). Motility of LPS mutants follows a similar pattern; *gal*E mutants swim normally, while heptose-less strains are essentially nonmotile (Komeda et al., 1977). The motility defect appears to be the result of a reduction in flagella number rather than an effect on flagellar function.

Heptose-less strains have few flagella visible on their surfaces (Irvin et al., 1975) and contain little envelope-associated flagellin (Ames et al., 1974). Reduction of the amount of outer membrane proteins and loss of motility occurs even in mutants missing only a portion of the core polysaccharide. Boman and Monner (1975) characterized a series of LPS mutants that lacked progressively more of the core polysaccharide. Each mutant, even that with the mildest defect (no rhamnose, no galactose, reduced glucose) lacks 65% of the usual amount of λ receptor (Randall, 1975) and is almost nonmotile (G. Hazelbauer and L. Randall, unpublished observations).

Studies by Komeda et al. (1977) demonstrate that nomotile, LPS-defective mutants contain little flagellin, little flagellin-specific mRNA, and probably no assembled basal ends. This is the same phenotype exhibited by *fla* mutants. The complex process of biosynthesis and assembly of flagella depends on many genes and the lack of any one of a large number of *fla* products appears to shut off the whole process (Iino, 1977). The LPS defects also shut down the bulk of flagellar synthesis and assembly, perhaps by means of a linkage that involves the product of *flaH*. Komeda et al. (1977) showed that supressor mutations that restored motility to *galU* (LPS-defective) strains mapped in *flaH*, implying that the normal *flaH* product mediates the effect of outer membrane alterations on *fla* gene activity. Since *flaH*-supressed *galU* mutants are motile, as are, to varying extents, LPS mutants (Tamaki et al., 1971) of the NS series (NS1, Komeda et al., 1977; NS1-NS5, G. Hazelbauer and L. Randall, unpublished observations), the effect of LPS mutations on motility is probably not simply a structural one. Thus the motility defect is not simply the result of faulty functioning of a flagellum inserted into a defective outer membrane, but is more likely a reflection of a complex control matrix that guides flagellar synthesis and assembly.

Heptose-less mutants are missing a significant proportion of the major outer membrane proteins (Ames et al., 1974; Koplow and Goldfine, 1974), but loss of those proteins by specific mutations does not reduce motility. Mutants missing matrix protein, ompA protein, both matrix protein and ompA protein (Henning and Haller, 1975), or lipoprotein (Hirota et al., 1977) all swim normally (Henning and Haller, 1975; E. T. Palva and G. Hazelbauer, unpublished observation).

This is an appropriate place to mention the relation of peptidoglycan to motility. Complete spheroplasts lacking peptidoglycan after treatment with penicillin or lysozyme–EDTA (Vaituzis and Doetsch, 1965, 1966) do not swim, probably because the rigid layer is required for flagella to exert their thrust (Stocker, 1956). Incomplete penicillin-induced spheroplasts continue to swim until the rod-shaped "rabbit ears" of the "prepenicillin" cell have disappeared. Synthesis of flagellin and assembly of flagella appear to require active synthesis of peptidoglycan (McGroarty et al., 1973).

3.2 Outer Membrane as a Barrier Between Stimulus and Receptor

Normal passage of maltose across the outer membrane requires the product of *lamB* (Szmelcman and Hofnung, 1975; Hazelbauer, 1975b; Szmelcman et al., 1976), an outer membrane protein with an apparent molecular weight on the order of 44,000 (Palva and Randall, 1978; Braun and Krieger-Brauer, 1977) that was initially identified as the receptor for bacteriophage λ (Randall-Hazelbauer and Schwartz, 1973). It appears that the *lamB* product mediates the passage of maltose by providing an aqueous pore (von Meyenburg and Nikaido, 1977; Benz et al., 1978). Outer membrane containing normal amounts of *lamB* product does not pose a significant barrier for maltose, as indicated by the correspondence of the 1 μM K_d of maltose-binding protein *in vitro* with the apparent K_M of maltose transport *in vivo* (Szmelcman et al., 1976). However, outer membrane lacking the protein seriously limits the access of external maltose to the periplasm, as documented by 100- to 500-fold increased K_M values for maltose transport by *lamB* mutants, yet does not constitute an inpenetrable barrier for the sugar since *lamB* mutants exhibit essentially normal V_{max} values for maltose transport (Szmelcman et al., 1976).

This pattern is paralleled by the tactic behavior of *lamB*$^+$ and *lamB* mutant strains. In a wild-type strain the stimulus–response relationship for maltose is an exact reflection of occupancy of maltose-binding protein by maltose (Hazelbauer, 1977), while the stimulus–response curve of null mutants in *lamB* is shifted to hundredfold higher concentrations of maltose (Hazelbauer, 1975b). The simplest hypothesis is that *lamB* null mutants exhibit a reduction in apparent affinity of the maltose transport system and a decrease in apparent sensitivity of the taxis system as a result of the same single defect, limited permeability of maltose through an outer membrane lacking the lamB protein. This implies that the lamB protein has no active role in the actual processes of transport or chemoreception but simply provides free passage for maltose between the exterior and periplasm, the site of functional interaction between sugar and recognition protein. However, it is difficult to test the possibility that, in addition to its function as a maltose pore, the lamB protein also has a more direct role in chemoreception (or for that matter, transport), but an indirect argument can be made against such direct participation. If the *lamB* product were directly involved in maltose reception, then one would expect a particular site on the protein to be critical for such participation, and that site would not be identical to either the phage absorption site or the areas critical for pore function. Thus missense mutants selected for phage resistance could be unperturbed in transport but defective in taxis, or even *vice versa*. In Figure 3 the maltose taxis response of a collection of *lamB* missense mutants selected as λ-resistant clones (Hofnung, et al., 1976) is plotted versus maltose uptake activity (Hazelbauer, 1975b). Activities of the missense mutants range from normal levels to ones as defective as *lamB* null mutants. There is a good correlation between the extent of the taxis defect and the severity of the transport defect, supporting the hypothesis that *lamB* mutations influence the tactic response to maltose only by reducing passage of maltose into the receptor-containing compartment.

At relatively high concentrations any di- or trisaccharide can cross the outer membrane by using the rather general pore formed by the matrix protein (Nakae and Nikaido, 1975; Nakae, 1976; Bavoil et al., 1977). For maltose that concentration is about 1 m*M*. For maltotriose it is higher and for higher maltodextrans it is even higher (Szmelcman et al., 1976; Braun and Krieger-Brauer, 1977; Silhavy et al., 1977). Thus the tactic response of a *lamB* mutant to maltodextrans should be more drastically reduced as the size of the attractant increases, as seen in Figure 4.

Maltose is rather an exception among the sugars metabolized by *E. coli*. It is one of the few sugars for which there is only one transport system and

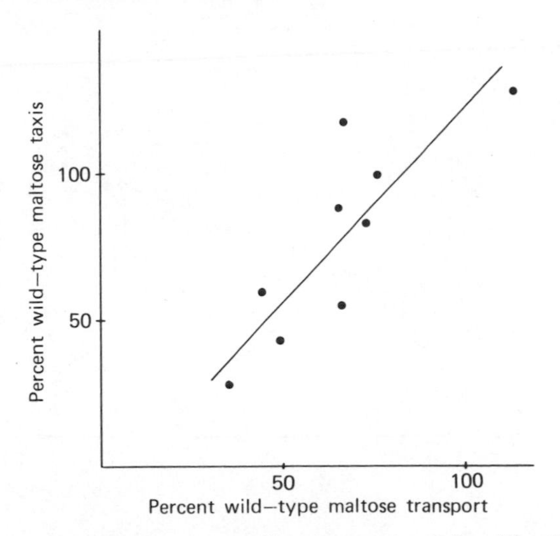

Figure 3 Maltose taxis response versus maltose transport activity of *lamB* missense mutants. Transport activity is initial rate of uptake at 10 μM maltose (Table 2, Hazelbauer 1975b); taxis response is total response in capillary assay (Table 1, Hazelbauer, 1975b).

the only sugar known to have its own specialized system for bridging the outer membrane. Thus the influence of the outer membrane barrier in the absence of its special pore is striking and easy to observe. Other sugars appear to pass the outer membrane through the matrix protein pore and thus one would predict that *ompB* mutants, missing all forms of that porin (Sarma and Reeves, 1977), should exhibit reduced sensitivity to those attractant sugars.

The possibility of a barrier to free exchange between the external medium and the local environment of receptors on the cell surface is frequently a concern of electrophysiologists and others studying receptor systems in higher organisms. The importance of that concern is that biochemical identification of a "physiological" receptor is often dependent on the correlation of *in vitro* affinity with *in vivo* sensitivity. If maltose taxis had been initially characterized in a *lamB* null mutant, then *in vitro* affinity of the maltose-binding protein would have appeared a hundredfold stronger than the value predicted by the *in vivo* response.

3.3 The Periplasm

The periplasmic space is created by the outer membrane. It is in this compartment that chemoreception occurs (see discussion in previous section), so it seems appropriate to include a consideration of the periplasm in this chapter.

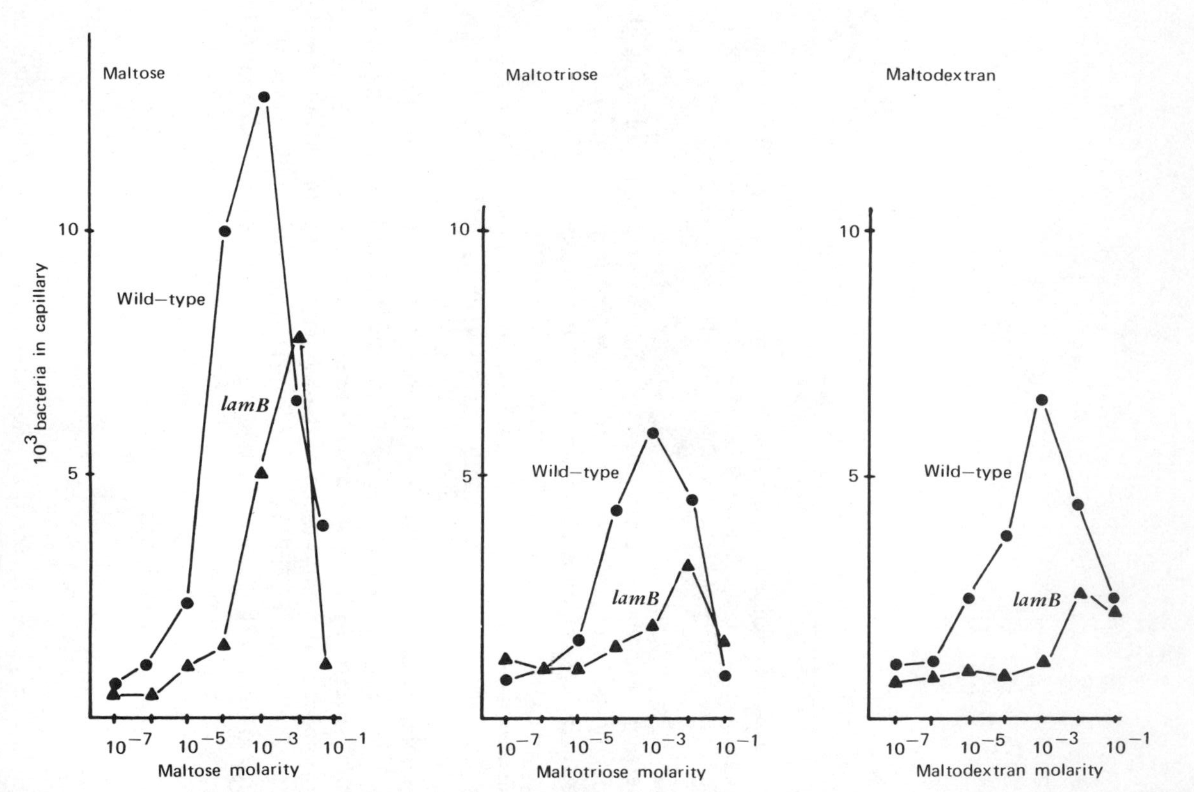

Figure 4 Chemotactic response of a *lamB* null mutant and its parent to maltose and higher maltodextrans. Wild-type parent is pop1021 and the mutant is pop1080, carrying lamB102, and an ochre mutation (Hazelbauer, 1975b; Hofnung et al., 1976). Maltose and maltotriose were purified by paper chromatography (Hazelbauer, 1975b), and maltodextrans (averaging seven glucose residues, a gift from G. Wöber, Marburg) were separated from low molecular weight compounds by ultrafiltration through a UM-2 membrane (Silhavy et al., 1977).

The volume of the periplasm is important in considerations of the functioning of periplasmic binding proteins (Silhavy et al., 1975), but attempts to estimate dimensions from electron micrographs of sectioned material suffer from the distinct possibility of artifactual changes [see the difference between cell dimension determined by phase contrast and by electron microscopy listed by Luria (1960)]. The elegant determinations of osmotic space by Stock et al. (1977) provide values for periplasmic volume expressed as microliters per milligram of dry weight of cells. Using a value of 0.25 mg dry weight/10^9 cells (Roberts et al., 1955), the volume can be expressed as femtiliters per cell (Table 1). The values of Stock et al. (1977) for periplasmic volume are 20–40% of total osmotically active space in the cell, numbers that may appear surprisingly large. Yet a significant proportion of the cytoplasm is occupied by ribosomes and nucleic acid, and that volume plus the volume of bound water does not contribute to osmotic space. It seems informative to express the periplasmic volume as a percentage of the total volume occupied by a bacterial cell (Table 1). Values in the range of 10% of total occupied cell volume seem quite reasonable. Previous to the availability of the measurements of Stock et al. (1977) I had attempted to estimate the periplasmic volume by using the dimensions of the flagellar basal end. The L ring is inserted into the outer membrane and the M ring is inserted into the cytoplasmic membrane. Reasoning that the distance between those rings should be relatively close to the usual *in vivo* separation of the two membranes, I assumed a periplasm width of 22 nm from the basal end dimensions determined by DePhamphilis and Adler (1971b and

Table 1 Volume of Cell Compartments

	Calculated Occupied Volume[a]		Measured Osmotically Active Volume[b]		
	Femtiliters	% Whole Cell	Femtiliters	%Whole Cell Osmotic Volume	% Whole Cell Occupied Volume
Whole cell	1.57	100	0.4–0.6	100	25–34
Cytoplasm	1.42	90.5	0.3–0.35	75–58	19–22
Periplasm	0.15	9.5	0.1–0.25	25–42	6–16

[a] Occupied volumes calculated assuming cell is a cylinder 1 μm in diameter, 2 μm high (Luria, 1960). Periplasmic and cytoplasmic volumes calculated as cylinder volumes using 22 nm as distance between outside of cytoplasmic membrane and inside of outer membrane (see text).

[b] From Stock et al. (1977). *S. typhimurium* cells in medium 63: lower values, Figure 6; higher values Table 1. Volumes in microliters per milligram of dry weight of cells converted to femtiliters per cell using 0.25 mg dry wt./10^9 cells (see text).

Figure 2). Calculations as described in footnote *a* of Table 1 yield a value in the same range as those of Stock et al. (1977). Estimates of the content of a particular binding protein in a fully induced cell are on the order of 1×10^4-5×10^4 copies/cell. Thus the concentration of binding protein in the periplasm would be 0.1–0.5 m*M*, significantly higher than the 0.1–1 μM range of dissociation constants for binding protein and ligand, but not as concentrated as previously estimated (Silhavy et al., 1975). In any case, a concentration of binding protein in the periplasm that is high in relation to the dissociation constant of the ligand–protein complex means that ligand molecules in the periplasm are almost always bound to a protein (Silhavy et al., 1975). Thus the number of ligands in the periplasm is essentially the same as the number of occupied binding proteins and the transducer system could directly count the number of attractant molecules by interactions with those occupied binding protein.

Stock et al. (1977) also make the interesting observation that there appears to be a Donnan equilibrium between the periplasm and the external solution that would create an electrical potential across the outer membrane of about 30 mV, negative inside, for cells suspended in minimal salts medium. It is not unreasonable to expect that this potential might be utilized in some way by processes that occur in the cell envelope (Schindler and Rosenbusch, 1978). A potential across the outer membrane might also be involved in unexplained differences between results obtained using a gram-negative (Szmelcman and Adler, 1976) and a gram-positive organism (Miller and Koshland, 1977) in studies of effects of chemotactic stimulation on membrane potential (see discussion above). In *B. subtilis*, tactic stimulation did not alter the distribution of triphenylmethylphosphonium ion (TPMP) between cell and medium, while in *E. coli* such stimulation resulted in transitory changes in cellular content of the ion. TPMP is known to distribute in response to membrane potential (Harold, 1976), so it would not be surprising if the outer membrane potential had some influence on the pattern of distribution.

4 CONCLUDING REMARKS

The purpose of this chapter is to provide a summary of our current understanding of the phenomenon of bacterial chemotaxis and to indicate how the structure and function of the outer membrane influences motility and taxis. I hope those interested in either or both areas will find this hybrid review useful. It seems prudent for those studying motility and taxis of gram-negative organisms to be aware of the distinguishing structure of those bacteria and it seems reasonable for adherents of the outer membrane to include motility and taxis among their expanding areas of involvement.

ACKNOWLEDGMENTS

Unpublished studies described here were supported by grant AI12858 from The National Institute of Allergy and Infectious Diseases and grants from the Swedish Natural Sciences Research Council. I thank L. L. Randall and H. C. Berg for critical reading of the manuscript and B. Berneholm for technical assistance.

REFERENCES

Adler, J. (1969), *Science*, **166,** 1588–1597.

Adler, J. (1975), *Annu. Rev. Biochem.*, **44,** 341–356.

Adler, J., and Epstein, W. (1974), *Proc. Natl. Acad. Sci. U.S.*, **71,** 2895–2899.

Adler, J., Hazelbauer, G. L., and Dahl, M. M. (1973), *J. Bacteriol.*, **115,** 824–847.

Aksamit, R., and Koshland, D. E., Jr. (1972), *Biochem. Biophys. Res. Commun.*, **48,** 1348–1353.

Aksamit, R. R., and Koshland, D. E., Jr. (1974), *Biochemistry*, **13,** 4473–4478.

Allweiss, B., Dostal, J., Carey, K., Edwards, T. F., and Freter, R. (1977), *Nature*, **266,** 448–450.

Ames, G. F-L., Spudich, E. N., and H. Nikaido (1974), *J. Bacteriol.*, **117,** 406–416.

Anraku, Y. (1968), *J. Biol. Chem.*, **243,** 3116–3122.

Armstrong, J. B., and Adler, J. (1969), *Genetics*, **61,** 61–66.

Armstrong, J. B., Adler, J., and Dahl, M. M. (1967), *J. Bacteriol.*, **93,** 390–398.

Bavoil, P., Nikaido, H., and von Meyenberg, K. (1977), *Mol. Gen. Genet.*, **158,** 23–33.

Bennet, U., Craig, S., Hollenberg, M. O., O'Keefe, E., Sahyoun, N., and Cuatrecasas, P. (1976), *J. Supramol. Struct.*, **4,** 99–120.

Benz, R., Janko, K., Boos, W., and Läuger, P. (1978), *Biochim. Biophys. Acta*, **511,** 505–519.

Berg, H. C. (1974), *Nature*, **249,** 77–79.

Berg, H. C. (1975a), *Annu. Rev. Biophys. Bioeng.*, **4,** 119–136.

Berg, H. C. (1975b), *Nature*, **254,** 389–392.

Berg, H. C., and Anderson, R. A. (1973), *Nature*, **245,** 380–382.

Berg, H. C., and Brown, D. A. (1972), *Nature*, **239,** 500–504.

Boman, H. C., and Monnor, D. A. (1975), *J. Bacteriol.*, **121,** 455–464.

Braun, V., and Krieger-Brauer, H. J. (1977), *Biochim. Biophys. Acta*, **469,** 89–98.

Brown, D. A., and Berg, H. C. (1974), *Proc. Natl. Acad. Sci. U.S.*, **71,** 1388–1392.

Carlile, M. S. (1975), in *Primitive Sensory and Communication Systems*, (M. S. Carlile, ed.) Academic Press, London, pp. 1–28.

Collins, A. L., and Stocker, B. A. D. (1976), *J. Bacteriol.*, **128,** 754–765.

Datta, D. B., Arden, B., and Henning, U. (1977), *J. Bacteriol.*, **131,** 821–829.

De Jong, M. H., van der Drift, C., and Vogels, G. D. (1976), *Arch. Microbiol.*, **111,** 7–11.

DePhamphilis, M. L. (1971), *J. Bacteriol.*, **105,** 1184–1199.

DePhamphilis, M. L., and Adler, J. (1971a), *J. Bacteriol.*, **105,** 376–383.

DePhamphilis, M. L., and Adler, J. (1971b), *J. Bacteriol.*, **105,** 384–395.

DePhamphilis, M. L., and Adler, J. (1971c), *J. Bacteriol.*, **105,** 396–407.

Diliberto, E. J., Jr., and Axelrod, J. (1976), *J. Neurochem.*, **26,** 1159–1165.

Diliberto, E. J., Jr., Viveros, O. H., and Axelrod, J. (1976), *Proc. Natl. Acad. Sci. U.S.*, **73,** 4050–4054.

Eckert, R. (1972), *Science*, **176,** 473–481.

Fraenkel, G. S., and Gunn, D. L. (1940), Clarendon Press, Oxford.

Goy, M. F., and Springer, M. S. (1978), in *Taxis and Behavior* (G. L. Hazelbauer, ed.), Chapman and Hall, London, pp. 3–33.

Goy, M. F., Springer, M. S., and Adler, J. (1977), *Proc. Natl. Acad. Sci. U.S.*, **74,** 4964–4968.

Harayama, S., Palva, E. T., and Hazelbauer, G. L. (1979), *Mol. Gen. Genet.*, **171,** 193–203.

Harold, F. M. (1976), *Curr. Top. Bioenerg.*, **6,** 83–149.

Hazelbauer, G. L. (1975a), *J. Bacteriol.*, **122,** 206–214.

Hazelbauer, G. L. (1975b), *J. Bacteriol.*, **124,** 119–124.

Hazelbauer, G. L. (1977), in *Olfaction and Taste* (J. LeMagnen, and P. MacLeod, eds.) Information Retrieval, London, pp. 47–54.

Hazelbauer, G. L., and Adler, J. (1971), *Nature, New Biol.*, **230,** 101–104.

Hazelbauer, G. L., and Harayama (1979), *Cell*, **16,** 617–625.

Hazelbauer, G. L., and Parkinson, J. S. (1977), in *Microbial Interactions* (J. L. Reissig, ed.) Chapman and Hall, London, pp. 61–98.

Hazelbauer, G. L., Mesibov, R. E., and Adler, J. (1969), *Proc. Natl. Acad. Sci. U.S.*, **64,** 1300–1307.

Henning, U., and Haller, I. (1975), *FEBS Lett.*, **55,** 161–164.

Hirota, Y., Suzuki, H., Nishimura, Y., and Yasuda, S. (1977), *Proc. Natl. Acad. Sci. U.S.*, **74,** 1417–1420.

Hofnung, M., Jezierska, a., and Braun-Breton, C. (1976), *Mol. Gen. Genet.*, **145,** 207–213.

Iino, T. (1977), *Annu. Rev. Genet.*, **11,** 161–182.

Irvin, R. T., Chatterjee, A. K., Sanderson, K. E., and Costerton, J. W. (1975), *J. Bacteriol.*, **124,** 930–941.

Kellermann, O., and Szmelcman, S. (1974), *Eur. J. Biochem.*, **47,** 139–149.

Kim, S., Wasserman, L., Lew, B., and Paik, W. K. (1975), *J. Neurochem.* **24,** 625–629.

Kleene, S. J., Toews, W. L., and Adler, J. (1977), *J. Biol. Chem.*, **252,** 3214–3218.

Komeda, Y., Icho, T., and Iino, T., (1977), *J. Bacteriol.*, **129,** 908–915.

Koplow, J., and Goldfine, H. (1974), *J. Bacteriol.*, **117,** 527–543.

Kort, E. N., Goy, M. F., Larsen, S. H., and Adler, J. (1975), *Proc. Natl. Acad. Sci. U.S.*, **72,** 3939–3943.

Koshland, D. E., Jr. (1977a), in *Advances in Neurochemistry* Vol 2, (B. W. Agranoff and W. H. Aprison, eds.), Plenum, New York, pp. 277–341.

Koshland, D. E., Jr. (1977b), *Science*, **196,** 1055–1063.

Larsen, S. H., Adler, J., Gargus, J. J., and Hogg, R. W. (1974a), *Proc. Natl. Acad. Sci. U.S.*, **71,** 1239–1243.

Larsen, S. H., Reader, R. W., Kort, E. M., Tso, W. W., and Adler, J. (1974b), *Nature*, **249,** 74–77.

Läuger, P. (1977), *Nature*, **268,** 360–361.

Lenger, J. (1975), *J. Bacteriol.*, **124,** 26–38.

Levy, S. B., and Leive, L. (1968), *Proc. Natl. Acad. Sci. U.S.*, **63,** 1435–1439.

Luria, S. E. (1960), in *The Bacteria* Vol. I (I. C. Gunsalus and R. Y. Stanier, eds.) Academic Press, New York, pp. 1–34.

McGroarty, E. J., Koffler, H., and Smith, R. W. (1973), *J. Bacteriol.*, **113,** 295–303.

Macnab, R. M. (1977), *Proc. Natl. Acad. Sci. U.S.*, **74,** 221–225.

Macnab, R. M., and Koshland, D. E., Jr. (1972), *Proc. Natl. Acad. Sci. U.S.*, **69,** 2509–2512.

Macnab, R., and Orston, M. K. (1977), *J. Mol. Biol.*, **112,** 1–30.

Manson, M. D., Tedesco, P., Berg, H. C., Harold, F. M., and van der Drift, C. (1977), *Proc. Natl. Acad. Sci. U.S.*, **74,** 3060–3064.

Matsumura, P., Silverman, M., and Simon, M. (1977), *J. Bacteriol.*, **132,** 996–1002.

Melton, T., Hartman, P. E., Stratis, J. P., Lee, T. L., and Davis, A. T. (1978), *J. Bacteriol.*, **133,** 708–716.

Mesibov, R., and Adler, J. (1972), *J. Bacteriol.*, **112,** 315–326.

Miller, J. B., and Koshland, D. E., Jr. (1977), *Proc. Natl. Acad. Sci. U.S.*, **74,** 4752–4756.

Mitchell, P. (1973), *J. Bioenerg.*, **4,** 63–91.

Muskavitch, M. M., Kort, E. N., Springer, M. S., Goy, M. F., and Adler, J. (1978), *Science*, **201,** 63–65.

Naitoh, Y., and Eckert, R. (1969), *Science*, **164,** 963–965.

Nakae, T. (1976), *Biochem. Biophys. Res. Commun.*, **71,** 877–884.

Nakae, T., and Nikaido, H. (1975), *J. Biol. Chem.*, **250,** 7359–7365.

Neu, H. C., and Heppel, L. A. (1976), *J. Biol. Chem.*, **240,** 3685–3694.

Nikaido, H. (1976), *Biochim. Biophys. Acta*, **433,** 118–132.

O'Dea, R. F., Viveros, O. H., Axelrod, J., Aswanikuman, S., Schiffman, E., Corcoran, B. A. (1978), *Nature*, **272,** 462–464.

Ordal, G. W. (1976), *J. Bacteriol.*, **126,** 706–711.

Ordal, G. W. (1977), *Nature*, **270,** 66–67.

Ordal, G. W., and Adler, J. (1974), *J. Bacteriol.*, **117,** 517–526.

Ordal, G. W., and Fields, R. B. (1977), *J. Theor. Biol.*, **68,** 491–500.

Palva, E. T., and Randall, L. L. (1978), *J. Bacteriol.*, **133,** 279–286.

Parkinson, J. S. (1976), *J. Bacteriol.*, **126,** 758–770.

Parkinson, J. S. (1977), *Annu. Rev. Genet.*, **11,** 397–414.

Parkinson, J. S. (1978), *J. Bacteriol.*, **135,** 45–53.

Pike, M. C., Kredich, N. M., Snyderman, R. (1978), *Proc. Natl. Acad. Sci. U.S.*, **75,** 3928–3932.

Randall-Hazelbauer, L. L., and Schwartz, M. (1973), *J. Bacteriol.*, **116,** 1436–1446.

Randall, L. L. (1975), *J. Bacteriol.*, **123,** 41–46.

Ridgeway, H. F., Silverman, M., and Simon, M. I. (1977), *J. Bacteriol.*, **132,** 657–665.

Roberts, R. B., Abelson, P. H., Cowie, D. B., Bolton, E. T., and Britten, R. J. (1955), in *Studies of Biosynthesis* in Escherichia coli, Carnegie Institution of Washington Publication 607, Washington D.C.

Rosenbusch, J. P. (1974), *J. Biol. Chem.*, **249,** 8019–8029.

Rubik, B. A., and Koshland, D. E. Jr. (1978), *Proc. Natl. Acad. Sci. U.S.*, **75,** 2820–2824.

Sarma, V., and Reeves, P. (1977), *J. Bacteriol.*, **132,** 23–27.

Schindler, H., and Rosenbusch, J. P. (1978), *Proc. Natl. Acad. Sci. U.S.*, **75,** 3751–3755.

Silhavy, T. J., Szmelcman, S., Boos, W., and Schwartz, M. (1975), *Proc. Natl. Acad. Sci. U.S.*, **72,** 2120–2124.

Silhavy, T. J., Shuman, H. A., Beckwith, J., and Schwartz, M. (1977), *Proc. Natl. Acad. Sci. U.S.*, **74,** 5411–5415.

Silverman, M., and Simon, M. (1973), *J. Bacteriol.*, **116,** 114–122.

Silverman, M., and Simon, M. (1974), *J. Bacteriol.*, **117,** 73–79.

Silverman, M., and Simon, M. (1977a), *J. Bacteriol.*, **130,** 1317–1325.

Silverman, M., and Simon, M. (1977b), *Proc. Natl. Acad. Sci. U.S.*, **74,** 3317–3321.

Silverman, M., and Simon, M. (1977c). *Annu. Rev. Microbiol.*, **31,** 397–419.

Springer, W. R., and Koshland, D. E., Jr. (1977), *Proc. Natl. Acad. Sci. U.S.*, **74,** 533–537.

Springer, M. S., Kort, E. N., Larsen, S. H., Ordal, G. W., Reader, R. W., and Adler, J. (1975), *Proc. Natl. Acad. Sci. U.S.*, **72,** 4640–4644.

Springer, M. S., Goy, M. F., and Adler, J. (1977a), *Proc. Natl. Acad. Sci. U.S.*, **74,** 183–187.

Springer, M. S., Goy, M. F., and Adler, J. (1977b), *Proc. Natl. Acad. Sci. U.S.*, **74,** 3312–3316.

Spudich, J. L., and Koshland, D. E., Jr. (1975), *Proc. Natl. Acad. Sci. U.S.*, **72,** 710–713.

Stock, J. B., and Koshland, D. E., Jr. (1978), *Proc. Natl. Acad. Sci. U.S.*, **75,** 3659–3663.

Stock, J. B., Rauch, B., and Roseman, S. (1977), *J. Biol. Chem.*, **252,** 7850–7861.

Stocker, B. (1956), *Symp. Soc. Gen. Microbiol.*, **6,** 19–40.

Strange, P. G., and Koshland, D. E., Jr. (1976), *Proc. Natl. Acad. Sci. U.S.*, **73,** 762–766.

Szmelcman, S., and Adler, J. (1976), *Proc. Natl. Acad. Sci. U.S.*, **73,** 4387–4391.

Szmelcman, S., and Hofnung, M. (1975), *J. Bacteriol.*, **124,** 112–118.

Szmelcman, S., Schwartz, M., Silhavy, T. J., and Boos, W. (1976), *Eur. J. Biochem.*, **65,** 13–19.

Tamaki, S., Sato, T., and Matsuhashi, M., (1971), *J. Bacteriol.*, **105,** 968–975.

Tamm, S. L. (1978), *J. Cell Biol.*, **78,** 76–92.

Tsang, N., Macnab, R., and Koshland, D. E., Jr. (1973), *Science*, **181,** 60–63.

Tso, W. W., and Adler, J. (1974), *J. Bacteriol.*, **118,** 560–576.

Vaituzis, Z., and Doetsch, R. N. (1965), *J. Bacteriol.*, **89,** 1586–1593.

Vaituzis, Z., and Doetsch, R. N. (1966), *J. Bacteriol.*, **91,** 2103–2104.

Van der Werf, P., and Koshland, D. E., Jr. (1977), *J. Biol. Chem.*, **252,** 2793–2795.

Von Meyenburg, K., and Nikaido, H. (1977), *Biochem. Biophys. Res. Commun.*, **78,** 1100–1107.

Warrick, H. M., Taylor, B. L., and Koshland, D. E., Jr. (1977), *J. Bacteriol.*, **130,** 223–231.

Willis, R. C., and Furlong, C. W. (1974), *J. Biol. Chem.*, **249,** 6926–6929.

Zukin, R., Hartig, P. R., and Koshland, D. E., Jr. (1977), *Proc. Natl. Acad. Sci.*, **74,** 1932–1936.

Chapter **14**

Pathogenic Aspects of Outer Membrane Components of Gram-Negative Bacteria

THOMAS M. BUCHANAN, M.D.

Associate Professor of Medicine and Pathobiology, The University of Washington, Seattle, Head, Immunology Research Laboratory, U.S. Public Health Service Hospital, Seattle, Washington

and

WILLIAM A. PEARCE, M.S.P.H.

Department of Pathobiology, The University of Washington, Seattle, Immunology Research Laboratory, U.S. Public Health Service Hospital, Seattle, Washington

1 INTRODUCTION

This chapter is limited to pathogenic aspects of structures that comprise the outer membrane of gram-negative organisms or are located in contact with, but just external to, that membrane. It includes, therefore, proteins, carbohydrates, and lipids structurally comprising, or extending beyond, the outer membrane that are thought to be instrumental in the pathogenesis of infections (Figure 1).

2 OUTER MEMBRANE COMPONENTS THAT CONTRIBUTE TO PATHOGENESIS

2.1 Attachment to Human Cells

Pili are hairlike structures approximately 70Å in diameter and 0.5–4 μm long that project outwards from the outer membrane of many gram-negative pathogenic organisms (Figure 2). When studied, piliation usually has been found to influence attachment of the organisms to cell surfaces and, in several instances, to affect phagocytosis. This chapter first examines pili of *Neisseria gonorrhoeae*, and the related investigations on pili of other microbes.

Pili were first reported on gonococci in 1971 by Jephcott, Reyn, and Birch Anderson, and independently by Swanson, Kraus, and Gotschlich. Within a short period it was demonstrated that piliated gonococci were

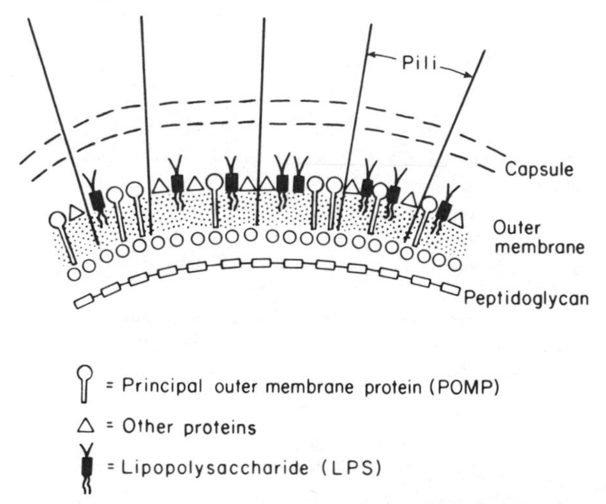

Figure 1 Schematic of the structures external to peptidoglycan in gram-negative bacteria. Surface components likely to contribute to the pathogenesis of infection and microbial virulence include pili, capsule, LPS, principal outer membrane protein, and peptidoglycan. In addition, proteolytic or glycosidic enzymes found in the periplasmic space may elute through holes in the outer membrane during infection to affect pathogenesis.

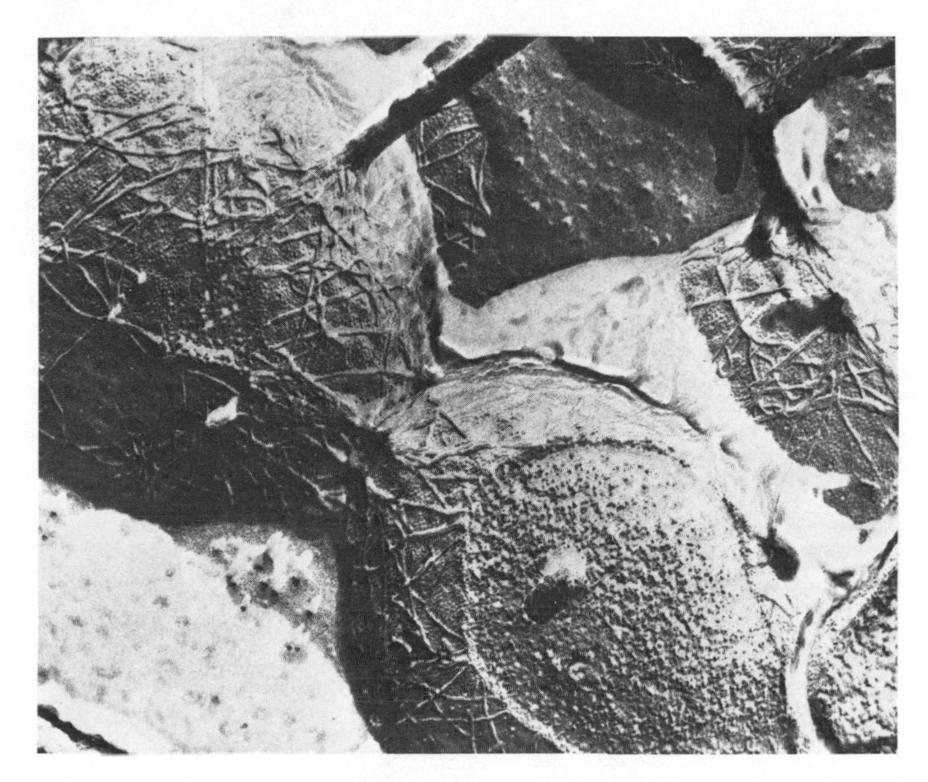

Figure 2 Freeze-fracture, freeze-etch prepared gonococci with pili on surface and cleaved outer membrane (×142,500). Courtesy of Dr. John Swanson.

more virulent than nonpiliated organisms for the chimpanzee (Brown et al. 1972) and the chick embryo (Buchanan and Gotschlich, (1973); Bumgarner and Finkelstein, 1973), increasing interest in pili as a virulence factor. The enhanced virulence of piliated gonococci appeared to correlate with greater attachment of piliated than nonpiliated organisms to human cells and with *resistance* of piliated gonococci to *phagocytosis*.

Many investigators have observed that piliated gonococci attach better to human cells than nonpiliated organisms (Tables 1 and 2). Punsalang and Sawyer reported enhanced attachment of piliated gonococci to rabbit, guinea pig, sheep, chicken, and human blood group O, Rh+ erythrocytes and to human buccal cells as compared to nonpiliated organisms (Punsalang and Sawer, 1973). Antiserum to partially purified pili inhibited both hemagglutination and epithelial cell adherence. This inhibition was removed only by adsorption of the antiserum with piliated gonococci, suggesting a role for pili in facilitating attachment to epithelial cells. Punsalang

Table 1 Hemagglutination Mediated by Pili

Organism	Characteristics	Agglutination of Human RBC[a]	Mannose Sensitivity	Reference
	Pili Size: 70–90 Å Diameter × 0.5–4.0 μm Length			
Neissera gonorrhoeae	Isolated pili will hemagglutinate	+++	No	Punsalang and Sawer, 1973; Waitkins, 1974; Koransky et al., 1975; Buchanan and Pearce, 1976; Pearce and Buchanan, 1978
Escherichia coli	Isolated pili will hemagglutinate	+	Yes	Duguid et al., 1955; Salit and Gotschlich, 1977a, Collier and Miranda, 1955
Enterotoxigenic *E. coli*	Colonization factor	+++	No	Evans et al., 1977; Orskov and Orskov, 1977
Other *Enterobacteriaceae* (*Shigella*, *Salmonella*, *Klebsiella*, *Serratia*)	Common pili	+	Yes	Duguid et al., 1966; Duguid, 1959; Duguid and Gillies, 1957, 1958; Old, 1972; Cowan et al., 1960
Vibrio cholerae	Not certain bacterial hemagglutination is due solely to pili; most inhibition by L-fucose and L-fucosides	+++	Partial	Jones, 1977; Tweedy et al., 1968; Jones and Freter, 1976

	Pili Size: 40–60 Å Diameter × 0.5–4.0 μm Length			
Proteus species		++	No	Duguid et al, 1966; Duguid and Gillies, 1957
Klebsiella, *Serratia*	"Thin-type" pili: hemagglutinate only tannic acid-treated RBCs	No	No	Duguid, 1959
Nonpathogenic *Neisseria*		No	No	Wistreich and Baker, 1971
Pseudonomas aeruginosa	Isolated hemagglutinins (mannose sensitive and resistant) reflect bacterial hemagglutination; unknown if they are pilus structures; most inhibition of mannose-resistant hemagglutinin by α-methyl-D-galactoside	+++	No/Yes	Bradley, 1972; Gilboa-Garber, 1972; Gilboa-Garber et al., 1977

[a] Hemagglutinating ability of bacteria for human erythrocytes: ++++, strong; ++, moderate; +, weak.

Table 2 Proposed Pili-Mediated Adherence to Host Tissue Correlating with Virulence in vivo

Organism	Characteristics	Possible Chemical Receptor	Reference
	70–90 Å × 0.5–4.0 μm Size Pili		
N. gonorrhoeae	Binding specificity of isolated pili reflects cellular and species specificity of localization *in vivo*	Similar to: Galβ1→3GalNAcβ1→4Gal	Brown et al., 1972; Buchanan and Gotschlich, 1973; Bumgarner and Finkelstein, 1973; Punsalang and Sawyer, 1973; Swanson, 1973; Swanson et al., 1975; James-Holmquest et al., 1974; James et al., 1976; Mårdh and Westrom, 1976; Ward et al., 1974; Tebbut et al., 1976; Johnson et al., 1977; Tramont, 1977; Pearce and Buchanan, 1978; Buchanan et al., 1978b
E. coli K-12	Presence of pili correlates with symptomatic infection and adherence to urinary tract epithelium	α-Methyl-D-mannoside	Salit and Gotschlich, 1977; Eden et al., 1976, 1977; Ofek et al., 1977
Enteropathogenic *E. coli*	Colonization factor provides specificity for small intestine and host specificity	Not D-mannose	Evans et al., 1977, 1978; Orskov and Orskov, 1977; McNeish et al., 1975

V. cholerae	Adherence by pili not confirmed; partially mannose-sensitive hemagglutinin may be distinct from L-fucose-sensitive adhesin	L-Fucose > D-mannose	Jones, 1977; Tweedy et al., 1968; Jones and Freter, 1976
Proteus mirabilus	Pili predominate during *in vivo* growth; enhance virulence; mediate adherence to urinary tract and renal epithelium; determine site of localization	Unknown	Silverblatt, 1974; Silverblatt and Ofek, 1976
Shigella flexneri	Pili mediate adherence to human intestinal epithelium	D-Mannose	Duguid and Gillies, 1957
	40–60 Å × 0.5–4.0 μm Size Pili		
Proteus mirabilis	40 Å pili may also contribute to adherence in pyelonephritis	Not mannose	Silverblatt, 1974
P. aeruginosa	Nature and function *in vivo* of the D-galactose and D-mannose-specific hemagglutinins not determined	α-Methyl-D-galactose, D-galactose, lactose, D-galNac, D-mannose	Gilboa-Garber, 1972; Gilboa-Garber et al., 1977

and Sawyer also found that a partially purified pili preparation could bind buccal epithelial cells and that this attachment was not blocked by simple sugars, including D-glucose, maltose, D-mannitol, dulcitol, D-sorbitol, raffinose, saccharose, lactose, D-fructose, D-galactose, D-mannose, and inulin. They also observed no inhibition of attachment of piliated gonococci at pH 7.4 to erythrocytes after treatment of the RBCs with trypsin (1 mg/ml) or neuraminidase (Punsalang and Sawer, 1973). The hemagglutination observations of Punsalang and Sawyer were confirmed, modified, and extended by Waitkins (1974), Koransky et al. (1975), and Buchanan and Pearce (1976). Swanson (1973) and Swanson et al. (1975a) noted that piliated gonococci attached more readily than nonpiliated organisms to human amnion tissue culture cells, human foreskin cells, and HeLa cells. James-Holmquest et al. (1974) and James et al. (1976) demonstrated enhanced attachment of piliated gonococci to human sperm. This increased attachment was blocked by antiserum to purified pili, suggesting that pili were responsible for the enhanced attachment. Mårdh and Westrom demonstrated that piliated organisms adhered more readily to human vaginal epithelial cells than nonpiliated gonococci (Mårdh and Westrom, 1976), and subsequently P. A. Mårdh, S. Colleen and B. Hovelius ("Attachment of Bacteria to Exfoliated Cells From the Urogenital Tract," submitted for publication) also noted enhanced attachment to uroepithelial cells by piliated gonococci. Ward et al. (1974) demonstrated enhanced attachment of piliated gonococci to fallopian tube epithelium as compared to nonpiliated organisms. Using scanning and transmission electron microscopy, they also were able to visualize pili anchoring gonococci to the membrane of the fallopian tube epithelial cell layer within 3 hr after perfusing the tube with gonococci. These findings suggested that pili were involved in the initial attachment of gonococci to human fallopian tube epithelial cells. Tebbutt et al. (1976) confirmed that piliated gonococci adhered better than nonpiliated organisms to fallopian tube mucosa, as well as to human endocervix and ectocervix. Interestingly, this same adherence advantage for piliated organisms was not present for guinea pig epithelial surfaces or for human bronchial mucosa, perhaps suggesting species and tissue specificity for the presence of the receptor for gonococcal pili. This species specificity was again recently confirmed by Johnson et al. (1977), who found rapid attachment of piliated gonococci to human fallopian tube mucosa but little, if any, attachment to oviducts from the rabbit, pig, or cow. Finally, antibodies to pili, as measured by their ability to inhibit attachment of piliated gonococci to human buccal cells, have been quantitated in human genital secretions following gonorrhea (Tramont, 1977). This antibody was IgG and IgA (11S) in immunoglobulin class, was quite specific for the infecting strains of gonococci, and persisted for at least a short period of time.

Pearce and Buchanan (1978) investigated the attachment of isolated gonococcal pili to human cells. They found that isolated pili attached in ten to forty-fold higher amounts to human cervical–vaginal or buccal mucosal cells, sperm, and fallopian tube mucosa (1–10 pili/μm^2, 10^4 pili binding sites per cervical–vaginal cell) than to human red blood cells (0.1 pilus/μm^2) or white blood cells (0.3 pilus/μm^2), as summarized in Table 3. The attachment of gonococcal pili to human cells demonstrated cellular specificity as seen from Table 3.

These data suggest that more binding sites for pili exist on the surface of cervical–vaginal, sperm, and buccal cells than on the other cells studied. This observation of the higher number of pili binding sites on cells that are histologically the most similar to the actual sites of human gonococcal infection is consistent with the hypothesis that pili-mediated attachment of gonococci to human cells may determine, in part, the sites of eventual gonococcal infection (Pearce and Buchnan, 1978).

Isolated pili attachment was pH ($3.5 < 4.5 > 5.5 > 7.5$) and temperature dependent (37°C > 20°C > 4°C). Tryptophan alterations produced inhibition of pili binding, as did heat (85°C × 1 hr) and antibody to pili (homologous > heterologous). Inhibition of attachment with low concentrations (< 1 μM) of ganglioside molecules or by treatment of human cells with selective exoglycosidases suggests that the structure for the pili receptor is similar to the underlined portion of the ganglioside structure: *Galβ1→3GalNAcβ1→4Gal*(NANA)β1→4Glc→ceramide (Buchanan et al. 1978b). Isolated pili were shown to be capable of producing hemagglutination as well (Buchanan and Pearce, 1976).

Many studies of the pili of Enterobacteriaceae have been reported and recently reviewed (Ottow, 1975; Jones, 1977). Pili that function as adhesive organelles in *Enterobacteriaceae* can be classified by their morphology

Table 3 Attachment of Gonococcal pili to Human Cells at 37°C, pH 7.4

Cell Tested	Pili/Cell[a]	Surface Area (μm^2)	Pili/μm^2
Buccal mucosal epithelium	25,000	5116	4.91
Cervical–vaginal epithelium	10,100	4056	2.48
Red blood cells	16	138	0.11
Polymorphonuclear leukocytes	68	237	0.29
Sperm	86	84	1.03
HeLa "M" cells	189	600	0.32
Fetal tonsil fibroblasts	824	1400	0.59

[a] Geometric mean values (for methods see Pearce and Buchanan, 1978).

(especially their diameter, as length is often variable), hemaglutinating (Table 1) and attachment (Table 2) ability, host cell species specificity, and whether facilitation of attachment by these pili is inhibited by D-mannose (mannose sensitive) or not (mannose resistant). Pili-mediated attachment of gonococci to human cells is mannose resistant (Buchanan et al., 1978b). This review primarily considers those pili in *Enterobacteriaceae* that are 70–90 Å in diameter and 0.5–4 μm long and that have been associated with virulence, attachment, and hemagglutinating functions. These pili have been given the name "common pili" and are classified as Type 1 by Duguid et al. (1966); many are mannose sensitive. Other pili with a smaller diameter (40–50 Å) have not been associated with enhanced attachment to human cells (Duguid, 1959), with the exception of *Proteus* pili (Type 4, Duguid classification (Duguid et al., 1966); see below). Examples of this "thin type" of pili are found on *Klebsiella* and *Serratia* (Type 3 pili, (Duguid et al., 1966)), in addition to common pili, and are mannose resistant but agglutinate only tannic acid-treated red cells (Duguid, 1959). Similar pili of this smaller diameter and mannose resistance have been found on nonpathogenic *Neisseria* (Wistreich and Baker, 1971), but their size and ability to agglutinate untreated, nonhuman erythrocytes may indicate a closer similarity to the smaller pili of *Proteus* (Type 4) described below. They are nearly uniform in failing to agglutinate human red cells (Wistreich and Baker, 1971) and no evidence exists to implicate their function in adherence to human cells.

Bacterial pili were first recognized on *Escherichia coli*, *Pseudomonas aeruginosa* (*pyocyanea*), and *Proteus mirabilis* in 1950 by Houwink and Van Iterson. The occurrence of these common pili on *E. coli* and the hemagglutination they mediate were defined in 1955 by Duguid et al. (1955), (Table 1). Subsequently, similar pili that facilitate hemagglutination were reported on *Shigella* (Duguid and Gillies, 1957), *Salmonella* (Duguid and Gillies, 1958; Duguid et al. 1966), *Klebsiella* and *Serratia* (Duguid and Gillies, 1958; Duguid, 1959). Pili hemagglutinins were observed for *Proteus* in 1958 (Duguid and Gillies, 1958), but they had a different spectrum of activity for untreated red cells. However, these pili agglutinated human erythrocytes and may contribute to the ability of *Proteus* to cause pyelonephritis (Silverblatt, 1974). Hoeniger (1965) found the width of these pili (about 40 Å) to be much smaller than that of the common pili of *Enterobacteriaceae*. Brinton (1967), and later, Silverblatt (1974) reported the additional occurrence on *Proteus* with 70 Å diameter pili that are indistinguishable in appearance from common pili. Significantly, these larger pili predominate during *in vivo* growth and appear to be the principal mediators of adherence during infection, as their presence significantly enhances virulence (Silverblatt, 1974). For *Enterobacteriaceae*, pili are distributed around the entire cell, whereas in *Pseudomonadacea*, pili most often have a polar or bipolar distribution and have retractile capabilities. (Houwink and Van Iterson, 1950; Bradley, 1972; Bradley and Pitt, 1974). Also, pili on *Pseudonomas aeruginosa* are thinner (about 60 Å; Bradley, 1972) than common pili. Pili have also been observed on some *Vibrio* species (Tweedy et al., 1968), although their pathogenic role is less well characterized. These pili are of similar dimensions (e.g., *Vibrio cholerae* pili are 72 Å in average diameter) to common pili and facilitate agglutination of human erythrocytes (Tweedy et al., 1968).

Most studies of pili function in *Enterobacteriaceae* have indicated that pili mediate hemagglutination and may facilitate attachment of these organisms to human cells. This hemagglutinin activity of *E. coli* has been conclusively demonstrated to result, at least in part, from pili, since purified pili alone were able to produce the hemagglutination (Salit and Gotschlich, 1977a). Monosaccharides and oligosaccharides containing α-linked D-mannose inhibit some to most pili-mediated hemagglutination of *E. coli* (Salit and Gotschlich, 1977a, Collier and Miranda, 1955), *Shigella* (Duguid and Gillies, 1958; Old, 1972) *Salmonella* (Duguid et al., 1966; Old, 1972), *Klebsiella*, and *Serratia* (Duguid, 1959; Cowarn et al., 1960). These substances do not inhibit a hemagglutinin of *Pseudomonas aeruginosa* Gilboa-Garber 1972) that has been partially purified and shown to possess an agglutinating capacity for human erythrocytes equivalent to intact *Pseudomonas* bacteria. Specific inhibition of this adhesive reaction is obtained with α-D-galactosides > D-galactose > lactose > *N*-acetyl-D-galactosamine (Gilboa-Garber, 1972). Recently, a second mannose-specific hemagglutinin was identified on a *P. aeruginosa* strain grown in the presence of acetylcholine or choline (Gilboa-Garber et al., 1977). Inhibition by D-mannose far exceeded that produced by α-methyl-D-mannoside. Both of these hemagglutinins are glycoproteins with estimated subunit molecular weights of 13,000–13,700 (galactose specific) and 11,000 (mannose specific). These sizes are conceivably in the correct range for pili subunits (Pearce and Buchnan, 1978; Brinton, 1967). However, the morphological characteristics of these hemagglutinins have not been determined, nor has an attachment role for them in the pathogenesis of infection been confirmed. Also, Type 4 pili of *Proteus* cause mannose-resistant hemagglutination (Duguid et al., 1966), but the mannose sensitivity of the adhesive properties of the 70 Å diameter *Proteus* pili is unknown. Thus the role of α-linked D-mannose-containing substances as potential receptors for *Proteus* adherence has not been clarified (Jones, 1977; Silverblatt, 1974). The hemagglutination produced by *Vibrio* species is inhibited by L-fucose (Jones and Freter, 1976) more than by D-mannose (Tweedy et al., 1968). The attachment of *Vibrio cholerae* to brush border epithelial cells is best inhibited by L-fucosides and L-fucose, and the specific attachment organelles and their receptors require further clarification. Also, many human-associated enteropathogenic *Escherichia coli* produce a colonization factor that is morphologically indistinguishable from common pili (Evans et al., 1977, 1978). However, the hemagglutination of human cells produced by the pili that constitute colonization factor is not inhibited by D-mannose (Evans et al., 1977; Orskov and Orskov, 1977), suggesting that the receptor for this virulence factor that facilitates host-specific intestinal attachment and colonization is different from the receptor for other pili of *E. coli*.

The inhibition of pili-mediated hemagglutination by D-mannose and related compounds was investigated in some depth by Old (1972). He demonstrated the importance of the α configuration at the C-1 position and of the hydroxyl positions at C-2, C-3, C-4, and C-6 for maximal inhibition of hemagglutination by piliated *Salmonella* or *Shigella*. The strongest inhibitors in this study were D-mannose, α-methyl-D-mannoside, 1,5-anhydromannitol, D-mannoheptulose, and α-D-mannose-1-phosphate. Two other compounds, D-fructose and yeast mannan, were moderately inhibitory at slightly higher concentrations (0.05–0.2% w/v). These results were comparable to those found by Salit and Gotschlich (1977a) for isolated *E. coli* common pili where the order of inhibition of hemagglutination was α-methyl-D-mannoside > D-mannose > yeast mannan > D-fructose.

Pili of nongonococcal gram-negative organisms also mediate adherence to host tissues, as summarized in Table 1. Mannose-sensitive pili facilitated attachment of *Shigella flexneri* to human colonic epithelium (Duguid and Gillies, 1958) and of strain K-12 of *E. coli* to monkey kidney cells in tissue culture (Salit and Gotschlich 1977b). *E. coli*, *Proteus*, *Klebsiella*, and *Pseudomonas* isolated from patients with urinary tract infections are usually piliated (Brinton, 1967). Further, *E. coli* isolated from patients with symptomatic urinary tract infection adhere in larger numbers to epithelial cells from the urinary tract than do organisms from patients with asymptomatic infection, and the presence of pili relates to this attachment ability (Eden et al., 1976, 1977). Also, these piliated organisms adhere more readily to human buccal mucosal cells (Ofek et al., 1977). These attachments are inhibited by α-methyl-D-mannoside for strains of K-12 *E. coli* (Salit and Gotschlich, 1977b; Ofek et al., 1977). Other pili of *E. coli*, termed colonization factor, appear important for enteropathogenic *E. coli* to selectively attach to the human small intestinal mucosal epithelium and cause diarrheal disease (McNeish et al., 1975, Evans et al., 1977, 1978). Attachment to intestinal epithelium was unaffected by 0.5% D-mannose (McNeisch et al.,

1975). These colonization factor pili cause hemagglutination that is not inhibited by D-mannose (Evans et al., 1977; Orskov and Orskov, 1977) and appear morphologically similar to pilus structures that facilitate selective colonization and neonatal diarrhea in piglets (K88 antigen: Jones and Rutter, 1972; Wilson and Hohmann, 1974; Nagy et al., 1977; Jones and Rutter, 1974; Hohmann and Wilson, 1975) or in calves and lambs (K99 antigen: Orskov et al., 1975; Burrows et al., 1976; Isaacson, 1977). These colonization-promoting antigens provide host specificity (Wilson and Hohmann, 1974; Orskov et al., 1975) and are mannose-resistant hemagglutinins (Jones and Rutter, 1974; Orskov et al., 1975; Burrows et al., 1976). In the rat model of pyelonephritis, Silverblatt (1974) determined that piliated (70 Å diameter) *Proteus* organisms more frequently produced infection than non-piliated bacilli, and electron micrographs demonstrated organisms that appeared to attach to epithelial cells of the renal pelvis by means of their pili. The role of *Proteus* pili in determining the site of localization in renal infection is further supported by evidence that they mediate attachment to rabbit urinary tract epithelial cells *in vitro* (Silverblatt and Ofek, 1976).

It is probable that structural differences among pili will explain their affinity for different cell surface receptors and differences in antigenicity. At present, no pili have been completely sequenced and only a few reports of the amino-terminal amino acid sequence structure are available. These reports indicate that pili from four different species whose attachment is not inhibited by mannose possess pili with similar amino-terminal amino acid sequences. However, the Type 1 pili of *E. coli*, whose attachment is inhibited by α-methyl-D-mannoside, possess a totally different amino-terminal amino acid sequence. These findings are summarized in Table 4.

The single most striking finding for the amino-terminal amino acid sequences given in Table 4 is the extreme conservation of sequence for pili from organisms quite different from gonococci. Each of four antigenically distinct gonococcal pili and the meningococcal pilus had identical sequences (Hermodson et al. 1978). Also a fifth gonococcal pilus type from strain P9 was obtained from Michael Ward in Southhampton, England and recently sequenced. The P9 pili were purified by a different method (Robertson et al., 1977) and they also had the same sequence (our unpublished results). The *Pseudomonas* pili had only four substitutions at positions 10 (Val for Ile), 13 (Ile for Val), 19 (Ile for Val), and 21 (Leu for Ile). The *Moraxella* pili had only two substitutions compared to *Neisseria* pili, and these were identical to the *Pseudomonas* pili substitutions as positions 13 and 19 (Table 4). Each of these substitutions can result from a single base change in DNA sequence, and none affect the extremely hydrophobic nature of this portion of the pilus. This highly conserved and hydrophobic sequence at the amino

Table 4 Pili Amino-Terminal Amino Acid Sequences

	1				5					10					15
Neisseria[a]	MePhe	Thr	Leu	Ile	Glu	Leu	Met	Ile	Val	Ile	Ala	Ile	Val	Gly	Ile
Pak[b]	MePhe	Thr	Leu	Ile	Glu	Leu	Met	Ile	Val	Val	Ala	Ile	Ile	Gly	Ile
Moraxela[c]	X	Thr	Leu	Ile	Glu	Leu	Met	Ile	Val	Ile	Ala	Ile	Ile	Gly	Ile
E. coli[d]	Ala	Ala	Thr	Thr	Val	Asn	Gly	Gly	Thr	Val	His	Phe	Lys	Gly	Glu
					20					25					30
Neisseria	Leu	Ala	Ala	Val	Ala	Leu	Pro	Ala	Tyr	Gln	Asp	Tyr	Thr	Ala	
Pak	Leu	Ala	Ala	Ile	Ala	Ile	Pro								
Moraxella	Leu	Ala	Ala	Ile	Ala	Leu	Pro	Ala	Tyr	Gln	Asp	Tyr	Ile	Ala	(Arg)
E. coli	Val	Val	Asn	Ala	Ala	?	Ala	Val	Asp						

[a] Amino-terminal amino acid sequence of antigenically distinct pili proteins from four different strains of *Neisseria gonorrhoeae* and from strain 13090 of *Neisseria meningitidis* (Hermodson et al., 1978).
[b] Amino acid sequence for pili protein from *Pseudomonas aeruginosa K* (Paranchych et al., 1978).
[c] Amino acid sequence of the pili protein of *Moraxella nonliquefaciens* by Froholm and Sletten (1977).
[d] Amino acid sequence of the pili protein of *E. coli* strain $B_{am}P+$, (Hermodson et al., 1978).

terminus of pili subunits from both near (meningococcus) and distant (*Moraxella* and *Pseudomonas*) relatives of gonococci suggests that this common sequence is important for pilus function or assembly.

An understanding of the macromolecular mechanisms of microorganism attachment to human cells may suggest ways to interrupt the attachment. For example, an excess of receptor for a given pilus might inhibit its binding to human cell surface receptors. An excess of α-methyl-D-mannoside might block attachment of *E. coli* to human cells, and its use in patients with indwelling urinary catheters might prevent *E. coli* urinary tract infections, providing mannose-sensitive pili mediate attachment for pathogenic *E. coli*. Alternatively, one might induce antibody by means of immunization that would inhibit pili attachment and prevent infection. Antiserum raised to purified pili in rabbits was effective at inhibiting attachment of isolated gonococal pili (Pearce and Buchanan, 1978) or of piliated gonococci (James-Holmquest et al., 1974) to human cells. Also, antibodies that inhibit attachment of *N. gonorrhoeae* are present in genital secretions of patients recovering from gonorrhea (Tramont, 1977). Thus immunization of humans with purified gonococcal pili (Brinton et al., 1978) may provide some protection against gonorrhea and such studies are currently underway. However, the antigenic variation of gonococcal pili is extensive (Buchanan and Pearce, 1976; Pearce and Buchanan, 1978; Brinton et al., 1978; Buchanan, 1975; Novatny, Turner, 1975; Buchanan, 1978), and inhibition of pili-mediated attachment is maximal only when antibody to the homologous pili serotype is used (Buchanan and Pearce, 1976; Pearce and Buchnan, 1978.). Thus the ultimate effectiveness of a pili vaccine to prevent attachment and perhaps infection is likely to depend on the extent of pili antigenic variation and/or the amount of attachment inhibition produced by heterologous antibody to pili antigenic determinants.

2.2 Invasion

Though not all gram-negative bacteria invade tissues to produce disease, many, such as meningococci, gonococci, *Salmonella*, *Shigella*, and *Pseudomonas* penetrate human epithelial cells as a usual occurrence in the course of infection. It is likely that outer membrane surface components or, perhaps, periplasmic molecules that diffuse through gaps in the outer membrane during infection influence the capacity of an organism to invade host tissues. Mechanisms of invasion are poorly understood, but some possibilities have been suggested. It has been reported that penetration of intestinal epithelial cells by *Shigella flexneri* 2a is dependent on an *O*-polysaccharide group antigenic determinant containing an *N*-acetyl-glucosamine-rhamnose-rhamnose repeat unit (Gremski et al., 1972). Only hybrids expressing chemically similar somatic antigens from *E. coli* donors retained the ability to invade. It is possible that this *Shigella* surface antigen reacts with lectin-like moieties on the intestinal cell surface, triggering the initiation of active phagocytosis of the *Shigella* organisms by the intestinal cell. Alternatively, the genetic transformation may have altered a separate component that is critical for invasion, and the change in *O*-polysaccharide antigen may be coincidental. *Shigella*, *Salmonella*, and *E. coli* all invade epithelial cells by an active phagocytic process (Takeuchi, 1967; Staley et al., 1969). Gonococci are actively phagocytized by epithelial cells such as human fallopian tube mucosa (Watt et al., 1976, Johnson et al., 1977). It has not been determined what stimulates the epithelial cell to phagocytize gonococci or other organisms. Pili might induce surface fluidity changes in the epithelial cells, triggering the phagocytic process. Alternatively, enzymes secreted from the organisms might remove hydrophilic groups (endopeptidases, exo and endoglycosidases) and the increased hydrophobicity might facilitate approximation of organisms and cell membrane leading to eventual invasion. Known enzymes produced by gram-negative bacteria that may enhance invasion are given in Table 5. Gonococci and meningococci

Table 5 Bacterial Enzymes That May Facilitate Invasion

Organism	Enzyme	Reference
N. gonorrhoeae, *N. meningitidis*	Alkaline endopeptidase	Chen and Buchanan, 1978
	L-Asparaginase	Chen and Buchanan, 1978
N. gonorrhoeae	Endoglycosidase	Apicella et al., 1978
	Phospholipase A, lysophospholipase	Wolf-Watz et al., 1975; Senff et al., 1976
E. coli	Two neutral serine proteases (cytoplasmic and periplasmic) Phospholipase A, lysophospholipase	Pacaud and Uriel, 1971; Pacaud and Richaud, 1975; Scandella and Kômberg, 1971; Nishijima et al., 1977
S. typhimurium	Phospholipase A, lysophospholipase	Osborn et al., 1972
Serratia species	Alkaline endopeptidase	Miyata et al., 1970
	Collagenase	McQuade and Crewther, 1969
P. mirabilis	Alkaline endopeptidase	Hampson et al., 1963
B. melaninogenicus	Collagenase	Haussmann and Kaufman, 1969
P. aeruginosa	Alkaline endopeptidase	Morihara, 1963
	Collagenase	Carrick and Berk, 1975
	Neutral elastase	Morihara et al., 1965

produce an endopeptidase and an asparaginase (Chen and Buchanan, 1978), as well as a probable endoglycosidase (Apicella et al., 1978), and these may participate in invasion. the *Neisseria* endopeptidases (Chen and Buchanan, 1978) may be representative of a type of enzyme distributed in gram-negative bacteria. Extracellular alkaline endopeptidases are also produced by *Proteus* (Hampson et al., 1963), *Serratia* (Miyata et al., 1970), and *Pseudomonas* (Morihara, 1963). A neutral elastase produced by *Pseudomonas aeruginosa* has also been purified (Morihara et al., 1965) that is distinct from the alkaline proteinase and has both elastolytic and proteolytic activity. *P. aeruginosa* strains that can produce both enzymes have been shown to be more pathogenic than those lacking these enzymes (Okada et al., 1976). *E. coli* (Pacaud and Richaud, 1975; Pacaud and Uriel, 1971), on the other hand, possess two distinct neutral serine proteases that are not extracellular. Protease I (Pacaud and Uriel, 1971) is periplasmically localized and chymotyrpsin-like in specificity, while protease II (Pcaud and Richaud, 1975) is cytoplasmic and resembles trypsin. Collagenase-like enzymes have also been reported for *Serratia marcescens* (McQuade and Crewther, 1969), *Pseudomonas aeruginosa* (Carrick and Berk, 1975), and *Bacteroides melaninogenicus* (Hausman and Kaufman, 1969), but the first enzyme hydrolyzes native collagen only slightly. *Salmonella* (Takeuchi, 1967) and *E. coli* (Staley et al., 1969) need only approach small intestine epithelium within a critical proximity of 350 and 250 Å, respectively, to trigger a sudden local degeneration of microvilli on brush border cells that they subsequently invade; the enzyme or other substance producing these lesions has not be characterized. *Pseudomonas aeruginosa* produces ulcerative corneal lesions in humans and rabbits and disrupts the stromal proteoglycan ground substance of the anterior chamber (Gray and Kreger, 1975). These lesions may result from the neutral proteinase with elastase activity mentioned above since this enzyme in highly purified form has reproduced a corneal ulcerative condition essentially identical to the lesions produced by the intact organism (Kreger and Gray, 1978). Gonococci also produce corneal invasion, and enzymes extracted from gonococci produce damage to human corneal tissue cultures (T. M. Buchanan, J. C. Newland, and K. C. S. Chen, unpublished data).

Calcium-dependent phospholipase A enzymes have been characterized in several gram-negative bacteria including *E. coli* Scandella and Kornberg, 1971), *Salmonella* (Osborn et al., 1972), and *N. gonorrhoeae* (Wolf-Watz et al., 1975; Senff et al., 1976). All are localized in the outer membrane. *Salmonella* (Osborn et al., 1972) and gonococci (Wolf-Watz et al., 1975) may also possess lysophospholipase activity. The *E. coli* phospholipase A_1 enzyme purified by Scandella and Kornberg (1971) has more recently been shown to also possess both A_1 and A_2 as well as lysophospholipase activities

(Nishijima et al., 1977). Phospholipases have not been correlated with virulence (e.g., avirulent phenotypic mutants of gonococci are unaltered in activity, Senff et al., 1976), but a function in pathogenesis can be speculated. It is known, for example, that phospholipase A and cleavage products from phospholipase A activity can affect mammalian membrane integrity. These effects include the release of cytoplasmic enzymes (Robinson and Wilkinson, 1973) and the activation of membrane-bound glycosyltransferases (Graham and Wood, 1974; Mookerjea and Yung, 1974). Bacterial phospholipases may prove helpful for the fusion of microbial and mammalian membranes or as a means to produce fluidity changes in the host membrane that result in stimulation of endocytosis by the mammalian cell with coincidental microbial invasion. Most or all of the above enzymes (Table 5) almost certainly are instrumental in microbial cell wall metabolism. What remains to be determined is the role, if any, that they play in invasion by microorganisms of human cells and tissues.

If the mechanisms of microbial invasion can be elucidated, this information may suggest ways to block the invasion and possibly prevent the infection.

2.3 Resistance to "Natural Immunity"

Natural host immunity might be defined as any condition capable of preventing or resisting infection by a pathogenic microorganism that is present despite no previous exposure of the host to that microorganism. These conditions at mucosal surfaces might include pH, lack of nutrients, lysozyme, cilia, fluid movement (e.g., peristalsis, micturition), and mucus secretions. Within tissues and the circulatory system, conditions comprising natural immunity might include phagocytic activity by macrophages and polymorphonuclear leukocytes, bactericidal activity resulting from direct complement activation by the microorganism, or complement activation by "cross-reactive" or "natural antibodies" formed initially against some organism other than the invading pathogen.

Components of the outer membrane of gram-negative bacteria are almost certainly important for the resistance to local natural immunity by pathogens. For example, the tight integrity of the outer membrane protects the organism against the penetration of lysozyme that would digest its peptidoglycan and protects against unfavorable pH or osmotic conditions. *E. coli* possess at least three outer membrane iron-carrier complex binding proteins that participate in independent high affinity systems (Braun, 1978) that allow these organisms to successfully compete for Fe^{3+} with communal organisms and mammalian cells. It is likely that similar iron-binding proteins, as well as molecules to facilitate the uptake and transport of other nutrients, are present within the outer membrane of most gram-negative pathogens. Attachment organelles of gram-negative bacteria help them to anchor to mammalian cell surfaces and thus avoid removal from the site of infection by local ciliary action or fluid movement. Mucus provides a protective coating over the eye, in the gastrointestinal tract, in the tracheobronchial tree, and in the female genital tract. Many gram-negative pathogens that infect these areas must first overcome the mucus barrier in order to reside and multiply on or within the host. Biochemically speaking, mucus is a glycoprotein containing more than 50% carbohydrate. There are hundreds of short (<25 monosaccharides) oligosaccharide chains usually linked through an *O*-glycosidic bond between *N*-acetyl galactosamine and a serine or threonine located on the protein backbone (Reid and Clamp, 1978). The freshly produced molecule has a molecular weight of several million and has oligosaccharide units usually of 8–10 monosaccharide residues linked to about every third amino acid of a protein backbone. These units project out from the backbone and maintain the extended nature of the molecule that contributes to the viscosity and barrier characteristics of mucus. These carbohydrate moieties also suffer considerable protection to the mucus against proteolytic enzymes. Thus microorganisms that penetrate a mucus barrier must either proteolytically cleave peptide stretches devoid of oligosaccharides or remove oligosaccharides by glycosidic enzymes. The most rapid cleavage of oligosaccharides would result from the production of endoglycosidases, and these enzymes are now being found in gram-negative pathogens (Apicella et al., 1978). It remains to be determined whether these enzymes are bound to the outer membrane or are confined to the periplasmic space in these pathogens.

2.4 Interference With Immune Response

Microorganisms may interfere with the human immune response by directly inhibiting lymphocyte responsiveness, using molecular mimicry, inducing immune tolerance, or evoking antigenic plasticity. Potential examples of these mechanisms are summarized in Table 6.

Direct inhibition of lymphocyte blastogenesis has been measured as decreased responses to phytohemagglutinin or Con A-stimulated lymphocyte blastogenesis produced by L-asparaginase from *E. coli* (Ohno and Hersh, 1970) and by gonocosin (Chen and Buchanan, 1978) produced by *N. gonorrhoeae* and *N. meningitidis*, an endopeptidase of narrow specificity cleaving primarily on the amino side of an Ala-Ala sequence. This gonocosin in highly purified form also reduced antigen-stimulated lympho-

Table 6 Proposed Mechanisms of Bacterial Interference with the Humoral Immune Response

Bacterial Product (Mediator)	Organism	Reference
A. Immunosuppression		
L-Asparaginase	*E. coli*	Ohno and Hersh, 1970; Schwartz, 1969; Chakrabarty and Friedman, 1970
	N. gonorrhoeae, *N. meningitidis*	Chen and Buchanan, 1978; T. M. Buchanan and K. C. S. Chen, unpublished observations
Iga protease	*N. gonorrhoeae*, *N. meningitidis*	Plaut et al., 1975, 1977; Mulks and Plaut, 1978
Peptidoglycan	Mycobacteria	Chedid et al., 1976
B. Molecular mimicry		
Sialic acid homopolymer	*E. coli* K-1	Artenstein et al., 1973
(capsular polysaccharide)	*N. meningitidis* Group B	Artenstein et al., 1973; Gotschlich, 1973; Gold et al., 1977
Blood group activity (*O*-polysaccharide)	*Enterobacteriaceae*	Robinson et al., 1971; Springer et al., 1961; Springer, 1973
	E. coli (urinary infections)	Cruz-Coke et al., 1965
	N. gonorrhoeae	Foster and Labrum, 1976
C. Toleragenicity		
Capsular polysaccaride	*N. meningitidis*	Gold et al., 1977
	S. pneumoniae	Felton et al., 1955; Baker et al., 1971
D. Antigenic variation		
Superficial heat-labile antigen (e.g., microcapsule)	*C. fetus*	Corbiel et al., 1975
Pilus	*N. gonorrhoeae*	Tramont, 1977; Tramont et al., 1978

cyte blastogenesis (T. M. Buchanan and K. C. S. Chen, unpublished data), as has been demonstrated for *E. coli* L-asparaginase (Ohno and Hersh, 1970). *In vivo* studies have further shown a severe depression of antigen-stimulated immune responses by *E. coli* L-asparaginase as measured by splenic antibody plaque-forming cells (Schwartz, 1969; Chakrabarty and Friedman, 1970) and serum antibody levels (Chakrabarty and Friedman, 1970). It is yet to be determined whether these lymphocyte active enzymes, are associated with the outer membrane of the organism or are primarily confined to the periplasmic space. However, during natural infection the outer membrane of gonococci, and perhaps other organisms, is disrupted, and it is likely that periplasmic enzymes become accessible to host cells and tissues. Another enzyme produced by *N. gonorrhoeae* and *N. meningitidis* acts directly on an immunologlobulin product, IgA. This IgA protease cleaves human serum and secretory IgA immunoglobulins of the IgA1 subclass, into Fabα and Fcα fragments (Plaut et al., 1975). It has been proposed that this enzyme might enhance the pathogenicity of meningococci and gonococci by decreasing the antibody activity of IgA in human secretions (Plaut et al., 1977). Possible mechanisms of IgA protective activity might include inhibition of attachment, agglutination of organisms, or some other as yet undefined protective effect. Involvement of this enzyme in pathogenesis is supported by a recent study (Mulks and Plaut, 1978) in which 8 different nonpathogenic *Neisseria* species lacked the IgA protease that was found in all of 60 clinical isolates of gonococci and meningococci. The precise pathogenic role of the IgA protease requires further study since IgA2 subclass immunoglobulins may represent a significant percentage of the IgA in local secretions, and since Fab fragments would be expected to retain some antigen-binding activity.

The research of Chedid and others (Chedid et al., 1976) has elucidated the adjuvant activity of the compound *N*-acetylmuramyl-L-alanyl-D-isoglutamine of the peptidoglycan of mycobacteria. These workers have also shown that slight alterations of this sequence, for example, substituting D-alanine for L-alanine, produces immunosuppression rather than an enhanced immune response. It is possible that some bacteria possess peptidoglycan structures that are immunosuppressive and that these structures facilitate their ability to persist in the infected host. The peptidoglycan is closely associated in some areas with the outer membrane of gram-negative organisms, and during natural infection the infected host is very likely to be exposed to peptidogylcan fragments of the infecting organism.

Microorganisms may also interfere with the human immune response by molecular mimicry (Table 6). This term refers to organisms that possess surface antigens that are identical to, or sufficiently similar to host antigens that they cannot be recognized by the host immune system. Two gram-negative organisms that cause a disproportionately high amount of bacteremia and meningitis are *Escherichia coli* K-1 and Group B *Neisseria meningitidis*. Approximately 84% of all *E. coli* isolated from neonates with meningitis is of the K-1 antigenic type (Robins et al., 1974; Glode et al., 1977). Group B meningococci are a dominant cause of meningococcal meningitis and meningococcal bacteremia, primarily in adults in the United States. The capsules of both the *E. coli* K-1 and Group B meningococci are homopolymers of sialic acid (Artenstein et al., 1973) and are antigenically cross-reactive (Artenstein et al., 1973), and the sialic acid residues are $\alpha 2 \rightarrow 8$ linked (Bhattacharjee et al., 1975; Jennings et al., 1977). Group B meningococcal capsular polysaccharide is a poor immunogen and thus far has failed as a basis for vaccination against Group B meningococcal meningitis (Gotschlich, 1973; Gold et al., 1977). This failure to respond to this homopolymer of sialic acid may be due to the close antigenic similarity of organisms coated with this capsule and human cells that also contain terminal sialic acids. In many human cells this sialylation covers other sugars that are recognized by lectin-like receptors on liver or spleen cells, and desialylation results in removal of these cells from the circulation (Aminoff et al., 1977; Ashwell and Van Lenten, 1972). Many gram-negative organisms have been reported to contain *O*-polysaccharide side chain or other antigens with blood group activity (Robinson et al., 1971; Springer et al., 1961; Springer 1973). These organisms might conceivably be more pathogenic for individuals of the same blood group (which they mimic) and such an association has been reported for urinary pathogens (Cruz-Coke et al, 1965) and gonococcal infections (Foster and Labrum, 1976). *Neisseria gonorrhoeae* have galactose as a dominant antigenic determinant of the core oligosaccharide (Perry et al., 1978) and the association of higher infection rates in blood group B patients (Foster and Labrum, 1976) who have terminal galactose antigenic dterminants in their blood group antigen might be related to the lack of anti-blood group B antibodies in these patients. However, other investigators (Miller et al., 1977) have not confirmed the greater prevalence of gonococcal infections in persons with blood group B, so further investigations are needed.

Some gram-negative microorganisms produce external membrane or capsular products that circulate in large antigenic excess during the infection (Table 6). This is particularly true for poorly biodegradable high molecular weight polysaccharides, such as the capsular substances of the pneumococcus or *Coccidioides*. The phenomenon of "immunological paralysis" or high zone tolerance, a condition of a poor or absent immune response following immunization with high amounts of polysaccharide antigen, is easily demonstrable for mice immunized with pneumococcal polysaccharides (Felton et al., 1955; Baker et al., 1971). Similarily, meningococcal polysaccharide vaccine produced lower antibody levels following a booster immunization than after a single injection, suggesting a toleragenic effect, when given to human infants (Gold et al., 1977). For these reasons, vaccine doses for the currently successful meningococcal and pneumonococcal capsular polysaccharide vaccines are limited to 50 μg per capsular serotype.

Another mechanism by which microorganisms interfere with the human immune response is antigenic plasticity (Table 6). This refers to the observation that microorganisms, particularly trypanosomes, *Plasmodium malariae*, *Borrelia*, and schistosomes frequently alter their surface antigenicity, at a rate rapid enough to avoid elimination by the host immune system. Some gram-negative bacteria also develop different surface antigenicity. *Campylobacter* (*Vibrio*) *fetus* (which produces endometritis and sterility in cattle) and gonococci (which may produce endometritis, salpingitis and sterility in humans) may alter their surface antigenicity, allowing persistent infection despite development of some measurable immunity. The surface antigenicity of *C. fetus* has been observed to vary during the chronic 6 month endometrial infection characteristic in cattle (Corbiel et al., 1975), and recurrent gonococcal infection in humans has been observed with gonocci that possess apparently identical antigens as the strain causing the initial infection except from antigenically different pili (Tramont, 1977; Tramont et al., 1978). Thus gonococci with different pili antigens are still able to attach to human cervical–vaginal cells and to resist phagocytosis (Buchanan et al., 1978a).

2.5 Resistance to Phagocytosis, Interaction With Opsonins

Several outer membrane components facilitate the resistance of gram-negative bacteria to phagocytosis, including pili, capsules, Vi antigens, and long *O*-polysaccharide chains of lipopolysaccharide as summarized in Table 7. In most instances, antibody to these components is opsonic.

Whether pili help a given organism resist phagocytosis may be related to the density of the receptor on phagocytes for the pili of that bacterium or to other effects such as phagocyte membrane fluidity alteration. The density of gonococcal pili binding sites ("receptors") is low on human polymorphonuclear leukocytes as compared to cervical–vaginal cells, buccal cells, and sperm (Pearce and Buchanan, 1978), and piliated gonococci resist phagocytosis (Jones and Buchanan, 1978). In contrast, Silverblatt and Ofek reported that piliated *Proteus mirabilis* organisms were more readily ingested by a monolayer of human granulocytes than were nonpiliated

Table 7 Outer Membrane Structures That Facilitate Resistance to Phagocytosis

Component	Example	Antibody to Component is Opsonic	Reference
Pilus	*N. gonorrhoeae*	Yes	Punsalang and Sawyer, 1973; Buchanan et al., 1978a; Jones and Buchanan, 1978; Thonghai and Sawyer, 1973; Thomas et al., 1973; Ofek et al., 1974; Dilworth et al., 1975; Gibbs and Roberts, 1975; Blake and Swanson, 1975; Swanson et al., 1975b; Brooke et al., 1976; Witt et al., 1976
Capsular (K) polysaccharide	*E. coli*	Yes	Howard and Glynn, 1971
	P. aeruginosa	Yes	Schwarzmann and Boring, 1971
	B. fragilis subspecies fragilis	Unknown	Onderdonk et al., 1977
	N. gonorrhoeae	Unknown	Richardson and Sadoff, 1977
	N. menigitidis	Yes	Roberts, 1970
Vi polysaccharide	*S. typhi*	Yes	Felix and Bhatnagar, 1935
O-Polysaccharide (LPS)	*S. typhimurium*	Yes	Jann and Westphal, 1975; Biozzi et al., 1963

Note: The Vi antigen of *Salmonella typhi* is also a polysaccharide and has been chemically characterized as a repeating polymer of NAc D-galactosamineuronic acid. This antigen is located on the cell surface protruding from the outer membrane and is antiphagocytic (Felix and Bhatnagar, 1935). Antibody to the Vi antigen would be expected to be opsonic.

organisms (Silverblatt and Ofek, 1978) and that lightly piliated organisms persisted and multiplied within rat kidneys after intravenous injection, whereas heavily piliated *Proteus* organisms that reached the kidney in equal numbers were more readily eliminated over the same time period (Silverblatt and Ofek, 1978). Many investigators have observed that piliated gonococci are more resistant to phagocytosis than nonpiliated organisms. Thonghai and Sawyer (1973), Punsalang and Sawyer (1973), and Thomas et al. (1973) first reported this in 1973. Subsequently, these observations were extended and confirmed in several animal systems, including humans, by Ofek et al. (1974), Dilworth et al. (1975), Gibbs and Roberts (1975), Blake and Swanson (1975), Swanson et al. (1975b), Brooks et al. (1976), Witt et al. (1976), and Jones and Buchanan (1978). Using a quantitative assay for phagocytosis (Jones and Buchanan, 1978) and antisera to purified gonococcal surface components whose antibody levels to pili principal outer membrane protein on LPS had been quantitated, Jones and Buchanan investigated which surface antigens of gonococci stimulate opsonins (Buchanan et al., 1978a). They found that the major surface component to stimulate opsonins was pili. Even more striking, an affinity column containing covalently bound purified gonococcal pili effectively removed nearly all opsonic antibodies from an antiserum to whole organisms, whereas purified antibody to pili eluted with 2*M* NaI from the same washed column was highly opsonic (Table 8, Buchanan et al., 1978a).

The capsules of many gram-negative bacteria have been shown to confer resistance to phagocytosis. It has been speculated that this antiphagocytic effect may be related to the hydrophobicity, charge, and visiosity properties of the polysaccharides comprising capsules (Schwarzmann and Boring, 1971). Howard and Glynn (1971) showed a direct relationship between the quantity of K (capsular) antigen on *E. coli* and their resistance to phagocytosis. Different species of *Enterobacteriaceae* are routinely more virulent for animal models if they are encapsulated, and some of this virulence is thought to be related to the resistance to phagocytosis provided by capsular antigens (Jann and Westphal, 1975). The virulence of *Bacteroides* appears closely related to the possession of a capsule. *Bacteroides* are the obligate anaerobes most frequently isolated from clinical infections (Gorbach et al., 1974). Up to 78% of all *Bacteroides* isolated from the blood stream are *Bacteroides fragilis* (Wilson et al., 1972). There are five subspecies of

Table 8 Inhibition of Opsonization and Adsorption of Opsonic Antibody With Homologous Pili (strain F62 of N. gonorrhoeae)

Antiserum	Pili (μg)	Phagocytosis[a]
Phosphate-buffered saline	—	0.136 ± 0.010
Anti-F62	—	0.836 ± 0.045
Anti-F62	150	0.396 ± 0.051
Anti-F62 (adsorbed affinity column)	—	0.178 ± 0.038
Purified antibody to F62 pili (from affinity column)	—	1.014 ± 0.215

[a] Average number of colony-forming units per macrophage. Mean plus or minus standard deviation.

Bacteroides fragilis. One, *Bacteroides fragilis* subspecies fragilis, accounts for few of the fecal isolates of *Bacteroides*, but for most of the systemic isolates from patients with infections (Wilson et al., 1972). Strains of this subspecies fragilis are encapsulated, whereas strains of each of the other subspecies are nonencapsulated (Kasper et al., 1977). In the rat model of intraabdominal sepsis, only encapsulated *Bacteroides* routinely produced abscesses. Alternatively, implantation of 200 μg of capsular polysaccharide alone or in conjunction with unencapsulated strains also produced abscesses (Onderdonk et al., 1977). This abscess-promoting ability of the *Bacteroides* capsule was not reproduced by comparable amounts of *E. coli* 07:K1(L):NM capsular polysaccharide (Onderdonk et al., 1977). Gonococci have been reported to contain capsules as demonstrable by India ink exclusion (Hendley et al., 1977; Richardson and Sadoff, 1977; James and Swanson, 1977) and by electron microscopic examination with alcian blue staining (Demarco de Hormalche and Thornley, 1978), and encapsulated gonococci were more resistant to phagocytosis than nonencapsulated organisms (Richardson and Sadoff, 1977). Antibody to capsule usually overcomes the antiphagocytic effect and is opsonic. For example, antibody to purified capsular polysaccharide from Group A or Group C meningococci is opsonic for each of these groups of organisms, respectively (Roberts, 1970).

Long *O*-polysaccharide chains on the LPS of gram-negative bacteria (smoothness) permit more prolonged survival of organisms injected intravenously, and some of this survival may be related to the greater resistance to phagocytosis of smooth as compared to rough organisms (Jann and Westphal, 1975). Antibody to O-antigens is often opsonic, as for example the immune-enhancement of phagocytosis produced by antibody to the O-antigen of *Salmonella typhimurium* (Biozzi et al., 1963).

2.6 Resistance to Intraphagocytic Killing

There are many reports of the survival and, in some instances, multiplication of some bacteria within phagocytes. These organisms include *M. tuberculosis*, *M. leprae*, *Listeria monocytogenes*, *Brucella abortus*, *Salmonella typhi*, *Toxoplasma gondii*, and *Chlamydia psittaci* (Table 9) (Wyrick and Brownridge, 1978). Unfortunately, the mechanisms by which these organisms are able to resist intracellular killing are poorly understood. However, it seems likely that surface or outer membrane components of these organisms are responsible for this intraphagocytic survival, since antibody-coated organisms are often killed and digested normally (Wyrick and Brownridge, 1978; Jones, 1975). The resistance of *Mycobacteria* to intracellular killing may relate to failure of fusion of lysosomes with the

Table 9 Examples of Resistance to Intraphagocytic Killing

Organism	Mechanism	Component Responsible	Reference
M. tuberculosis	Failure of lysosome–phagosome fusion	Unknown	Armstrong and Hart, 1971
Mycobacteria	Resistance to digestion by lysosomal enzymes	Cell surface waxes, hydrophobic mycosides	Imada et al., 1969
C. psittaci	Failure of lysosome–phagosome fusion: antibody protection	Cell surface proteins	Wyrick and Brownridge, 1978; Moulder et al., 1976
T. gondii	Failure of lysosome–phagosome fusion; antibody removes protection	Cell surface proteins	Jones, 1975; Jones and Hirsch, 1972

phagosomes (Armstrong and Hart, 1971), plus protection from cell surfaces waxes, including hydrophobic mycosides (Imada et al., 1969), that provide resistance to digestion by lysosomal enzymes. Both *Chlamydia* and *Toxoplasma* have also been shown to prevent fusion of lysosomes and phagosomes, and this property is likely to be dependent on cell surface proteins (Moulder et al., 1976; Jones and Hirsch, 1972).

2.7 Resistance to or Interaction With Antibody–Complement Mediated Bactericidal Activity

Resistance to serum killing by gram-negative bacteria may be associated with encapsulation, long *O*-polysaccharide side chains of LPS, or occasionally with specific outer membrane proteins. Examples of these resistance conferring components are given in Table 10. Often, but not always, antibody to the surface or outer membrane component associated with serum resistance results in enhanced bactericidal activity when complement is present. In some instances, particularly with encapsulation, bactericidal activity does not result from antibody–antigen complement complexes, presumably because the distance from the capsule surface to the outer membrane is too great so that a lethal lesion does not result.

A relationship between virulence and resistance to bactericidal activity has been noted for many gram-negative organisms (Rowley, 1954; Muschel,

Table 10 Examples of Resistance to the Antibody–Complement Mediated Bactericidal Reaction

Bacterial Mediator of Resistance	Organism	Reference
Capsular polysaccharide	*E. coli*	Glynn and Howard, 1970; Glynn et al., 1971; Schiffer et al., 1976
	N. meningitidis Groups A and C	Gotschlich et al., 1969
O-Polysaccharide (smooth LPS)	*S. typhimurium*	Reynolds and Pruul, 1971
Outer membrane protein	*N. gonorrhoeae*	Hildebrandt and Buchanan, 1978; Shubin and Weil, 1963

1960). Roantree and Rantz (1960) and Roantree and Pappas (1960) found that bacteremic strains of enteric bacilli were almost always serum resistant, but the resistance to bactericidal activity was much less common in strains isolated from stool or from the urine. More recently, this same higher frequency of serum resistance in strains isolated from the blood stream as compared to these from stool, urine, or mucosal surface isolates has been confirmed for *E. coli* (Vosti and Randall, 1970), *Pseudomonas aeruginosa* (Young and Armstrong, 1972), and *N. gonorrhoeae* (Schoolnick et al., 1976).

For *E. coli*, serum resistance was found to correlate directly with the quantities of K (capsular) antigen (Glynn and Howard, 1970; Table 10). Organisms causing pyelonephritis were encapsulated more often than bacteria isolated from the lower urinary tract (Glynn et al., 1971). Recently, a specific capsular polysaccharide, the K-1 antigen, has been recognized as the polysaccharide most frequently present on invasive *E. coli* (Schiffer et al., 1976). Antibody to specific capsular polysaccharides may be bactericidal, and this is a basis for the successful pure capsule vaccines to prevent meningitis due to Group A and Group C meningococci (Gotschlich et al., 1969).

Strains of *Enterobacteriaceae* with long *O*-polysaccharide side chains on their LPS ("smooth strains") are generally more resistant to killing by normal human serum than are rough bacteria (Reynolds and Pruul, 1971).

The resistance of *N. gonorrhoeae* to killing has been correlated with possession of a principal outer membrane protein (POMP) of specific antigenicity and molecular weight (Hildebrandt and Buchanan, 1978). Seventy-five (88%) of 85 serum-resistant strains isolated from patients with disseminated gonococcal infection (DGI) had a single POMP antigenic type (1 of approximately 10 POMP serotypes) when tested for POMP antigen in an enzyme-linked immunosorbent assay (ELISA) (Hildebrandt and Buchanan, 1978). Only 10 (33%) of 33 serum-sensitive strains of *N. gonorrhoeae* possessed this same POMP antigenic type. Each of 20 serum-resistant DGI strains with the characteristic POMP antigenic serotype had a POMP subunit molecular weight of approximately 36,500 as compared to POMP molecular weights of 36,000–39,000 for serum-sensitive strains when examined by SDS–polyacrylamide slab gel electrophoresis (Hildebrandt and Buchanan, 1978; Hildebrandt et al., 1978). Finally, when serum-sensitive gonococci were transformed to serum resistance using DNA from serum-resistant donors, each of 10 serum-resistant transformants acquired a POMP with a subunit molecular weight of 36,500, and with POMP antigenicity characteristic for DGI strains (Hildebrandt et al., 1978). Subsequently, the LPS serotype of these strains has been tested and the donor LPS serotypes differed from those of the recipients; the serum-resistant transformants retained the LPS serotype of the recipient, indicating that LPS was not transformed at the same time as the POMP in these 10 transformants (T. M. Buchanan, unpublished data).

2.8 Toxin Production

The most potent bacterial toxins are produced by gram-positive bacteria (*Clostridium tetani*, *C. diphtheriae*, *C. botulinum*, *Bacillus anthracis*), and these undoubtably contribute significantly to the pathogenesis of the diseases that they cause. Some gram-negative bacteria also produce exotoxins, and these include *Shigella dysenteriae*, *Pseudomonas aeruginosa*, *V. cholerae*, *E. coli*, *Pasteurella pestis* and *Bordetella pertussis* (Table 11). Most of these are released from the bacterium during the course of infection and contribute to disease progression. A further toxin, endotoxin or LPS, is found in nearly all gram-negative bacteria and, when released into the human bloodstream, produces profound physiological effects of "septic shock" characterized by fever, hypotension, and peripheral vasodilation (Shubin and Weil, 1963).

The exotoxin produced by *Shigella dysenteriae* is comparable in potency to tetanus toxin and is the most potent of the toxins produced by gram-negative bacteria. It is a heat labile protein that has a subunit molecular weight of 75,000 (Keusch and Grady, 1972), and the cell surface receptor for this toxin appears to be oligosaccharides of $\beta 1 \rightarrow 4$ linked NAc glucosamine (Keusch and Jacewicz, 1977). The precise mechanism of action is

Table 11 Toxins of Gram-Negative Bacteria[a]

Microorganism and Toxin	Toxin Properties	Receptor Properties	Mechanism of Action
S. dysenteriae/enterotoxin	Protein, heat labile; mol. wt. 75,000 (Keusch and Grady, 1972)	Oligosaccharides of $\beta 1 \rightarrow 4$ NAc glucosamine (Keusch and Jacewicz, 1977)	Unknown, induces fluid loss from intestine and vascular endothelial damage in brain (Mims, 1977)
P. aeruginosa/exotoxin	Protein, heat labile (56°C × 10 min) (Callahan, 1974); subunit mol. wt. a fragment-27,000, b fragment 45,000; pI = 5.0 (Vasil et al., 1977)	?	Cytotoxic, inhibits protein synthesis in a manner similar to diphtheria toxin, ADP-ribosylates the same amino acid of elongation factor 2 (Iglewski et al., 1977)
V. cholerae/cholera toxin	B subunits, protein, 8000 daltons, 6/molecule, binds to receptor (Cuatrecasas et al., 1973) A subunit, protein, 30,000 daltons, activates adenyl cyclase (Finkelstein, 1975; Lönnroth and Holmgren, 1973)	GM_1 ganglioside (Holmgren et al., 1973; Cuatrecasas, 1973; Van Heyningen, 1974)	Activated adenyl cyclase: ↑ cyclic AMP → net secretion of Cl^- into bowel lumen and ↓ Na^+ absorption (Field, 1971; Kimberg et al., 1971)
E. coli/enterotoxin	Protein heat labile (Gyles, and Barnum 1969) (65°C × 30 min); subunit mol. wt. 102,000, pI = 6.9 (Donner, 1975)	GM_1 ganglioside (Finkelstein, 1975)	Activated adenyl cyclase: ↑ cyclic AMP → loss of fluid + electrolytes into bowel (Finkelstein, 1975)

E. coli/enterotoxin	Heat stable (Kohler, 1968; Sack et al., 1975), (65°C × 30 min) may be heat labile toxin complexed with endotoxin (Smith and Gyles, 1970)	Probably GM_1 ganglioside	Same as above + LPS effects
Yersinia pestis/plague toxi	Protein, subunit mol. wt. 24,000 (two chains with mol. wt. of 12,000) (Montie et al., 1975; Montie and Montie, 1971, 1973)	?	Inhibits O_2 uptake, NADH-Coenz Q-reductase activity blocked in electron transport system (Kadis and Ajl, 1970)
B. pertussis/whooping cou; toxin	Protein, heat labile	?	Slows cilia action in respiratory tract, dermal necrosis in rabbits
Gram negative/endotoxin	Lipopolysaccharide (Lüderitz, 1977), lipid A, core oligosaccharide, and *O*-polysaccharide, heat stable	?	Endogenous pyrogen, noradrenalin, serotonin, and histamine release (Guenter et al., 1969; Gilbert, 1960), most biologic effects 2° to lipid A (Nowotny, 1977; Galanos et al., 1977)

[a] Listed in approximate order of the potency of each toxin.

unknown, but the toxin acts to induce profound fluid loss from the intestine and vascular endothelial damage in the brain. The exotoxin of *Pseudomonas aeruginosa* is also quite potent. It is a protein that is inactivated at 56°C for 10 min in the purified form (Callahan, 1974), and the active portion (fragment a) has a subunit molecular weight of 27,000 (Vasil et al., 1977). Its fragment b has a molecular weight of 45,000 and the toxin has a *p*I of 5.0 (Vasil et al., 1977). The receptor for this toxin is unknown. The toxin is cytotoxic and its mechanisms of action is similar to that of diphtheria toxin. It inhibits protein synthesis by adenosine-5′-diphosphate-ribosylation of the same amino acid in elongation factor 2 that is affected by diphtheria toxin (Iglewski et al., 1977). The role of cholera toxin in cholera is one of the best studied exampled of bacterial toxins in pathogenesis. Cholera toxin is a protein comprised of A and B subunits. The B or binding subunits have a subunit molecular weight of approximatley 8000, and approximately six of these subunits aggregate to form a complex that allows binding of the toxin to the surface of intestinal cells (Cuatrecasas et al., 1973; Finkelstein, 1975). The A subunit has a molecular weight of approximately 30,000 and there is one A subunit for each six B subunits in the functioning toxin (Lönnroth and Holmgren, 1973). The functioning toxin binds specifically to GM_1 ganglioside receptors on the cell surface by means of its B subunits (Holmgren et al., 1973; Cuatrecasas, 1973; Van Heyningen, 1974). After it is bound, the A subunit is inserted into the cell membrane, resulting in activation of adenyl cyclase, which leads to increased levels of cyclic AMP and a net secretion of chloride ion into the bowel lumen and a net reduction in the absorption of sodium (Field, 1971; Kimberg et al., 1971). The exotoxin of *Escherichia coli* resembles cholera toxin in many ways, including antigenic cross-reactivity, binding to GM_1 ganglioside, and activation of adenyl cyclase leading to increased cyclic AMP and fluid accumulation in the bowel lumen (Finkelstein, 1975). However, in other ways it is quite different. The heat-labile protein toxin (Gyles and Barnum, 1969) is destroyed by 65°C treatment for 30 min and has a subunit molecular weight of 102,000, with a *p*I of 6.9 (Donner, 1975). A heat-stable toxin of *E. coli* has been described by several investigators (Kohler, 1968; Sack et al., 1975). Since all the other exotoxins of gram-negative bacteria are heat-labile proteins this would be unusual. An explanation for this discrepancy is that the toxin is the same in each case, but the heat stability is found in impure preparations and is contributed by a complexing of the heat-labile toxin with heat-stable contaminating endotoxin rendering the whole complex heat stable (Donner, 1975; Smith and Gyles, 1970).

The plague toxin produced by *Yersinia pestis* is a protein with a subunit molecular weight of 24,000, comprised of two chains of 12,000 daltons each

(Montie et al., 1975; Montie and Montie 1971, 1973). The receptor for the plague toxin has not been characterized, but its action is to inhibit cellular O_2 uptake. In one study the site of action that produced this effect was a block in the electron transport chain at the site of reduced nicotinamide adenine dinucleotide-coenzyme Q reductase activity (Kadis and Ajl, 1970). The whooping cough toxin produced by *B. pertussis* is a heat-labile protein. Its cell surface receptor(s) has not been chemically characterized, and it produces dermal necrosis in rabbits and impaired ciliary activity in cells of the respiratory tract (Munoz and Bergman, 1977).

The endotoxin of gram-negative bacteria is the lipopolysaccharide (LPS) that constitutes a major portion of the outer membrane. The "septic shock" state produced in man during gram-negative bacteremia can be reproduced in monkeys by injection of isolated, purified endotoxin (Gaenter et al, 1969). This shock state probably arises from an effect on the vascular system mediated through endogenous pharmacological agents such as noradrenalin, histamine, and serotonin, whose release has been stimulated by LPS (Gilbert, 1960). The molecule of LPS is comprised of three parts: an outer *O*-polysaccharide chain, an intermediate core oligosaccharide, and an inner Lipid A moiety that is imbedded into the lipid layer of the outer membrane (Lüderitz, 1977). Most of the biologic effects of LPS require the lipid A moiety, though antigencity is mediated through the *O*-polysaccharide and, to some extent, through the core oligasaccharide (Nowotny, 1977; Galanos et al., 1977).

3 CONCLUSIONS

The era of understanding the pathogenesis of infectious diseases at a molecular and macromolecular level is just beginning. This chapter reviews some aspects of how surface components of gram-negative bacteria contribute to the pathogenesis of infections. Individual components facilitate attachment, invasion, resistance to natural immunity, interference with the host immune response, resistance to phagocytosis or interaction with opsonins, resistance to intraphagocytic killing, resistance to or interaction with antibody–complement mediated bactericidal activity, and host tissue damage by means of toxin production. These surface components that facilitate the pathogenesis of infection do so usually by specific and precise interactions with specific host cell surface or other molecules, and these interactions are beginning to be understood biochemically. It is expected that in the next 10 years major advances will occur toward better understanding the biochemical and molecular basis of the interactions between bacteria and humans that lead to infection. Hopefully, a clearer understanding of the molecular basis for the pathogenesis of infections will suggest improved methods for the diagnosis and prevention of infectious diseases.

REFERENCES

Aminoff, D., Vorder Bruegge, W. F., Bell, W. C., Sarpolis, K., and Williams, R. (1977), *Proc. Natl. Acad. Sci. U.S.*, **74,** 1521–1524.

Apicella, M. A., Breen, J. F., and Gagliardi, N. C. (1978), in *Immunobiology of* Neisseria gonorrhoeae (GF Brooks et al., eds.), American Society for Microbiology, Washington, D.C., pp. 108–112.

Armstrong, J. A., and Hart, P. D., (1971), *J. Exp. Med.*, **134,** 713–740.

Artenstein, M. S., Kasper, D. L., Brandt, P. L., and Zollinger, W. D., (1973), in *New Approaches for Inducing Natural Immunity to Pyogenic Organisms*, (J. B. Robbins et al., eds.), DHEW publication no. (NIH) 74-553, pp. 57–59.

Ashwell, G. and Van Lenten, L., (1972), *J. Biol. Chem.*, **247,** 4633–4640.

Baker, P. J., Stashak, P. W., Amsbaugh, D. F., and Prescott, B., (1971), *Immunology*, **20,** 469–480.

Bhattacharjee, A. K., Jennings, J. H., Kenny, C. P., Martin, A., and Smith, I. C. P. (1975), *J. Biol. Chem.*, **250,** 1926–1932.

Biozzi, G., Stiffel, C., LeMinor, L., Mouton, D., and Bouthillier, Y. (1963), *Ann. Inst. Pasteur*, **105,** 635–660.

Blake, M., and Swanson, J. (1975), *Infect. Immun.*, **11,** 1402–1404.

Bradley, D. E. (1972), *Genet. Res.*, **19,** 30–51.

Bradley, D. E., and Pitt, T. L. (1974), *J. Gen. Virol.*, **24,** 1–15.

Braun, V. (1978), in *Relations Between Structure and Function with Prokaryotic Cell* (R. Y. Stanier, H. J. Rogers, J. B. Ward, eds), Cambridge University Press, Cambridge, pp. 111–138.

Brinton, C. C., Jr. (1967), in *The Specificity of Cell Surfaces* (B. D. Davis and L. Warren, eds), Prentice-Hall, Englewood Cliffs, N.J., pp. 37–70.

Brinton, C. C., Bryan, J., Dillon, J. A., Guerina, N., Jacobsen, L. J., Labik, A., Lee, S., Levine, A., Lim, S., McMichael, J., Polen, S., Rogers, K., To, C-AC and To M-SC 1978), in *Immunobiology of* Neisseria gonorrhoeae (G. F. Brooks et al., eds.), American Society for Microbiology, Washington, D.C., pp. 155–178.

Brooks, G. F., Israel, K. S., and Petersen, B. H. (1976), *J. Infect. Dis.*, **134,** 450–462.

Brown, W. J., Lucas, C. T., and Kuhn, U. S. G. (1972), *Br. J. Vener. Dis.*, **48,** 177–178.

Buchanan, T. M. (1975), *J. Exp. Med.*, **141,** 1470–1475.

Buchanan, T. M. (1978), *J. Infect. Dis.*, **138,** 319–325.

Buchanan, T. M., and Gotschlich, E. C. (1973), *J. Exp. Med.*, **137,** 196–200.

Buchanan, T. M., and Pearce, W. A. (1976), *Infect. Immun.*, **13,** 483–1489.

Buchanan, T. M., Chen, K. C. S., Jones, R. B., Hildebrandt, J. F., Pearce, W. A., Hermodson, M. A., Newland, J. C., and Luchtel, D. L. (1978a), in *Immunobiology of* Neisseria gonorrhoeae, (G. F. Brooks et al., eds.), American Society for Microbiology, Washington, D. C., pp. 145–154.

Buchanan, T. M., Pearce, W. A., and Chen, K. C. S. (1978b), in *Immunobiology of* Neisseria gonorrhoeae, (G. F. Brooks et al. eds.), American Society for Microbiology, Washington, D.C., pp. 242–249.

Bumgarner, L. R., and Finkelstein, R. A. (1973), *Infect. Immun.*, **8,** 919–924.

Burrows, M. R., Sellwood, R. and Gibbons, R. A. (1976), *J. Gen. Microbiol.*, **96,** 269–275.

Callahan, L. T. III (1974), *Infect. Immun.*, **9,** 113–118.

Carrick, L. Jr., and Berk, R. S. (1975), *Biochim. Biophys. Acta*, **391,** 422–434.

Chakrabarty, A. K., and Friedman, H. (1970), *Science*, **167,** 869–870.

Chedid, L., Audibert, F., Lefraneier, P., Choay, J., and Lederer, E. (1976), *Proc. Natl. Acad. Sci. U.S.*, **73,** 2472–2475.

Chen, K. C. S., and Buchanan, T. M. (1978) in *Immunobiology of* Neisseria gonorrhoeae, (G. F. Brooks et al., eds.), American Society for Microbiology, Washington, D.C., pp. 30–34.

Collier, W. A. and DeMiranda, J. C. (1955), *Antionie Van Leeuwenhoek J. Microbiol. Serol.*, **21,** 133–140.

Corbiel, L. B., Schurig, G. G. D., Bier, P. J., and Winter, A. J. (1975), *Infect. Immun.*, **11,** 240–244.

Cowan, S. T., Steel, K. J., Shaw, C., and Duguid., J. P., (1960), *J. Gen. Microbiol.*, **23,** 601–612.

Cruz-Coke, R. L., Parcades, L., and Montenegro, A. (1965), *J. Med. Genet.*, **2,** 185–188.

Cuatrecasas, P. (1973), *Biochemistry*, **12,** 3558–3566.

Cuatrecasas, P., Parikh I., and Hollenberg, M. D. (1973), *Biochemistry*, **12,** 4253–4264.

Demarco de Hormaeche, R., and Thornley, M. T. (1978), *J. Gen. Microbiol.*, **106,** 81–91.

Dilworth, J. A., Hendley, J. O., and Mandell, G. L. (1975), *Infect. Immun.* **11,** 512–516.

Donner, F. (1975), in *Microbiology* (D. Schlessinger ed.), American Society for Microbiology, Washington, D.C., pp. 242–251.

Duguid, J. P. (1959), *J. Gen. Microbiol.*, **21,** 271–286.

Duguid, J. P., and Gillies, R. R. (1957), *J. Pathol. Bacteriol.*, **74,** 397–411.

Duguid, J. P., and Gillies, R. R. (1958), *J. Pathol. Bacteriol.*, **75,** 519–520.

Duguid, J. P., and Smith, I. W., Dempster, I. G., and Edmunds, P. N. (1955), *J. Pathol. Bacteriol.*, **70,** 335–348.

Duguid, J. P., Anderson, E. S. and Campbell I. (1966), *J. Pathol. Bacteriol.*, **92,** 107–138.

Eden, C. S., Hanson, L. A., Jodal, U., Lindberg, U., and Akerlund, A. S. (1976) *Lancet*, **II,** 490–492.

Eden, C. S., Eriksson, B., and Hanson, L. A. (1977), *Infect. Immun.*, **18,** 767–774.

Evans, D. G., Evans, D. J. Jr., and Tjoa, W. S. (1977), *Infect. Immun.*, **18,** 330–337.

Evans, D. G., Evans, D. J. Jr., Tjoa, W. S., and DuPont, H. L. (1978), *Infect. Immun.*, **19,** 727–736.

Felton, L. D., Kauffmann, F., Prescott, B., and Ottinger, B. (1955), *J. Immun.*, **74,** 17–26.

Felix, A., and Bhatnager, S. S. (1935), *Br. J. Exp. Pathol.*, **16,** 422–434.

Field, M. (1971), *N. Engl. J. Med.*, **284,** 1137–1144.

Finkelstein, F. A. (1975), in *Microbiology* (D. Schlessinger ed.), American Society for Microbiology, Washington, D.C., pp. 236–241.

Foster, M. T., and Labrum, A. N. (1976), *J. Infect. Dis.*, **133,** 329–330.

Froholm, L. O., and Sletten, K. (1977), *FEBS. Lett.*, **73,** 29–32.

Galanos, C., Freudenberg, M., Hase, S., Jay, F., and Ruschmann, E. (1977), in *Microbiology* (D. Schlessinger ed), American Society for Microbiology, Washington, D.C., pp. 269–276.

Gemski, P. Jr., Sheahan, D. G., Washington, O., and Formal, S. B. (1972), *Infect. Immun.*, **6,** 104–111.

Gibbs, B. L., and Roberts, R. B. (1975), *J. Exp. Med.*, **141,** 155–171.

Gilbert, R. P. (1960), *Physiol. Rev.*, **40,** 245–279.

Gilboa-Garber, N. (1972), *FEBS Lett.*, **20,** 242–244.

Gilboa-Garber, N., Mizrahi, L., and Garber, N. (1977), *Can. J. Biochem.*, **55,** 975–981.

Globe, M. P., Sutton, A., Robbins, J. B., MacCracken, G. H., Gotschlich, E. C., Kaijser, B., and Hanson, L. A. (1977), *J. Infect. Dis.*, **136,** S93–S97.

Glynn, A. A., and Howard, C. J. (1970), *Immunology*, **18,** 331–346.

Glynn, A. A., Brumfitt, W., and Howard, J. C. (1971), *Lancet*, **I,** 514–516.

Gold, R., Lepow, M. L., Goldschneider, I., and Gotschlich, E. C. (1977), *J. Infect. Dis.*, **136,** 531–535.

Gorbach, S. L., Thadepalli, H. and Norsen, J. (1974), in *Anaerobic Bacteria: Role in Disease*, (A. Balows et al. eds.), Charles C. Thomas, Springfield, Illinois, pp. 399–407.

Gotschlich, E. C. (1973), in *New Approaches for Inducing Natural Immunity to Pyogenic Organisms* (J. R. Robbins et al. eds.), DHEW publication no. (NIH) 74-553, pp. 67–72.

Gotschlich, E. C., Goldschneider, I., and Artenstein, M. S. (1969), *J. Exp. Med.*, **129,** 1367–1384.

Graham, A. B., and Wood, G. C. (1974), *Biochim. Biophys. Acta*, **370,** 431–440.

Gray, L. D., and Kreger, A. S. (1975), *Infect. Immun.*, **12,** 419–432.

Guenter, C. A., Fiorica, V., and Hinshaw, L. B. (1969), *J. Appl. Physiol.*, **26,** 780–786.

Gyles, c. L., and Barnum, D. A. (1969), *J. Infect. Dis.*, **120,** 419–426.

Hampson, S. E., Mills, G. L., and Spencer, T. (1963), *Biochim. Biophys. Acta*, **73,** 476–487.

Hausmann, E., and Kaufman E. (1969), *Biochim. Biophys. Acta.*, **194,** 612–615.

Hendley, J. O., Powell, K. R., Rodewald, R., Holzgrefe, H. H., and Lyles, R. (1977), *N. Engl. J. Med.*, **296,** 608–611.

Hermodson, M. A., Chen, K. C. S., and Buchanan, T. M. (1978), *Biochemistry.*, **17,** 442–445.

Hildebrandt, J. L., and Buchanan, T. M. (1978), in *Immunobiology of* Neisseria gonorrhoeae (G. F. Brooks et al., eds.), American Society for Microbiology, Washington, D.C., p. 138.

Hildebrandt, J. L., Mayer, L. W., Wang, S. P., and Buchanan, T. M. (1978), *Infect. Immun.*, **20,** 267–273.

Hoeniger, J. F. M. (1965), *J. Gen. Microbiol.*, **40,** 29–42.

Hohmann, A., and Wilson, M. R. (1975), *Infect. Immun.*, **12,** 866–880.

Holmgren, J., Lönnroth, I., and Svennerholm, L. (1973), *Infect. Immun*, **8,** 208–214.

Houwink, A. L., and Van Iterson, W. (1950), *Biochim. Biophys. Acta*, **5,** 10–44.

Howard, C. J., and Glynn, A. A. (1971), *Immunology*, **20,** 767–777.

Iglewski, B. H., Liu, P. V., and Kabat, D. (1977), *Infect. Immun.*, **15,** 138–144.

Imada, T., Kanetsuna, F., Rieber, F., Galindo, B., and Cesari, I. M. (1969), *J. Med. Microbiol.*, **2,** 181–186.

Issacson, R. E. (1977), *Infect. Immun.*, **15,** 272–279.

James, A. N., Knox, J. M., and Williams, R. P. (1976), *Br. J. Vener. Dist.*, **52,** 128–135.

James, J. F., and Swanson, J. (1977), *J. Exp. Med.*, **145,** 1082–1086.

James-Holmquest, A. N., Swanson, J., Buchanan, T. M., Wende, R. D., and Williams, R. P. (1974), *Infect. Immun.*, **9,** 897–902.

Jann, K., and Westphal, O. (1975), in *The Antigens*, Vol. 3 (M. Sela, ed.), Academic Press, New York, pp. 1–125.

Jennings, H. J., Bharracharjee, A. K., Bundle, D. R., Kenny, C. P., Martin, A. and Smith, I. C. P. (1977), *J. Infect. Dis.*, **136,** 578–583.

Jephcott, A. E., Reyn, A., and Birch-Andersen, A. (1971), *Acta Pathol Microbiol. Scand. Sec. B*, **79,** 437–439.

Johnson, A. P., Taylor-Robinson, D., and McGee, Z. A. (1977), *Infect. Immun.*, **18,** 833–839.

Jones, G. W. (1977), in *Receptors and Recognition*, (Microbial Interactions, Series B, Vol. 3), (J. L. Reissing, ed.), Chapman and Hall, London, pp. 139–176.

Jones, G. W., and Freter, R. (1976), *Infect. Immun.*, **14,** 240–245.

Jones, G. W., and Rutter, J. M. (1972), *Infect. Immun.*, **6,** 918–927.

Jones, G. W., and Rutter, J. M. (1974), *J. Gen. Microbiol.*, **84,** 135–144.

Jones, R. B., and Buchanan, T. M. (1978), *Infect. Immun.*, **20,** 732–738.

Jones, T. C. (1975), in *Mononuclear Phagocytes; Immunity, Infection, and Pathology*, (Van Furth R, ed.), Blackwell Scientific Publications, Oxford, pp. 595–607.

Jones, T. C., and Hirsch, J. C. (1972), *J. Exp. Med.*, **136,** 1173–1194.

Kadis, S., and Ajl, S. J. (1970), in *Microbial Toxins*, Vol. 3 (T. C. Montie et al. eds.), Academic Press, New York, pp. 39–67.

Kasper, D. L., Hayes, M. E., Reinap, B. G., Craft, F. O., Onderdonk, A. B., and Polk, B. F. (1977), *J. Infect. Dis.*, **136,** 75–81.

Keusch, G. T., and Grady, G. F. (1972), *J. Clin. Invest.*, **51,** 1212–1218.

Keusch, G. T., and Jacewicz, M. (1977), *J. Exp. Med.*, **146,** 535–546.

Kimberg, D. V., Field, M., Johnson, J., Henderson, A., and Gershon, E. (1971), *J. Clin. Invest.*, **50,** 1218–1230.

Kohler, E. M. (1968), *Am. J. Vet. Res.*, **29,** 2263–2274.

Koransky, J. R., Scales, R. W., and Kraus, S. J. (1975), *Infect. Immun.*, **12,** 495–498.

Kreger, A. S., and Gray, L. D. (1978), *Infect. Immun.*, **19,** 630–648.

Lönnroth, I., and Holmgren, J. (1973), *J. Gen. Microbiol.*, **76,** 417–427.

Lüderitz, O. (1977), in *Microbiology* (D. Schlessinger, ed.), American Society for Microbiology, Washington, D.C., pp. 239–246.

McNeish, A. S., Fleming, J., Turner, P., and Evans, N. (1975), *Lancet*, **II,** 946–948.

McQuade, A. B., and Crewther, W. G. (1969), *Biochim. Biophys. Acta*, **191,** 762–764.

Mårdh, P.-A., and Westrom, L. (1976), *Infect. Immun.*, **13,** 661–666.

Miller, J. J., Novotny, P., Walker, P. D., Harris, J. R. W., and MacLennan, I. P. B. (1977), *Infect. Immun.*, **15,** 713–719.

Mims, C. A. (1977), Academic Press, New York, pp. 60–64.

Miyata, K., Maejima, K., Tomoda, K., and Isono, M. (1970), *Agr. Biol. Chem.*, **34,** 310–318.

Montie, T. C., and Montie, D. B. (1971), *Biochemistry*, **70,** 2094–2100.

Montie, T. C., and Montie, D. B. (1973), *Biochemistry*, **12,** 4958–4965.

Montie, T. C., Montie, D. B., and Wennerstrom, D. (1975), in *Microbiology* (D. Schlessinger ed.), American Society for Microbiology, Washington, D.C., pp. 278–282.

Mookerjea, S., and Yung, J. M. W. (1974), *Can. J. Biochem.*, **52,** 1053–1066.

Morihara, K. (1963), *Biochim. Biophys. Acta*, **73,** 113–124.

Morihara, K., Tsuzuki, H. Oka, T., Inoue, H., and Ebata, M. (1965), *J. Biol. Chem.*, **240,** 3295–3304.

Moulder, J. W., Hatch, T. P., Byrne, G. I., and Kellogg, K. R. (1976), *Infect. Immun.*, **14,** 277–289.

Mulks, M. H., and Plaut, A. G. (1978), *N. Engl. J. Med.*, **299,** 973–976.

Munoz, J. J., and Bergman, R. K. (1977), Dekker, New York, pp. 1–235.

Muschel, L. H. (1960), *Ann. N.Y. Acad. Sci.*, **88,** 1265–1272.

Nagy, B., Moon, H. W., and Isaacson, R. E. (1977), *Infect. Immun.*, **16,** 344–352.

Nishijima, M., Nakaike, S., Tamori, Y., and Nojima, S. (1977), *Eur. J. Biochem.* **73,** 115–124.

Novotony, P., and Turner, W. H. (1975), *J. Gen. Microbiol.*, **89,** 87–92.

Nowotny, A. (1977), in *Microbiology* (D. Schlessinger ed.), American Society for Microbiology, Washington, D.C., pp. 247–252.

Ofek, I., Beachey E. H., and Bisno, A. L. (1974), *J. Infect. Dis.*, **129,** 320–315.

Ofek, I., Mirelman, D., and Sharon, N. (1977), *Nature*, **265,** 623–625.

Ohno, R., and Hersh, E. M. (1970), *Blood*, **35,** 250–262.

Okada, K., Kawaharajo, K., Homma, J. Y., Aoyama, Y., and Kubota, Y. (1976), *Jap. J. Exp. Med.*, **46,** 245–256.

Old, D. C. (1972), *J. Gen. Microbiol.*, **71,** 149–157.

Onderdonk, A. B., Kasper, D. L., Cisneros, R. L., and Bartlett, J. G. (1977) *J. Infect. Dis.*, **136,** 82–89.

Orskov, I., and Orskov, F. (1977), *Med. Microbiol. Immunol.*, **163,** 99–110.

Orskov, I., Orskov, F., Smith, H. W., and Sojka, W. J. (1975), *Acta Pathol. Microbiol. Scand. B.*, **83,** 31–36.

Osborn, M. J., Gander, J. E., Parisi, E. and Carson, J. (1972), *J. Biol. Chem.*, **247,** 3962–3972.

Ottow, J. C. G. (1975), *Annu. Rev. Microbiol.*, **29,** 79–108.

Pacaud, M., and Richaud, C. (1975), *J. Biol. Chem.*, **250,** 7771–7779.

Pacaud, M., and Uriel, J. (1971), *Eur. J. Biochem.*, **23,** 435–442.

Paranchych, W., Frost, L. S., and Carpenter, M. (1978), *J. Bacteriol.*, **134,** 1179–1180.

Pearce, W. A., and Buchanan, T. M. (1978), *J. Clin. Invest.*, **61,** 931–943.

Perry, M. B., Daoust, V., Johnson, K. G., Diena, B. B., and Ashton, F. E. (1978), in *Immunobiology of* Neisseria gonorrhoeae (G. F. Brooks et al. eds.), American Society for Microbiology, Washington, D.C., pp. 101–107.

Plaut, A. G., Gilbert, J. V. Artenstein, M. S., and Capra, J. D. (1975), *Science* **190,** 1103–1105.

Plaut, A. G., Gilbert, J. V., and Wistar, R. J. (1977), *Infect. Immun.*, **17,** 130–135.

Punsalang, A. P. Jr., and Sawyer, W. D. (1973) *Infect. Immun.*, **8,** 255–263.

Reid, L., and Clamp, J. R., (1978), *Br. Med. Bull.*, **34,** 5–8.

Reynolds, B. L., and Pruul, H. (1971), *Infect. Immun.*, **4,** 764–771.

Richardson, W. P., and Sadoff, J. C. (1977), *Infect. Immun.*, **15,** 663–664.

Roantree, R. J., and Pappas, N. C. (1960), *J. Clin. Invest.*, **39,** 82–88.

Roantree, R. J., and Rantz, L. A. (1960), *J. Clin. Invest.*, **39,** 72–81.

Robbins, J. B., McCracken, G. H. Gotschlich, E. C., Orskov, F., Orskov, I., and Hanson, L. A., (1974), *N. Engl. J. Med.*, **290,** 1216–1220.

Roberts, R. B. (1970), *J. Exp. Med.*, **131,** 499–513.

Robertson, J. N., Vincent, P., and Ward, M. E. (1977), *J. Gen. Microbiol.*, **102,** 169–177.

Robinson, J. M., and Wilkinson, J. H. (1973), *Clin. Chim. Acta*, **47,** 347–356.

Robinson, M. G., Rolchin, D., and Halpern, C. (1971), *Am. J. Hum. Genet.*, **23,** 135–145.

Rowley, D. (1954), *Br. J. Exp. Pathol.*, **35,** 528–538.

Sack, D. A., Merson, M. H., Wells, J. C., Sack, R. B., and Morris, G. K. (1975), *Lancet*, **II,** 239–241.

Salit, I. E., and Gotschlich, E. C. (1977a), *J. Exp. Med.*, **146,** 1169–1181.

Salit, I. E. and Gotschlich, E. C. (1977b), *J. Exp. Med.*, **146,** 1182–1194.

Scandella, C. J., and Kornberg, A. (1971), *Biochemistry*, **10,** 4447–4456.

Schiffer, M. S., Oliveira, E., Glode, M. P., McCracken, G. H., Sarff, L. M., and Robbins, J. B. (1976), *Pediatr. Res.*, **10,** 82–87.

Schoolnick, G. K., Buchanan, T. M., and Holmes, K. K. (1976), *J. Clin. Invest.*, **58,** 1163–1173.

Schwartz, R. S. (1969) *Nature* (*Lond.*), **224,** 275–276.

Schwarzmann, S., and Boring, J. R. III (1971), *Infect. Immun.*, **3,** 762–767.

Senff, L. M., Wegener, W. S., Brooks, G. F., Finnerty, W. R., and Makula, R. A. (1976), *J. Bacteriol.*, **127,** 874–880.

Shubin, H., and Weil, M. H. (1963) *JAMA*, **185,** 850–853.

Silverblatt, E. J. (1974), *J. Exp. Med.*, **140,** 1696–1711.

Silverblatt, F. J., and Ofek, I. (1976), *Clin. Res.*, **24,** 454A.

Silverblatt, F. J., and I. Ofek (1978a), in *Infections of the Urinary Tract*. Proceedings of the 3rd International Symposium on Pyelonephritis, London, 1975 (E. H. Kass ed.), University of Chicago Press, Chicago, in press.

Silverblatt, F. J., and Ofek, I. (1978b), *J. Infect. Dis.*, **138,** 664–667.

Smith, H. W., and Gyles, G. L. (1970), *J. Med. Microbiol.*, **3,** 387–401.

Springer, G. F. (1973), in *New Approaches for Inducing Natural Immunity to Pyogenic Organisms* (J. B. Robbins et al., eds.), DHEW publication no. (NIH) 74-553, pp. 27–39.

Springer, G. F., Williamson, P. C., and Brandes, W. C. (1961), *J. Exp. Med.*, **113,** 1077–1093.

Staley, T. E., Jones, E. W., and Corley, L. D. (1969), *Am. J. Path.*, **56,** 371–392.

Swanson, J. (1973), *J. Exp. Med.*, **137,** 571–589.

Swanson, J., Kraus, S. J., and Gotschlich, E. C. (1971), *J. Exp. Med.*, **134,** 886–906.

Swanson, J., King, G., and Zeligs, B. (1975a), *Infect. Immun.*, **11,** 453–459.

Swanson, J., Sparks, E. Young, E., and King, G. (1975b), *Infect. Immun.*, **11,** 1352–1361.

Takeuchi, A. (1967), *Am. J. Pathol.*, **50,** 109–136.

Tebbut, G. M., Veale, D. R., Hutchinson, J. G. P., and Smith, H. (1976), *J. Med. Microbiol.*, **9,** 263–273.

Thomas, D. W., Hill, J. C., and Tyeryar, J. Jr. (1973), *Infect. Immun.*, **8,** 98–104.

Thongthai, C., and Sawyer, W. D. (1973), *Infect. Immun.*, **7,** 373–379.

Tramont, E. C. (1977), *J. Clin. Invest.*, **59,** 117–124.

Tramont, E. C., Hodge, W., and Ciak, J. (1978), *Clin. Res.*, **26,** 407A.

Tweedy, J. M., Park, R. W. A. and Hodgkiss, W. (1978), *J. Gen. Microbiol.*, **51,** 235–244.

Van Heyningen, S. (1974), *Science*, **183,** 656–657.

Vasil, M. L., Kabat, D. and Iglewski, B. H. (1977), *Infect. Immun.*, **16,** 353–361.

Vosti, K. L., and Randall, E. (1970), *Am. J. Med. Sci.*, **259,** 114–119.

Waitkins, S. A. (1974), *Br. J. Vener. Dis.*, **50,** 272–278.

Ward, M. E., Watt, D. J., and Robertson, J. N. (1974), *J. Infect. Dis.*, **129,** 650–659.

Watt, P. J., Ward, M. E., and Robertson, J. N. (1976), in *Sexually Transmitted Diseases*. (R. D. Catterall and C. S. Nicol eds.), Academic Press, New York, pp. 89–101.

Wilson, M. R., and Hohmann, A. W. (1974), *Infect. Immun.*, **10,** 776–782.

Wilson, W. R., Martin, W. J., Wilkowske, C. J. and Washington, J. A. (1972), *Mayo Clin. Proc.*, **47,** 639–646.

Wistreich, G. A., and Baker, R. F. (1971), *J. Gen. Microbiol.*, **65,** 167–173.

Witt, K., Veale, D. R., Finch, H. Penn, C. W., Sen, D., and Smith, H. (1976), *J. Gen. Microbiol.*, **96,** 341–350.

Wolf-Watz, H., Elmros, T. Normark, S. and Bloom, G. D. (1975), *Infect. Immun.*, **11,** 1332–1341.

Wyrick, P. B., and Brownridge, E. A. (1978), *Infect. Immun.*, **19,** 1054–1060.

Young, L. S., and Armstrong, D. (1972), *J. Infect. Dis.*, **126,** 257–276.

Index